Edward John Routh

A Treatise on Analytical Statics with Numerous Examples

Volume 2

AF808801

Elibron Classics
www.elibron.com

Elibron Classics series.

© 2005 Adamant Media Corporation.

ISBN 1-4021-7320-2 (paperback)
ISBN 1-4021-2766-9 (hardcover)

This Elibron Classics Replica Edition is an unabridged facsimile of the edition published in 1908 by the University Press, Cambridge.

Elibron and Elibron Classics are trademarks of Adamant Media Corporation. All rights reserved.

This book is an accurate reproduction of the original. Any marks, names, colophons, imprints, logos or other symbols or identifiers that appear on or in this book, except for those of Adamant Media Corporation and BookSurge, LLC, are used only for historical reference and accuracy and are not meant to designate origin or imply any sponsorship by or license from any third party.

A TREATISE ON

ANALYTICAL STATICS

CAMBRIDGE UNIVERSITY PRESS WAREHOUSE,
C. F. CLAY, MANAGER.
London: FETTER LANE, E.C.
Edinburgh: 100, PRINCES STREET.

Leipzig: F. A. BROCKHAUS.
Berlin: A. ASHER AND CO.
New York: G. P. PUTNAM'S SONS.
Bombay and Calcutta: MACMILLAN & CO. LTD.

[*All Rights reserved.*]

A TREATISE ON
ANALYTICAL STATICS

WITH ILLUSTRATIONS TAKEN FROM THE THEORIES OF
ELECTRICITY AND MAGNETISM

BY

EDWARD JOHN ROUTH, Sc.D., LL.D., M.A., F.R.S., ETC.

VOLUME II

CAMBRIDGE:
AT THE UNIVERSITY PRESS.
1908

First Edition 1892.
Reprinted 1902, 1908.

PREFACE.

IN the first edition of this treatise the subject of Attractions was presented only in its gravitational aspect. This limitation was formerly customary, when electricity was less studied than now, but the result has become somewhat unsatisfactory. When lecturing on the subject the Author found that some of the most striking examples of Attraction were those derived from the theory of electricity. While it was impossible wholly to pass these over, it appeared that the interest in them was sensibly diminished if they were discussed without explanations of their meaning. Examples on the attractions of thin layers of matter, subject to what appeared to be arbitrary laws, seemed to have no real applications.

For these reasons a selection has been made of those propositions in Magnetism and Electricity which appeared most forcibly to illustrate the theory of Attraction. These have been joined together, with brief introductions, so as to form a continuous story which could be understood without reference to any other book.

These illustrations have been so far separated from the rest of the volume that any portion of them may be omitted by a reader who desires to confine his attention chiefly to gravitational problems.

Some theorems, which it was not deemed expedient to include in the text, have been shortly discussed in the notes at the end of the volume. These are not always closely connected with the theory of attractions, yet, being natural developments of the text, will probably assist the reader in following the argument.

The general arrangement of the gravitational part of "Attraction" has been only slightly altered. New theorems have, however, been introduced and the demonstrations of some of the old ones simplified.

The second part of this volume is on the stretching and bending of rods. The investigation of the stretching, and consequent thinning, of a rod is founded on Hooke's law. The fact that (with certain restrictions) the stress couple is proportional to the bending is assumed as an experimental result and applied to determine the bending of rods and springs under various circumstances. The problem, when put into this form, is properly included in a treatise on Statics. Although this chapter is not a treatise on the theory of Elasticity, it did not seem proper wholly to omit the theoretical considerations by which the truth of the fundamental law is confirmed. Accordingly some simple examples which had been briefly discussed in the last edition have been retained.

The theory of Astatics occupies the third part of this volume. It was discussed with sufficient fulness in the first edition and only very slight alterations have now been made.

A separate index to each of the three chapters has been given. So many results are included under the head of Attraction that it was found impossible to mention them all without unduly lengthening the list. It was also necessary to classify some theorems only under one heading.

Finally, I desire to express my thanks to Mr J. D. H. Dickson of Peterhouse for the very great assistance he has given me in correcting most of the proof-sheets and for his many valuable suggestions.

EDWARD J. ROUTH.

PETERHOUSE,
December, 1901.

CONTENTS.

ATTRACTIONS.

Introductory remarks.

Attraction of Rods, Discs, &c.

The Potential.

Spherical Surface.

Laplace's, Poisson's and Gauss' Theorems.

Theorems on the Potential.

Attraction of a Thin Stratum.

Green's Theorem.

Given the Potential, find the Body.

Method of Inversion.

Circular Rings and Anchor Rings.

Attraction of Ellipsoids.

Rectilinear Figures.

Laplace's Functions and Spherical Harmonics.

Magnetic Attractions.

Electrical Attractions.

Magnetic Induction.

THE BENDING OF RODS.

Introductory Remarks.

The Stretching of Rods.

The Bending of Rods.

Rods in Three Dimensions.

ASTATICS.

Astatic Couples.

The Central Ellipsoid.

The Central Plane and the Central Point.

The Confocals.

Arrangement of Poinsot's Central Axes.

Reduction to Three and to Four Forces.

NOTES.

ATTRACTIONS.

Introductory remarks.

1. Law of attraction. If two particles of matter are placed at any sensible distance apart, they attract each other with a force which is directly proportional to the product of their masses and inversely proportional to the square of the distance.

Let m, m' be the masses of two particles, r their distance apart; if F be the mutual attraction which each exerts upon the other, then F is given by the equation $F = \kappa \frac{mm'}{r^2}$.

If f be the acceleration produced by the attraction of m at the distance r, then $f = \kappa \frac{m}{r^2}$.

The quantity κ is called the *constant of attraction*. Its magnitude depends on the particular units in which the masses m, m', the distance r and the force F are measured. To avoid the continual recurrence of this constant running through every equation, it is usual to so choose the units that $\kappa = 1$. When this is done the units are called *theoretical* or *astronomical units*.

Putting $\kappa = 1$ in the equations, we see that when m and r are both unity the acceleration f is also unity. We infer that the astronomical unit of mass is that mass which, when collected into a particle, produces by its attraction at a unit of distance the unit of acceleration. The expression for F shows that the unit of force is the attraction which a particle whose mass is the astronomical unit of mass exerts on an equal particle at a unit of distance.

To avoid the continual repetition of the same set of words, we shall use the phrase *attraction at a point* to mean the attraction on a unit of mass collected into a particle and placed at that point.

It is convenient to use different systems of units for different purposes. The astronomical units should be used in analytical investigations. In any numerical applications we may choose such units of space and time as we may find convenient, and then introduce into our formulæ the factor κ with its appropriate value.

It may be noticed that in using different units for different purposes we are following the analogy of other mathematical sciences. In practical trigonometry we measure angles in degrees, in theoretical trigonometry we adopt that unit by which our analytical formulæ are most simplified. Also in algebra we have one base in logarithms for use in calculations and another for theoretical investigations; and so on through all the sciences.

2. Numerical estimate. To obtain a numerical estimate of the magnitude of the force of attraction, we must determine by experiment the mutual attraction of some two bodies. We may exhibit the result in either of two forms: (1) we may determine the value of κ when the units of space, mass, &c. have been chosen; (2) we may determine the magnitude of the astronomical unit of mass by expressing it as a multiple of some other known mass.

The two bodies on which the experiment should be tried are obviously the earth and some body at its surface. Regarding the earth as a sphere, whose strata of equal density are concentric spheres, it will be shown further on that its attraction on all external bodies is the same as if its whole mass were collected into a particle and placed at its centre. If then m be the mass of the earth and a its radius, the acceleration of a body at its surface is $\kappa m/a^2$. Let g be the acceleration actually produced by the attraction of the earth on any body placed at its surface. We thus form the equation $\kappa m/a^2 = g$.

Several experiments have been made to determine the mean density of the earth. One of these is the Cavendish experiment, but there have been others conducted on different plans. The result is that the mean density has been variously estimated to be from $5\frac{1}{2}$ to 6 times that of water. According to Baily's repetition of the Cavendish experiment the ratio is 5·67. Representing this ratio by β, we learn that the attraction of a sphere of water, of the same size as that of the earth, will produce in a body, placed at its surface, an acceleration equal to g/β.

3. *To find the value of κ when the units of space, mass, and time are the centimetre, the gramme and the second.* Since the mass of a cubic centimetre of water is one gramme nearly, the mass m of a sphere of water of the same size as the earth is $\frac{4}{3}\pi a^3$ grammes, where the radius a is measured in centimetres. By the experiment just described $\frac{\kappa m}{a^2} = \frac{g}{\beta}$; taking $\beta = 5{\cdot}67$, $g = 981$ (see Vol. I. Art. 11), we find $\kappa = \frac{3g}{4\pi a\beta} = \frac{1}{1543 \times 10^4}$.

If therefore the attracting masses are measured in grammes and the distances in centimetres, the expression for F with this value of κ gives the attraction in dynes.

Let m be the mass, measured in grammes, of a particle which produces by its attraction at the distance of one centimetre a unit of acceleration. Then m is the astronomical unit of mass. The formula $f = \kappa m/r^2$ gives $1 = \kappa m$, $\therefore m = 1543 \times 10^4$ grammes.

Let F be the force measured in dynes which one astronomical unit of mass exerts on another at the distance of one centimetre. The formula $F = \kappa mm'/r^2$ gives $F = 1/\kappa$ since $m = m'$ and $m\kappa = 1$. The force F is 1543×10^4 dynes.

4. *To find the value of κ when the units of space and time are the foot and the second, and those of mass and force are the pound and the poundal.* Since the weight of a cubic foot of water is the same as that of $\gamma = 61$ pounds nearly, the mass m of a sphere of water of the same size as the earth is $\frac{4}{3}\pi a^3\gamma$, where the radius a is measured in feet. By the experiment just described $\frac{\kappa m}{a^2} = \frac{g}{\beta}$. If we take $a = 20926000$ feet this gives $\kappa = \frac{1}{93 \times 10^7}$.

If therefore the attracting masses are measured in pounds and the distances in feet, the expression for F with this value of κ gives the attraction in poundals.

The astronomical unit of mass, when the foot and the second are the units of space and time, is 93×10^7 pounds and the astronomical unit of force is 93×10^7 poundals. A poundal is roughly equal to the weight of half an ounce. See Vol. I. Art. 11.

5. *Dimensions of κ and m.* When the unit of mass is arbitrarily chosen the attraction F of a particle of mass m on a particle of equal mass is $F = \kappa m^2/r^2$. It follows that the dimensions of κ are the same as $FL^2\mu^{-2}$ or $L^3\mu^{-1}t^{-2}$ where F, L, μ, t stand for force, length, mass, and time. When the factor κ is omitted the dimensions of astronomical mass include those of κ and become the same as those of $\mu\kappa^{\frac{1}{2}}$ or, which is the same thing, $F^{\frac{1}{2}}L$ or $L^{\frac{3}{2}}\mu^{\frac{1}{2}}t^{-1}$. This also follows at once from the formula $F = m^2/r^2$. These dimensions are the same as those of the electrostatic measure of electricity. See Maxwell's *Electricity*, Arts. 41 42.

6. Ex. 1. Prove that the mass of the particle which at the distance of one centimetre from a particle of equal mass attracts it with the force of one dyne is 3928 grammes. Everett's *Units and Physical Constants.*

Ex. 2. Show that a cubic foot of water, collected into a particle, attracts an equal particle placed at the distance of one foot with a force equal to the weight of $1/(8 \times 10^6)$ pounds.

7. Law of the direct distance. There are other laws besides that of the inverse square which may govern the attraction of bodies in special cases. Some of these will be mentioned as we proceed. But the most useful is that in which the attraction varies as the distance. In this case the attraction of two particles, each on the other, is represented by $F = mm'r$, where m, m' are their masses, and r, the distance between them.

8. When the attraction obeys the law of the direct distance, the resultant attraction of any body at any point is found at once by using Art. 51 of Vol. I. Let O be any point, A_1, A_2, &c. the positions of the attracting particles; let m_1, m_2, &c. be their masses. The component attractions at O are then given by $X = \Sigma mx = \bar{x}\Sigma m$, $Y = \bar{y}\Sigma m$, $Z = \bar{z}\Sigma m$, where $\bar{x}$, $\bar{y}$, $\bar{z}$ are the coordinates of the centre of gravity of the body or system of attracting points.

It immediately follows that the resultant attraction at O is the same as if the whole mass Σm of the attracting system were collected into a single particle placed at the centre of gravity. *The resultant force on a particle at O tends therefore towards the centre of gravity of the attracting system, and is proportional to the distance of the attracted point from it.*

9. In what follows, when no special law of force is mentioned, it is to be understood that the law meant is that of the inverse square. This is often called the Newtonian law.

When the law of attraction is said to be $f(r)$, it is meant that the mutual attraction of two particles whose masses are m, m' placed at a distance apart equal to r is $mm'f(r)$.

Attraction of rods, discs, &c.

10. **Attraction of a rod.** *To find the attraction of a uniform thin straight rod AB at any external point P.*

Let m be the mass of a unit of length, then m is called the *line density* of the rod. Let p be the length of the perpendicular PN from P on the rod. Let QQ' be any element of the rod, $NQ = x$; let also the angle $NPQ = \theta$, then $x = p \tan\theta$.

The attraction at P of the element QQ' is

$$\frac{mdx}{PQ^2} = \frac{md\,(p\tan\theta)}{(p\sec\theta)^2} = \frac{md\theta}{p}.$$

Let X, Y be the resolved attractions at P parallel and perpendicular to the length AB. Let the angles NPA, NPB be α, β,

then $$X = \int m\frac{d\theta}{p}\sin\theta = \frac{m}{p}(\cos\alpha - \cos\beta) \quad\quad (1),$$

$$Y=\int m\frac{d\theta}{p}\cos\theta=\frac{m}{p}(\sin\beta-\sin\alpha)\ \ldots\ldots\ldots\ (2).$$

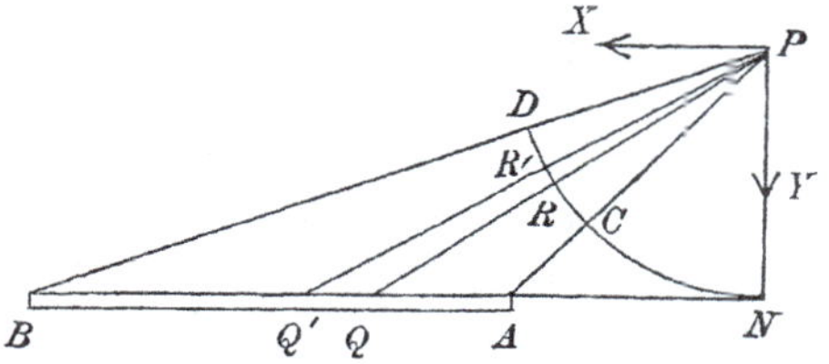

11. Substitute for $\cos\alpha$, $\cos\beta$ their values obtained from the triangles PNA, PNB; the resolved attraction parallel to the rod takes the useful form
$$X=\frac{m}{PA}-\frac{m}{PB}\ \ldots\ldots\ldots\ldots\ (3).$$

It should be noticed that this is the attraction at P of the rod AB resolved in the direction from A towards B.

12. Describe a circle with centre P and radius PN and let the portion CD included between the distances PA, PB represent a thin circular rod of the same material and section as the given rod AB.

The attraction at P of the element RR' of the circular rod is therefore $\dfrac{m\,.\,RR'}{PR^2}=m\dfrac{pd\theta}{p^2}=m\dfrac{d\theta}{p}$. But this has just been proved to be the same as the attraction of the element QQ'. Thus each element of the rod AB attracts P with the same force as the corresponding element of the rod CD. *The resultant attraction of the straight rod AB is therefore the same in direction and magnitude as that of the circular rod CD.*

13. The resultant attraction at P of the circular rod CD must clearly bisect the angle CPD. It immediately follows that *the direction of the resultant attraction at P of a straight rod AB bisects the angle APB.*

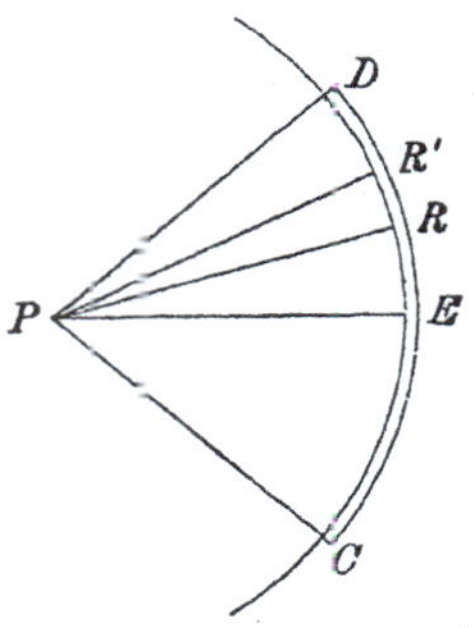

To find the magnitude of the resultant attraction at P of the circular arc CD, we draw PE bisecting the angle CPD. Let the angle any radius PR makes with PE be ψ. Let 2γ be the angle CPD. Since $RR'=pd\psi$ the attraction of the whole circular arc when resolved along PE is

$\int \frac{mpd\psi}{p^2} \cos \psi = \frac{m}{p} . 2 \sin \gamma$, the limits of the integral being $\psi = -\gamma$ and $\psi = \gamma$. *The magnitude F of the resultant attraction at P of a straight rod AB is given by* $F = \frac{2m}{p} \sin \frac{APB}{2}$.

14. When the rod AB is infinite in both directions the angle APB is equal to two right angles. *The resultant attraction of an infinite rod at any point P is equal to $2m/p$, and it acts along the direction of the perpendicular p drawn from P to the rod.*

This proposition leads to *a useful rule* which helps us *to find the attraction of any cylindrical surface or solid* which is infinitely extended in both directions. We pass a plane through the attracted point P perpendicular to the generating lines and cutting the cylinder in a cross section. If the attracting body be composed of elementary rods of line density m, each of these attracts P as if a mass $2m$ were collected into its cross section and *the law of attraction were changed to the inverse distance. The attraction of the whole cylinder is then equal to that of this cross section.* If the cylinder be solid and of volume density ρ, the cross section is an area of surface density 2ρ; if the cylinder is a surface of surface density σ, the cross section is a curve of line density 2σ. The same rule will apply to a heterogeneous cylinder provided the density along each generator is uniform.

Three laws of attraction are therefore especially useful. These are (1) the law of the inverse square, (2) that of the inverse distance, and (3) that of the direct distance.

15. When the point P moves about and comes to the other side of the attracting rod AB, crossing AB produced but not passing through any portion of the attracting rod, the components X, Y remain continuous functions of the coordinates of P, and will continue to represent the component attractions. When P lies in AB produced Y takes the singular form $0/0$, but it is evident that it changes sign through zero. The resultant attraction is then given by (3) which is free from singularity.

When P passes through the material of the rod the case is somewhat different. When P approaches the thin rod, the angles β and α become ultimately $\frac{1}{2}\pi$ and $-\frac{1}{2}\pi$, the Y component becomes infinite while X remains finite. The attraction is therefore ultimately perpendicular to the rod and finally changes sign through infinity. When P is inside the indefinitely thin rod the Y component is zero by symmetry and the X component represents the attraction.

In the preceding analysis we have regarded the linear dimensions of the transverse section of each element QQ' as infinitesimal when compared with the distance from P. This however is not true for any material rod when P approaches very closely to any point of it. The rod (or at least the portion which is near to P) must then be regarded as a cylindrical solid.

16. Ex. 1. If two forces be applied at P acting along AP, PB taken in order, and each equal to m/p, prove that their resultant is equal in magnitude to the attraction of the rod AB and acts in a direction perpendicular to that attraction.

Ex. 2. The sides of a triangle are formed of three thin uniform rods of equal density. Prove that a particle attracted by the sides is in equilibrium if placed at the centre of the inscribed circle.

If one side of the triangle repel while the other two attract the particle, prove that the centre of an escribed circle is a position of equilibrium. [Math. T.]

This follows at once from Art. 12. Draw straight lines from the centre I of the inscribed circle to the corners A, B, C of the triangle, cutting the circle in A', B', C'. The attractions of the sides AB, BC, CA are the same as those of the arcs $A'B'$, $B'C'$, $C'A'$, that is their resultant attraction is the same as that of the whole circle on the centre. This attraction is clearly zero.

Ex. 3. Four uniform straight rods of equal density form a quadrilateral, and their lengths are such that the sum of two opposite sides is equal to the sum of the other two opposite sides. Find the position of equilibrium of a particle under the attraction of the four sides.

Ex. 4. Every particle of three similar uniform rods of infinite length, lying in the same plane, attracts with a force varying inversely as the square of the distance; prove that a particle will be in equilibrium if it be placed at the centre of gravity of the triangle ABC enclosed by the rods. [Math. Tripos, 1859.]

The attractions at P are perpendicular to the sides of the triangle and therefore, when P is in equilibrium, their magnitudes are proportional to those sides. Hence by Art. 14 the areas APB, BPC, CPA are equal and therefore P is the centre of gravity.

Ex. 5. A particle is placed at any point P on the bisector of the angle C of a triangle. Show that the direction of the resultant attraction of the three sides at P bisects the angle APB and is equal in magnitude to $2m\left(\frac{1}{\gamma}-\frac{1}{\alpha}\right)\sin\frac{APB}{2}$, where α and γ are the perpendiculars from P on the sides BC, AB respectively.

Describe a circle centre P to touch the sides AC, BC. The resultant attraction of these two sides is equal and opposite to that of the arc of the circle which lies between the straight lines AP, BP on the side remote from C (Art. 12).

Ex. 6. Two uniform parallel straight rods AB, CD attract each other: show that the components of their mutual attraction, respectively perpendicular and parallel to the rods, are

$$Y=\frac{mm'}{p}(BC-BD-AC+AD), \qquad X=mm'\log\frac{BC'+BC}{AC'+AC}\cdot\frac{AD'+AD}{BD'+BD},$$

where C', D' are the projections of C, D on the rod AB, p the distance between the rods, and m, m' the masses per unit of length.

Ex. 7. P is a particle in the diagonal AC of a square $ABCD$, and within the square; show that the attraction of the perimeter of the square upon P is equal to $M\cdot\frac{OP}{PA\,.\,PB\,.\,PC}$; where M is the mass of the perimeter, O the centre of the square.

[Trin. Coll., 1882.]

Ex. 8. Let the finite rod AB be produced both ways to infinity and let the portion beyond A attract and the portion beyond B repel P, the portion between A

and B exerting no force at P. Prove that the resultant force at P bisects the angle external to APB and is equal to $\frac{2m}{p}\cos\frac{APB}{2}$.

Describe a circle, centre P, to touch AB and intersect PA in C and BP produced in H. The resultant force at P is therefore equal to the attraction of the arc CH. Art. 12.

Ex. 9. The law of attraction of a uniform thin straight rod is the inverse κth power. Prove that the components of attraction at a point P parallel and perpendicular to the rod are respectively

$$X = \frac{m}{\kappa-1}\left(\frac{1}{PA^{\kappa-1}} - \frac{1}{PB^{\kappa-1}}\right), \qquad Y = \frac{m}{p^{\kappa-1}}\int_\alpha^\beta (\cos\theta)^{\kappa-1}\,d\theta,$$

the latter integral can be found by a formula of reduction in the usual way.

Ex. 10. The law of attraction of a cylinder infinitely extended in both directions is the inverse κth power. Prove that the attraction at a point P is equal to that of the cross section provided (1) the law of attraction of the section is the inverse $(\kappa-1)$th power and (2) ratio of its density to the cylindrical density is $2\int(\cos\theta)^{\kappa-1}\,d\theta$, the limits being 0 to $\frac{1}{2}\pi$, (see Art. 14).

17. Curvilinear rods. The method by which the attraction of the straight rod AB is replaced by that of the circular arc CD in Art. 12 may be extended to other curves.

Two curvilinear rods AB, CD are so related that if any two radii vectores OAC, OBD are drawn, the attractions of the intercepted arcs AB, CD at the origin O are the same in direction and magnitude. It is required to find the relation between the densities of the rods.

Since the attractions are equal for all arcs, they are equal for infinitesimal arcs. Let OQR, $OQ'R'$ be two consecutive radii vectores; ds, ds' the arcs QQ', RR'; m, m' the masses at Q, R per unit of length. Then if the law of attraction is the inverse κth power of the distance we have

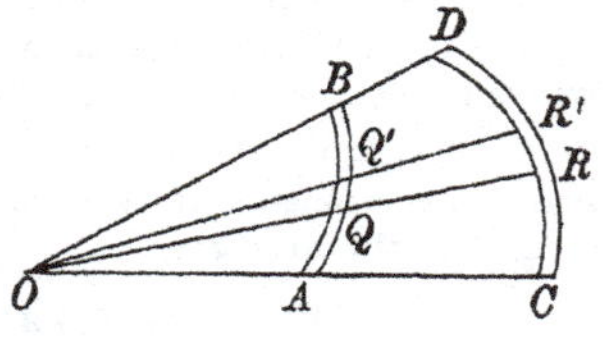

$$\frac{m\,ds}{r^\kappa} = \frac{m'ds'}{r'^\kappa},$$

where $r = OQ$, $r' = OR$. If ϕ, ϕ' be the angles the radius vector OQR makes with the tangents at Q and R, this gives

$$\frac{m}{r^{\kappa-1}\sin\phi} = \frac{m'}{r'^{\kappa-1}\sin\phi'} \quad\text{.....................} \quad (1).$$

The densities of the curvilinear rods at corresponding points must therefore be proportional to $r^{\kappa-1}\sin\phi$.

If the law of attraction be the inverse square, two curvilinear rods in one plane equally attract the origin, if the densities at corresponding points in the two rods are proportional to the perpendiculars from the origin on the tangents.

18. If the two curves are so related that *each is the inverse of the other*, we have $OQ \cdot OR = OQ' \cdot OR'$. A circle can therefore be described about the quadrilateral $QRR'Q'$. In the limit when QQ', RR' become tangents this gives $\sin\phi = \sin\phi'$. If also $\kappa = 1$, we see that $m = m'$. It follows therefore that when the law of attraction is the inverse distance, any curvilinear rod and its inverse, if of equal uniform line density, equally attract the origin.

19. Ex. 1. Let the law of attraction be the inverse distance and let P be any point attracted by a uniform straight rod AB. Draw PN perpendicular to the rod and describe a circle on PN as diameter. Prove that the attraction of AB at P is the same as that of the corresponding arc CD of the circle intercepted between the straight lines PA, PB, if the line densities are equal. Compare Art. 12.

Ex. 2. Two rigid and equal semicircular arcs of matter with uniform section and density are hinged together at both extremities. The matter attracts according to the law of gravitation. If equal and opposite forces applied along the line joining the middle points of the semicircles keep them apart with their planes at right angles, the magnitude of each force will be $4m^2 \log(1+\sqrt{2})$, where m is the mass of unit length of arc. [Math. Tripos, 1874.]

20. Some inverse problems. Ex. 1. A uniform rod is bent into the form of a curve such that *the direction of the attraction of any arc PQ at the origin O bisects the angle POQ*. Show that the curve is either a straight line or a circle whose centre is O.

The data lead to the differential equation $\int \frac{ds}{r^2}\sin\theta = \tan\frac{\theta}{2}\int\frac{ds}{r^2}\cos\theta$. The limits of the integrals being 0 and θ. The equation may be solved by differentiation.

Ex. 2. Find the law of density of a curvilinear rod of given form that *the direction of the attraction at a given point O of any arc PQ may bisect the angle POQ*. If the law of attraction be the inverse κth power of the distance, the result is that the line density m at P must be proportional to $pr^{\kappa-2}$ where $r = OP$ and p is the perpendicular on the tangent at P.

Draw any circle, centre O, intersecting OP, OQ in C, D. The attraction of CD (regarded as a uniform rod) at O is by hypothesis the same in direction as that of PQ and may (by giving CD the proper density) be made the same in magnitude also. Include the additional elements QQ', DD'. It is clear that unless their attractions at O are equal the attraction of PQ' cannot coincide in direction with that of CD'. The attractions at O of corresponding elements of the two rods are therefore equal. Hence as in Art. 17 the density m at every point of PQ varies as $pr^{\kappa-2}$. The proposition may also be proved analytically as indicated in the last example.

Ex. 3. A uniform rod is bent into a curve such that the direction of the attraction at the origin of any arc PQ passes through the centre of gravity of the arc. Prove that, either the law of attraction is the direct distance, or the curve is a straight line which passes through the origin.

Ex. 4. If any uniform arc of an equiangular spiral attract a particle placed at the pole, prove that the resultant attraction acts along the line joining the pole to the intersection of the tangents at the extremities of the arc.

Prove also that if any other given curve possess this same property, the law of attraction must be $F=\frac{\mu}{p^2}\frac{dp}{dr}$, where p is the perpendicular drawn from the attracted particle on the tangent at the point of which the radius vector is r.

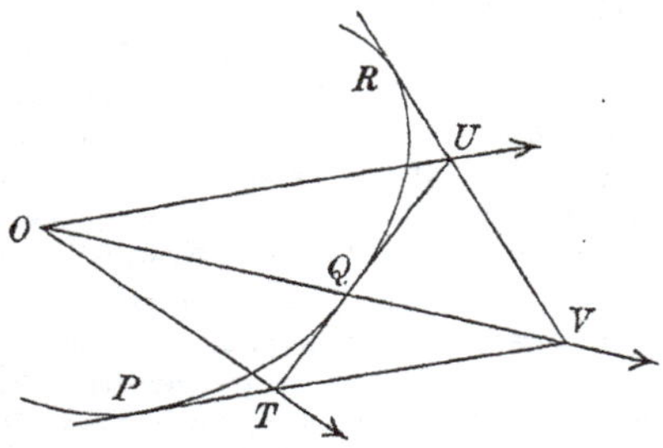

Reversing the attracting forces, we may regard the rod as acted on by a centre of repulsive force. Since the resultant force on any arc PQ acts along OT, where T is the intersection of the tangents at P and Q, we may resolve that force into two components which act along TP and TQ. It follows that the resultant force on any arc PQ may be balanced by two forces or tensions acting along the tangents at P and Q.

To complete the analogy of the force at P to a tension, we must show that that force is always the same whatever the length of the arc PQ may be. To prove this let PQ, QR be two contiguous arcs, and let the tangents at P, Q meet in T, those at Q, R in U, those at P, R in V. Resolving the forces at T, U, V as before, the components along PT, QT and RU, QU must together be equivalent to the components along PV, RV. We have to deduce from this that the components along PT and PV are equal. This follows at once by taking moments about U.

The conditions of equilibrium of the rod are therefore the same as those of a string acted on by a central force. Referring to Art. 474, Vol. I., the tension is obviously $T=A/p$ and the force $f(r)$ has the value given above. See the *Solutions of the Senate House problems for the year* 1860, page 61. The analytical solution leads to an interesting differential equation which can be solved without great difficulty.

21. Attraction of a circular disc. *To find the attraction of a uniform thin circular disc at any point in its axis.*

Let O be the centre, ABA' the disc seen in perspective; OZ the axis, i.e. a straight line drawn through O perpendicular to the plane of the disc. Let a be the radius of the disc, m the mass per unit of area, usually called *the surface density*. Let P

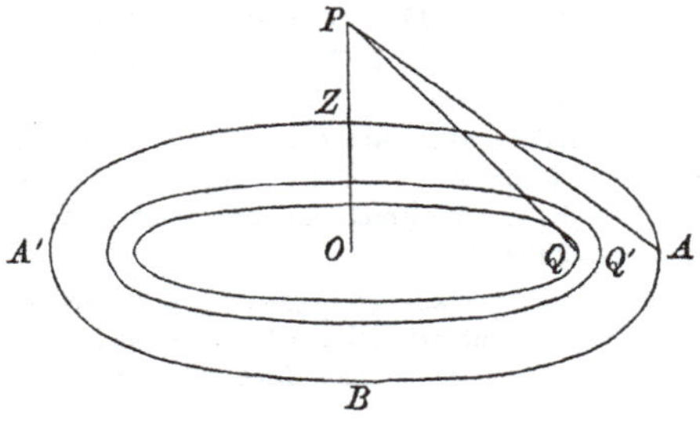

be the point at which the attraction is required $OP = p$, and the angle $OPA = \alpha$.

Describe an elementary annulus, represented in the figure by QQ'. Let x, $x + dx$ be its radii, and let θ be the angle OPQ. The resultant attraction of the disc at P is $F = \int \frac{2\pi x dx \cdot m}{QP^2} \cos \theta$, where the limits of the integral are $x = 0$ and $x = a$. Since $x = p \tan \theta$ and $QP = p \sec \theta$, we find

$$F = 2\pi m \int \sin \theta d\theta = 2\pi m (1 - \cos \alpha).$$

Here α is the *acute angle* subtended at the attracted point by the radius of the disc.

It appears from this that the attraction of a uniform thin circular disc at a point P in its axis depends only on the surface density and on the angle 2α subtended at P by a diameter of the disc. It will be presently seen that if ω be the solid angle subtended at P by the disc, the attraction is $m\omega$, (Art. 26).

22. From this we deduce the attraction of an infinite thin plate or disc by putting $\alpha = \frac{1}{2}\pi$. We thus find that *the attraction of an infinite thin plate at any point P is $2\pi m$ and is therefore independent of the distance of P from the plate.*

We also infer that the attraction of a *circular disc of finite radius a at a point P on the axis very near the disc is ultimately $2\pi m$.* The attraction is $A = \int 2\pi (1 - p/r) \rho dp$ where $r = PA$, ρ is the density, t the thickness, $m = \rho t$, and the limits are $p = p$ to $p + t$. After integration this reduces to $2\pi m$, *provided p/a and t/a are ultimately zero.*

At first sight it may appear anomalous that the attraction of an infinite plate should be independent of the distance of the particle from the plate, but we may understand how it can happen by considering what elements of the disc are effective in producing the attraction. Each element of an annulus QQ', whose centre is O, attracts P with a force acting along the straight line joining P to that element, and the component of force along PO is obtained by multiplying this attraction by $\cos OPQ$. When the point P is near O, this cosine is small and therefore it is only the portion of the disc near O which exerts any sensible attraction in the direction PO. As P recedes from O, the cosine for each annulus gets larger and the resolved attraction becomes greater. Thus the area of effective attraction increases in size as the particle recedes. At the same time as the particle P recedes from O the actual attraction of each annulus on it decreases. It follows from the analysis in the last article that the increase of effective area just balances the decrease of attraction due to increased distance, so that on the whole the attraction is independent of the distance.

23. If g, g' be the attractions due to gravity on two table-lands whose difference of level is x, show that $g' = g\left(1 - \frac{5}{4}\frac{x}{a}\right)$ approximately, where a is the radius of the earth.

To obtain this result, we regard the attraction of the table-land as sensibly the same as that of an infinite plate, Art. 22. The attraction is therefore $2\pi\rho x$, where ρ is the density of the table-land or flat mountain. If ρ' be the mean density of the earth, its attraction, viz. g', is $\frac{4}{3}\pi\rho' a$. There are reasons for believing that the mean surface density of the earth is about half the mean density of the whole earth; when therefore the true density of the table-land is unknown we may as an approximation put $\rho = \frac{1}{2}\rho'$. The attraction of the table-land is thus approximately $\frac{3}{4}gx/a$. The attraction of the earth at the altitude x is $g\left(\frac{a}{a+x}\right)^2 = g\left(1 - 2\frac{x}{a}\right)$ approximately. Adding this to the attraction of the table-land we arrive at the result given.

This theorem was first used by Bouguer in his *Figure de la Terre.* A short account of this treatise is given in Art. 363 of Todhunter's *History of Attractions,* &c. A similar result is also given by Poisson in Art. 629 of his *Traité de Mécanique.* See also Clarke's *Geodesy.* It is often called Dr Young's rule. According to *Nature,* Feb. 10, 1898, a good account of the controversy about the second term of Bouguer's formula is given by G. R. Putnam in the scientific work of the Boston party on the sixth Peary expedition to Greenland. Report A.

Ex. Sir W. Siemens invented an instrument to measure the depth of the sea under a ship on the principle of balancing gravitation by the force of a spring. If the mean surface density of the earth be three times that of sea water, and the mean density of the whole earth five and a half times that of sea water, show that at a depth h in the sea, the diminution of gravity is $\frac{8}{11}hg/a$, where a is the radius of the earth.

24. Attraction of a Cylinder. Ex. 1. Find the attraction of a uniform solid right circular cylinder at a point P on its axis.

Let ρ be the density of the cylinder, a its radius. Let O be the centre of gravity, $OP = p$. Let us take as the element of volume the slice of the cylinder between two planes drawn perpendicular to the axis at distances x and $x + dx$ from O, measured positively towards P.

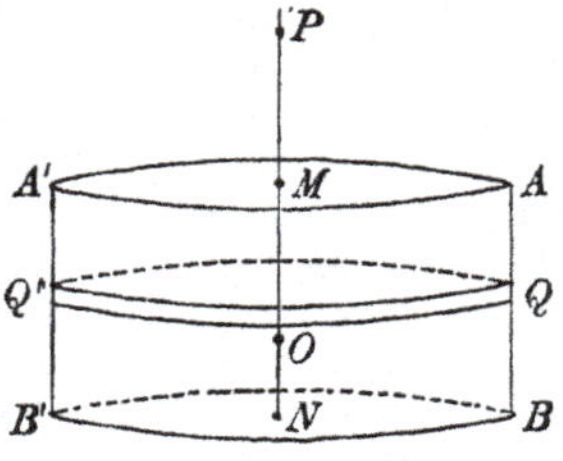

First, let P be outside the cylinder. Let 2θ be the angle subtended at P by any diameter QQ' of the slice, and let $PQ = r$. Since the mass per unit of area of the slice is $m = \rho dx$, the attraction at P is $2\pi\rho dx(1 - \cos\theta) = 2\pi\rho dx\left(1 - \frac{p-x}{r}\right)$. But $(p-x)^2 + a^2 = r^2$, $\therefore (x-p)\,dx = r dr$. The whole attraction of the cylinder at P is therefore $F = 2\pi\rho\int(dx + dr)$, where the limits of integration are $x = -\frac{1}{2}AB$ to $x = \frac{1}{2}AB$ and $r = PB$ to $r = PA$. The resulting attraction is therefore $F = 2\pi\rho(AB + PA - PB)$, where AB is any generating line and A is the extremity nearest to P. We notice that AB is equal to the difference of the distances from the plane sections passing through A and B.

Next, let P be inside the cylinder, but nearer to the plane section $A'A$ than to $B'B$. Since θ is the *acute angle* subtended at P by the radius of the attracting slice, we must equate $\cos\theta$ to $\pm(p - x)/r$, the sign being different on opposite sides of P. To avoid this discontinuity we draw a plane $C'C$ perpendicular to the axis so that P lies midway between the sections $A'A$ and $C'C$. The resultant attraction of the matter between $A'A$ and $C'C$ at P is therefore zero. The resultant attraction of the rest of the cylinder is given by

$$F = 2\pi\rho(CB + PC - PB)$$
$$= 2\pi\rho(CB + PA - PB).$$

Here CB is equal to the difference of the distances of P from the plane sections through A and B, measured positively in opposite directions.

Another Solution. We may also find the attraction by dividing the cylinder into elementary columns or filaments parallel to the axis. We find that the resolved force parallel to the axis is therefore the difference between the values of the integral $\int \rho d\sigma/r$ for the two plane faces, where r is here the distance of $d\sigma$ from P. Since $d\sigma = 2\pi r dr$, and the limiting values of r for the faces AA', BB' respectively are PM to PA and PN to PB, we easily arrive at the same result as before.

Ex. 2. Find the ratio of the radius of the base to the height of a right circular cylinder of given volume so that the attraction at the centre of one of the circular ends may be the greatest possible. The required ratio is $\frac{1}{8}(9 - \sqrt{17})$. Playfair's problem. See Todhunter's *History of Attractions*, Art. 1585.

Ex. 3. A right circular cylinder is of infinite length in one direction and is homogeneous, the finite extremity being perpendicular to the generators. Prove that the attraction at the centre of this end is $2M/a$, where M is the mass per unit of length, and a is the radius.

If the cylinder be elliptic, of the same density and mass per unit of length as before, and of eccentricity e, then the attraction will be n times the former value,

where $n = \frac{2}{\pi}(1 - e^2)^{\frac{1}{4}} \int_0^{\frac{\pi}{2}} \frac{d\theta}{\sqrt{1 - e^2\sin^2\theta}}$. [St John's Coll., 1887.]

Ex. 4. A solid right circular cylinder of uniform density ρ stands on the plane of xy and is infinite in the positive direction of the axis of z. Show that the z component of its attraction at a point P of its base is ρl, where l is the perimeter of an ellipse having the base for the auxiliary circle and P for one focus. See Art. 11.

Ex. 5. A vertical solid cylinder of height h, radius a, and density ρ, bounded by plane ends perpendicular to the axis, is divided by a plane through the axis into two parts. Show that the horizontal attraction of either part at the centre of the base is

$$2h\rho \log \frac{a + \sqrt{(a^2 + h^2)}}{h}.$$ [Coll. Ex., 1888.]

25. Attraction of a surface. *All sections of a uniform cone which are of the same thickness, and have their plane faces parallel to a given plane, exert equal attractions at the vertex.*

Let AB, $A'B'$ be two thin parallel laminæ of the same thickness dt. Let ρ be the density of the cone. With the same vertex O describe an elementary cone cutting the laminæ in QR, $Q'R'$. The attractions of QR, $Q'R'$ at O are to each other as their masses divided by the squares of the distances. Since the thicknesses are equal, the masses are proportional to the areas, and these by similar figures are proportional to the squares of the distances OQ, OQ'. Thus the attractions of the elements QR, $Q'R'$ at O are equal. Hence the attractions of the laminæ AB, $A'B'$ at O are the same both in direction and magnitude.

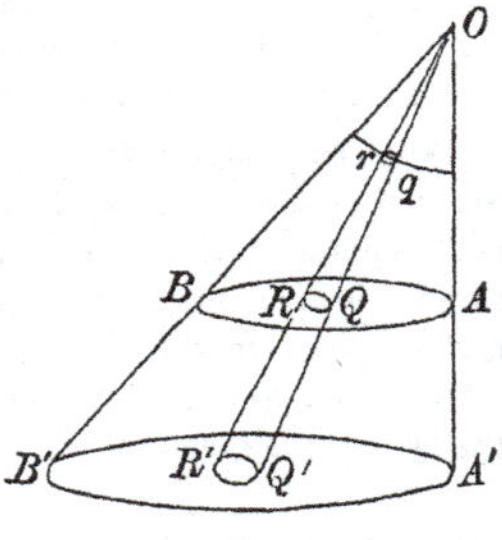

This being true for all thin laminæ must, by integration, be also true for all thick sections. And in general *any two parallel slices of the same cone, whether thick or thin, attract the vertex in the same direction with forces proportional to their thicknesses.*

26. As the attraction of the element QR at any point O is wanted in several theorems further on, it is convenient to determine an expression for its magnitude.

Let $d\sigma$ be the area of the element QR, m its mass per unit of area, r its distance from O; the attraction at O is then $md\sigma/r^2$.

To simplify this expression, we use the solid angle subtended at O by the area. Just as in plane trigonometry an angle is measured by the arc subtended in a circle of unit radius, so the solid angle contained by any cone is measured by the surface cut off by the cone from a sphere of unit radius with its centre at the vertex.

Let the elementary cone whose base is QR intercept on the unit sphere an elementary area qr, and let this area be $d\omega$, then $d\omega$ measures the solid angle subtended at O. Let ψ be the angle the normal to the elementary area QR makes with the radius vector OQ, then $d\sigma \cos\psi$ is the area of a section of the cone made by a plane drawn through Q perpendicular to OQ. Hence by similar figures

$$d\sigma \cos\psi/r^2 = \text{area } qr = d\omega.$$

The attraction of the element is therefore $m \sec\psi \,.\, d\omega$. If p be the perpendicular from O on the plane of the element, then $r\cos\psi = p$, and the attraction of the element at O may also be written in the form $mrd\omega/p$. If r, θ, ϕ are the Eulerian polar coordinates of a point referred to any axes with the origin at O, it is clear that $d\omega = \sin\theta d\theta d\phi$.

27. It follows from this result that the attraction at O of an element $d\sigma$ when resolved perpendicular to its plane is $md\omega$.

Hence we may deduce by integration that *the attraction at O of a plane uniform lamina of any form when resolved perpendicular to the plane is $m\omega$, where m is the mass of a unit of area of the lamina, and ω is the solid angle subtended at O by the lamina.* This theorem is due to Playfair, *Edin. Trans* Vol. VI., 1812.

Ex. If l, m, n be the direction cosines of the radius vector of an element of a surface, and if l, m, n can be expressed in terms of two parameters a and b, show that the normal attraction of the element on the origin is $\Delta da\,db\,dk$, where dk is the thickness of the element and Δ is the determinant in the margin. [Caius Coll.]

$$\begin{vmatrix} l, & m, & n \\ \frac{dl}{da}, & \frac{dm}{da}, & \frac{dn}{da} \\ \frac{dl}{db}, & \frac{dm}{db}, & \frac{dn}{db} \end{vmatrix}$$

28. The method explained in Art. 17 by which the attraction at the origin of one thin rod may be replaced by that of another of a more convenient form may be extended to surfaces.

Let the law of attraction be the inverse κth power of the distance. Refer to the figure of Art. 17 and equate the attractions of the elementary areas QR, $Q'R'$, we have $\frac{md\sigma}{r^\kappa} = \frac{m'd\sigma'}{r'^\kappa}$.

By Art. 26 $d\sigma \cos\psi = r^2 d\omega$, hence $\frac{m}{r^{\kappa-2}\cos\psi} = \frac{m'}{r'^{\kappa-2}\cos\psi'}$.

It follows that, *if two curvilinear laminæ are so related that their masses per unit of area, at points on the same radius vector drawn from a point O, are connected by the above equation, then the attractions at O of the portions included within any conical surface whose vertex is O are the same in direction and magnitude.*

For example, if the law of attraction be the inverse cube, the attraction at a point O of any portion of a thin plane area is the same in direction and magnitude as that of the corresponding portion of a spherical surface having its centre at O, and touching the plane, the masses per unit of area of the plane and sphere being equal. This corresponds to the theorem in Art. 12, which connects the attraction of a straight rod with that of a circle.

29. *If the plane area be bounded by an ellipse (the law of attraction being the inverse cube) the resultant attraction at any point O acts along the axis of the enveloping cone whose vertex is O.*

To prove this we notice that the enveloping cone is a quadric cone and that therefore the portion of the spherical surface (centre O) enclosed within it is symmetrical about the internal axis of the cone. The resultant attraction of the spherical surface at O must therefore act along that axis. By a known theorem in geometry this axis is normal to the ellipsoid which passes through O and has the given ellipse for a focal conic.

30. Ex. 1. Show that the attraction at a point O of any portion of a thin plane disc is the same in direction and magnitude as that of the corresponding portion of a spherical surface having for a diameter the perpendicular ON drawn from O to the plane. The two attracting surfaces are supposed to be homogeneous and of equal mass per unit of area.

Ex. 2. A tetrahedron is constructed of thin metal, the faces being of equal and uniform density. Prove that if the law of attraction were the inverse cube of the distance, a particle would be in equilibrium if placed at the centre of the inscribed sphere. See Art. 16, Ex. 2.

Ex. 3. Prove that the ratio of the attractions of a solid right cone at the centre of the base and at the vertex is $\frac{\sqrt{2}-\log(\sqrt{2}+1)}{\sqrt{2}-1}$, the angle at the vertex being a right angle.

Ex. 4. An infinite lamina is bounded by two parallel straight lines. Prove that its component attractions X and Y respectively parallel and perpendicular to its plane are $X=2m\log r'/r$ and $Y=2m\,.\,\theta$, where r', r are the distances of the attracted point P from the two edges, θ the angle these distances make with each other and m the surface density. See Art. 14.

Ex. 5. Prove that the resultant attraction of a uniform rectangular plate at a point P on its axis is $4m\sin^{-1}(\sin\alpha\sin\beta)$ where α, β are the angles subtended at P by perpendiculars drawn from the centre on the sides and m is the surface density. Playfair, *Edin. Trans.* 1812.

Ex. 6. Prove that the attraction of a uniform elliptic disc at the focus is $\frac{2\pi m}{e}\{1-\sqrt{(1-e^2)}\}$ where m is the surface density.

The attraction is $X=\iint mrd\theta\,dr\cos\theta/r^2$. Describe a circle of arbitrary radius c with its centre at the focus: the attraction of the enclosed area is zero. Integrate from $r=c$ to the elliptic rim and from $\theta=0$ to 2π. In this way we avoid the infinite $\log r$ at the origin.

31. The solid of greatest attraction. *To find the solid of revolution of given mass which exerts the greatest attraction at a point O situated on the axis.*

Let us trace the surface such that the attraction at the given point O, of a particle of given mass m placed at any point of the surface, when resolved along the given axis, is equal to a given constant C. Taking O for origin and the given axis as the axis of reference, the equation of that surface is clearly $\frac{m}{r^2}\cos\theta=C$. By giving C different values we obtain a system of surfaces. It is evident from the definition that the surface defined by any value of C lies outside that defined by a greater value of C. It follows that the resolved attraction of a particle lying on any one surface is greater than that of an equal particle situated on any external surface.

It is evident from the equation that all these surfaces are similar and similarly situated, and that they all touch a plane drawn through O perpendicular to the given axis.

Let us select that surface whose volume would just contain the given mass. The solid of greatest attraction must coincide with the surface thus selected; for if any portion lies outside the selected surface, the attraction would be increased by moving that portion into the vacant places within the selected surface and thus filling them up.

The solid of greatest attraction has therefore such a form that the attraction at the given point of a given particle placed at any point of the surface when resolved along the given axis is always the same.

The problem of finding the solid of greatest attraction was proposed and solved by Silvabelle. The principle used above, that the resolved attraction must be constant over the surface, is due to Playfair, *Edin. Trans.* 1812. The following example is also due to him.

32. Ex. Supposing the law of attraction to be the inverse κth power of the distance, find the form of an infinitely long cylinder so that the attraction may be a maximum at an external point.

Take the point for origin; pass a plane through it perpendicular to the generating lines of the cylinder. Let r be the radius vector of any point on this section, θ the angle made by r with the direction of the resultant attraction. The equation of the curve is included in $\cos\theta = Cr^{\kappa-1}$. *When the law of attraction is the inverse square the required cylinder is right circular.*

33. Attraction of mountains. It is a matter of some importance to determine by direct experiment the effect of the attraction of a neighbouring mountain on the direction of the plumb line. This was attempted by Bouguer in Peru but without any great success. In 1772 Maskelyne, then Astronomer Royal, proposed to repeat the experiment. He pointed out a mountain in Yorkshire as suitable for the purpose. He suggested also that the defect of matter in the valley between Helvellyn and Saddleback might produce an effect of an opposite character which would be sensible. The mountain Schehallien in Scotland was finally chosen. It is a narrow ridge running east and west in a comparatively flat country and is about 2000 feet above the general level.

Let f, f' be the horizontal attractions of the mountain at two stations north and south. The angular deviations of the plumb line from the direction of gravity will then be $\alpha = f/g$, and $\alpha' = f'/g$. The meridional distance between the two stations was found by a survey over the mountain to be 4364·4 feet. By dividing this by the radius of the earth, the difference of latitude of the two stations was found to be 42″·9. By observing the zenith distance of the same star at both stations the difference of the angles which the direction of the star made with the directions of the plumb line at the two stations was found to be $54''\cdot6$. The difference between these two angles, viz. $11''\cdot7$, is evidently equal to the sum of the angular deviations α, α' produced on the plumb lines by the attraction of the mountain.

34. To find the attraction f at a station A, a contour map of the country was made. This was divided into twenty rings by circles having A for a common centre, their radii being in arithmetical progression. These rings were subdivided into rectangular spaces by radii vectores drawn from A. The mountain was thus theoretically divided into elementary columns placed on these rectangular bases. Let GP be a vertical drawn through the centre of gravity G of any base cutting the surface of the mountain in P. Let z be the angle PAG subtended by PG at A. The attraction of this column is nearly equal to $2m\sin\frac{1}{2}z/AG$ and its direction bisects the angle PAG, where m is the line density of the column (Art. 13).

Let r, θ be the polar coordinates of G referred to A as origin and the meridian AM as axis of x. Let Δr be the difference of two consecutive radii, and $\Delta\theta$ the angle between two consecutive radii vectores. Then $m=\mu r.\Delta r.\Delta\theta$ nearly, where μ is the density of the column. The resolved attraction of the column along the meridian is therefore

$$X=\frac{2m}{r}\sin\frac{z}{2}\cos\frac{z}{2}\cos\theta=\mu\sin z.\Delta r.\Delta\sin\theta$$

nearly. The constant difference Δr was taken to be $666\frac{2}{3}$ feet. The radii vectores were drawn according to the following law. The first being directed along the meridian, the others were drawn making with the meridian angles whose sines were successively 1/12, 2/12, 3/12, &c. There were therefore 48 columns over each ring. Also $\Delta\sin\theta$ was constant and equal to 1/12. It is now evident that the attraction f of the mountain may be found by forming the sum

$$\sin z_1+\sin z_2+\sin z_3+\ldots$$

for all the columns and multiplying the result by $\frac{1}{12}\mu.\Delta r$. The twenty rings drawn round each station included 960 columns. This space was bounded by a circle of radius $2\frac{1}{2}$ miles. It was assumed that the attraction of the matter beyond this distance might be neglected.

35. By such processes as these the sum of the two opposite attractions at the two stations was found as a known multiple of the density μ of the hill. If R be the radius of the earth, ρ its density, we have $g=\frac{4}{3}\pi\rho R$, Art. 77. We thus have $\alpha+\alpha'$ expressed as a known multiple of μ/ρ. By equating this result to the circular measure of $11''\cdot7$, we find that the mean density of the hill is $\frac{5}{9}$ths of that of the earth.

A geological survey was subsequently made by Playfair to discover the average density of the hill. After many corrections Hutton gave 4·95 as the mean density of the earth, that of water being unity.

Other mountains also have been used for this purpose. The observations of James and Clarke on Arthur's Seat gave 5·316, while those of Mendenhall in Japan led to 5·77 as the mean density of the earth.

36. There are two other methods of finding the mean density one by observations in mines and the other by processes analogous to the Cavendish experiment. These have been used many times and lead to results which differ slightly, in excess or defect, from $5\frac{1}{2}$.

A short history of the older experiments may be found in Airy's *Figure of the Earth*. Much however has been done since 1830, the date of this treatise. An account of the experiments up to the year 1894 is given in Poynting's essay on the mean density of the Earth.

37. At the end of a paper on the Schehallien experiment (*Phil. Trans.* 1821) Hutton suggested that one of the great pyramids of Egypt might be used instead of a mountain to find the mean density of the earth. He calls to mind the great size of one of these, its height being nearly double that of St Paul's Cathedral. Its regular figure and known composition would, he says, yield facilities in the calculation of its attraction. Observations could then be made at four stations, one on each face, and these could be placed much nearer to the centre of gravity of the attracting mass than was possible in an irregular mountain. Such was his enthusiasm, that he declared that even his age of eighty years would hardly prevent him from joining an expedition for this purpose.

38. Ex. 1. The tide in the Bay of Fundy rises 100 feet from low to high water mark. It has been proposed to find the density of the earth by determining the attraction of the tide-wave on a plumb-line at high and low tide on the same principle as Maskelyne's experiment at Schehallien. Supposing the attraction of the tide-wave at a point O on the shore to be represented by that of the water within a cylinder whose axis is the vertical at O, whose height l is 100 feet and radius r, show that the deviation of the plumb-line is $\frac{3l}{2\pi RD}\log\frac{2r}{l}$, where R is the radius of the earth, D its mean density, and r is large compared with l.

Show that this expression increases slowly compared with r, and that if r be taken between 2 and 4 miles, the deviation to be observed will be about two-fifths of a second. This is much smaller than the deviation to be observed in Maskelyne's experiment, which was about eleven seconds. On the other hand the attracting mass is a homogeneous body instead of a heterogeneous mountain.

Ex. 2. The section of a long wedge-shaped mountain is an equilateral triangle having BC for base. If P be the point on the face AB at which the horizontal attraction is greatest, prove that $\log\frac{y(a-x)}{x^2}=\sqrt{3}\sin^{-1}\frac{a\sqrt{3}}{2y}$ where $x=BP$, $y=CP$ and a is the length of any side of the section. The equation is nearly satisfied by $x=\frac{1}{4}a$.

The Potential.

39. Let A_1, A_2, &c. be the positions of any number of fixed attracting particles; m_1, m_2, &c. their masses. The potential of these particles* at any proposed point P is defined to be

$$V = \frac{m_1}{r_1} + \frac{m_2}{r_2} + \&c. = \Sigma \frac{m}{r},$$

where r_1, r_2, &c. are their distances from P regarded as positive quantities. For the sake of distinction this is sometimes called the Newtonian Potential. See Art. 9.

This may be called the geometrical definition of the potential. Another definition founded on the principle of work will be given a little further on. In discussing the attractions of geometrical figures the former is the more convenient for use, but in many physical applications the latter will be found the more satisfactory.

We may notice that *as the point P moves in space the potential is, by the definition, a continuous function of the position of P.* We must however except the case in which any one of the distances r_1, r_2, &c. vanishes or changes sign, for then the term m/r ceases to represent the potential of the particle from which r is measured. *The potential is also a one-valued function of the coordinates of P.*

40. If m be the mass of any one of the attracting particles, A its position, r its distance from a point P, the potential of m at P is m/r. Let P' be any point adjacent to P, and let $PP' = ds$. The difference of the potentials of m at P and P' is then

$$\frac{d}{ds}\left(\frac{m}{r}\right) ds = -\frac{m}{r^2}\frac{dr}{ds} ds.$$

* The function now called the Potential was used by Legendre in 1784 who refers to it when discussing the attraction of a solid of revolution. Legendre however expressly ascribes the introduction of the function to Laplace and quotes from him the theorem connecting the components of attraction with the differential coefficients of the function. M. Bianco in the *Rivista di Matematica*, 1893, gives quotations from Bist (*Institut Paris*, 1806) and from Baltzer (*Geschichte des Potentials*, 1878) showing that Lagrange used the same function in 1777 when discussing the motion of several bodies mutually attracting each other (*Academy of Berlin*, 1777). See also "Il problema Meccanico della figura della Terra" (*Torino*, 1880) by M. Bianco. The name, Potential, was first used by Green in his *Essay on the application of Mathematical Analysis to the theories of Electricity and Magnetism*, published in 1828. Green gave many of the theorems on this function now in continual use, which have been since associated with the names of others who have discovered them a second time. Gauss also uses the name in Art. 3 of his memoir on *Forces acting inversely as the square of the distance*, Leipsic 1840, translated in the third volume of Taylor's *Scientific Memoirs*. The reader may also consult Todhunter's *History*, Arts. 790, 1138, and Thomson and Tait's *Treatise on Natural Philosophy*, Art. 483.

If ϕ be the angle $AP'P$, we have $\cos\phi = dr/ds$. The attraction of m at P acts in the direction PA, and is equal to m/r^2; hence its resolved part in the direction PP' is $\frac{m}{r^2}\cos APP' = -\frac{m}{r^2}\frac{dr}{ds}$. Comparing this with the above result we see that if P, P' be two adjacent points, the excess of the potential at P' over that at P, divided by the distance PP' is equal to the resolved attraction in the direction PP'.

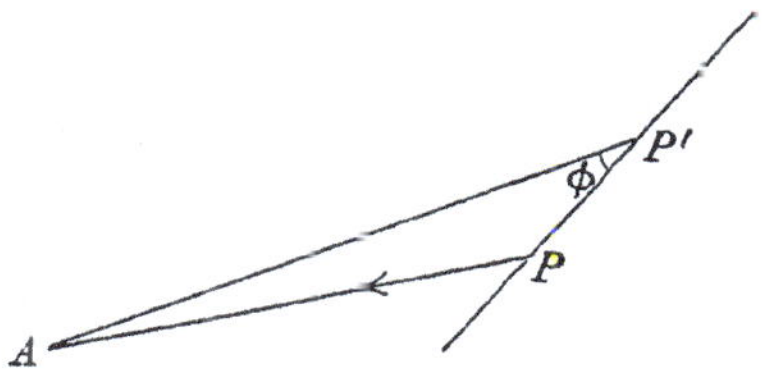

This, being true for every particle of an attracting system is necessarily true for the whole. We have therefore the following theorem. *If V, V' be the potentials of a system at two neighbouring points P, P', the attraction at P resolved in the direction PP' in which s is measured is the limit of* $\frac{V'-V}{PP'} = \frac{dV}{ds}$.

So long as the point P is situated outside the attracting mass the potentials V and V' are both finite and this proof is free from ambiguity. The case in which P lies within the attracting mass will be considered a little further on.

41. By taking the displacement PP' parallel to the axes of x, y, z in turn, we see that the components of the attraction in the *positive directions* of the axes are respectively

$$X = \frac{dV}{dx}, \quad Y = \frac{dV}{dy}, \quad Z = \frac{dV}{dz}.$$

In the same way the components of the attraction in polar coordinates may be expressed. Let r, θ, ϕ be the polar coordinates of any point P, let F, G, H be the components at P in the directions in which dr, $rd\theta$, $r\sin\theta d\phi$ are drawn, then

$$F = \frac{dV}{dr}, \quad G = \frac{dV}{rd\theta}, \quad H = \frac{dV}{r\sin\theta d\phi}.$$

In the theory of gravitation *the attraction* of one particle on another is taken to be mm'/r^2 (Art. 3), and *repulsion* is then represented by supposing that the mass of one of the particles is negative. In other theories, for example in that of electricity, repulsion is taken as the standard case and then attraction occurs when the masses have opposite signs. *In both cases the geometrical definition of the potential is $V = \Sigma m/r$* (Art. 39). When therefore repulsion is taken as the standard the signs of the forces given above must be changed. Thus *the force in the positive direction of the axis of x is $X = -dV/dx$*, and so on.

42. It appears from this proposition that, when the potential V of a body fixed in space is given, its resolved attractions at any point P can be found by simply differentiating the potential with regard to the coordinates of that point. It follows that, if two different bodies have equal potentials throughout any space, they equally attract any particle placed in that space. Thus the attraction of a body is determined by the single function V instead of the three components X, Y, Z.

One chief reason for the use of the potential is that a body, so far as its quality of attraction is concerned, is analytically given by a single function without the necessity of stating either the form or the structure of the attracting body.

When the potential is used merely to find the forces, it is obvious that we may add an arbitrary constant to its value as defined in Art. 39. We then have $V = \Sigma\, m/r + C$, where C is the constant added. When the attracting bodies are finite, it is convenient to choose C so that V is zero at an infinite distance; this assumption makes $C = 0$. When the attracting bodies extend to infinity, the potential, as defined in Art. 39, is sometimes found to contain an infinite constant. It may then be preferable to keep C arbitrary and to absorb into its value all constants not immediately required. There is a certain inconvenience in having different definitions of the potential for finite and infinite bodies, especially when we wish to proceed from one to the other as a limit. In stating the results therefore for the Newtonian law of force we shall adhere to the definition of Art. 39. In special cases such a constant may then be added as may most simplify the expression for V.

43. Potential for other laws of force. When the law of force is the inverse κth power, the potential is $V = \dfrac{1}{\kappa - 1} \Sigma \dfrac{m}{r^{\kappa-1}}$. We then find by the same reasoning as in Art. 40 that dV/ds is the resolved force at P in the direction in which ds is measured.

When the law of force is the inverse distance, the potential is $V = C - \Sigma m \log r$. This is sometimes called *the logarithmic potential.*

44. Work and potential. A definition of the potential may also be given founded on the principle of work. Referring to the figure of Art. 40, let a particle of unit mass travel along the elementary arc PP'. It has been already shown that the resolved attraction in the direction PP' is dV/ds. The work done by the attraction is therefore $(dV/ds)\, ds$. If the particle continue its journey along any curve, starting from some point P and arriving at some other point Q, the work done by the attraction is $\int dV = V_2 - V_1$, where V_1 and V_2 are the potentials at P and Q. Thus the excess of the potential at Q over that at P is the work done by the attraction on a particle of unit mass as it travels by *any path* from P to Q.

If the attracting body is finite in all directions, the potential at a point P infinitely distant is zero. It follows that, *the potential at any point Q is the work done by the attracting forces on a particle of unit mass, as it travels from an infinite distance along any path to the point Q.* In the same way the potential at Q is the work which must be done against the attraction by some external cause to move a unit particle from Q to an infinite distance.

The several particles of the attracting mass are supposed to remain fixed in space while the attracted particle makes its journey from P to Q.

45. Level surfaces. The locus of points at which the potential has any one given value is called *a level surface.* It is also called *an equipotential surface.*

At any point of a level surface the resultant force acts along the normal to the surface. To show this, let P_1 be a point on a level surface, and let P_2 be any neighbouring point also on the surface. If V_1, V_2 be the potentials at these points, the component force in the direction of any tangent P_1P_2 will be the limit of $\dfrac{V_2 - V_1}{P_1P_2}$. This is zero since $V_1 = V_2$. The resultant force must therefore act along the normal at P_1.

46. Let two neighbouring level surfaces be drawn at which the potentials are respectively $V_1 = c$ and $V_2 = c - \delta c$. *The normal attraction at any point P of either surface is inversely proportional to the length of the normal at that point intercepted between these level surfaces.* To prove this, let the normal at any point P_1 on the first surface intersect the second surface in P_2. The normal force at P_1 is then ultimately $F = \dfrac{V_2 - V_1}{P_1P_2} = \dfrac{\delta c}{P_1P_2}$. It is therefore evident that F varies inversely as P_1P_2.

If a rigid surface were constructed having the form of a level surface and coincident with it, it is clear that a particle, placed at any point of the surface, would be pulled by the attracting body in a direction normal to the surface. The particle, if placed on the proper side, would therefore be in equilibrium. Level surfaces are therefore also called *surfaces of equilibrium.*

47. *A Line of force* is a curve such that the direction of the resultant force at any point is a tangent to the curve. It is evident that the whole system of level surfaces is cut orthogonally by the system of lines of force.

48. Ex. 1. A free particle placed at rest at any point of a line of force will move along the curve in such a direction that the potential increases.

Ex. 2. Show that, if attracting matter be arranged so that the direction of the resultant attraction at any external point P shall always pass through a fixed point O, the magnitude of the resultant attraction will be a function only of the distance OP, and will not depend on the angular coordinates of OP.

To prove this we notice that the level surfaces are spheres, because the normal at every point P passes through O. Hence the potential is a function of r only, Art. 45.

49. **Potentials of rods.** *To find the level surfaces and the potential of a thin uniform straight rod AB at any point P.*

It has already been proved that the direction of the attraction of a rod AB at any point P bisects the angle between the distances PA, PB (Art. 13). It follows from Art. 45 that *the level surfaces are prolate spheroids having their foci at the extremities of the rod.*

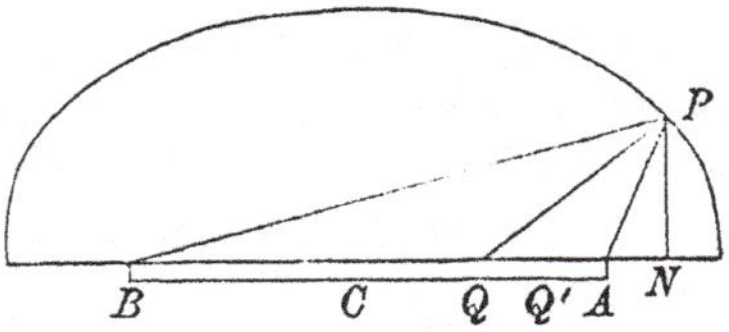

To find the potential we notice that at all points on the same level surface the potentials are equal. It is therefore sufficient to find the potential at some convenient point on each spheroid.

Let C be the middle point of the rod, $2l$ its length, m the line density. Let r, r' be the distances of P from A, B. Let $2a$ and e be the major axis and eccentricity of the spheroid, then $ae = l$, $2a = r + r'$. The potential at the extremity of the major axis and therefore at any point on the spheroid is

$$V = \int \frac{m dx}{a - x} = m \log \frac{a + l}{a - l} = m \log \frac{r + r' + 2l}{r + r' - 2l},$$

the limits of the integral being $x = -l$ to l.

50. When the rod is infinite in both directions the potential is easily deduced from the attraction already found in Art. 14.

Since the magnitude of the attraction is $2m/p$ and its direction is PN, it is evident that the potential must be $V = C - 2m \log p$, where C is a constant and p is the distance of P from the rod.

We may also deduce this result from the expression for the potential of a finite rod. Suppose the point P to be situated in the straight line drawn through C perpendicular to the rod.

Then $$r = (l^2 + p^2)^{\frac{1}{2}} = l + \tfrac{1}{2}\frac{p^2}{l} \text{ and we have}$$

$$V = m \log \frac{r+l}{r-l} = 2m \log 2l - 2m \log p.$$

We thus see that *the constant C is really infinite and equal to* $2m \log 2l$ when we adhere to the definition of Art. 39.

51. Ex. 1. Let the rod AB be produced both ways to infinite distances. Let the portion beyond A attract and that beyond B repel P, the part between A and B exerting no force. Prove that the level surfaces are hyperboloids having A and B for foci and that the potential at P is $m \log \dfrac{2l + r' - r}{2l - r' + r}$. Prove also that if the portion AB is evanescent the level surfaces are right cones and that the potential is $2m \log \cot \frac{1}{2}\psi$ where ψ is the angle of the cone.

Ex. 2. Show that the potential of a thin rod AB at any point P is

$$V = m \log (\cot \tfrac{1}{2} PAB \,.\, \cot \tfrac{1}{2} PBA).$$

Ex. 3. A thin uniform rod AB is attracted by a body of any form: show that the component of the attraction along the length BA of the rod is $m(V_A - V_B)$, where V_A and V_B are the potentials of the body at A and B, and m is the mass of the rod per unit of length.

By Art. 11 this theorem is true when the rod is attracted by a single particle; it is therefore true by summation when attracted by any body.

Ex. 4. A uniform thin chain AB is enclosed in a smooth curvilinear tube which it just fits, and is attracted by a body of any form. Show that the force urging the chain to move in the tube is $m(V_A - V_B)$. Hence show that the position of equilibrium may be found by equating the potentials of the body at the extremities of the chain.

That the force depends only on the positions of the extremities of the chain, and not on its length or form, may also be shown by another kind of reasoning. Let the chain be completed into a circuit by uniting two chains in different tubes at their extremities. If the forces were not equal the chain would begin to move round the circuit, and thus a perpetual motion would be caused by the mere presence of an attracting body.

Ex. 5. When the law of attraction is the inverse cube, the potential of a uniform thin rod AB at any point P is $m\gamma/2p$, where γ is the angle APB, p the perpendicular from P on the rod, and m is the line density.

When the law is the inverse fourth power, the potential is $m(\sin\beta - \sin\alpha)/3p^2$ where β, α are the angles p makes with PB and PA.

Ex. 6. A plane lamina is bounded by two parallel straight lines whose distance apart is $2l$. The surface density at any point Q is $\beta(QM \,.\, QN)^\lambda$ where QM, QN are the distances of Q from the bounding straight lines and $2\lambda=\kappa-3$. The law of attraction is the inverse κth power of the distance. Prove that the level surfaces are confocal elliptic cylinders having the foci on the bounding straight lines. Find also the potential for any cylinder.

First find the attraction at P of an elementary strip Q whose breadth is dx. Put $x=p\tan\theta$ where $p=PO$ is the perpendicular from P to the lamina. The attraction of the strip can then be put into the form $\frac{A}{p}\left(\frac{\sin MPQ \,.\, \sin NPQ}{\sin OMP \,.\, \sin ONP}\right)^{\lambda}\left(\frac{\cos\theta}{p}\right)^{\mu} d\theta$, where A is a constant and $\mu=\kappa-3-2\lambda$. If then κ and λ are so related that the exponent $\mu=0$ the attraction of the elementary strip at Q is a symmetrical function of the angles PQ makes with PM, PN. The elements on each side of the bisector of the angle MPN will then equally attract P. The direction of the attraction therefore bisects the angle MPN. The magnitude of the attraction is found by resolving along the bisector and the potential by using the method of Art. 49. In this proof the plane $PMQN$ is taken to be perpendicular to the boundaries.

52. Ex. A number of infinite straight attracting rods are arranged at equal distances on the surface of a circular cylinder of radius a. If n be the number of rods, m the mass of each per unit of length, prove that their potential at any point P is given by

$$V=C-m\log(r^{2n}-2a^nr^n\cos n\theta+a^{2n}),$$

where r is the distance of P from the axis of the cylinder and θ the angle r makes with a plane through the axis and one of the attracting rods.

By making n infinite while the whole mass is given, show that the potential of a uniform thin cylindrical shell at the point P is $C-4\pi aM\log a$ or $C-4\pi aM\log r$ according as P is inside or outside the cylinder, the mass per unit of area being M.

These expressions follow from Art. 50 by using De Moivre's property of the circle.

These results are of considerable interest because they help us to understand how the potential of a thin cylindrical shell is a discontinuous function of the coordinates, being constant at all points within the cylinder and depending on the logarithm of the distance from the axis at points outside. Supposing the number of rods to be very great but not infinite, the potential at any point P is represented by a continuous function of the coordinates of P, i.e., as P travels from the interior to the exterior through the interstices between the rods the potential is always the same function of the coordinates. When P is inside the cylinder, r/a is less than unity, and by expanding the logarithm in powers of r/a we see that

$$V=C-2mn\log a+2m\left(\frac{r}{a}\right)^n\cos n\theta+\&c.$$

It follows that when n is large the potential is sensibly constant throughout the interior except in the immediate neighbourhood of the surface of the cylinder on which the rods lie. When P is outside, a/r is less than unity and by expanding the logarithm in powers of a/r we find $V=C-2mn\log r+2m\left(\frac{a}{r}\right)^n\cos n\theta+\&c.$ It appears that, except in the immediate neighbourhood of the surface of the cylinder, the potential when n is large does not sensibly differ from $C-2mn\log r$ at any point outside.

As n increases, the small space within which the potential differs from the first term of these series gets continually less, and in the limit is zero, so that we may say that the potential is constant throughout the interior of the cylinder and, except for C, varies as the logarithm of the distance throughout external space.

53. Potentials of discs and cylinders. *To find the potential of a circular disc at any point P situated on its axis.*

Referring to the figure of Art. 21, the potential at P of the annulus QQ' is $2\pi mxdx/PQ$, where x and $x+dx$ are the radii of the annulus, and m the mass of the disc per unit of area. If p be the perpendicular from P on the disc and r the distance PQ, we have $r^2 = x^2 + p^2$ and $rdr = xdx$. Substituting, we find that the potential V of the disc is $V = 2\pi m \int dr = 2\pi m(r_1 - p)$, where r_1 is the distance from P of any point on the perimeter.

If a be the radius of the disc, we may also write this in the form

$$V = 2\pi m \{\sqrt{a^2+p^2} - p\}.$$

When the radius a of the disc is infinite we expand the radical and retain only the lowest power of p/a. We thus find $V = A - 2\pi mp$ where A is an infinite constant.

54. Ex. 1. The law of force being the inverse κth power of the distance, prove that the potential of an infinite disc at a point distant p from its plane is $C - \dfrac{2\pi mp^{3-\kappa}}{(\kappa-1)(3-\kappa)}$ where C is infinite or zero according as $\kappa<3$ or $\kappa>3$. When $\kappa=3$ the potential is $C - \pi m \log p$, where C is infinite.

Ex. 2. Show that the potential of a circular cylinder of density ρ, radius a, and small thickness h at an external point P on the axis close to the cylinder is $2\pi\rho h(a-p)$, where p is the mean of the distances of P from the two plane faces of the cylinder. See Art. 9.

55. Infinite Cylinders. *An indefinitely thin homogeneous layer of attracting matter of surface density m is placed on an infinitely long right circular cylinder. It is required to find the potential and the attraction at any internal or external point P.*

We replace the cylinder by a fine ring of line density $m' = 2m$ which occupies the position of the cross section through P and attracts according to the law of the inverse distance, Art. 14.

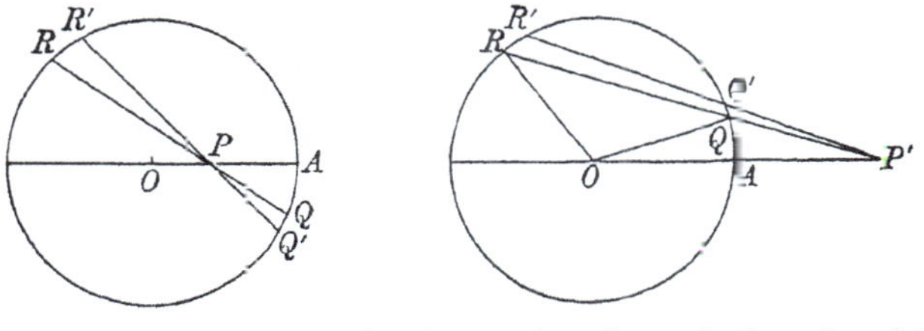

Let QR, $Q'R'$ be two chords passing through P and making a small angle $d\theta$ with each other. Let $PQ = u_1$, $PR = u_2$, $QQ' = ds_1$, $RR' = ds_2$. Let ϕ be either of the equal angles OQR, ORQ.

The attractions at P of the elements QQ', RR' are respectively $m'ds_1/u_1$ and $m'ds_2/u_2$. But since $u_1d\theta/ds_1$ and $u_2d\theta/ds_2$ are each equal to $\cos\phi$, we see that each of these attractions is equal to $m'd\theta/\cos\phi$. The attractions are therefore equal.

If P is inside the circle, these attractions balance each other. *The resultant attraction of the whole circle is zero. The potential is therefore constant and equal to that at the centre.*

56. If P is outside the circle as at P', let $\theta = OP'Q$, $r' = OP'$; then $r'\sin\theta = a\sin\phi$. The attraction of each of the elements at Q and R being $m'd\theta/\cos\phi$, the resolved attraction at P' of the whole circle along $P'O$ is

$$X = 2\int \frac{m'd\theta}{\cos\phi}\cos\theta = 2m'a\int \frac{\cos\theta\, d\theta}{\sqrt{(a^2 - r'^2\sin^2\theta)}} = \frac{2m'a}{r'}\left[\sin^{-1}\frac{r'\sin\theta}{a}\right].$$

The limits of θ are found by drawing two tangents from P' to the circle; the integral is to be taken from $\sin\theta = -a/r'$ to $+a/r'$. We therefore find $X = M/r'$ where $M = 2\pi am'$.

The attraction therefore of the ring is the same as if its whole mass were collected into its centre. *The attraction of the cylindrical layer at an external point is the same as if its whole mass were equally distributed along the axis.*

The potential is deduced from the attraction by integrating $dV/dr = -M/r$. *The potential at an external point is therefore* $V = C - M\log r$. We know by Art. 50 that the constant C is really infinite.

57. *The attraction of a solid cylindrical shell bounded internally and externally by coaxal right circular cylinders* may be deduced from the preceding results.

By dividing the body into elementary cylindrical shells we see at once that the attraction and potential at an external point are the same as if the whole mass were equally distributed along the axis.

At an internal point the attraction is zero and the potential is an infinite constant.

Lastly, let P be any point in the substance of the shell, r its distance from the axis. Let us describe a coaxal cylinder passing through P and dividing the whole attracting body into two shells. The attraction at P of the outer shell is zero; the attraction of the inner is the same as if its mass were arranged along the axis.

The line density is $\rho\pi(r^2 - a^2)$ where a is the radius of the inner boundary of the attracting cylinder and ρ the density. *The attraction* is by Art. 14

$$2\pi\rho(r^2 - a^2)/r.$$

58. Heterogeneous cylinder. *An indefinitely thin layer of attracting matter is placed on an infinitely long cylinder of radius a, so that the surface density m is uniform along any generating line but varies from one generating line to another. It is required to find the potential at any point P.*

We replace the cylinder by a fine ring, of line density $m' = 2m$, which occupies the cross section through P, the law of attraction of the ring being the inverse distance, Art. 14.

Let the plane of the circle be that of xy, the centre O being the origin. Let the polar coordinates of P be r, ϕ. Let QQ' be an element of the ring, the angle xOQ being q; let $q - \phi = \psi$.

The line density m' at Q is some given function of q, thus we expand (using Fourier's rule if necessary) in a series of the form

$$m' = \Sigma(A_n \cos nq + B_n \sin nq) \qquad (1),$$

where Σ implies summation for integral values of n from $n=0$ to ∞. We write this in the form

$$m' = \Sigma(E_n \cos n\psi + F_n \sin n\psi) \qquad (2),$$

where $E_n = A_n \cos n\phi + B_n \sin n\phi$, $F_n = -A_n \sin n\phi + B_n \cos n\phi$.

The element of mass at Q being $m'ad\psi$, and the distance PQ being u, the potential at P of the whole circle is

$$V = -\int ad\psi m' \log u + C = -\tfrac{1}{2}\int ad\psi m' \log(a^2 - 2ar\cos\psi + r^2) + C,$$

where the limits are $\psi = 0$ to 2π and C is a constant.

By writing $2\cos\psi = \xi + 1/\xi$ where ξ is an imaginary exponential we have

$$\log(1 - 2h\cos\psi + h^2) = \log(1 - h\xi) + \log(1 - h/\xi)$$
$$= -2\{h\cos\psi + \tfrac{1}{2}h^2\cos 2\psi + \tfrac{1}{3}h^3\cos 3\psi + \&c.\}.$$

This series is convergent when h is less than unity.

To obtain a convergent series for V, we expand the logarithm in powers of r/a or a/r according as P is inside or outside the circle. We therefore write the potential in the forms

$$V = -\int ad\psi \cdot \frac{m'}{2} \log\left\{1 - 2\frac{r}{a}\cos\psi + \left(\frac{r}{a}\right)^2\right\} - \int ad\psi \cdot \frac{m'}{2} \log a^2 + C,$$

$$V = -\int ad\psi \cdot \frac{m'}{2} \log\left\{1 - 2\frac{a}{r}\cos\psi + \left(\frac{a}{r}\right)^2\right\} - \int ad\psi \cdot \frac{m'}{2} \log r^2 + C,$$

according as P is inside or outside the circle.

Suppose first that $m' = E_n \cos n\psi$. Then by remembering that $\int \cos n\psi \cos m\psi d\psi = 0$ or π according as m and n are unequal or equal, the limits being 0 and 2π, we easily find that $V = E_n \frac{\pi a}{n}\left(\frac{r}{a}\right)^n + C$ or $E_n \frac{\pi a}{n}\left(\frac{a}{r}\right)^n + C$ according as P is inside or outside, except when $n = 0$.

Next suppose that $m' = F_n \sin n\psi$, then since $\int \cos n\psi \sin m\psi d\psi = 0$, the limits being 0 and 2π, we find by the same reasoning that the potential at P is constant, whether P is inside or outside.

When $n = 0$, we have $m' = E_0$ and the potentials take the form $-E_0\pi a \log a^2 + C$, or $-E_0\pi a \log r^2 + C$, according as the point P is inside or outside.

When the line density at Q is given by *a single term* of the series (1), it is evident from (2) that E_n *represents the line density of the ring at the point where the radius vector OP cuts the ring.*

Finally, the potential for the whole ring is found by adding together the potentials for the separate terms of the series (1).

Ex. The density of a thin stratum, on a right circular cylinder of radius a, is proportional to the distance from a plane through the axis, and its greatest value is D. Prove that the potential at any point P is $2\pi a^2 D\xi/r^2$ or $2\pi D\xi$ according as P is outside or inside, where ξ and r are the distances of P from the given plane and from the axis respectively.

59. Systems of particles. If a particle of mass m_1' travel from a position at which the potential is zero along any path to any assigned position B_1, it is clear from what precedes that the work done by the attracting forces is $V_1 m_1'$, where V_1 is the potential at B_1. If a second particle m_2' travel from a position of zero potential to the position B_2, it is clear that the additional work is $V_2 m_2'$, where V_2 is the potential at B_2 of the same attracting forces.

Generalizing this, let there be two systems of particles; let the masses of the first be m_1, m_2, &c., and let these be situated at the points A_1, A_2, &c. Let the masses of the second be m_1', m_2', &c. and let these be situated at the points B_1, B_2, &c. Let V_1, V_2, &c. be the potentials of the first system at B_1, B_2, &c.; V_1', V_2', &c. the potentials of the second system at A_1, A_2, &c. Let us also suppose that each particle of either system acts on all the particles of the other, but does not attract any particle of its own system. The work done by the attracting forces in moving the particles of the second system from positions of zero potential to their assigned positions is
$$W' = V_1 m_1' + V_2 m_2' + \ldots$$
In the same way the work of bringing the particles of the first system from positions of zero potential to the positions A_1, A_2, &c. under the influence of the attracting forces of the second system is
$$W = V_1' m_1 + V_2' m_2 + \ldots$$

If r_{12} be the distance between the particles m_1, m_2', and r_{21} that between the particles m_2, m_1', and so on, the values of the potentials V_1, V_1' are
$$V_1 = \frac{m_1}{r_{11}} + \frac{m_2}{r_{21}} + \&\text{c.}, \qquad V_1' = \frac{m_1'}{r_{11}} + \frac{m_2'}{r_{12}} + \&\text{c.}$$

Substituting, we find that each of the expressions W, W' is equal to
$$\frac{m_1 m_1'}{r_{11}} + \frac{m_1 m_2'}{r_{12}} + \frac{m_2 m_1'}{r_{21}} + \ldots = \Sigma \frac{mm'}{r}.$$

If the forces were repulsive instead of attractive this formula expresses the work the system would do if the particles (under the influence of their repulsions) retired to infinite distances.

This symmetrical expression is called *the mutual potential energy* or *the mutual work* of the two systems according as the standard of force is repulsion or attraction (Art. 41).

The work done when either system moves from one given position to another under the influence of the attractions of the other system is the difference of their mutual works in the two positions. *If both systems are moved, each from one given position to another, under the influence of their mutual attractions*, it easily follows, by moving them one at a time, that *the work done is the excess of their mutual work in their final positions over that in their initial positions.*

60. If the particles are elements of a solid body the argument is still the same. Let dv' be an element of the volume of any finite mass M', ρ' its density, V the potential of any fixed system of attracting bodies at the element dv'; the work of collecting together the mass M' is $\int V\rho' dv'$.

This formula may be put into the form of a rule. *To find the mutual work of two attracting masses in assigned positions, we multiply the mass of each element of one body by the potential of the other at that element, and then integrate the result throughout the volume of the first body.*

61. The particles of a system mutually attract each other and are in assigned positions. Supposing them to have been originally at distances so far apart that their mutual attractions were zero, it is required to find the work done by their attractions as they are collected together and brought each into its assigned position.

Let us suppose that the particles $m_1, m_2, \ldots m_{n-1}$ have been brought into their proper places. We now bring m_n from infinity into its place under the attraction of $m_1 \ldots m_{n-1}$. The work is $m_n \left\{\frac{m_1}{r_{1n}} + \frac{m_2}{r_{2n}} + \ldots \frac{m_{n-1}}{r_{n-1,\,n}}\right\}$. Thus m_n is taken once with each of the masses $m_1, m_2, \ldots m_{n-1}$. When we bring in succession m_{n+1}, m_{n+2}, &c. from infinity we obtain a similar series for each and therefore m_n is taken once with each of these masses as it is brought in. Thus m_n is taken once with every mass except itself. If m, m' are the masses of any two particles, r their distance apart in the final arrangement, the work of the attractions when collecting the system is $W = \Sigma \frac{mm'}{r}$.

Let V_1 be the potential in the given final arrangement at the particle m_1 of all the particles except m_1; V_2 the potential at m_2 of all the particles except m_2 and so on. Then

$$V_1 = \frac{m_2}{r_{12}} + \frac{m_3}{r_{13}} + \&c., \qquad V_2 = \frac{m_1}{r_{12}} + \frac{m_3}{r_{23}} + \&c.$$

Let us consider how often the mass m_n occurs in the expression $V_1 m_1 + V_2 m_2 + \dots$. It occurs once in $V_1 m_1$ combined with m_1, once in $V_2 m_2$ combined with m_2 and so on. Again it occurs in $V_n m_n$ combined with every other mass. Thus on the whole m_n occurs *twice* combined with every other mass. It follows that the work of collecting the system is

$$W = \tfrac{1}{2}(V_1 m_1 + V_2 m_2 + \dots) = \tfrac{1}{2}\Sigma Vm.$$

We thus arrive at the following rule. *To find the work done by the attractions of a system of particles brought from infinite distances to any assigned positions we multiply the mass of each element by the potential at that element, integrate throughout the volume, and halve the result.*

This rule, when the final sign is reversed, also gives the work when the particles move from any assigned positions to infinite distances. To find the work when the particles move from one assigned arrangement to another, we add together the work when taken from the first arrangement to infinite distances and the work when brought from infinite distances to the second arrangement. If the system be moved, like a rigid body, from one place to another so that the relative positions of the particles in the two places are the same, it is clear that no work is done by the mutual attractions of the particles.

62. In this investigation we have treated the masses m_1, m_2, &c. as if their linear dimensions were infinitely small compared with their distances apart. In the case of a continuous body the portions of matter not in contact can be divided into elements so small that the above assumption is correct, but the argument might be supposed to fail for two elements which finally become contiguous.

We notice however that in finding the potential of any solid mass at a point P we may omit the matter within any indefinitely small element of volume enclosing P if its density is finite. For since potential is "mass divided by distance" and the mass varies as the cube of the linear dimensions, it follows that the potentials of similar bodies at points similarly situated must vary as the square of the linear dimensions. The potential must therefore vanish when the mass becomes elementary and the distance indefinitely small. In applying therefore the form $W = \frac{1}{2}\Sigma Vm$ to a solid body we write ρdv for m and take V to be the potential of the whole mass at the element dv.

In the same way, in finding the potential of a surface at a point P on the surface we may omit the element contiguous to P if the surface density is finite. For, the potentials of similar areas at similarly situated points vary as the linear dimensions, and are zero when the areas become elementary.

63. It appears from the definition of potential that its dimensions are not the same as those of work. The potential of a particle whose mass is m at a point P distant r is m/r. If a particle of mass m' is situated at the point P, the mutual potential energy or work of these two particles is mm'/r. *The dimensions of potential are therefore mass divided by distance, those of work are mass squared divided by distance.*

Spherical Surface.

64. *To find the potential of a thin uniform spherical shell at any point.*

Let O be the centre of the shell, a the radius of either bounding surface, m the mass per unit of area. Let P be the point at which the potential is required, $OP = r$.

Taking on the surface of the shell an annulus QQ' whose axis is OP, let the angle

$POQ = \theta$, and $QP = u$.

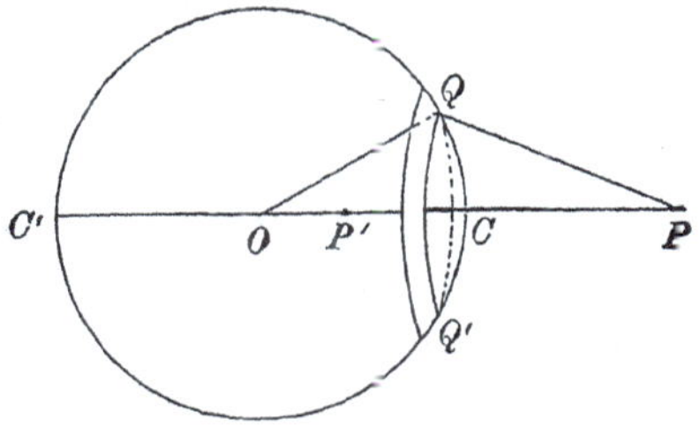

Since the mass of the annulus is $m \cdot ad\theta \cdot 2\pi a \sin\theta$ by Pappus' theorem (Vol. I. Art. 413), the potential at P of the whole shell is

$$V = \int \frac{2\pi m a^2 \sin\theta \, d\theta}{u}.$$

Since $u^2 = r^2 + a^2 - 2ar\cos\theta$, we have $udu = ar\sin\theta\, d\theta$.

Substituting, we find $V = \frac{2\pi ma}{r}\int du$. If the point P is external to the surface as shown in the figure, the limits of u are $u = PC$ to $u = PC'$, i.e. $u = r - a$ to $r + a$. In this case $V = \frac{4\pi ma^2}{r}$. If the point P is inside the shell as at P', the limits of u are $u = P'C$ to $u = P'C'$, i.e. $u = a - r$ to $a + r$. In this case $V = \frac{4\pi ma^2}{a}$.

If M be the whole mass of the shell, $M = 4\pi ma^2$, and these expressions take the form $V = \frac{M}{r}$ or $V = \frac{M}{a}$ according as the attracted point P lies outside or inside the shell.

When the point P is at the centre, u is constant and cannot be properly taken as the independent variable. But since every element of the attracting mass is equally distant from P, it is evident that the potential at the centre is equal to the mass divided by the radius, and this agrees with the above result.

65. Since the potential is the same at all points within the spherical shell, it follows that its differential coefficient with regard to each of the coordinates is zero. Thus *the attraction of a thin uniform spherical shell at an internal point is zero.*

Since a thick shell bounded by concentric spheres may be regarded as composed of a sufficient number of thin shells, it follows that *the attraction of a thick shell bounded by concentric spheres at an internal point is zero.*

This theorem is also true for *a heterogeneous thick shell provided the strata of equal density are concentric spheres.* For in this case each of the thin shells into which it is analysed is homogeneous.

66. Since the potential at an external point of a uniform thin shell is M/r, we see that the force at an external point P resolved in the direction OP is equal to $-M/r^2$. *The attraction therefore acts in the direction from P towards the centre, and is the same as if the whole mass were collected at its centre.*

As before, since a thick shell may be analysed into elementary thin shells, it follows that *the attraction of a thick shell bounded by concentric spheres or of a solid sphere at any external point is the same as if the whole mass were collected into its centre. Also this is true for heterogeneous shells provided the strata of equal density are concentric spheres.*

These theorems on the attraction of a spherical shell as well as that of a spheroid at an internal point are due to Newton.

67. It remains to find the attraction of a thin uniform shell on an elementary area which is part of itself. Let the attracted point P be at C, then $CQ = u$, $r = a$, and $\cos QCO = u/2a$. Proceeding as before, we find the attraction X at C is

$$X = \int \frac{2\pi ma^2 \sin\theta\, d\theta}{u^2} \cdot \frac{u}{2a} = \frac{\pi m}{a} \int du,$$

the limits of u being 0 and $2a$. This gives $X = 2\pi m = \frac{1}{2}\frac{M}{a^2}$.

The attraction of a thin uniform shell on an element of its surface is the same as if half the mass of the shell were collected at its centre.

68. That the attraction of a thin uniform shell bounded by concentric spheres at an internal point P is zero, may be shown by an elementary geometrical argument which applies also to the case of some ellipsoidal shells.

With P as vertex describe an elementary cone cutting the surfaces of the shell in $QQ'qq'$, $RR'rr'$ respectively. Let $PQ = r$, $Qq = dr$; $PR = r'$, $Rr = dr'$. If $d\omega$ be the solid angle of the elementary cone, the volumes of the elementary solids at Q and R will be respectively $r^2 d\omega dr$ and $r'^2 d\omega dr'$. Their attractions at P are therefore $\rho d\omega dr$ and $\rho d\omega dr'$, where ρ is the density. These attractions will balance each other whenever the form of the shell is such that the intercepted parts Qq, Rr of the chord $qQRr$ are equal. This being true for all chords through P, the attraction of

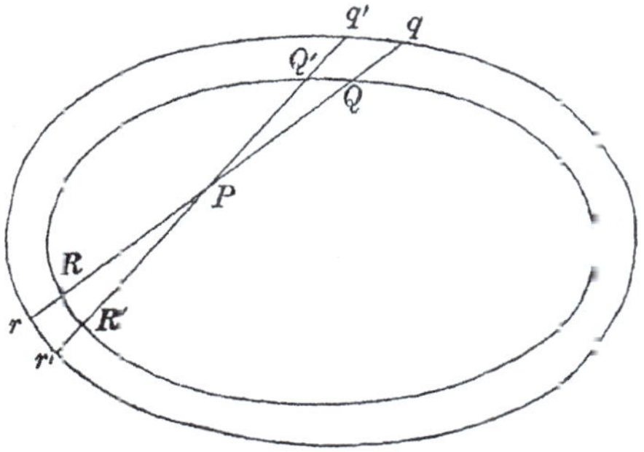

every element is balanced by that of the opposite element, and the resultant attraction on P is zero.

When the shell is bounded by concentric spheres these intercepted parts are evidently equal. The resultant attraction on any internal point is therefore zero.

When the shell is bounded by similar and similarly situated concentric ellipsoids the same is also true. To prove this we notice that, since the chords parallel to QR have in the two ellipsoids a common diametral plane, the chords QR and qr must have the same middle point. It follows that the intercepted parts Qq and Rr are equal.

Since a thick shell may be analysed into elementary thin shells, it follows that *the attraction of any homogeneous shell bounded by similar and similarly situated concentric ellipsoids at any internal point is zero.*

69. If P is on the outside of a thin ellipsoidal shell, bounded by similar concentric ellipsoids, we may show by similar reasoning that *the enveloping cone whose vertex is P divides the surface into two portions whose attractions at P are the same in direction and magnitude.*

When P is indefinitely close to the outer margin of the shell, the infinitely small portion on the nearer side of the polar plane exerts the same attraction at P as all the rest of the shell. If the thin shell is spherical, the resultant attraction is known to be the same as if the whole mass were collected at its centre. Putting m for the mass per unit of area, the attraction at P of each of the portions on the two sides of the polar plane is $2\pi m$.

70. We may apply these results to the solid bounded by two concentric similar and similarly situated hyperboloids. If one sheet attract and the other repel, the attraction on P is zero, provided both sheets are on the same side of P.

Also a paraboloidal shell bounded by two equal paraboloids having their axes coincident but their vertices separate exerts no attraction at an internal point.

71. If the thin shell is ellipsoidal and P is very close to the outer margin, the distance of P from the polar plane is infinitely smaller than the linear dimensions of the curve of contact. The attraction at P of the portion on the nearer side of the polar plane is therefore the same as that of an infinite plate of the same thickness, see Art. 22. The attraction at P of each of the portions on the two sides of the polar plane is therefore $2\pi m$, where m is the mass of the shell in the neighbourhood of P per unit of area. The attraction of the whole shell at a point P, just outside the shell, is therefore twice that of an infinite plate of the same thickness as that of the shell at P, i.e. *the attraction is $4\pi m$.* It also follows that *the direction of the attraction* is the same as that of the infinite plate and *is normal to the shell.* This line of argument will be more fully considered further on.

Let a, b, c be the semi-axes of the inner boundary of the shell,

p a perpendicular drawn from the centre to the tangent plane, ρ the uniform density of the shell, then $m = \rho dp$. The volume v of the ellipsoid is $\frac{4}{3}\pi a^3 (bc/a^2)$, and the volume dv of the shell (being the differential obtained on the supposition that b/a and c/a are constants) is $4\pi bcda$ and the mass M of the shell is ρdv. Also since the bounding surfaces are similar $da/a = dp/p$. *The resultant attraction of a thin ellipsoidal shell bounded by similar ellipsoids at an external point close to its surface is therefore equal to* $\dfrac{Mp}{abc}$ *and its direction is normal to the surface.*

72. **Cylindrical elliptic shell.** By making one axis of the ellipsoidal shell infinite, we deduce that *the attraction of any homogeneous shell bounded by similar and similarly situated concentric elliptic cylinders at any internal point is zero.*

Let μ' be the mass per unit of length of *a thin cylindrical shell*, and let the infinite axis be c; then the whole mass of the shell is $\mu' c$. *The resultant attraction at any point just outside is equal to* $\mu' p/ab$ *and its direction is normal to the surface.*

73. Ex. 1. Prove that, if the attraction of a shell is zero at all internal points and the inner surface is an ellipsoid, the outer surface is a similar and similarly situated concentric ellipsoid.

If possible let the outer surface have some other form. Describe a similar ellipsoid to *enclose* and touch the outer surface at some point T. The difference between the ellipsoidal shell thus formed and the given shell possesses also the property that the attraction is zero. This shell has no thickness at the point T of contact, and the surface density m at T is zero.

Let P be a point inside this shell very near T, draw a plane through P parallel to the tangent plane at T. The attraction of the matter on one side of this plane balances that on the other. But the attraction of the matter on one side is ultimately zero (being in fact $2\pi m$), hence the attraction of the other is unbalanced and the particle P cannot be in equilibrium. [Todhunter's *History*, &c. Art. 1473.]

Ex. 2. If the matter composing a thin shell bounded by concentric spheres attract according to the inverse κth power of the distance, prove that the resultant force on an internal particle P acts towards or from the centre according as κ is less or greater than 2. Cavendish, *Phil. Trans.* 1771.

The plane section whose centre is P is such that the longer segment PQ of every chord QPR of the sphere is on the same side of the plane as the centre of the sphere. Since the masses of the elements Q, R are as PQ^2 to PR^2, the attractions are as $PQ^{2-\kappa}$ to $PR^{2-\kappa}$. The first is greater or less than the second according as $\kappa < 2$ or > 2.

Ex. 3. If matter attracting according to the law of gravitation be uniformly distributed upon the circumference of a circle, show that the chord of contact of tangents drawn to the circle from any external point divides the circle into two arcs, such that the potentials at the point due to each arc are the same.

[Math. Tripos.]

74. Potential of an annulus. We may use the method of Art. 64 to find the potential of an annulus of a thin uniform spherical shell at a point P on its axis.

Let $DD'EE'$ be the annulus; let $PD = u_1$, $PE = u_2$, $OP = r$.

The potential of an elementary annulus QQ' being the same as before, the potential V of the whole annulus is

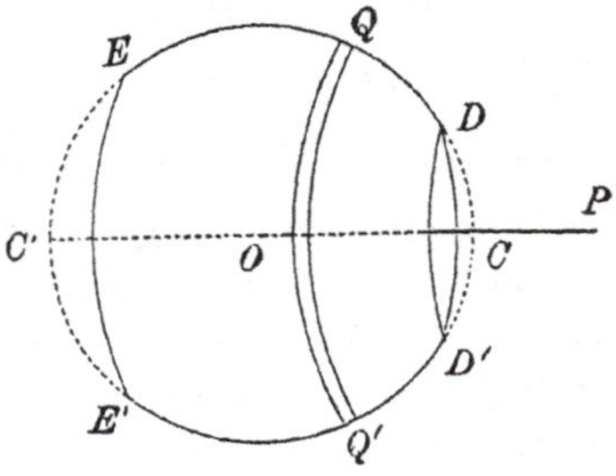

$$V = \frac{2\pi ma}{r}\int du = \frac{2\pi ma}{r}(u_2 - u_1),$$

since in our case the limits of integration are $u = PD$ and $u = PE$. In the same way the mass M of the annulus is

$$M = \frac{2\pi ma}{r}\int udu = \frac{\pi ma}{r}(u_2^2 - u_1^2).$$

The potential of the whole annulus is $V = \dfrac{M}{\frac{1}{2}(u_1 + u_2)}$.

75. If we suppose the annulus to form a complete sphere except for two small holes DD', EE', we have an expression for the potential which applies equally to points inside and outside the shell, provided they lie on the axis. Let y be the radius of either hole. When P is inside the shell the sum of the distances u_1 and u_2 differs from the diameter only by small quantities of the order y^2 and the potential is therefore sensibly constant. When P passes through the hole DD' the distance u_1 has a minimum value equal to y and then begins to increase without vanishing or changing sign. When P is outside the shell the sum of the distances u_1 and u_2 differs from twice the distance of P from the centre by quantities of the order y^2, so that the potential sensibly follows the law of the inverse distance. See Art. 39.

76. Ex. 1. The internal and external radii of a thin spherical shell of density ρ are $a - t$ and a. Prove that the difference of the potentials at two points, one inside and the other outside, both close to the surface, is $2\pi\rho t^2$. We notice that this difference is of the second order of the small quantity t.

Ex. 2. A thin spherical shell of radius a attracts an internal particle P at a distance r from the centre. If the shell be divided into two parts by a plane through P perpendicular to the radius the resultant attraction of each part at P is $\dfrac{2\pi ma}{r^2}\{a - (a^2 - r^2)^{\frac{1}{2}}\}$ where m is the surface density. [Todhunter's *History*, 1615.]

77. A solid sphere. *To find the attraction of a solid uniform sphere at an internal point P.*

Describe a sphere concentric with the given surface to pass through P. The attraction at P of the matter between this sphere and the given surface is zero; Art. 65. The attraction at P of the matter within this sphere is the same as if it were collected at the centre, Art. 66. If r be the distance of P from the centre O, the attraction is $\frac{4}{3}\pi\rho r^3/r^2$, where ρ is the density. It follows that *the attraction of a solid homogeneous sphere at an internal point distant r from the centre is* $\frac{4}{3}\pi\rho r$.

If (x, y, z) be the coordinates of P referred to the centre as origin, X, Y, Z the components of attraction, we have also

$$X = -\tfrac{4}{3}\pi\rho x, \qquad Y = -\tfrac{4}{3}\pi\rho y, \qquad Z = -\tfrac{4}{3}\pi\rho z.$$

These are obtained by resolving the resultant attraction, viz. $\frac{4}{3}\pi\rho r$, parallel to the axes.

78. *To find the potential of a solid sphere at an internal point P.*

If x and $x + dx$ are the radii of an elementary shell, taken within the sphere passing through P, its potential at P is $4\pi\rho x^2 dx/r$, Art. 64. In the same way, if y and $y + dy$ are the radii of an elementary shell outside the same sphere, its potential at P is $4\pi\rho y^2 dy/y$. The potential at P of the whole sphere is therefore

$$V = \int_0^r \frac{4\pi\rho x^2 dx}{r} + \int_r^a \frac{4\pi\rho y^2 dy}{y}.$$

If the density ρ of the sphere is uniform, this integral becomes

$$V = \frac{2\pi\rho}{3}(3a^2 - r^2).$$

If the density is any function of the distance from the centre, the integration can be effected when the function is given.

79. Ex. 1. A portion of a homogeneous spherical shell is cut off by a cone whose vertex is at the centre and whose solid angle is $d\omega$. Show that the attraction, per unit of mass, of the rest of the shell on this portion is

$$\pi\rho(b-a)\frac{b^2 + 2ab + 3a^2}{b^2 + ab + a^2},$$

where a and b are the internal and external radii of the shell. Hence show that when the shell is indefinitely thin the attraction is *half* that just outside.

Since the resultant attraction of a body on itself is zero the attraction of the rest of the shell is the same as that of the whole shell. The attraction on the portion included is $\int \frac{4\pi}{3}\rho^2 \cdot \frac{r^3 - a^3}{r^2} r^2 dr d\omega$; dividing this by the mass attracted, viz. $\frac{1}{3}\rho(r^3 - a^3)\,d\omega$, we have the result above given.

Ex. 2. Prove that the pressure per unit of length, on any normal section of a spherical shell of mass M and radius a, due to the mutual gravitation of the particles tends to the limit $M^2/16\pi a^3$, as the thickness of the shell is indefinitely diminished.

[Math. Tripos.]

Ex. 3. A solid homogeneous sphere is divided by a plane through its centre into two hemispheres. These being placed with their plane faces coincident, show that the force required to pull them apart is $\frac{3}{16}M^2/a^2$, where M is the mass of the sphere and a its radius.

Ex. 4. The density of a solid sphere varies as the nth power of the distance from the centre. Show that the potential at an internal point is

$$V=\frac{4\pi\rho}{(n+2)(n+3)}\left\{(n+3)a^2-\frac{r^{n+2}}{a^n}\right\},$$

where ρ is the density at the surface and $n+2$ is positive.

Ex. 5. A homogeneous sphere is divided into two parts by a plane QNR bisecting OP at right angles, P being any point within the sphere and O the centre. If a be the radius of the sphere and $c=OP$, prove that the attraction at P of the larger part of the sphere cut off by the plane QNR is n times the attraction at P of the whole sphere, where $n=(3a-c)/4c$.

Ex. 6. If I be an external point and C the centre of a sphere, prove that the sphere on IC as diameter, the sphere with centre I and radius IC or the polar plane of I will divide the sphere into two parts exerting equal attractions at I, according as the law of attraction is the inverse square, the inverse cube, or the inverse fourth power of the distance. [St John's Coll., 1885.]

If the law be the inverse nth power, and a radius vector from I as origin cut the sphere in Q, R and the dividing surface in S, then $2(IS)^{3-n}=(IQ)^{3-n}+(IR)^{3-n}$ except when $n=3$. The results given follow at once.

Ex. 7. If a homogeneous solid hemisphere of radius a and density ρ be referred to the centre of the complete sphere as origin, the bounding plane circle as plane of xy and the radius of the hemisphere perpendicular to the plane of xy as axis of z, then the attraction at the origin is along the axis of z and is equal to $\pi\rho a$. Further show that if V be the potential at a point xyz near the origin, then

$$V=\pi\rho a^2+\pi\rho az-\tfrac{1}{3}\pi\rho\{x^2+y^2+4z^2\} \text{ (within the hemisphere),}$$

and $$V=\pi\rho a^2+\pi\rho az-\tfrac{1}{3}\pi\rho\{x^2+y^2-2z^2\} \text{ (without the hemisphere).}$$

[St John's Coll., 1886.]

Ex. 8. The potential of a solid hemisphere of radius a and unit density, at an external point P situated on the axis at a distance ξ from the centre, is

$$V=\frac{2\pi a^3}{3\xi}\pm\frac{2\pi}{3\xi}\{(\xi^2+a^2)^{\frac{3}{2}}-\xi^3-\tfrac{3}{2}a^2\xi\},$$

the upper or lower sign being taken according as P is on the convex or plane side of the body. The potential at an internal point may be found by subtracting from the potential of the complete sphere, that of the missing half.

Ex. 9. A point P is situated very near to the rim of a thin hemispherical shell on a prolongation of a radius of the rim. Prove that the component of attraction at P of the shell in a direction perpendicular to the plane of the rim is ultimately $2m\log 8a/x$, where a is the radius, x the infinitely small distance of P from the rim, and m the surface density.

Ex. 10. Two mutually attracting spheres are placed at rest in a vacuum. The radius of each is one foot and the distance apart (surface to surface) is 1/4th inch, and the density the same as the mean density of the earth. Prove that they will meet in less than 250 seconds. This problem is due to Newton who gave a wrong numerical result. [Todhunter, *History*, &c. Art. 725.]

80. Other laws of force. Ex. 1. Let the law of force be the inverse κth power of the distance. Let the potential of a thin homogeneous spherical surface at a point P be V_κ' or V_κ according as P is external or internal. Prove the following results

$$V_\kappa' = \frac{M}{(\kappa-1)(\kappa-3)} \cdot \frac{(r-a)^{3-\kappa} - (r+a)^{3-\kappa}}{2ar}, \quad V_\kappa = \frac{M}{(\kappa-1)(\kappa-3)} \cdot \frac{(a-r)^{3-\kappa} - (a+r)^{3-\kappa}}{2ar},$$

$$rV_{\kappa+1}' = \frac{-(\kappa-1)}{\kappa(\kappa-2)} \frac{d}{dr}(rV_\kappa'), \quad ar^{\kappa-3}V_{\kappa+1} = \frac{\kappa-1}{\kappa(\kappa-2)} \frac{d}{dr}(r^{\kappa-2}V_\kappa),$$

where M is the mass and a the radius.

Ex. 2. Prove that the potential of a homogeneous solid sphere of unit density at an internal point P distant r from the centre is

$$V_\kappa = \frac{4\pi}{(\kappa-1)(\kappa-3)} \left\{ \frac{(a-r)^{5-\kappa} - (a+r)^{5-\kappa}}{2(5-\kappa)r} + \frac{(a-r)^{4-\kappa} + (a+r)^{4-\kappa}}{2(4-\kappa)} \right\}.$$

To this we add an infinite constant when $\kappa > 4$. The integral takes another form when $\kappa = 4$.

Ex. 3. Let the law of attraction be the inverse cube. Prove that the potential of a thin spherical shell at a point P distant r from the centre is V_3' or V_3 according as P is external or internal, where

$$V_3' = \frac{M}{4ar} \log \frac{r+a}{r-a}, \qquad V_3 = \frac{M}{4ar} \log \frac{a+r}{a-r}.$$

Prove also that the potential of a solid sphere of unit density at an internal point is

$$V = \frac{\pi}{r} \left\{ \frac{a^2 - r^2}{2} \log \frac{a+r}{a-r} + ar \right\}.$$

81. *To find the potential of a shell bounded by any two non-intersecting spheres.*

Let A and B be the centres of the spheres, a and b their radii. Let ρ be the density of the attracting matter which fills the space between these spheres.

The potential at any point P is evidently the difference of the potentials of the spheres each regarded as a solid sphere of density ρ. If r, r' be the distances of P from A and B respectively, the potential at P is

$$V = \tfrac{4}{3}\pi\rho \left(\frac{a^3}{r} - \frac{b^3}{r'} \right) \text{ or } \frac{2\pi\rho}{3} (3a^2 - 3b^2 - r^2 + r'^2),$$

according as P is outside or inside both spheres. If P lie between the spheres

$$V = \tfrac{2}{3}\pi\rho \left(3a^2 - r^2 - \frac{2b^3}{r'} \right).$$

82. We may use the same principle *to find the attraction of a shell bounded by two non-intersecting spheres.*

Suppose, for example, that the attracted point lies within both spheres. The force at P is evidently the resultant of two forces, (1) an attraction equal to $\tfrac{4}{3}\pi\rho \,.\, PA$ acting along PA, and (2) a repulsion equal to $\tfrac{4}{3}\pi\rho \,.\, BP$ acting along BP. By the triangle of forces, the resultant of these is equal to $\frac{4}{3}\pi\rho \,.\, BA$ acting parallel to BA. Thus *the attraction at all internal points is the same in direction and magnitude.* The attraction at an external point may be found in the same way.

83. Ex. Two spheres touch at a point O, and the space between is filled with homogeneous attracting matter. Show that, when the radii differ by an infinitely small quantity, the attractions at two external points, one at O and the other at the opposite extremity of the diameter through O, are as 1 : 5. What is the ratio if the points are inside both spheres?

84. A theorem of Gauss. *The mean value of the potential of any attracting system, taken for all points on any spherical surface, is equal to the potential at the centre due to that part of the attracting system which lies outside the sphere plus the quotient of the mass inside the sphere by the radius.*

Let $d\sigma$ be any element of surface of the sphere, V the potential of all the attracting mass at this element. Let M be the mass inside the sphere and M' that outside, and let V_1 be the potential of the latter at the centre C. Let a be the radius of the sphere, then we have to prove that $\dfrac{\int V d\sigma}{4\pi a^2} = V_1 + \dfrac{M}{a}$.

Let m be the mass of any particle of the attracting system, and let it be situated at a point A. Its potential at any point Q of the sphere is therefore m/AQ. The part of the integral $\int V d\sigma$ due to this mass is therefore $\int m d\sigma/AQ$. The integral $\int d\sigma/AQ$ is evidently the potential at A of a thin stratum placed on the sphere, of unit surface density, and is therefore equal to $4\pi a^2/AC$ or $4\pi a^2/a$ according as the point A is situated outside or inside the sphere.

Taking all the particles of the attracting system, every particle m outside the sphere contributes a term $4\pi a^2 \,.\, m/AC$ to the integral $\int V d\sigma$ while every particle m' inside contributes a term $4\pi a^2 \,.\, m'/a$. We therefore have $\dfrac{\int V d\sigma}{4\pi a^2} = \Sigma \dfrac{m}{AC} + \dfrac{\Sigma m'}{a}$. Since V_1 is the potential of the external mass at the centre of the sphere, the result follows at once.

Ex. Prove that the mean value of the potential of a body, taken for all points equally distributed throughout the volume of a sphere which is external to the body, is equal to the potential of the body at the centre. This theorem was given by Poisson for *the component of attraction* on any given direction. *Comptes Rendus*, vol. VII., 1838.

85. Heterogeneous spherical shells. The potential of a heterogeneous spherical shell may be found by the help of Laplace's functions more easily than by any other method. Although there are several cases of heterogeneous shells whose attractions may be found by special artifices, it does not seem useful to stop over these when they can all be treated by one comprehensive method. We must however postpone the discussion of this method until after we have reached Laplace's equation. In the meantime there are some general theorems on heterogeneous shells which are independent of Laplace's functions, and to these we shall now turn our attention.

86. *The potential of a thin heterogeneous spherical shell being supposed known at all internal points, it is required to find the potential at all external points.*

Let O be the centre, a the radius of the sphere. Let P, P' be two points on the same radius, one inside and the other outside, such that

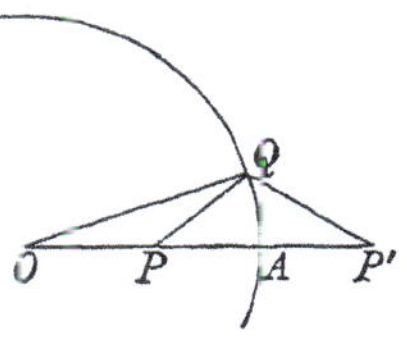

$$OP \,.\, OP' = a^2.$$

The points P, P' are called *inverse points.* Let $OP = r$, $OP' = r'$.

Let Q be any point on the surface, then since $OP \,.\, OP' = OQ^2$ the triangles QOP, $P'OQ$ are similar. It follows that the ratio QP/QP' is constant for all points on the sphere, and that this ratio is equal to a/r'.

Let V, V' be the potentials of the whole shell at P, P'. If m be an element of mass at Q, the potentials of m at P and P' are respectively m/QP and m/QP'. Since these have a constant ratio for all positions of Q and all values of m, the potentials V, V' must have the same ratio. We therefore have $V' = V\dfrac{a}{r'}$.

If the law of force is the inverse κth power of the distance, the potentials of the mass m at P and P' respectively are in the ratio $1/(QP)^{\kappa-1}$ to $1/(QP')^{\kappa-1}$. We therefore have $V' = V\left(\dfrac{a}{r'}\right)^{\kappa-1}$. If the law of force is the inverse distance we find in the same way that $V' - V = M \log a/r'$ where M is the whole mass of the shell.

We notice that *these theorems do not require the shell to be homogeneous or the sphere to be complete.* They apply to any distributions of attracting matter on the surface of the sphere.

Ex. The potential at an internal point of a thin homogeneous shell of radius a being $V = M/a$, prove that the potential V' at an external point distant r' from the centre is $V' = M/r'$.

87. A theorem of Stokes. Let X, X' be the radial components of the attractions at P, P', estimated positively when directed from the centre. Then since

$$rr'=a^2, \qquad X=\frac{dV}{dr}, \quad X'=\frac{dV'}{dr'}=-\frac{dV}{dr}\frac{a^3}{r'^3}-V\frac{a}{r'^2}=-X\frac{a^3}{r'^3}-V\frac{a}{r'^2},$$

when the points P, P' approach indefinitely near to the surface $r'=a$, and this equation reduces to $$X'+X=-V/a.$$
We therefore have the following theorem. *The sum of the inward normal attractions at two points on the same radius, one just inside and the other just outside a thin heterogeneous spherical shell, is equal to the potential at either point divided by the radius.* This theorem is given by Sir G. Stokes in his article on the *Figure of the Earth*, and is there proved by the use of Laplace's functions.

88. Let Y, Y' be the components of the attraction at P, P' perpendicular to OPP'. Let the radius vector OPP' turn round O through an angle $d\theta$. Then

$$Y'=\frac{dV'}{r'd\theta}=\frac{dV}{d\theta}\cdot\frac{a}{r'^2}=\frac{dV}{rd\theta}\left(\frac{a}{r'}\right)^3=Y\left(\frac{a}{r'}\right)^3.$$

When the points P, P' approach indefinitely near to the surface we have $Y'=Y$.

89. A converse problem. *To determine the law of force when it is given that the attraction of every thin uniform spherical shell at every external point is the same as that of an equal particle placed at the centre.* Laplace, *Méc. Céleste*, vol. I. p. 163.

Let the potential of a particle m at a distance u be $mf(u)$. The potential of the shell at a point P is $\frac{2\pi ma}{r}\int uf(u)\,du$, the limits being $r-a$ to $r+a$ or $a-r$ to $a+r$ according as P is external or internal, Art. 64. Since the attractions are equal, the potentials of the shell and the central point must differ by a quantity independent of r. Hence

$$\frac{2\pi ma}{r}\int uf(u)\,du=4\pi ma^2f(r)+2\pi maA \quad\text{.......................(1)},$$

where A may be a function of a but is independent of r. *If the potentials of the shell and the central point are also to be equal we must have $A=0$.*

Put $uf(u)=F'(u)$, the equation (1) then becomes

$$F(r+a)-F(r-a)=2aF'(r)+Ar \quad\text{.........................(2)},$$

where $r>a$. Since the equality is to hold for shells of all radii, we may differentiate this equation with regard to a. Differentiate twice with regard to r and twice with regard to a, we then have $$F^{iv}(r+a)=F^{iv}(r-a) \quad\text{................................(3).}$$
Since r and a are independent variables this equation cannot hold unless each side is a constant, for if we write $r=a$, we have $F^{iv}(2a)=$ a constant. We therefore have

$$F(r)=\alpha+\beta r+\gamma r^2+\delta r^3+\epsilon r^4 \quad\text{............................(4)},$$

where α, β, γ, δ, ϵ are constants. Since (3) has been obtained from (2) by differentiation, this value of $F(r)$ may not satisfy (2). Substitute in (2) and we find $\delta=0$, $A=8a^3\epsilon$. We thus have

$$f(u)=\frac{\beta}{u}+2\gamma+4\epsilon u^2 \quad\text{..................................(5).}$$

The only laws of force therefore which can make the attraction of every shell equal to that of its central point are the inverse square, the direct distance and any combination of these. It is unnecessary to include the case in which the potential is constant, since the attraction is then zero. *If the potentials also are to be equal* we must have $A=0$ and therefore $\epsilon=0$. *The potential must then vary as the inverse distance.* If the potential of m is $-\frac{1}{2}mu^2$ the force varies as the distance. It is easy to prove by direct integration that the potential of the shell cannot be equal to that of the central point, but exceeds it by $-\frac{1}{2}Ma^2$.

90. *We may also enquire the law of force when it is given that the attraction of every thin spherical shell is zero at all internal points.* We then find

$$F(a+r) - F(a-r) = Ar \quad \text{..........................} (6).$$

Differentiate twice with respect to r, we find $F''(a+r) = F''(a-r)$. Since both a and r are independent variables, this as before requires that each side should be equal to some constant β. We then find

$$f(u) = \frac{a}{u} + \beta \quad \text{..........................} (7),$$

where $2\beta = A$. *The only law of force is therefore that of the inverse square.*

91. We have assumed in this investigation that the law of force is required to be independent of the radius of the spherical shell. *If we remove this restriction, there may be other laws of force which make the attraction of a given shell at all external points equal to that of a central mass*.*

To determine these laws we must solve equation (2) without differentiating it with regard to a, because a is no longer arbitrary. Since (2) is linear, we follow the rule in differential equations and put $F(r) = er^4 + Me^{pr}$, where the first term represents a particular integral introduced to clear (2) of the term Ar. Substituting this value of $F(r)$ in (2) we arrive at the equation $e^{pa} - e^{-pa} = 2pa$. This equation gives all the possible values of p.

This equation has three roots equal to zero and has no other real values of pa. These lead to the value of $F(r)$ given in (4). To find the imaginary roots we put $pa = \alpha + \beta i$, we then have $\cos\beta \,.\, \sinh\alpha = \alpha$, $\sin\beta \,.\, \cosh\alpha = \beta$.
By roughly tracing these curves (regarding α and β as coordinates) we find that there is an intersection between $\beta = 2n\pi$ and $(2n+\frac{1}{2})\pi$, where n is any integer except zero. There is therefore a possible law of potential which however is a function of the radius of the spherical surface.

We may obtain a simpler result if we enquire when *the potential* of a thin shell can be equal to that of *a central particle whose mass is μ times that of the shell.* The right-hand side of (2) must then be multiplied by μ and we have $A = 0$. We then find $e^{pa} - e^{-pa} = 2\mu pa$. This equation determines μ when p has any given real value. The law of potential is $f(r) = (Be^{pr} + Ce^{-pr})/r$. This law is the same for all spheres but the ratio of the central mass to that of the shell depends on the radius. That this law of potential satisfies the conditions given above is easily verified by actual integration.

92. Method of differentiation. Let the potential of a homogeneous body of density ρ at any point P, (ξ, η, ζ), be $V = \phi(\xi, \eta, \zeta)$. If we move the body a small distance $d\xi$, the point P remaining fixed, the potential at P of the body in its new position is $V - (dV/d\xi)\, d\xi$. Let us now construct a composite body whose density at any point Q is the difference of the densities at Q of the given body in its two positions. Since the boundaries are not the same, the composite body consists solely of a thin layer of matter placed on the boundary of the given body. The surface

* It is stated in *Nature*, No. 1572, Dec. 1899, that Dr Bakker has written a paper on this subject in the Proceedings of the Royal Academy of Sciences of Amsterdam. The author has not been able to see this memoir.

density at any point R is $\rho \cos\phi d\xi$, where ϕ is the angle the outward normal at R makes with the axis of ξ. We therefore arrive at the following rule; *if $V = \phi(\xi, \eta, \zeta)$ is the potential at P of a solid homogeneous body, the potential at P of a layer on its boundary of surface density $A\rho\cos\phi$ is $-AdV/d\xi$, or, which is the same thing, AX where X is the ξ component of attraction at P.* Here A is a constant for all elements of the attracting body.

If the body is heterogeneous, let its density be $\rho' = \psi(x', y', z')$; the interior of the composite body is not now vacant, its density is $Ad\rho'/dx'$, while the surface density at R is, as before, $A\rho\cos\phi$, where ρ is the density at R of the given body. We notice that *when the density of the given body is zero along the bounding surface, the potential of a body of density $d\rho'/dx'$ is $dV/d\xi$.*

93. Ex. 1. As an example consider the case of a *homogeneous solid sphere.* The ξ components of attraction at P are $\frac{4}{3}\pi a^3 \rho\xi/r^3$ or $\frac{4}{3}\pi\rho\xi$ according as P is external or internal. Hence these are also the potentials of a surface layer of density $\rho\cos\phi$, or $\rho x'/a$ if x' is measured from the centre.

Ex. 2. If V be the potential at P of a homogeneous body, prove that the potential at the same point of a thin layer on its surface of surface density $A(x\mu - y\lambda)$ is $A\left(x\frac{dV}{dy} - y\frac{dV}{dx}\right)$ where λ, μ, ν are the direction cosines of the normal. [Turn the body round the axis of z through an angle $\delta\phi$.]

Ex. 3. The surface density at any point Q of an infinitely extended plane is m, E is a given point distant $EO = z$ from the plane. The potential of the plane at any point P on the side of the plane opposite to E is V. Let $EQ = r'$, $EP = r$ and let θ be the angle EO makes with EP. Assuming the first of the following theorems deduce the others.

If $m = \dfrac{\mu}{r'^3}$ then $V = \dfrac{2\pi\mu}{zr}$,

$m = \dfrac{3\mu}{r'^5}$,, $V = \dfrac{2\pi\mu}{z^2 r}\left\{\dfrac{1}{z} + \dfrac{\cos\theta}{r}\right\}$,

$m = \dfrac{3 \,.\, 5\mu}{r'^7}$,, $V = \dfrac{2\pi\mu}{z^3 r}\left\{\dfrac{3}{z^2} + \dfrac{3\cos\theta}{zr} + \dfrac{3\cos^2\theta - 1}{r^2}\right\}$,

$m = \dfrac{Ax' + By'}{r'^5}$,, $V = \dfrac{2\pi}{3}\,\dfrac{A\xi + B\eta}{zr^3}$.

To deduce the second result from the first we perform the operation $\dfrac{-1}{z}\dfrac{d}{dz}$ on both m and V. The third is similarly deduced from the second and so on. To obtain the fourth we refer E to fixed coordinates x, y, z and operate on the first with d/dx and d/dy.

The first result for a point P on the axis EO produced is obtained by an easy integration. It follows by a theorem of Legendre on the attraction of solids of revolution (to be proved presently) that this result being true for a point P on the axis is necessarily also true when P does not lie on the axis.

94. Similar solids. Let dv, dv' be the volumes of two corresponding elements Q, Q'; ρ, ρ' their densities; r, r' their distances from two corresponding points P, P'. The lines QP, $Q'P'$ are parallel and the forces have the ratio $\rho dv/r^2$ to $\rho' dv'/r'^2$, which is the same as the constant ratio ρr to $\rho' r'$. The resultant attractions of similar and similarly situated solids at corresponding points are therefore parallel and have the ratio ρr to $\rho' r'$.

In the same way the attractions of similar surfaces at corresponding points are in the ratio of their surface densities.

Heterogeneous bodies. Let the density of a solid body at any point Q be $\rho = \psi(x, y, z)$, where ψ is a homogeneous function of the coordinates of s dimensions. Let the potential at a point P be $V = \phi(\xi, \eta, \zeta)$.

Increase the dimensions of the body and the distance of P from the origin O in any given ratio $1 : \beta$. We thus have two bodies *bounded by similar surfaces* S, S' attracting two points P, P' similarly situated. Since the potentials at the points P, P' of corresponding elements at Q, Q' are proportional to the masses divided by the distances, the potential at P' of the enlarged body is $V' = \beta^{s+2} \phi(\xi, \eta, \zeta)$.

The potential at P' of a thin shell bounded by the surfaces β and $\beta + d\beta$ may be found by differentiating V' with regard to β on the supposition that the coordinates of P' (viz. $\beta\xi$, &c.) are constant. If we finally put $\beta = 1$, this shell will become a thin layer placed on the surface S. Since $d\xi/\xi = -d\beta/\beta$, &c. we have for the potential

$$dV = \left\{(s+2) V - \xi \frac{dV}{d\xi} - \eta \frac{dV}{d\eta} - \zeta \frac{dV}{d\zeta}\right\} d\beta\,*$$

where $V = \phi(\xi, \eta, \zeta)$. Since this shell is bounded by similar surfaces, and its density is $\psi(x, y, z)$, its *surface density* σ at x, y, z, is $\sigma = p\psi(x, y, z)\, d\beta$, where p is the perpendicular on the tangent plane. Also if M, M' be the masses of the original body and the stratum, $M' = M(s+3)\, d\beta$. We may substitute for $d\beta$ one or other of these values according as we wish to express the potential in terms of the surface density or the mass.

Laplace's, Poisson's and Gauss' theorems.

95. Laplace's theorem†. Let (ξ, η, ζ) be the coordinates of any particle A of the attracting matter, and let m be the mass of that particle. Let (x, y, z) be the coordinates of any point P.

* This formula for the potential of a heterogeneous stratum placed on the surface of a known body was given by Ferrers in *Q. J.* vol. xv. 1877.

† *Mécanique Céleste*, T. II.

Taking the particle m apart from the rest of the matter, its potential at P is $V_1 = m/r$,

where $$r^2 = (x-\xi)^2 + (y-\eta)^2 + (z-\zeta)^2 \text{.................} (1).$$

Since $$r\frac{dr}{dx} = x - \xi,$$

we find $$\frac{dV_1}{dx} = -m\frac{x-\xi}{r^3}, \quad \therefore \frac{d^2V_1}{dx^2} = -\frac{m}{r^3} + \frac{3m(x-\xi)^2}{r^5}.$$

Also, $$\frac{d^2V_1}{dy^2} = -\frac{m}{r^3} + \frac{3m(y-\eta)^2}{r^5}, \quad \frac{d^2V_1}{dz^2} = -\frac{m}{r^3} + \frac{3m(z-\zeta)^2}{r^5}.$$

Adding these three expressions and remembering equation (1) we find $$\frac{d^2V_1}{dx^2} + \frac{d^2V_1}{dy^2} + \frac{d^2V_1}{dz^2} = 0.$$

Let now V be the potential of the whole attracting matter at P. Then, since V is the sum of the potentials of the several particles, it immediately follows that $\dfrac{d^2V}{dx^2} + \dfrac{d^2V}{dy^2} + \dfrac{d^2V}{dz^2} = 0.$

In this investigation we have assumed that the point P does not coincide with any one of the attracting particles. If it did the meaning of the potential of that particle would require some further consideration. *The theorem has therefore been proved to be true only for a point external to the attracting matter.* It will be presently shown that the right-hand side is not zero when the attracted particle forms a part of the attracting mass.

Laplace's equation is a differential equation which must be satisfied by the potential of every body at all points not occupied by attracting matter. If a general solution of the equation could be found, that solution would comprise within its compass the potential and therefore the component attractions of all bodies.

Laplace's function $\dfrac{d^2V}{dx^2} + \dfrac{d^2V}{dy^2} + \dfrac{d^2V}{dz^2}$ is often written in the abbreviated form $\nabla^2 V$.

96. When the law of attraction is the inverse κth power of the distance we have $V_\kappa = \dfrac{1}{\kappa-1}\Sigma\dfrac{m}{r^{\kappa-1}}$, (Art. 43). We may then prove that

$$\left(\frac{d^2}{dx^2} + \frac{d^2}{dy^2} + \frac{d^2}{dz^2}\right) V_\kappa = (\kappa-2)\,\Sigma\frac{m}{r^{\kappa+1}} = (\kappa-2)(\kappa+1)\,V_{\kappa+2}.$$

When therefore the potentials of a body at an external point P are known functions of the coordinates of P for the laws of the inverse cube and the inverse fourth power, this theorem enables us to find by simple differentiation the potentials of the same body for any higher inverse power. [Jellett, *Brit. Assoc.*, Dublin 1857.]

Ex. If the point P is internal and the body is homogeneous and of density ρ, prove that the left-hand side of Jellett's equation should be increased by the constant $4\pi\rho 0^{2-\kappa}$. [See Arts. 80, 105.]

97. When the attracting body is a heterogeneous spherical surface we find that

$$V_{\kappa+2}=\frac{1}{\kappa+1}\,\frac{1}{a^2-r^2}\left\{2r\frac{dV_\kappa}{dr}+(\kappa-1)\,V_\kappa\right\},$$

where a is the radius and r the distance of the point P from the centre. This result holds whether P is external or internal.

We have $V_\kappa=\dfrac{1}{\kappa-1}\displaystyle\int\frac{m\,d\sigma}{u^{\kappa-1}}$, $u^2=a^2+r^2-2apr$, where $d\sigma$ is an element of area at Q, $u=QP$ and $p=\cos QOP$. To obtain the result, substitute this value of V_κ on the right-hand side and eliminate p.

This theorem may be used to find the potential of a circular ring or of any curve which can be drawn on a sphere.

98. When the attracting body is a lamina of finite extent, not necessarily homogeneous, and the potentials at points in the plane of the lamina only are required, the formula takes either of the forms

$$V_{\kappa+2}=\frac{1}{\kappa^2-1}\left(\frac{d^2}{dx^2}+\frac{d^2}{dy^2}\right)V_\kappa=-\frac{1}{\kappa+1}\,\frac{d^2V_\kappa}{dz^2}.$$

When the potential of the lamina is required at a point P not in its plane, we notice that the component of force at P normal to the plane due to any particle m of the attracting plane is $-mz/r^{\kappa+1}$ where r is the distance of m from P. Summing up for all the particles we find $V_{\kappa+2}=-\dfrac{1}{\kappa+1}\,\dfrac{dV_\kappa}{zdz}$. [James Roberts, *Quarterly Journal*, 1881.]

99. *The potential V_κ of a body when the law of force is the inverse κth power cannot be constant throughout any finite space unoccupied by matter unless the law of force is the inverse square.* It is assumed that all the m's have the same sign, that is *every particle must attract or every particle must repel.* For if $V_\kappa=0$, we have by Jellett's theorem either $V_{\kappa+2}=0$ or $\kappa=2$. But $V_{\kappa+2}$ is by definition the sum of a number of terms all of which have the same sign, and therefore cannot vanish. In the same way *the potential of a lamina cannot be constant throughout any finite area in its plane unless the law of force is the inverse distance.*

100. Another important theorem should be noticed. If we transform the coordinates from one system of *rectangular* Cartesian axes x, y, z to another x', y', z' according to the scheme in the margin, it is well known that

	x	y	z
x'	a_1,	a_2,	a_3
y'	b_1,	b_2,	b_3
z'	c_1,	c_2,	c_3

$$x'=a_1x+a_2y+a_3z,\qquad \frac{d}{dx'}=a_1\frac{d}{dx}+a_2\frac{d}{dy}+a_3\frac{d}{dz}.$$

Thus x, y, z and d/dx, d/dy, d/dz are transformed by the same rules. It immediately follows that since $x^2+y^2+z^2=x'^2+y'^2+z'^2$

$$\frac{d^2V}{dx^2}+\frac{d^2V}{dy^2}+\frac{d^2V}{dz^2}=\frac{d^2V}{dx'^2}+\frac{d^2V}{dy'^2}+\frac{d^2V}{dz'^2}.$$

This is an analytical proof of the invariant property of Laplace's equation. The result follows more simply from Poisson's theorem (Art. 105), for each side of the equation is there proved to be equal to $-4\pi\rho$.

101. Potential at an internal point. The potential at a point P of any particles situated at the points A_1, A_2, &c. has already been defined in Art. 39

to be $\Sigma m/r$. It is evident from this definition that, if a *finite quantity* of matter be situated at any one of the points A_1, A_2, &c. in a condensed form, the potential at a point P in the immediate neighbourhood of that point is very great, and at that point itself this definition would make the potential infinite. But if the attracting matter is so distributed in space that the mass which occupies any elementary volume dv is ρdv where ρ is finite, we shall now show that the potential in this portion of space need not be infinite.

The potential at a point P in the interior of a body of finite density may be found by taking P as the origin of polar coordinates and integrating all round throughout the body. *In this way we make r positive for every particle*, Art. 39. Describe a small surface S enclosing P and let its equation be $r=\epsilon f(\theta, \phi)$, where ϵ is a small constant factor. An element dv of the volume distant r from P is equal to $r^2 d\omega dr$, where $d\omega$ is the solid angle subtended at P. If then V_2 be the potential at P of the matter filling this surface, we have

$$V_2=\int \frac{\rho dv}{r}=\iint \rho d\omega r dr \quad\text{........(1)},$$

where the limits of integration for r are 0 and $\epsilon f(\theta, \phi)$. It is evident therefore that V_2 is of the order ϵ^2.

It follows that when ϵ is evanescent the value of V_2 is zero. Thus the matter filling the surface may be removed without altering the potential of the whole attracting mass. *In finding therefore the potential of a body at any internal point P we may regard P as situated in an infinitely small cavity, and determine the potential as if P were an external point.*

Let us consider next the *resolved attraction* at the point P of the matter filling the small surface described above. Let X_2 be the component parallel to the axis of x, then

$$X_2=\int \frac{\rho dv}{r^2}\cos\theta=\iint \rho\cos\theta d\omega dr \quad\text{........................(2)},$$

where θ is the angle the radius vector r makes with the axis of x. *It is evident that X_2 is at least of the order ϵ of small quantities, and therefore vanishes when the size of the surface is evanescent.* Since $\cos\theta$ is negative when $\theta>\pi/2$ the order of the term may be higher than ϵ.

Lastly let us find the order of dV_2/dx. To simplify the integrations let us suppose that the surface is spherical, so that we may use the formula for the potential already obtained in Art. 78. Let the radius of the sphere be ϵ, let the coordinates of its centre be (a, b, c) and those of P be (x, y, z). Then

$$V_2=\tfrac{2}{3}\pi\rho\{3\epsilon^2-(x-a)^2-(y-b)^2-(z-c)^2\} \quad\text{...................(3)}.$$

It follows at once that

$$\frac{dV_2}{dx}=-\frac{4\pi\rho}{3}(x-a), \qquad \frac{d^2V_2}{dx^2}=-\frac{4\pi\rho}{3} \quad\text{...................(4)}.$$

Since $x-a$ is less than ϵ, it is clear that *dV_2/dx is a small quantity of at least the order ϵ, and vanishes when ϵ is evanescent.* In the same way the first differential coefficients of V_2 with regard to y and z are evanescent with ϵ. The second differential coefficients of V_2 with regard to x, y or z are however not small.

We have supposed the density of the matter within the evanescent sphere to be uniform. It is however clear that, if we substituted for ρ an expression of the form

$$\rho=\rho_0+A(x-a)+\&c.$$

we should merely add to the expression for V_2 terms of the order ϵ^3.

102. To prove that *the relation $X=dV/dx$* which has been established for an external point *also holds for an internal point.* Let V_1, V_2 and X_1, X_2 be the potentials and components of force at P due respectively to the matter outside and inside a small spherical surface S. Then $V=V_1+V_2$ and $X=X_1+X_2$. Since P is external to that part of the body which is outside S, $X_1 = dV_1/dx$. We have just proved that dV_2/dx and X_2 are each equal to zero when the size of the surface is evanescent. Hence $X = dV/dx$.

Ex. 1. Let the point P be situated at the middle point of the axis of a right circular cylindrical cavity of altitude $2h$, and let x be measured along the axis. Prove that $\frac{dX_2}{dx} = -4\pi\rho\left(1 - \frac{h}{l}\right)$ where l is the distance of P from any point of either rim. Thence show that in a flat cylindrical cavity dX_2/dx is $-4\pi\rho$ and in a long cylinder is zero.

Ex. 2. Let the law of force be the inverse κth power of the distance. Prove that for a homogeneous body the relation $X = dV/dx$ holds for an internal point P.

It is sufficient to prove this for a sphere enclosing the point P. Take P for origin, we then find by an easy polar integration the value of X. The value of V at the same point has been given in Art. 80. The integrations are shortened by taking P near the centre.

103. We shall now prove that, *when a point P passes from the interior of a body of finite density into external space, both the potential and the attraction undergo no sudden change of magnitude, but the second differential coefficients of the potential are discontinuous in value.*

Describe round the point A of emergence a small surface S of any convenient form. Since both the potential and the attraction due to the matter within S are zero, the points near A may be regarded as both external and internal.

All that is meant is that there is *a numerical continuity* in the potential. The potentials of a solid sphere, for example, are represented by different analytical expressions at points inside and outside, but at the surface both these have the same numerical value, viz. M/a, Art. 78.

104. *When P traverses an infinitely thin stratum whose surface density is finite* the volume density is not finite. It will be shown further on that the potential is continuous, but that the attraction does undergo a sudden change of value, and an expression will be found for the change.

It is at once evident from Arts. 15 &c. that *when P arrives at an infinitely thin line of finite line density, both the attraction and the potential are infinite.*

105. **Poisson's theorem*.** If V be the potential of a body at an internal point P at which the density ρ is finite, then

$$\frac{d^2V}{dx^2} + \frac{d^2V}{dy^2} + \frac{d^2V}{dz^2} = -4\pi\rho.$$

Describe a spherical surface of radius ϵ enclosing the point P, let (a, b, c) be the coordinates of its centre, (x, y, z) those of P. Let the radius ϵ be so small that the matter enclosed by the sphere may be regarded as of uniform density.

Let V_2 be the potential at P of the matter within the sphere,

* This theorem is given by Poisson in the third volume, page 388, of the *Nouveau Bulletin des Sciences par la Société Philomathique de Paris*, 5ᵉ Année 1812. He proceeds very nearly as in Art. 105.

V_1 that of the rest of the body, then $V = V_1 + V_2$. But by Laplace's theorem $\nabla^2 V_1 = 0$, hence

$$\nabla^2 V = \nabla^2 V_2 = \frac{d^2 V_2}{dx^2} + \frac{d^2 V_2}{dy^2} + \frac{d^2 V_2}{dz^2} = \frac{dX_2}{dx} + \frac{dY_2}{dy} + \frac{dZ_2}{dz},$$

where X_2, Y_2, Z_2 are the resolved attractions at P of the matter within the sphere. But by Art. 77

$$X_2 = -\tfrac{4}{3}\pi\rho\,(x-a), \quad Y_2 = -\tfrac{4}{3}\pi\rho\,(y-b), \text{ \&c.}$$

It easily follows by substitution that $\nabla^2 V = -4\pi\rho$. Another proof of this theorem founded on Gauss' theorem is given a little further on.

We may notice that the centre of the sphere, though arbitrary in position, must not be taken coincident with P. The reason is that we differentiate V_2 with regard to the coordinates of P, i.e. we make P travel from the point (x, y, z) to a neighbouring point $(x + dx$, &c.). But since the centre of the sphere is fixed, it cannot be made to coincide with both the positions of P.

Ex. When the law of force is the inverse distance and the attracting body is a lamina attracting particles of its own substance prove that $\frac{d^2V}{dx^2} + \frac{d^2V}{dy^2} = -2\pi\rho$. [Deduce this from the attraction of a cylinder (Art. 14) or from that of a circular area (Art. 57) by the method of Art. 105.]

106. Gauss' theorem. *Let S be any closed surface, and let M_1 be the sum of the attracting masses which lie within the surface,*

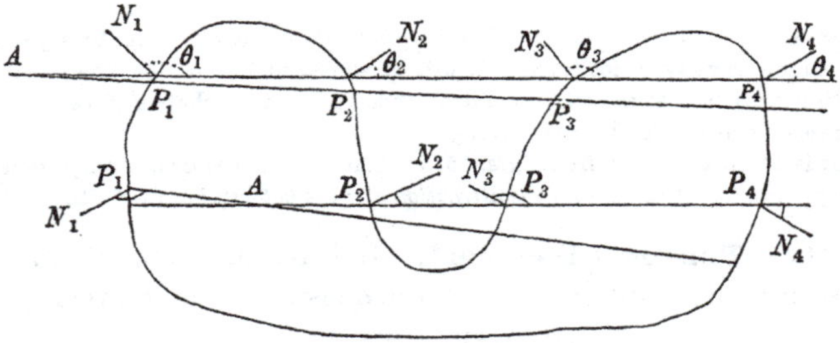

M_2 the sum of the masses outside. Let $d\sigma$ be any element of area of this surface, F the normal resolute at this element of the attraction of the whole mass both internal and external. Then $\int F d\sigma = \pm 4\pi M_1$ where the integration extends over the whole surface of S and the upper or lower sign is taken according as F is estimated positive or negative when the normal force acts inwards.*

* This theorem was given by Gauss in 1839, his paper is translated in Vol. III. of Taylor's *Scientific Memoirs*. It was also given by Sir W. Thomson in 1842 in his papers on *Electrostatics and Magnetism*. The demonstration given by Sir G. Stokes in 1849 has been followed here. He also deduces the Cartesian form of Poisson's equation from Gauss' theorem. See his *Mathematical and Physical Papers*.

Let m be the mass of any particle of the attracting system, and let it be situated at the point A. A straight line drawn through A to intersect the surface S in any point will also intersect it in some other point, but, if the surface is re-entrant, it may enter and issue from the surface any even number of times. Let the points of intersection, taken in order, be P_1, P_2, &c., and let the direction P_1P_2, &c. be called the positive direction of the straight line.

Let θ_1, θ_2, &c. be the angles the positive direction of P_1P_2, &c. makes with the normals P_1N_1, P_2N_2, &c. drawn *outwards*. It is evident that where the line enters the surface $\cos\theta$ is negative, and where it issues from the surface $\cos\theta$ is positive, thus the angles θ_1, θ_2, &c. are alternately obtuse and acute.

With A for vertex describe about this straight line an elementary cone whose solid angle is $d\omega$, and let it intersect the surface S in the elementary areas $d\sigma_1$, $d\sigma_2$, &c. If the distances $AP_1 = r_1$, $AP_2 = r_2$, &c., these elementary areas by Art. 26 are

$$d\sigma_1 = r_1^2 d\omega \sec(\pi - \theta_1), \quad d\sigma_2 = r_2^2 d\omega \sec\theta_2, \text{ \&c.} \ldots\ldots (1).$$

If the point A is external to the surface as in the upper part of the figure, the normal resolutes taken positively when acting outwards are $F_1 = \dfrac{m}{r_1^2}\cos(\pi - \theta_1)$, $F_2 = -\dfrac{m}{r_2^2}\cos\theta_2$, &c......(2).

Since the signs of these terms are alternately positive and negative, it follows that when A is external

$$F_1 d\sigma_1 + F_2 d\sigma_2 + \text{\&c.} = 0 \ldots\ldots\ldots\ldots\ldots\ldots (3).$$

If the point A is internal and lies between P_1 and P_2, as represented in the lower part of the figure, the sign of the force F_1 must be changed. We therefore have

$$F_1 d\sigma_1 + F_2 d\sigma_2 + \text{\&c.} = -2m\,d\omega \ldots\ldots\ldots\ldots\ldots (4).$$

If the point A lie between P_2 and P_3, the signs of the first two terms in the series (2) are changed, and the equation (4) resumes the form (3), and so on.

If we now let the straight line AP_1P_2 &c. revolve round A into all positions, all the elements of the surface will be included in the integration. We therefore find for an external point $\int F d\sigma = 0$ (5).

For an internal point the integration of the right-hand side of (4) is limited to a hemisphere of the unit sphere, Art. 26. We therefore have $\int F d\sigma = -4\pi m \ldots\ldots\ldots\ldots\ldots\ldots\ldots (6).$

Let now the system consist of any number of particles m_1, m_2,

&c. inside, and m_1', m_2', &c. outside the surface S. The particles outside contribute nothing to the integral $\int F d\sigma$, while the particles inside contribute respectively $-4\pi m_1$, $-4\pi m_2$, &c. On the whole, when F is measured positively outwards, we have

$$\int F d\sigma = -4\pi M_1 \quad \text{.........................(7)},$$

where M_1 stands for the sum of the internal particles m_1, m_2, &c.

The truth of the theorem is not affected if some of the matter, instead of being attractive, be repulsive. Such matter must however be regarded as having a negative mass.

107. The product $F d\sigma$ represents the product of the normal resolute of the attraction at an element multiplied by the area of the element across which it is supposed to act. This product is sometimes called the *flux* or *flow* of the attraction across the elementary area $d\sigma$ in the direction in which the component F is measured. When the particles of the body attract, the proposition asserts that the whole inward flux across any closed surface is equal to 4π multiplied by the mass inside. The product $F d\sigma$ is also called the *induction through the element*; see Maxwell's *Electricity*.

We sometimes require the flux or induction across a *portion only* of the surface S instead of across the whole. Let this portion subtend a finite solid angle ω at any one attracting point m. Then by what precedes *the flux or induction across this portion due to the attraction of m is $m\omega$.* If there are several attracting points we may find the flux due to each and add the results together.

108. *To deduce Laplace's and Poisson's theorems from Gauss' theorem.* To effect this we take as the closed space to which we apply Gauss' theorem the element suited to the coordinates we intend to use. Let P be any point of space and let $d\xi$, $d\eta$, $d\zeta$ be the lengths of the three edges which intersect at P.

In Cartesian coordinates the element has its edges parallel to the coordinate axes and therefore $d\xi = dx$, $d\eta = dy$, $d\zeta = dz$. The sides of the polar element are $d\xi = dr$, $d\eta = r d\theta$, $d\zeta = r \sin\theta d\phi$, while those for cylindrical coordinates are $d\xi = dR$, $d\eta = R d\phi$, $d\zeta = dz$.

It should be noticed that in all these cases the three edges which meet at any corner of the element are at right angles. The mass inside the element is $M = \rho d\xi d\eta d\zeta$ in every case.

Let V be the potential at P. Consider first the two faces perpendicular to the edge at $d\xi$; the inward flux for the one and the outward flux for the other are

$$F = \frac{dV}{d\xi} d\eta d\zeta, \qquad F + \frac{dF}{d\xi} d\xi.$$

The total outward flux for these two is therefore $\frac{dF}{d\xi} d\xi$. Treat the two other pairs of faces in the same way and equate the whole flux to $-4\pi M$. We then have

$$\frac{d}{d\xi}\left(\frac{dV}{d\xi} d\eta d\zeta\right) d\xi + \frac{d}{d\eta}\left(\frac{dV}{d\eta} d\xi d\zeta\right) d\eta + \frac{d}{d\zeta}\left(\frac{dV}{d\zeta} d\xi d\eta\right) d\zeta = -4\pi\rho d\xi d\eta d\zeta.$$

If we now substitute for $d\xi$, $d\eta$, $d\zeta$ their values in Cartesian coordinates and divide by the product $dx\,dy\,dz$, this becomes

$$\frac{d^2V}{dx^2}+\frac{d^2V}{dy^2}+\frac{d^2V}{dz^2}=-4\pi\rho.$$

If we substitute the polar values of $d\xi$, $d\eta$, $d\zeta$ and divide by $r^2dr\sin\theta\,d\theta\,d\phi$ we find

$$\frac{1}{r^2}\frac{d}{dr}\left(r^2\frac{dV}{dr}\right)+\frac{1}{r^2\sin\theta}\frac{d}{d\theta}\left(\sin\theta\frac{dV}{d\theta}\right)+\frac{1}{r^2\sin^2\theta}\frac{d^2V}{d\phi^2}=-4\pi\rho.$$

If we substitute the cylindrical values we have

$$\frac{1}{R}\frac{d}{dR}\left(R\frac{dV}{dR}\right)+\frac{1}{R^2}\frac{d^2V}{d\phi^2}+\frac{d^2V}{dz^2}=-4\pi\rho.$$

Poisson's equation in oblique Cartesian coordinates takes the following form. Let α, β, γ be the *sines* of the angles between the axes; A, B, C the angles between the coordinate planes, then

$$\nabla^2V=D^2\left\{\alpha^2\frac{d^2V}{dx^2}+\beta^2\frac{d^2V}{dy^2}+\gamma^2\frac{d^2V}{dz^2}-2\alpha\beta\cos C\frac{d^2V}{dx\,dy}-2\beta\gamma\cos A\frac{d^2V}{dy\,dz}-2\gamma\alpha\cos B\frac{d^2V}{dz\,dx}\right\},$$

where $$D=\alpha\beta\sin C=\beta\gamma\sin A=\gamma\alpha\sin B.$$

109. Orthogonal and elliptic coordinates. Let the equations of three surfaces which intersect at right angles be

$$\alpha=f_1(x,y,z),\quad \beta=f_2(x,y,z),\quad \gamma=f_3(x,y,z)\ldots\ldots\ldots\ldots\ldots\ldots(1),$$

where α, β, γ are three parameters whose values determine which surface of each system is taken. *These parameters may be regarded as the coordinates of the point P of intersection of the three surfaces.*

Let $P\xi$, $P\eta$, $P\zeta$ be normals to the surfaces α, β, γ at the point P of intersection. Let the direction cosines of these normals be $(\lambda_1\ \mu_1\ \nu_1)$, $(\lambda_2\ \mu_2\ \nu_2)$, $(\lambda_3\ \mu_3\ \nu_3)$. We therefore have $\lambda_1=\frac{\alpha_x}{h_1}$, $\mu_1=\frac{\alpha_y}{h_1}$, $\nu_1=\frac{\alpha_z}{h_1}$; $h_1^2=\alpha_x^2+\alpha_y^2+\alpha_z^2$ where suffixes denote partial differential coefficients.

Let $PQ=d\xi$ be an element of the normal $P\xi$ and let (xyz), $(x+dx$, &c.) be the coordinates of the extremities of $d\xi$. Then

$$\frac{d\alpha}{h_1}=\frac{1}{h_1}(\alpha_x dx+\alpha_y dy+\alpha_z dz)=\lambda_1 dx+\mu_1 dy-\nu_1 dz.$$

The right-hand side represents the sum of the projections of dx, dy, dz on the normal and this is $d\xi$. Hence

$$d\xi=\frac{d\alpha}{h_1},\qquad d\eta=\frac{d\beta}{h_2},\qquad d\zeta=\frac{d\gamma}{h_3}.$$

The general equation of flux for the orthogonal element $d\xi\,d\eta\,d\zeta$ is by Art. 108

$$\frac{d}{d\xi}\left(\frac{dV}{d\xi}d\eta\,d\zeta\right)d\xi+\text{\&c.}=-4\pi\rho\,d\xi\,d\eta\,d\zeta.$$

Substitute the values of $d\xi$, $d\eta$, $d\zeta$, and we find after division by $d\alpha\,d\beta\,d\gamma$

$$\frac{d}{d\alpha}\left(\frac{h_1}{h_2h_3}\frac{dV}{d\alpha}\right)+\frac{d}{d\beta}\left(\frac{h_2}{h_1h_3}\frac{dV}{d\beta}\right)+\frac{d}{d\gamma}\left(\frac{h_3}{h_1h_2}\frac{dV}{d\gamma}\right)=\frac{-4\pi\rho}{h_1h_2h_3}.$$

The quantities $h_1h_2h_3$ are given by

$$h_1^2=\alpha_x^2+\alpha_y^2+\alpha_z^2,\quad h_2^2=\beta_x^2+\beta_y^2+\beta_z^2,\quad h_3^2=\gamma_x^2+\gamma_y^2+\gamma_z^2,$$

and are supposed to be expressed in terms of the orthogonal coordinates $\alpha\beta\gamma$, the Cartesian coordinates xyz being eliminated by using the equations of the orthogonal surfaces. This equation is sometimes called *Lamé's transformation of Poisson's equation.*

110. Since V is regarded as a function of α, β, γ, we have

$$\frac{dV}{dx} = \frac{dV}{d\alpha}\alpha_x + \frac{dV}{d\beta}\beta_x + \frac{dV}{d\gamma}\gamma_x,$$

with similar expressions for dV/dy and dV/dz. These we differentiate again and substitute in Poisson's equation. Since the surfaces α, β, γ are orthogonal the coefficients of $d^2V/d\alpha d\beta$ &c. are zero. We therefore have

$$-4\pi\rho = \frac{d^2V}{d\alpha^2}(\alpha_x^2 + \alpha_y^2 + \alpha_z^2) + \frac{dV}{d\alpha}(\alpha_{xx} + \alpha_{yy} + \alpha_{zz}) + \&c.$$

Let the arbitrary functions α, β, γ be so chosen that they satisfy Laplace's equation. The Poisson equation then becomes

$$-4\pi\rho = h_1^2\frac{d^2V}{d\alpha^2} + h_2^2\frac{d^2V}{d\beta^2} + h_3^2\frac{d^2V}{d\gamma^2} \quad\text{..........................(1).}$$

Let α be the potential of a thin ellipsoidal shell of unit mass, such as that described in Art. 68. Let (abc) be its semiaxes. It will be shown in the chapter on the attraction of ellipsoids that the level surfaces of the shell are the confocal ellipsoids. Let $(a'b'c')$, $(a''$ &c.), $(a'''$ &c.) be the semiaxes of the three confocals which pass through any external point P.

Since $h_1^2 = \alpha_x^2 + \alpha_y^2 + \alpha_z^2$, it is evident that h_1 *is the component of force at* P due to the shell in a direction normal to the ellipsoid $(a'b'c')$. It will also be shown that this force is $h_1 = \frac{d\alpha}{dp'} = -\frac{p'}{a'b'c'}$, where p' is the perpendicular from the centre on the tangent plane. Similar expressions must hold for the hyperbolic confocals by the principle of continuity.

If D_1, D_2, are the semi-diameters of the confocal ellipsoid respectively parallel to the normals at P to the confocal hyperboloids we know that $p'D_1D_2 = a'b'c'$ by the properties of conjugate diameters. Also by the properties of confocal quadrics $D_1^2 = a'^2 - a''^2$, $D_2^2 = a'^2 - a'''^2$ and $p'dp' = a'da'$. By using these expressions, we put the equation (1) into the form

$$(a''^2 - a'''^2)\frac{d^2V}{d\alpha^2} + (a'''^2 - a'^2)\frac{d^2V}{d\beta^2} + (a'^2 - a''^2)\frac{d^2V}{d\gamma^2} = 4\pi\rho\,(a''^2 - a'''^2)(a'''^2 - a'^2)(a'^2 - a''^2).$$

Since $p'dp' = a'da'$ the potentials α, β, γ are to be found from

$$\frac{d\alpha}{da'} = -\frac{1}{b'c'}, \qquad \frac{d\beta}{da''} = -\frac{1}{b''c''}, \qquad \frac{d\gamma}{da'''} = -\frac{1}{b'''c'''}.$$

This form of Poisson's equation agrees with that given by Lamé.

Theorems on the Potential.

111. *The potential of any attracting system cannot be an absolute maximum or minimum at any point unoccupied by matter*.*

If V be the value of the potential at any point P whose

* The theorems in this section may for the most part be found in Gauss' memoir on *Forces varying inversely as the square of the distance*, 1840. In the *Cambridge and Dublin Mathematical Journal*, Vol. IV. 1849, there is an interesting collection of theorems on the potential by Sir G. Stokes. Most of these were already known, but the proofs were much improved and put into new and better forms. This paper is reprinted in his collected works Vol. I. p. 104. The reader may also refer to papers by Lord Kelvin in various volumes of the *Cambridge and Dublin Mathematical Journal*, 1842 and 1843, reprinted in his *Electricity and Magnetism*. There is also a memoir by Chasles in the additions to the *Connaissances des Temps* for 1845.

coordinates are x, y, z, the value V' of the potential at any neighbouring point P' whose coordinates are $x+\xi$, $y+\eta$, $z+\zeta$ will be given by

$$V' = V + V_x\xi + V_y\eta + V_z\zeta$$
$$+ \tfrac{1}{2}(V_{xx}\xi^2 + V_{yy}\eta^2 + V_{zz}\zeta^2 + 2V_{xy}\xi\eta + 2V_{yz}\eta\zeta + 2V_{zx}\zeta\xi) + \&c.,$$

where partial differential coefficients are represented as usual by suffixes.

If V were a maximum or minimum at the point x, y, z, the first differential coefficients V_x, V_y, V_z would each be zero, and the three second differential coefficients V_{xx}, V_{yy}, V_{zz} (besides fulfilling some other conditions) would have the same sign. But since the point P is unoccupied by matter, they must satisfy Laplace's equations, Art. 95. Their sum must therefore be zero. It is therefore impossible that all three should have the same sign.

It has not been assumed that the masses of all the particles have the same sign. *The theorem is still true if the forces due to some particles are attractive, and those due to others are repulsive.*

When the law of force is the inverse distance and the attracting body is a lamina, we have at all points in that plane $V_{xx}+V_{yy}=0$, Art. 105. It follows that in this case also the potential cannot be an absolute maximum or minimum at any point in the plane of the lamina unoccupied by matter. For other laws of force in which the sum of V_{xx}, V_{yy}, V_{zz} is not zero, the argument does not apply.

We have here assumed that we may apply Taylor's theorem to the potential. That we may do so follows from the definition given in Art. 39. It is clear that the potential at P of a single particle and therefore of a system of particles whose total mass is finite is a function of the coordinates of P which is continuous and finite as long as P does not traverse any attracting matter. We may however put the argument into another form which has the advantage of avoiding the use of series.

112. *Another proof.* With P as centre describe a sphere of small radius. If the potential V were an absolute maximum at P, the potential at any point Q of the sphere must be less than that at P. Thus V is decreasing for a displacement along every radius of the sphere. It follows from Art. 41 that the outward normal force F at Q is negative at every point of the sphere. But by Gauss' theorem $\int F d\sigma = 0$ (Art. 106), which requires that F should be positive for some elements of the sphere and negative for others. In the same way it may be shown that the potential cannot be an absolute minimum at P.

113. *If the point P be situated within the substance of a continuous attracting body of finite positive density ρ, the potential may be a maximum but cannot be a minimum at P.*

To prove this we observe that the potential function V here satisfies Poisson's equation instead of Laplace's. Since the sum of the three differential coefficients V_{xx}, V_{yy}, V_{zz} is negative, it is possible that each may be negative. In that case V is in general a maximum.

If we adopt the second proof, we notice that Gauss' theorem requires $\int F d\sigma$ to be equal to $-4\pi M$, where M is the mass inside the sphere of small radius. It follows that F may be negative, and therefore V be decreasing, for a displacement along every radius. The quantity V may therefore be a maximum at P.

114. If any arbitrary curve is drawn in space not intersecting any portion of the attracting matter, the potential may vary from point to point of the curve. At some points the potential may be a maximum and at others a minimum for displacements restricted to that curve. For example, if the curve touch a level surface the space differential coefficient of the potential is zero at the point of contact and the potential may be either a maximum or a minimum. What we have proved in Art. 111 is that the potential cannot be a maximum or minimum at any point for displacements in every direction.

If the curve is a line of force, it cuts the level surfaces at right angles and the space differential coefficient of the potential cannot vanish unless the *resultant force* is zero, (Art. 47). The potential at a point P which travels along *a line of force* always in the same direction must therefore *continually increase or continually decrease* until P arrives at a point of equilibrium.

At a point of equilibrium there are some directions in which the potential increases and others in which it decreases (see Art. 120). The point P may therefore resume its journey (though not necessarily in the same direction as before) so that the potential at P continues to increase or decrease. *The journey can be continued to an infinite distance unless stopped by arrival at a point of the attracting mass.*

115. *If the potential is equal to any given constant quantity A at all points of a closed surface S which does not contain any portion of the attracting mass, it must be constant and equal to A at all points of the space contained within the surface S.*

For if it were not constant, there would be some point at which either it is greater than at all the other points or less than at all other points. But this has just been proved to be impossible.

116. Ex. 1. As an example of this theorem consider the case of a spherical shell of uniform thickness and density. Describe a concentric sphere within the shell. By symmetry the potential must be the same at all points of its surface.

Since there is no attracting matter within this sphere, it follows that the potential is constant throughout its interior.

Ex. 2. If the potential is not constant throughout the superficies of any closed surface S, let A be the greatest and B the least value. Prove that the potential at all points within S lies between A and B. [Stokes.]

Ex. 3. A level surface S completely encloses all the attracting matter of a system. If the consecutive level surfaces extending from S to infinity be drawn, prove that the potential continually decreases outwards from each to the next until it vanishes at an infinite distance.

117. *If the potential is constant throughout any finite space, it is also constant throughout all external space which can be reached without passing through any portion of the attracting mass.* [Stokes.]

The external boundary of the space is necessarily a level surface. If possible let A be a point outside the space at which the potential is a little greater than within the space. Since the level surface through A cannot cut the boundary, the potential at all points in the neighbourhood of A is greater than within the space. We can therefore describe an indefinitely small sphere, passing through A and having its centre O within the space, such that the potential is increasing outwards along every radius drawn from O to any point on the sphere outside the space and is constant along every radius which lies wholly within the space. It follows that the normal force has the same sign at every element of this sphere. This however by Gauss' theorem is impossible. In the same way it may be shown that no point A can exist in the neighbourhood of the space at which the potential is less than within the space.

Another Proof. It has already been pointed out in Art. 39 that the potential at P is a continuous function of the coordinates of P. It follows that when an expression has been found which represents the potential throughout any finite empty space that expression must also represent the potential throughout all external space which can be reached without passing through any portion of the attracting mass.

118. Points of equilibrium. If an isolated particle placed at any point P be in equilibrium under the attraction of any system, that point is called a point of equilibrium. When every point of a curve is a point of equilibrium, the curve is called a line or curve of equilibrium.

When the potential of the attracting mass is known, the positions of the points of equilibrium are found by equating the first differential coefficients of the potential to zero, viz. dV/dx, dV/dy, &c.; for these represent the resolved parts of the forces parallel to the axes.

119. *The equilibrium of a free isolated particle attracted by fixed bodies cannot be stable for all displacements or unstable for all displacements, but must be stable with reference to some displacements and unstable with reference to others.* Earnshaw's theorem. *Camb. Transac.*, 1839.

If the equilibrium were stable when the particle occupied a position P, the potential must decrease in all directions from P, i.e. the potential would be an absolute maximum at P, which has been proved impossible. In the same way the equilibrium could not be unstable for all displacements.

120. *A particle is in equilibrium at a point P. It is required to find the equation of the cone which, having its vertex at P, separates the displacements for which the equilibrium is stable from those for which it is unstable.*

The level surface which passes through any given point has in general a tangent plane at that point, but when the given point is a point of equilibrium, such as P, the first differential coefficients V_x, V_y and V_z are zero, and the equation of the plane is nugatory.

Resuming the expression for the potential V' at any point $(x+\xi, \text{\&c.})$ neighbouring to (x, y, z), we have, (Art. 111)

$$V' - V = \tfrac{1}{2}V_{xx}\xi^2 + \text{\&c.} + V_{xy}\xi\eta + \text{\&c.} + \text{cubes}......(1).$$

For any small displacement from P which makes V' greater than V, the force on the particle will act from P, and the equilibrium will therefore be unstable (Art. 41). For any displacement from P which makes V' less than V, the equilibrium at P will be stable. To find the directions which separate the stable and unstable displacements, we put $V' = V$. The equation of the *separating cone* is therefore found by equating to zero the terms of the lowest order on the right side of equation (1).

The separating cone is therefore *a quadric cone*, unless all the differential coefficients of the second order are also zero. It is *a real cone*, since by Laplace's theorem V_{xx}, V_{yy} and V_{zz} cannot all have the same sign whatever rectangular axes it is referred to.

The level surfaces in the immediate neighbourhood of a point P unoccupied by matter are in general planes, but if P be a position of equilibrium, they are hyperboloids with the separating cone for a common asymptotic cone. If PQ be any radius vector of one of these hyperboloids, the force of restitution for a given small displacement along PQ varies inversely as PQ.

121. Ex. 1. Show that three straight lines at right angles can always be drawn through the vertex on the surface of the separating cone. There is an infinite number of such systems of straight lines.

Ex. 2. If the attracting body is symmetrical about an axis and the point of equilibrium lie on the axis, prove that the separating cone is a right circular cone of semi-vertical angle $\tan^{-1}\sqrt{2}$. [This follows at once from Laplace's theorem, Art. 95.]

Ex. 3. The lines of force in the immediate neighbourhood of a point of equilibrium, when referred to the principal diameters of the separating cone as axes, are $z^c = Mx^a = Ny^b$, where a, b, c are the reciprocals of V_{xx}, V_{yy}, V_{zz} at the point of equilibrium, and M, N are two arbitrary constants.

Ex. 4. If a number of mutually repelling particles are enclosed in a rigid boundary, show that when in stable equilibrium they all reside on the surface. [If any one were not on the surface, that particle would be in unstable equilibrium, the remaining particles being held at rest.] Kelvin, see *Papers on Electrostatics*, &c., p. 100.

Ex. 5. Three uniform thin rods AB, BC, CA, which form a triangle, attract a particle P placed at the centre of the inscribed circle. The particle is therefore in equilibrium. Show that the equilibrium is unstable for all displacements in the plane of the triangle.

122. If two sheets of a level surface intersect along a line, every point of that line is a point of equilibrium.

Let P be such a point, then at least three tangents can be drawn to the sheets of the level surface not all lying in one plane and making finite angles with each other. Since the force along each of these is zero, it follows that the particle is in equilibrium.

123. At every point of the curve of intersection of two sheets of a level surface, the tangent cone becomes two planes which are the tangent planes to the two sheets. The tangent cone may therefore be written in the form

$$(a\xi + b\eta + c\zeta)(a'\xi + b'\eta + c'\zeta) = 0.$$

Comparing this with the form already found (Art. 120), we have

$$aa' + bb' + cc' = V_{xx} + V_{yy} + V_{zz}.$$

This is zero by Laplace's theorem; the tangent planes are therefore at right angles. We therefore infer that, *if two sheets of a level surface intersect, they intersect at right angles.*

124. Ex. 1. The tangent cone becomes two planes whenever its discriminant is zero. Prove that in a level surface these planes cannot be imaginary. [If it were possible, the cone could be reduced to the form $(a\xi + b\eta + c\zeta)^2 + (a'\xi + b'\eta + c'\zeta)^2 = 0$. This would make $a^2 + a'^2 + b^2 + \&c. = 0$, by Laplace's theorem, which is impossible.]

Ex. 2. Show that an isolated line in free space cannot form part of a level surface.

If the potential at a point P were greater than that at some neighbouring point Q and less than that at R, it would follow from the principle of continuity that there must be some point between Q and R on every path from one to the other at which the potential is equal to that at P. If then an isolated line form part of a level surface, the potential must be either greater than at all neighbouring points not on the line or less than at all such points. On either alternative the second proof, by which it is shown that the potential cannot be an absolute maximum or minimum, is contradicted, Art. 112.

125. Rankine's theorem. If at any point of a level surface all the differential coefficients of V up to the nth inclusive with regard to x, y and z are zero, we know from solid geometry that there is a tangent cone of the $(n+1)$th order at that point. If $(n+1)$ sheets intersect along a line, the same thing will be true at every point of that line, and the tangent cone will be the product of the $(n+1)$ tangent planes.

Let us suppose that the level surface is such that at two consecutive points P, P' all the differential coefficients of V up to the nth are zero; let us examine the form of the surface in the immediate neighbourhood of those two points.

Taking P for origin and PP' for the axis of z, we have at the origin all the following differential coefficients equal to zero:

$$\frac{d^nV}{dx^n},\ \frac{d^nV}{dx^{n-1}dy},\ldots\ \frac{d^nV}{dy^n},\ \frac{d^nV}{dx^{n-1}dz},\ \frac{d^nV}{dy^{n-1}dz},\ \frac{d^nV}{dx^{n-2}dz^2},\ \&c. \quad\ldots\ldots\ldots(1).$$

These are also zero when z receives an increment dz; hence their differential coefficients with regard to z are all zero. It therefore follows that every differential coefficient of V of the $(n+1)$th order which has dz, dz^2, &c. in the denominator is zero at the origin. If therefore V' be the value of the potential at a point ξ, η, ζ, we find on making the expansion by Taylor's theorem

$$\left.\begin{array}{r} V'-V=A_0\xi^{n+1}+A_1\xi^n\eta+\ldots+A_{n+1}\eta^{n+1} \\ +\text{powers of } \xi,\ \eta,\ \zeta \text{ of } (n+2)\text{th order}\end{array}\right\}\quad\ldots\ldots\ldots\ldots\ldots\ldots\ldots\ldots(2),$$

where A_0, A_1, &c. are constants. It follows that the terms of the lowest order in the expansion do not contain ζ.

The level surface which passes through the origin is given by $V'-V=0$. This level surface has therefore $(n+1)$ tangent planes at the origin given by

$$U=A_0\xi^{n+1}+A_1\xi^n\eta+\ldots+A_{n+1}\eta^{n+1}=0\ldots\ldots\ldots\ldots\ldots\ldots\ldots\ldots(3).$$

All these tangent planes pass through the two given consecutive points P, P'.

We shall now prove that all these tangent planes are real, and that each makes the same angle with the next in order. The expression for V' given in (2) must satisfy Laplace's equation, hence the expression for U given in (3) must also satisfy that equation. Transforming to cylindrical coordinates, U becomes $U=Pr^{n+1}$, where P is some function of ϕ. By Art. 108, since z is absent from U, we have

$$\frac{d^2U}{dr^2}+\frac{1}{r}\frac{dU}{dr}+\frac{1}{r^2}\frac{d^2U}{d\phi^2}=0.$$

Substitute, and we find $$\{(n+1)\,n+n+1\}\,P+\frac{d^2P}{d\phi^2}=0.$$

$$\therefore\ P=A\cos\{(n+1)\,\phi+\alpha\}.$$

The equation (3) therefore reduces to $\cos\{(n+1)\,\phi+\alpha\}=0$, which gives $n+1$ planes, making equal angles, each with the next in order.

126. Tubes of force. If we draw a line of force through every point of a closed curve, we construct a tube which is called a *tube of force*. By choosing the closed curve properly we can make the section of the tube indefinitely small; it is then called a *filament*. It is evident that the resultant attraction at any point P of a filament acts in the direction of the tangent to the length of the filament.

127. *The magnitude of the attractive force at any point of the same filament is inversely proportional to the area of the normal section of the filament at that point.*

Let σ, σ' be the areas of the normal sections of the filament at any two points P, P'. Let F, F' be the attractive forces at the same points. These forces act along the tangents at P, P' to the length of the filament and are supposed to be measured positively in the same direction along the arc.

Let us apply Gauss' theorem to the space enclosed by the filament and the two normal sections. Since the filament contains no attracting matter the *total flow* of the attraction across the *whole surface* is zero. The flow across the sides of the tube is zero, because at each point the resultant force acts along the length of the tube. The flow across the two normal sections must therefore be zero, hence $F\sigma - F'\sigma' = 0$, that is $F\sigma$ is constant for the same tube.

128. Ex. 1. Let the attracting body be a sphere. The lines of force are by symmetry normals to the surface; the filaments are therefore conical surfaces of small angle. If r be the distance of P from the centre, $\sigma = r^2 d\omega$; hence Fr^2 is constant along any line of force. Thus it follows at once that the force of attraction at any external point varies inversely as the square of its distance from the centre.

Ex. 2. If F be the normal force at any point P of a level surface; ρ, ρ' the radii of curvature of the principal sections and ds an element of the arc of a line of force at the same point P, then will $\dfrac{d \log F}{ds} + \dfrac{1}{\rho} + \dfrac{1}{\rho'} = 0$.

Construct on the level surface an elementary rectangle $PQSR$ such that the sides PQ, PR are elements of the lines of curvature at P. Let the tube of force having this rectangle as base intersect a neighbouring level surface in $P'Q'S'R'$. If σ, σ' are the areas of these rectangles and $ds = PP'$, we have by the properties of similar figures

$$\frac{\sigma'}{\sigma} = \frac{(\rho + ds)(\rho' + ds)}{\rho\rho'} = 1 + \left(\frac{1}{\rho} + \frac{1}{\rho'}\right) ds.$$

If F, $F + dF$ be the forces at P, P', we know that $F\sigma = (F + dF)\sigma'$, Art. 127. This immediately reduces to the required result. See Bertrand *on isothermal surfaces*, *Liouville's J.* 1844, vol. IX.

129. *If two different bodies have equal potentials over the surface of any space not including any attracting matter, they have equal potentials throughout that space, and also at all external space which can be reached without passing through any of the attracting matter of either body.*

For let the attraction of one of the bodies be changed into repulsion. Then the potential due to both bodies is zero over the surface of the given space. That is, the united potential is *constant* over the surface; it is therefore also constant and zero throughout the enclosed space, and at all points of external space which can be reached without crossing any attracting matter;

Arts. 115 and 117. Returning then to the original supposition that both the bodies attract, it easily follows that their potentials are equal.

130. *If two different bodies have equal potentials over the whole boundary of any surface enclosing both, they have equal potentials throughout all external space.*

As before, changing the attraction of one body into repulsion, let us consider the potential of both bodies regarded as one system. Their united potential is therefore zero over the whole boundary of the surface. It is also zero over the boundary of an infinite sphere. Since the space between the surface and the sphere contains no attracting matter, the potential is also zero throughout that space, Art. 115. Returning to the original supposition, that both bodies attract, we see that their potentials must be equal.

Ex. An unknown body is surrounded by a sphere of radius a. The direction of the attraction at all points of this sphere is normal to the sphere and its magnitude is equal to a given constant F. Prove that the attraction at any external point is Fa^2/r^2.

The sphere is a level surface because the force is normal. The potential of the body at any point of the sphere is therefore equal to that of a particle whose mass is Fa^2 placed at the centre.

131. *If two different bodies have the same level surfaces throughout any empty space, their potentials throughout that space are connected by a linear relation.*

Let V and V' be the two potentials. Since when V is constant, V' is also constant, it follows that V' is some function of V, say $V'=f(V)$. Then by differentiation we easily find

$$\frac{d^2V'}{dx^2}+\frac{d^2V'}{dy^2}+\frac{d^2V'}{dz^2}=\frac{df}{dV}\left\{\frac{d^2V}{dx^2}+\frac{d^2V}{dy^2}+\frac{d^2V}{dz^2}\right\}+\frac{d^2f}{dV^2}\left\{\left(\frac{dV}{dx}\right)^2+\left(\frac{dV}{dy}\right)^2+\left(\frac{dV}{dz}\right)^2\right\}.$$

Since the space is external to both bodies, this, by Laplace's equation, reduces to $0=\frac{d^2f}{dV^2}$, unless V is constant throughout the space considered. If V is constant, the level surfaces for both bodies are indeterminate and therefore V' also is constant. We therefore have in both cases $V'=AV+B$, where A and B are two constants. Suppose the space considered includes the points at infinity, then when the attracting masses are finite in size and density both V and V' vanish at such points. We then have $B=0$. Again V and V' must vanish at infinity in the ratio of the attracting masses; we therefore find $V'/V=M'/M$ if M, M' be the masses of the attracting systems. We thus have the theorem; *if two finite bodies have the same external level surfaces and have equal masses, their attractions at all external points are the same in magnitude and direction. Quarterly Journal of Mathematics*, 1867.

When the space in which the two bodies have the same level surfaces encloses both bodies, this theorem follows at once from that proved in Art. 130. Since the two bodies have the innermost level surface common, we can by altering the mass of one of them make their potentials equal over that surface. The potentials of the changed bodies are then equal over all external space and the potentials of the original bodies have a constant ratio.

132. As an example of this theorem, consider the case of a spherical shell. The external level surfaces of such a shell and those of an equal mass placed at its centre are both spheres. Hence the attraction of a spherical shell at any external point is the same as that of an equal mass placed at its centre.

Again, the level surfaces of two equal and parallel infinite plates are both planes. Hence their *attractions* at any point are in a constant ratio. But at an infinite distance the attractions of two such plates when separated by a finite interval tend to equality, hence the ratio of the attractions is unity. It follows that the attraction of an infinite plate at an external point is independent of its distance. In the same way the attraction of an infinite circular cylinder is the same as if the whole mass were uniformly distributed along the axis.

133. The theorems in this section have been enunciated with special reference to the potential of an attracting system, but a little consideration will show that they have a more extended application.

If V be any continuous one-valued function which satisfies Laplace's equation and is not infinite within any given space, it follows from the argument in Art. 111 that V cannot be an absolute maximum or minimum at any point within that space.

Most of the other theorems are simple corollaries from this one general principle, and apply therefore to any finite continuous function which satisfies Laplace's equation.

For example, if such a function be constant over the boundary of any space and not infinite within that space, it must be constant throughout that space.

To take another example, let V be a finite continuous function which satisfies Laplace's equation, then $V=c$ is a system of surfaces. If any member of this system intersects itself in a singular line, the two sheets are at right angles. If several sheets intersect in a singular line, each tangent plane makes the same angle with the next in order.

Let V, V' be two continuous solutions which are both finite and one-valued at all points of space bounded by a surface S and are equal at every point of that surface, then they are equal throughout that space. The space considered may be external to S provided the functions are also equal at all points on the surface of some sphere of infinite radius enclosing S. This theorem shows that *when the values of a function V are known at all points of the boundary of a space, it is determinate throughout* *that space, provided it is known to satisfy Laplace's equation and to be finite throughout that space.*

134. To trace level curves and lines of force. Ex. 1. Three equal particles are placed at the corners ABC of an equilateral triangle. Trace the level curves and lines of force in the plane of the triangle.

We first search for the points of equilibrium. The centre of gravity G is evidently one such point. The level curves near G are conics (Art. 120) which must have GA, GB, GC for three principal diameters. The conics are therefore circles. The equilibrium is clearly stable for a displacement perpendicular to the plane and is therefore unstable for some (and therefore also for all) displacements in the plane (Art. 111). The potential is therefore a minimum at G for displacements in the plane ABC.

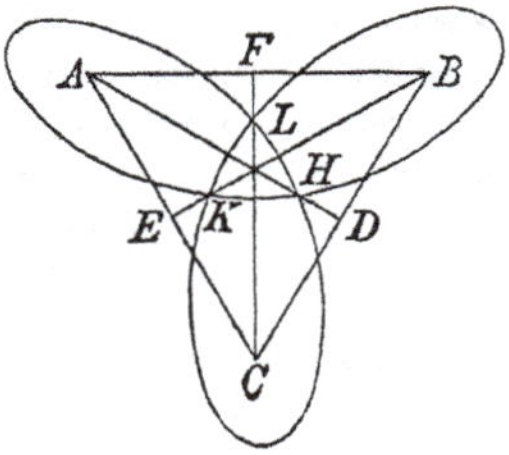

Let D, E, F be the feet of the perpendiculars from the corners. Since the force at F tends towards G and G is a point of minimum potential, there must be a point of equilibrium between G and F. There are therefore three points of equilibrium which are H, K, L. The level curve which passes through these points governs the whole sketch and is exhibited in the figure. Some of the other level curves fill up the four vacant areas and others surround the three loops.

To sketch the lines of force. It will be found convenient to mark the level curves or surfaces with small arrows to indicate the direction of the normal force. We then have the following rule; no line of force can pass from a point A on one level surface to a point B on another unless either the arrows at both A and B tend in the same direction along the line of force or the line of force passes through a point of equilibrium which lies between A and B.

The arrows on the sides of the curvilinear triangle HKL all tend outwards from G, while those on the three curvilinear triangles which surround A, B, C tend inwards towards those points. Hence a line of force beginning at A must either proceed to an infinite distance or cross KL. If it enter the triangle HKL it cannot emerge without passing through G. It must then proceed onwards to either B or C.

There are conical points at H, K and L. The level surfaces near G are not closed but bend over A, B, C, and surround the conical points.

Ex. 2. Two particles whose masses are m, m' are placed at A and B, both being attractive. Trace their level surfaces.

Ex. 3. Three equal particles are placed at three points A, B, C in a straight line. The particles A and C attract while B repels. Trace the level surfaces.

135. Potential at a distant point. *To find the potential of a body finite in all directions at any distant external point*.*

Let the origin O be a point not far from the body. Let Q be

* The expansion of the potential at a distant point is originally due to Poisson, but was put into a convenient form by MacCullagh, *R. Irish Trans.* 1855. Some of the following theorems were given by the author in the *Quarterly J.* 1857. The name *centrobaric* is due to Lord Kelvin, who gave several theorems on these bodies in the *Proc. R. S. E.* 1864. The results in Arts. 140, 141 are taken from Thomson and Tait, 1883.

the position of any particle of the body, m its mass, (x, y, z) its coordinates, r its distance from the origin. Let (ξ, η, ζ) be the coordinates of the point P, $OP = r'$, and the angle $POQ = \theta$.

To generalize the investigation we shall assume that the law of attraction is the inverse κth power of the distance. We then have

$$V = \frac{1}{\kappa - 1}\Sigma \frac{m}{(r'^2 - 2rr'\cos\theta + r^2)^{\frac{\kappa-1}{2}}}$$

$$= \Sigma \frac{m}{r'^{\kappa-1}}\left\{\frac{1}{\kappa-1} + \frac{r\cos\theta}{r'} + \frac{(\kappa+1)\cos^2\theta - 1}{2}\left(\frac{r}{r'}\right)^2 + \dots\right\} \dots\dots (1).$$

The first term of the series is $\frac{\Sigma m}{r'^{\kappa-1}}\frac{1}{\kappa-1}$. Hence *the attraction at a very distant point is ultimately the same as if the whole mass were collected into a single particle and placed at O.*

To find a closer approximation to the true attraction, let the point O be such that the second term of the series vanishes. This requires that $\Sigma mr\cos\theta = 0$. Since $rr'\cos\theta = x\xi + y\eta + z\zeta$, this gives $\xi\Sigma mx + \eta\Sigma my + \zeta\Sigma mz = 0$ for all values of ξ, η, ζ. *The point O will therefore be the centre of gravity of the body.*

We have now to consider the third term of the series. Let A, B, C be the moments of inertia of the body about any three straight lines at right angles meeting in O, I the moment of inertia about the straight line OP, then

$$2\Sigma mr^2 = A + B + C, \quad I = \Sigma m (r\sin\theta)^2.$$

Writing $1 - \sin^2\theta$ for $\cos^2\theta$ and making these substitutions we find for the third term

$$\frac{\kappa(A + B + C) - 2(\kappa+1) I}{4}\frac{1}{r'^{\kappa+1}} \dots\dots\dots\dots (2).$$

When the law of force is the inverse square and the centre of gravity is the origin we arrive at MacCullagh's expression for the potential, viz.

$$V = \frac{M}{r'} + \frac{A + B + C - 3I}{2r'^3} + \dots \quad \dots\dots\dots (3),$$

where M is the mass of the body.

Ex. When the law of attraction is the inverse distance, the potential of a single particle takes the form $C - m\log r'$. Prove that the potential of a body at a distant point is

$$V = C - M\log r' + \frac{A+B+C-4I}{4r'^2} + \dots \quad \dots\dots\dots\dots (4).$$

136. *If two bodies have equal potentials at all external points, their centres of gravity must coincide and their masses must be equal. If the law of force is the inverse κth power the bodies are* *equimomental, unless* $\kappa = -1$. *If the law is the inverse square, the difference of their moments of inertia about every straight line must be constant.*

The potential of each body can be represented by the series described in Art. 135 and these series must be equal, term to term. The equality of the first terms requires that the masses should be equal. Taking the origin O at the centre of gravity of one body, the second term must be missing for both series and therefore the centre of gravity of the second body must also be at the origin.

Comparing the third terms of the series we have

$$\kappa (A + B + C) - 2 (\kappa + 1) I = \kappa (A' + B' + C') - 2 (\kappa + 1) I' \quad \ldots (5),$$

where unaccented and accented letters refer to corresponding quantities in the two bodies. It follows that (unless $\kappa = -1$) $I - I'$ is the same for all axes passing through the common centre of gravity. The axes of maximum and minimum moments of inertia in the two bodies are therefore the same. Since these are the principal axes of inertia, *the two bodies must have the directions of their principal axes coincident.* Since $I - I'$ is the same for every axis, it follows that the four differences $A - A'$, $B - B'$, $C - C'$, and $I - I'$ are equal. The equation (5) then becomes

$$(\kappa - 2)(I - I') = 0 \quad \ldots\ldots\ldots\ldots\ldots\ldots (6).$$

Unless $\kappa = 2$, we have $I = I'$ and therefore *the moments of inertia of the two bodies about every axis are equal, each to each.* If however $\kappa = -1$ these conditions are not necessary. When κ has this value the law of attraction is the direct distance. In this case it has already been proved that a body, whatever be its form, attracts any particle as if it were collected into its centre of gravity (Art. 8).

These are necessary conditions that two bodies should be equipotential (unless $\kappa = -1$), but they are not sufficient. It is also necessary that all the subsequent terms of the potential series should be equal, each to each.

We have assumed here that the law of attraction is some one integral inverse power of the distance. If the law be represented by a series of inverse powers such as $\mu/r^{\kappa} + \mu'/r^{\kappa+1} + \&c.$, it is evident that so far as the series (1) in Art. 135 is concerned we need only consider the three lowest powers of r in the law of attraction. The remaining powers enter only into the terms of that series not included in our approximation. Proceeding in the same way we again arrive at the results stated in the enunciation.

137. Centrobaric bodies. When a body is such that its potential at every point is equal to that of a particle of mass M situated at some fixed point O, the body is said to be *centrobaric*. In other words, the body is equipotential to a particle. We infer immediately from Art. 136 that *M is equal to the mass of the body and that O is its centre of gravity.* Since for a particle $I'=0$, it follows that the moment of inertia I of the body about every axis is the same. *The body therefore cannot be centrobaric unless every axis at the centre of gravity is a principal axis.*

The condition (6) now becomes $(\kappa-2)I=0$. It appears therefore that the series (1) of Art. 135 cannot reduce to its first term unless $\kappa=2$ or $I=0$. The latter condition cannot be satisfied unless the masses of some of the particles are negative, that is unless some of the particles attract and others repel P. *Assuming that all the particles attract P, according to some inverse power of the distance, we see that the attraction of a body cannot be the same as if the whole mass were collected into its centre of gravity unless the law of force be either the direct distance or the inverse square of the distance.*

138. Ex. If the law of force be the inverse square, the potential of a body at all external points cannot be the same as that of two masses M_1 and M_2 placed at two points A, B fixed in the body unless (1) the body and masses have their centres of gravity coincident, (2) the moments of inertia of the body about every axis through the centre of gravity perpendicular to AB are equal.

139. Potential constant in a cavity. *In a similar manner, when a body has a cavity within its substance we may determine the necessary conditions that the potential should be constant throughout the cavity.* Taking the origin within the cavity, we have at all points close to the origin

$$V=\Sigma\frac{m}{r^{\kappa-1}}\left\{\frac{1}{\kappa-1}+\frac{r'\cos\theta}{r}+\frac{(\kappa+1)\cos^2\theta-1}{2}\left(\frac{r'}{r}\right)^2+\dots\right\};$$

the expansion is in powers of r'/r because r' is less than r.

This cannot be independent of r' unless the coefficient of each power of r' is zero. Equating the coefficient of r'^2 to zero, we have

$$\Sigma\frac{m}{r^{\kappa+1}}\{(\kappa+1)\cos^2\theta-1\}=0.$$

Writing α, β, γ for $\Sigma\frac{mx^2}{r^{\kappa+3}}$, $\Sigma\frac{my^2}{r^{\kappa+3}}$, $\Sigma\frac{mz^2}{r^{\kappa+3}}$ and putting the point P in succession on the axes of x, y, z we have $\kappa\alpha=\beta+\gamma$, $\kappa\beta=\gamma+\alpha$, $\kappa\gamma=\alpha+\beta$. These give $\kappa=2$, or $\kappa=-1$ and $\alpha+\beta+\gamma=0$, or

α, β, γ each zero. The two latter alternatives require that all the m's should not have the same sign. Hence *if every particle of the body be attractive, the potential cannot be constant throughout any cavity unless the law of attraction is the inverse square,* (Art. 99).

140. Assuming that a body attracts all points in external space as if the whole mass were collected into its centre of gravity, prove that (1) the centre of gravity is inside the external boundary, (2) the external boundary is a single closed surface.

If the centre of gravity O were in the same external space as the attracted point P, we could surround it by a small sphere, centre O, radius ϵ, which does not enclose any particle of the attracting mass. The flux across this sphere is therefore zero, Art. 106. But since the force on P tends always to O, the flux is also $4\pi M$. These results contradict each other unless the whole mass is equal to zero.

Again, if the attracting system consist of two separate portions, the centre of gravity O must lie inside one of them. Enclosing the other portion in a sphere, the flux across the surface is $4\pi M'$, if M' be the mass of this portion. But since O lies outside the sphere, it is also zero. These results cannot coexist unless the mass of that portion is zero.

141. A body B is such that the resultant attraction between it and a given body A is a force which always passes through the centre of gravity O of B, in whatever position A is placed. Prove that the resultant attraction between B and every body is a force which passes through the centre of gravity of B.

Let the body A be turned about a fixed point P sufficiently distant from B, that the body A in its motion never meets the fixed body B. In all these positions the resultant attraction of A on B is a force which passes through the centre of gravity of B. Hence if every particle of the mass of A be uniformly distributed over the surface of the sphere which that particle describes in its motions, the resultant attraction of the mass thus obtained is also a force which passes through the centre of gravity of B. The mass thus obtained is a spherical shell whose resultant attraction at any point of B is the same as if it were collected at the centre P. The resultant action between the body B and a particle placed at P is a force which passes both through P and the centre of gravity of B. The body B is therefore centrobaric for all points P beyond a certain distance and therefore for all points of space which can be reached from P without passing over any of the attracting mass, Art. 129.

Attraction of a thin stratum.

142. A theorem due to Green*. Let a thin heterogeneous stratum of attracting matter be placed on a surface which has no

* The theorem $X' - X = 4\pi m$ is of great importance in the theory of attraction. The principle of the demonstration given in Art. 142 was used for a spherical shell by Lagrange in 1759 and was afterwards applied by Coulomb to the case of a thin electrical film of any form (*Paris Mémoires*, 1788). Poisson gives a generalization of the theorem to any film (*Mém. de...l'Institut*, 1811, *Connaissance des Temps* for 1829, p. 375). Cauchy deduces the same result for any film from the general formulæ of attractions (*Bulletin...Soc. Philomathique*, 1815, p. 53). The theorem is commonly called Green's theorem (*Essay...on Electricity and Magnetism*, 1828). It was afterwards re-discovered by Gauss, 1840. A proof on the same general principle as that in Art. 142 was given by Kelvin in 1842, see the reprint of his papers on *Electrostatics and Magnetism*, 1842, and Thomson and Tait, Art. 478.

conical points or other singularities. Let ρ be the density and t the thickness at any point A of the surface, and let $m=\rho t$, so that m is the surface density at the point A. In what follows we shall regard m as finite and t as indefinitely small, so that ρ is very large.

Let P, P' be two points situated on the normal at A, one inside the surface and the other outside, both close to the stratum; it is required to find the attractions at P and P'.

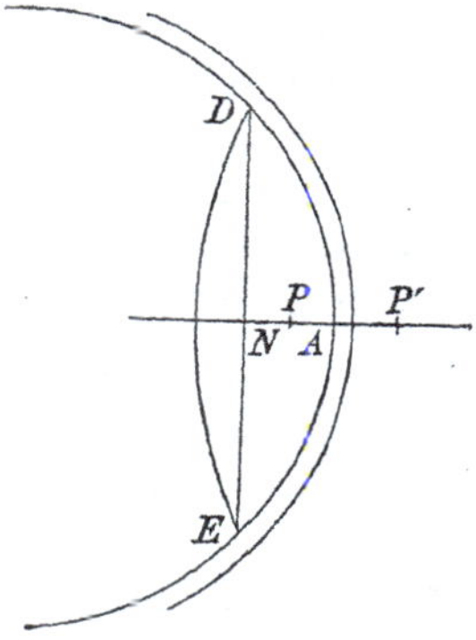

With centre A, and a small geodesic radius a, describe on the surface a circle whose circumference is DE, and let DN be a perpendicular on the normal at A drawn from any point D. The radius a of this circle is infinitely greater than either AN or the thickness t but infinitely less than either radius of curvature of the surface.

This circle divides the whole attracting stratum into two parts whose attractions at P and P' will be separately considered. Let us first find the attraction of the small portion DAE which we may suppose to lie in the tangent plane at A.

We take the attracted point P for origin and the normal PA for the axis of x, let $PA=p$. By Arts. 21, 22, the attraction is

$$2\pi\rho\int dx\left(1-\frac{x}{\sqrt{(a^2+x^2)}}\right)=2\pi\rho\left[t-\{a^2+(p+t)^2\}^{\frac{1}{2}}-\{a^2+p^2\}^{\frac{1}{2}}\right],$$

the limits being p to $p+t$. Now ultimately p/a and t/a are zero, while $\rho t=m$. We have therefore for the attraction

$$2\pi\rho t\left[1-\frac{2p+t}{2a}\right]=2\pi m.$$

The attractions at P, P' are therefore each equal to $2\pi m$. They are directed along the normal in opposite directions, and their difference is $4\pi m$.

We have supposed the stratum DAE to lie in the tangent plane at A. But the effect of the curvature would be simply to change the attraction $2\pi m$ of a plane disc into $2\pi m\,(1-\beta/r)$, where β is a quantity of the order a or p. These additional terms are zero because both a and p are infinitely smaller than r. That this is so may be made clearer by considering the case in which the surface is spherical. The disc DAE is then bounded by a right cone whose vertex is at the centre and whose

semi-angle is a/r. By using Art. 74 and retaining the first powers of p/a, t/a and a/r we find
$$\beta = \tfrac{1}{2}(a + 2p).$$

Consider next the attraction of the portion of the stratum remote from A. Let F, F' be the x-components of attraction at P, P'. Since these depend on the attracting mass, each contains the factor ρt or m. Also, since the distance PP' is infinitely smaller than the distance of either P or P' from the nearest attracting element, F' differs from F by $(dF/dx)\,t$. The difference is therefore of the order ρt^2 or mt. We may therefore regard F, F' as equal.

Taking both portions of the attracting stratum into the account and representing by X, X' the normal attractions of the whole system at P, P' we have
$$X = F - 2\pi m, \quad X' = F' + 2\pi m \quad\text{.............. (1)},$$
where X, X' are measured positively from P' to P. Since F, F' are ultimately equal, these give
$$X' - X = 4\pi m, \quad F = \tfrac{1}{2}(X' + X) \quad\text{.............. (2)}.$$
The equation $X' - X = 4\pi m$ shows that when attraction is taken as the standard case, $4\pi m$ is equal to the *sum* of the normal attractions at each side of the stratum, the attractions being measured *towards* the stratum. When repulsion is the standard case, $4\pi m$ is equal to the *sum* of the normal repulsions, the repulsion being measured on each side *from* the stratum.

If there are any other attracting bodies in the field which are at finite distances from the points P and P', their attractions at these points are ultimately equal. It follows that in both the formulæ (2) we may suppose X, X', and F to mean the normal components due to all causes.

143. The equation $F = \tfrac{1}{2}(X + X')$ enables us to find the normal attraction of a thin heterogeneous stratum on an elementary portion of itself.

Let the element be a small cylinder whose base is the area $d\sigma$ situated at A and whose altitude is the thickness t of the stratum. The normal attraction of the adjacent portion DAE on the cylindrical element is ultimately zero because it is the same as the normal attraction of an infinite plate on a portion of itself. The attraction of the remote portion of the stratum is $Fmd\sigma$. It follows therefore from (2) that *the whole normal force per unit of mass acting on the element is the arithmetic mean of the normal attractions just inside and just outside the stratum.* The normal force on the matter $md\sigma$ which covers an element of area $d\sigma$ is $Fmd\sigma$ and is therefore equal to $\frac{X'^2 - X^2}{8\pi} d\sigma$, where X, X' are the normal forces at each side of the element.

144. *We may also show that the parallel tangential components of attraction just inside and just outside the stratum are equal.* Let the axis of y be parallel to a tangent at A to either boundary of the stratum. Let Y, Y' be the components of attraction at P, P'. Considering first the adjacent portion DE of the stratum, it has already been shown that the resultant attractions at P, P' are each directed along the normal PP'; hence this portion contributes nothing to Y or Y'. Considering next the remote portion of the stratum, it may be shown as in Art. 142 that the components Y, Y' differ by terms of the order mt. In the limit therefore when t is very thin, we have $Y' = Y$.

145. *We shall now show that the potentials at P, P' are also equal.* The potentials due to the remote portion of the stratum for the same reasons as before can differ only by terms of the order mt. Consider next the portion of the stratum adjacent to A; the potentials at two points equally distant from the two faces of the stratum evidently differ by terms of an order higher than mt. See also Art. 76, Ex. 1. Taking both portions of the stratum, we see that the potentials at P and P' are ultimately equal.

146. It follows from this proposition that *if a point travel from a position P just within a thin stratum to another P' just outside, both on the same normal, the normal component of the attraction is increased by the quantity $4\pi m$, where m is the surface density. At the same time the tangential components of the attraction and the potential are unaltered.*

147. We may also deduce Green's theorem from the proposition, due to Gauss, that the flux of the attraction over a closed surface is 4π multiplied by the mass inside. See Art. 106.

Let the axis of x be a normal to the stratum, measured positively inwards, and let it cut the boundaries in the points A, A'. Let us consider the flux of the attraction across an element of volume whose edges parallel to the axes x, y, z are respectively $AA' = t$, dy and dz.

Proceeding as in Art. 108 we have

$$(X' - X)\,dydz + \frac{dY}{dy}\,tdydz + \frac{dZ}{dz}\,tdydz = 4\pi\rho t dydz,$$

where $X' - X$ has not been equated to $(dX/dx)\,dx$ because there is attracting matter on one side only of each of the two faces perpendicular to the axis of x. Substituting $m = \rho t$ in the equation, dividing by $dydz$ and taking the limit, we find $X' - X = 4\pi m$.

148. Ex. 1. A thin layer of heterogeneous attracting matter is placed on a sphere of radius a. If V be the potential and m the surface density at any point A, show that the normal attractions on each side of the stratum are $V/2a \pm 2\pi m$. Art. 87.

Ex. 2. Prove that, if matter attracting according to the law of the inverse square be so distributed over a closed surface that the resultant attraction on every external particle in the immediate neighbourhood is in the direction of the normal, the resultant *attraction* on every internal point is zero.

The outer boundary of the stratum is by definition a level surface. The inner boundary is therefore also a level surface. The result then follows from Art. 115 because there is no attracting matter within that surface.

Green's Theorem.

149. Let a portion of space be enclosed by a surface which we shall call S. Let V, P, Q, R be any one-valued finite functions of x, y, z, and let $dv = dxdydz$. Let us integrate

$$U = \iiint \left(P\frac{dV}{dx} + Q\frac{dV}{dy} + R\frac{dV}{dz} \right) dxdydz \quad \text{.........(1)},$$

throughout the given space S. The first term becomes by an integration by parts

$$\iint [PV]\,dydz - \iiint V\frac{dP}{dx}\,dxdydz \quad \text{...........(2)}.$$

We have here integrated all the elements which lie in a column parallel to the axis of x. Let AB be one of these columns and let it intersect the surface S at A and B in the elementary areas $d\sigma, d\sigma'$. If $(\lambda'\mu'\nu')$ be the direction cosines of the outward normal at the upper limit B we have $dydz = \lambda' d\sigma'$. In the same way if $(\lambda\mu\nu)$ be the direction cosines of the outward normal at the limit A, we have $dydz = -\lambda d\sigma$, since λ' is positive and λ negative. The quantity in the square brackets in the first term of (2) is to be taken between the limits A and B and is therefore

$$(PV)_B\,\lambda' d\sigma' - (PV)_A\,(-\lambda d\sigma) \quad \text{..............(3)},$$

where the suffix indicates the place at which the value of the quantity in brackets is to be taken. The two terms in (3) have now to be integrated, the first for all elements such as B on the right-hand side of the bounding curve CD and the second for all

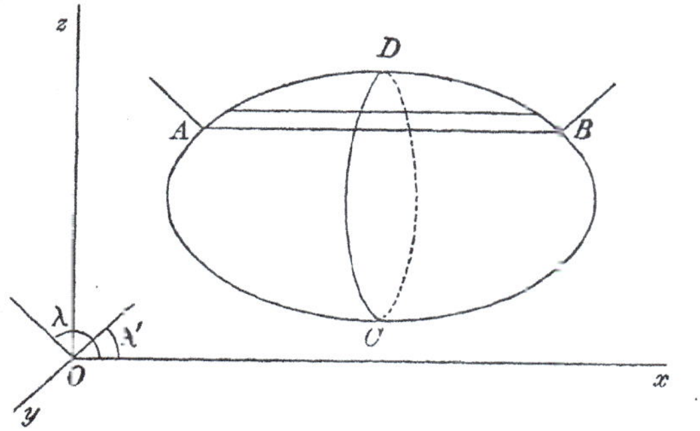

elements such as A on the left. These together are the same as $\int PV\lambda d\sigma$ taken for all elements of the surface, where λ now stands for the cosine of the angle the outward normal at $d\sigma$ makes with the axis of x.

Treating the other terms of (1) in the same way we have

$$U = \int V(P\lambda + Q\mu + R\nu)\, d\sigma - \int V\left(\frac{dP}{dx} + \frac{dQ}{dy} + \frac{dR}{dz}\right) dv \ldots (4).$$

Let P, Q, R be the components of a vector I and let $I \cos i$ be the normal component at the element $d\sigma$. The equation (4) then becomes

$$U = \int VI \cos i\, d\sigma - \int V\left(\frac{dP}{dx} + \frac{dQ}{dy} + \frac{dR}{dz}\right) dv \ldots\ldots\ldots (5).$$

In this way the volume integral (1) has been replaced by a surface integral, when the vector is such that

$$\frac{dP}{dx} + \frac{dQ}{dy} + \frac{dR}{dz} = 0.$$

Let this vector be the attractive force of some system whose potential is V'. To be more general, let $P = dV'/dx$, $Q = dV'/dy$, $R = dV'/dz$, then

$$U = \iiint \left(\frac{dV}{dx}\frac{dV'}{dx} + \frac{dV}{dy}\frac{dV'}{dy} + \frac{dV}{dz}\frac{dV'}{dz}\right) dx\,dy\,dz \ldots\ldots (6),$$

where V, V' are two arbitrary functions of xyz Let dn be an element of the outward normal at $d\sigma$, then

$$\frac{dV'}{dx}\lambda + \frac{dV'}{dy}\mu + \frac{dV'}{dz}\nu = \frac{dV'}{dn} \ldots\ldots\ldots\ldots\ldots (7).$$

Also let ρ, ρ' be such functions of xyz that

$$-4\pi\rho = \nabla^2 V, \quad -4\pi\rho' = \nabla^2 V'.$$

Then by (4) the symmetrical expression U takes either of the forms

$$U=\int V\frac{dV'}{dn}d\sigma+4\pi\int V\rho' dv \quad\ldots\ldots\ldots\ldots\ldots(9)$$

$$=\int V'\frac{dV}{dn}d\sigma+4\pi\int V'\rho dv \quad\ldots\ldots\ldots\ldots(10).$$

The equality of the expressions (6), (9) and (10) is usually called Green's theorem.

150. If the functions V, V' satisfy Laplace's equation we have $\rho=0$, $\rho'=0$, the equality then becomes

$$U=\int V\frac{dV'}{dn}d\sigma=\int V'\frac{dV}{dn}d\sigma \quad\ldots\ldots\ldots\ldots\ldots(11).$$

151. Let V, V' be the potentials of two attracting systems. Let W be the mutual work of the first and that portion of the second system which is internal to S; let W' be the mutual work of the second and that portion of the first which is internal to S. Then, by Art. 59, Green's equation becomes

$$U=\int VF'd\sigma+4\pi W=\int V'Fd\sigma+4\pi W' \quad\ldots\ldots\ldots\ldots\ldots(13),$$

where F, F' are the outward normal components of force at the element $d\sigma$.

The expression for U admits also of interpretation. Let (XYZ), $(X'Y'Z')$ be the components of force due to the two systems at any point (xyz) within S. Let R, R' be the resultant forces, ϕ the angle between the directions of R, R'. Then, by (1),

$$U=\int(XX'+YY'+ZZ')\,dv=\int RR'\cos\phi\, dv \quad\ldots\ldots\ldots\ldots\ldots(14).$$

If the two systems are the same, i.e. if the particles occupy the same positions in the two systems and have equal masses, Green's equation becomes

$$U=\int R^2dv=\int VFd\sigma+4\pi\int V\rho dv,$$

where F is the outward normal force at the element $d\sigma$.

152. Instead of considering the space internal to S *we may integrate through the space between S and a sphere of infinite radius enclosing S and having its centre at a finite distance from S.* We must then of course include this sphere in the surface integration over S. *Let V, V' be the potentials of some masses M, M' respectively*, then for points on the surface of the sphere $V=M/a$ and $dV'/dn=-M'/a^2$ also $d\sigma=a^2d\omega$, where a is the radius and $d\omega$ is the elementary solid angle subtended at the centre by $d\sigma$. We therefore have for the sphere

$$\int V\frac{dV'}{dn}d\sigma=-MM'\int\frac{d\omega}{a}=-4\pi\frac{MM'}{a},$$

and this is zero when a is infinite. We may therefore *in this case apply the equality* (9) *and* (10) *without further change to the space outside S.* We notice that dn is always to be measured outwards from the space over which the integration extends.

153. *To deduce Gauss' theorem.* Let us put unity for V'. Since this value satisfies Laplace's equation, we have $\rho'=0$. The equality (9) and (10) takes the form

$$\int\frac{dV}{dn}d\sigma=-4\pi\int\rho dv \quad\ldots\ldots\ldots\ldots\ldots(15).$$

Let the space of integration be the *finite space* enclosed by a surface S. We thus avoid the integration over the surface of a sphere of infinite radius. Supposing V to be the potential of any attracting mass, the function of x, y, z represented by ρ becomes, by Poisson's theorem, the density of the mass at the element dv. The right-hand side of this equation is therefore $-4\pi M$, where M is that portion of the mass which is inside S. Also dV/dn represents the outward normal force. The equation therefore asserts that the whole outward flux across any surface S is $-4\pi M$. This is Gauss' theorem.

154. Green's equivalent layer. Let $V' = 1/r'$ where r' is the distance of any point within the space of integration from some given point P. Let the integration extend throughout the space internal or external to S according as P is external or internal In this way we make $1/r'$ finite throughout the integration.

Since $4\pi\rho' = -\nabla^2 V'$, ρ' is now zero, and Green's equation becomes

$$\int V \frac{d}{dn}\left(\frac{1}{r'}\right) d\sigma - \int \frac{dV}{dn}\frac{d\sigma}{r'} = 4\pi \int \frac{\rho\, dv}{r'} \quad \text{.........(16).}$$

Here the r' on the right-hand side is the distance of P from dv and on the left-hand side r' is the distance of the same point from the element $d\sigma$ of the surface.

We shall now suppose that V is the potential of some attracting system, part of which may be inside S and part outside. The right-hand side of the equation is evidently $4\pi V_1$ where V_1 is the potential at P of that part of the attracting mass which is on the side of S opposite to P.

The equation asserts that the potential at P of that part of the system on the opposite side is equal to that of a thin layer placed on the surface S whose surface density D at any point Q, $(PQ = r')$ is given by

$$4\pi D = r' V \frac{d}{dn}\left(\frac{1}{r'}\right) - \frac{dV}{dn} = -\frac{1}{r'}\frac{d}{dn}(Vr'),$$

where V is the potential at Q of the whole system. To make D independent of the position of P we shall get rid of the terms which contain r'.

155. Let the surface S be such that the potential V of *the whole attracting system* is constant and equal to V_s over its area. Then S is a level surface, or a closed portion of a level surface, of the whole system. Since $1/r'$ is the potential at $d\sigma$ of a unit mass placed at P, we have by Gauss' theorem

$$\int V \frac{d}{dn}\left(\frac{1}{r'}\right) d\sigma = V_s \int \frac{d}{dn}\left(\frac{1}{r'}\right) d\sigma = 4\pi V_s \text{ or } 0,$$

according as P is within or without the finite space enclosed by the surface S. The plus sign is given to $4\pi V_s$ because we are integrating throughout the space on the side of S opposite to P, and *dn is therefore measured towards P*.

Lastly let us place on the surface S a thin layer of matter whose surface density ρ'' is given by $\rho'' = -\frac{1}{4\pi}\frac{dV}{dn'}$ where dn' is measured *outwards* from the finite space enclosed by S. Let V'' be the potential of this layer at P, then

$$V'' = \int \frac{\rho''\, d\sigma}{r'} = -\frac{1}{4\pi}\int \frac{dV}{dn'}\frac{d\sigma}{r'}.$$

In the equation (16), when P is internal dn is measured inwards and therefore $dn = -dn'$; when P is external $dn = dn'$. That equation therefore becomes

$$V_s - V'' = V_1, \text{ or } V'' = V_1 \qquad \text{.............. (17)},$$

according as P is internal or external. We deduce the three theorems enunciated in the next article.

156. Let S be a level surface of an attracting system. Let a thin layer of attracting matter be placed on the surface S such that its surface density ρ'' at any point Q is given by the equation

$$4\pi\rho'' = -\frac{dV}{dn'} \qquad \text{...................... (18)},$$

where V is the potential at Q due to the attracting system, and dn' is measured positively outwards from the finite enclosed space.

(1) The potential of the layer at any point P, external to the level surface S, is equal to the potential at the same point of that portion of the attracting system which is within S.

(2) The potential of the layer at any point P internal to S, increased by the potential at the same point of that portion of the attracting system which is external to S, is constant for all positions of P, and is equal to the potential V_s of the whole attracting system at the level surface S.

(3) The whole mass of the stratum is $\int\rho'' d\sigma$, and by Gauss' theorem, this is equal to the mass of that portion of the attracting system which is inside S.

If the surface S encloses all the attracting system, the second theorem asserts that the potential of the layer at all internal points is constant and equal to that of the attracting system at the level surface S.

This form of the theorem will enable us to find the law of distribution of a charge of electricity, given to any solid insulated conductor whose boundary is a level surface of some known attracting system.

157. Ex. It is known that a prolate spheroid is a level surface of a uniform thin attracting rod whose extremities are at the foci S, H of the spheroid, Art. 49. Find the surface density of the thin stratum which, when placed on the spheroid, has the same attraction at all external points as the rod.

The surface density ρ'' at any point Q of the spheroid is given by $4\pi\rho''=F$, where F is the resultant attraction at Q. Also $F=2m\sin\frac{1}{2}SFH/y$, where y is the distance of Q from the rod. By using some geometrical properties of conics this leads to the result that ρ'' is proportional to the perpendicular p from the centre on the tangent plane at Q. The whole mass of the stratum is equal to that of the rod.

158. Points at which V is infinite. If P be any arbitrary point taken in the interior of the space bounded by the surface S, it is evident that one of the columns of integration parallel to each coordinate axis will pass through P. It is necessary that in each of these three columns the subject of integration should be finite. We have therefore assumed in the proof given in Art. 149 that (1) both the functions V, V' are finite and continuous, (2) that their first and second differential coefficients with regard to x, y, z are each finite throughout the space considered. If any of the functions be infinite at some point A within S, we must surround that point by an infinitesimal sphere, and integrate only over the space between the sphere and the surface S.

159. Green's equation is

$$\int V\frac{dV'}{dn}\,d\sigma+4\pi\int V\rho'\,dv=\int V'\frac{dV}{dn}\,d\sigma+4\pi\int V'\rho\,dv\ldots\ldots\ldots\ldots\ldots(\text{I.}).$$

Let us suppose that one term of V' is $1/r'$, where r' is a distance measured from P. We shall substitute this term in Green's equation, and the space of integration shall be that between a small sphere, centre P, radius ϵ, and the surface S.

Consider first the integrals taken over the surface of the sphere. Since $d\sigma=\epsilon^2 d\omega$, we have by changing to polar coordinates

$$\int\frac{1}{r'}\frac{dV}{dn}\,d\sigma=\int\frac{1}{\epsilon}\frac{dV}{dn}\,\epsilon^2 d\omega=0,$$

$$\int V\frac{d}{dn}\left(\frac{1}{r'}\right)d\sigma=V_P\int\left(\frac{1}{\epsilon^2}\right)\epsilon^2 d\omega=4\pi V_P,$$

where dn has been measured from the space of integration, that is *inwards* on the sphere, and V_P has been written for the value of V at P.

Consider next the volume integrals. Since r' is finite throughout the space of integration, $\rho'=0$ and the term $\int V\rho'\,dv$ disappears. The integral $\int\rho\,dv/r'$ is to be taken only for the space outside the sphere, but since $dv=r'^2 dr' d\omega$ if we include the integral for the space within the sphere we have only added zero (see Art. 101).

Green's equation *for the term* $V'=1/r'$ takes the form

$$\int V\frac{d}{dn}\left(\frac{1}{r'}\right)d\sigma-\int\frac{dV}{dn}\frac{d\sigma}{r'}+4\pi V_P=4\pi\int\frac{\rho\,dv}{r'}\text{.................(II.)},$$

where the surface integrals are taken only over the surface S and the volume integrals throughout the space S, ignoring the sphere altogether.

Since V_P is the potential at an internal point P of the whole mass and $\int\rho dv/r'$ the potential of the mass inside S, this equation becomes identical with (16) of Art. 154 when we change the sign of dn.

Let V be the potential of some attracting system, part of which may be inside S and part outside. Also let

$$4\pi D=r'V\frac{d}{dn}\left(\frac{1}{r'}\right)-\frac{dV}{dn}=-\frac{1}{r'}\frac{d}{dn}(Vr')$$

give the surface density D at any point Q of a thin layer placed on S, where $r'=PQ$ and V is the potential at Q of the whole mass. The equation (II.) then asserts that the potential, at a point P inside S, of that part of the attracting system which is also inside S, exceeds the potential of the stratum by the potential V_P of the whole mass at P.

160. We may notice that if V or V' be the potential of a system of bodies of finite density, neither V nor its first differential coefficients are infinite at any point of the mass, see Art. 101.

If one term of V' were m/r' we may regard the particle m as the limit of a small sphere of radius ϵ and density ρ_0, where $\frac{4}{3}\pi\rho_0\epsilon^3=m$. The integrations in (I.) can then be made throughout the space enclosed by S without reference to the sphere. The integral $4\pi\int V\rho' dv$ will supply an additional term equal to $4\pi V_P m$. In this way we arrive at once at the final equation (II.).

161. Multiple-valued functions. It has been supposed in these theorems that the functions V, V' have only one value at the same point of space. If they are potentials of attracting masses, they are each of the form $\Sigma m/r$ and can have only one value. But if they are obtained as solutions of Laplace's equations, as in hydrodynamics, they may be many-valued functions. Thus let a fluid be running round in a ring-like vessel. If V be the velocity potential at any point P, we know by the principles of hydrodynamics that $dV/ds=u$, where s is the arc described, and u is the velocity at P. Since the velocity is always positive, the velocity potential V must always increase as P travels round the ring. When P has made a complete turn, it comes to the point it started from, and V has a different value. If we put Laplace's equation into cylindrical coordinates (Art. 108), we easily see that $V=\tan^{-1}y/x=\phi$ satisfies the equation and represents such a motion.

162. In order to apply Green's equation to a multiple-valued function by integrating throughout the space enclosed in a ring-shaped surface we must deprive the function of its multiple values by placing a barrier at any point and including this barrier as one of the boundaries. In this way the point P is prevented from making a complete circuit and the function is reduced to a single-valued form. It may be that the surface has several ring-like passages interlacing, and it may then be necessary to insert several barriers before the function is reduced to a single-valued form.

Taking the simpler case of a single ring-like surface, let us suppose that the potential V is always increased by the same quantity c when the point P starting from any position has made a complete circuit and has returned to the same position again. Similarly let V' be increased by c'. Let da be an element of the area of a barrier placed anywhere across the ring-like cavity. Let s be an arc measured from the barrier round the ring to the barrier again, say from $s=0$ to $s=l$. Consider the part of the boundary formed by the two sides of the barrier; remembering that dn is measured outwards, we have $dn=-ds$ for the side defined by $s=0$, and $dn=ds$ for the side $s=l$. We thus have, when we integrate over both sides of the barrier,

$$\int V\frac{dV'}{dn}\,da=-\int V\frac{dV'}{ds}\,da+\int (V+c)\,\frac{d(V'+c')}{ds}\,da=c\int\frac{dV'}{ds}\,da.$$

Supposing V and V' to be solutions of Laplace's equation, Green's theorem becomes

$$U=\int V\frac{dV'}{dn}\,d\sigma+c\int\frac{dV'}{ds}\,da=\int V'\frac{dV}{dn}\,d\sigma+c'\int\frac{dV}{ds}\,da,$$

where along the surface S, dn is measured outwards, and across the barrier ds is measured in the positive direction round the ring.

163. Ex. 1. Let V, V' represent as before any two functions of (x, y, z), and let a be a third finite function of the same variables. Beginning with

$$U=\iiint a^2\left(\frac{dV}{dx}\frac{dV'}{dx}+\frac{dV}{dy}\frac{dV'}{dy}+\frac{dV}{dz}\frac{dV'}{dz}\right)dx\,dy\,dz,$$

show, by the same succession of integrations as in Art. 149, that

$$U=\int a^2V'\frac{dV}{dn}\,d\sigma+4\pi\int V'\rho\,dv=\int a^2V\frac{dV'}{dn}\,d\sigma+4\pi\int V\rho'dv,$$

where
$$-4\pi\rho=\frac{d}{dx}\left(a^2\frac{dV}{dx}\right)+\frac{d}{dy}\left(a^2\frac{dV}{dy}\right)+\frac{d}{dz}\left(a^2\frac{dV}{dz}\right),$$

and $-4\pi\rho'$ represents a similar expression with V' written for V. This is Kelvin's extension of Green's theorem. See Thomson and Tait, Part I., p. 167.

Ex. 2. If V, V' be two solutions of the differential equation

$$\frac{d}{dx}\left(a^2\frac{dV}{dx}\right)+\frac{d}{dy}\left(a^2\frac{dV}{dy}\right)+\frac{d}{dz}\left(a^2\frac{dV}{dz}\right)=0,$$

and if also $V=V'$ at all points of a closed surface S, prove that $V=V'$ throughout the enclosed space.

Let $u=V-V'$, then u also is a solution of the differential equation. Writing u for both V, V' in the general theorem of Ex. 1, we have

$$\int a^2\left\{\left(\frac{du}{dx}\right)^2+\left(\frac{du}{dy}\right)^2+\left(\frac{du}{dz}\right)^2\right\}dv=\int a^2u\frac{du}{dn}\,d\sigma.$$

The right-hand side is zero since u vanishes at all points of the surface S. But the left-hand side is the sum of a number of positive quantities and cannot be zero unless each vanishes. Thus du/dx, du/dy, du/dz are each zero at all points inside S, i.e. the function u is a constant. Since it is given equal to zero at the surface S, it must be zero at all points within S. Lejeune Dirichlet uses a similar argument in *Crelle*, XXXII. 1844.

This differential equation is of great importance in the analytical theory of heat.

Ex. 3. Show in the same way that if $dV/dn=dV'/dn$ at all points of the surface S, then $V=V'$ throughout the space enclosed. [Here $du/dn=0$.]

Ex. 4. If both V and V', besides being solutions of the differential equation, also satisfy the equation $dV/dn=-kV$ at all points of S, where k is a function of the coordinates which is always *positive*, prove that $V=V'$. [Here the right-hand side of Ex. 2 would otherwise be negative.]

Ex. 5. If V be one solution of the differential equation in Ex. 2 such that $dV/dn = -kV$ at all points of a surface S, where k is always positive, prove that there is no other solution of that differential equation which satisfies this condition. [Use a proof similar to that in Art. 133.]

Given the potential, find the body.

164. Poisson's equation $4\pi\rho = -\nabla^2 V$ supplies a partial solution to this question. The potential V being given throughout all space we find ρ by differentiation. This value of ρ, if finite throughout space, determines the only body which could have the given potential. If the potential is given as a discontinuous function of the coordinates difficulties may arise in applying Poisson's equation at the points or surfaces of discontinuity. The following theorem will therefore be necessary.

165. Let the potential V throughout a given space S be the given function $\phi(x, y, z)$, throughout a neighbouring space S', let the potential be $\psi(x, y, z)$, and so on. In this way we regard all space as divided into compartments within each of which the potential is a different function of the coordinates. We suppose in the first instance that the given potentials are nowhere infinite.

As a point P moves in space, passing from one compartment to the next, we know by Art. 145 that there should be no sudden change in the numerical value of the potential. We therefore suppose that *the given potentials ϕ, ψ have equal values at all points of the common boundary.* This implies that the space rates of the potential tangential to the common boundary are equal. *The tangential components of force must therefore be equal.*

If the normal forces at the boundary are not also equal, there will be a film of attracting matter at the boundary (Art. 146) whose surface density σ is given by Green's equation

$$4\pi\sigma = \frac{d\phi}{dn} + \frac{d\psi}{dn'},$$

where dn, dn' are measured in directions outwards from the spaces S, S', and therefore, at points inside each space, towards the boundary.

We have now proved that the only arrangement of matter which could produce the given system of potential values is one consisting partly of solid matter given by Poisson's equation filling the compartments and partly of films on the boundaries. It remains to prove by integration that this arrangement does actually fulfil the given conditions. The results of these integrations are supplied by Green's theorem.

166. Let us write ϕ for the arbitrary function V in Green's theorem and let $D=\frac{1}{4\pi}\left\{r'\phi\frac{d}{dn}\left(\frac{1}{r'}\right)-\frac{d\phi}{dn}\right\}$ as defined in Arts. 154 and 159. Then the potential at P when P is outside S is equal to that of a stratum of surface density D placed on S, and when P is inside S, the potential at P exceeds that of the stratum by $\phi(x, y, z)$. Let this stratum be included (*with the sign of D changed*) as part of the attracting system, the potential at a point outside S is then zero and at a point inside S the potential is ϕ. The proposed conditions are satisfied for the space S.

Treating the neighbouring space S' in the same way, we obtain an internal density determined as before by Poisson's equation and a superficial density which, when its sign is changed, is the same as that given by D except that the function ϕ is replaced by ψ and the element dn of the normal is measured in the opposite direction.

Adding together the two superficial densities and remembering that ϕ and ψ are equal at those points of the boundary which are common to S and S', we observe that the first terms of each destroy each other. We therefore find for the density of the superficial stratum

$$\sigma'=\frac{1}{4\pi}\left\{\frac{d}{dn}\phi(x, y, z)+\frac{d}{dn'}\psi(x, y, z)\right\},$$

where dn and dn' inside each compartment are measured towards the boundary, so that $dn=-dn'$. We notice that this law of density is independent of the position of P.

If the given potential ϕ is infinite at any point A within the space S we must suppose that a finite quantity Q of attracting matter is situated at A (Art. 101). The quantity Q may be found by enclosing A within a small sphere and using Gauss' theorem, $4\pi Q=\int F d\sigma$. If the potential is infinite along a curve, the line density may be found by enclosing an elementary arc within the sphere.

167. Ex. 1. The potential at a point Q is $\phi=2\pi(b^2-a^2)$, $\psi=\frac{2}{3}\pi(3b^2-r^2-2a^3/r)$ or $\chi=\frac{4}{3}\pi(b^3-a^3)/r$, according as the distance r of Q from the origin is less than a, lies between a and b, or is greater than b. Find the attracting system.

Considering the space in which r is less than a, we see that both the volume density and the part of the surface density $d\phi/4\pi dn$ are zero.

Considering the space in which r lies between a and b, the volume density is found by substituting in $\rho=-\frac{1}{4\pi r}\frac{d^2\psi r}{dr^2}$, Art. 108, and is equal to unity. The part of the superficial density found by substituting in $d\psi/4\pi dn$ is zero at the inner boundary and $-(b^3-a^3)/3b^2$ at the outer.

Lastly in the space in which r is greater than b, the volume density is zero and the part of the superficial density $d\chi/4\pi dn=(b^3-a^3)/3b^2$.

Joining these together, we find that each of the two surface densities is zero and that the attracting body is a spherical shell of radii a and b and unit density.

Ex. 2. Find the attracting system whose potential V is equal to

$$\mu(1-Lx^2-My^2-Nz^2)$$

at all points within the ellipsoid $Lx^2+My^2+Nz^2=1$ and zero at all external points.

The system is a homogeneous ellipsoid whose density is $\mu(L+M+N)/2\pi$, together with a superficial stratum whose surface density at Q is $-\mu/2\pi p$, where p is the perpendicular on the tangent plane at Q.

Since this stratum is equivalent to a thin homogeneous confocal shell (see Vol. I. Art. 430), this result supplies a simple relation between the potential of a homogeneous solid ellipsoid and that of a homogeneous confocal shell. See Art. 224.

Method of Inversion.

168. Inversion from a point*. Let O be any assumed origin, and let Q be a point moving in any given manner. If on the radius vector OQ we take a point Q' so that $OQ.OQ'=k^2$, *then Q and Q′ are called inverse points.* If Q trace out a curve, Q' traces out the inverse curve; if Q trace out a surface or solid, Q' traces out the inverse surface or solid. The points Q, Q' are sometimes said to be *inverse with regard to a sphere* whose centre is O and radius k.

Let P', Q' be the inverse points of P, Q, then since the products $OP.OP'$, $OQ.OQ'$ are equal and the angles POQ, $P'OQ'$ are the same, the triangles POQ, $P'OQ'$ are similar. We therefore have

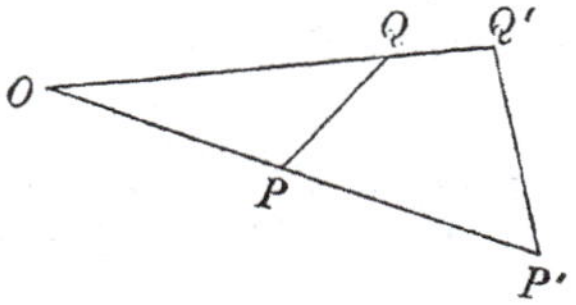

$$\frac{1}{P'Q'}=\frac{1}{PQ}\cdot\frac{OQ}{OP'} \quad \text{......(1).}$$

Let m, m' be the masses of two particles placed respectively at Q, Q', and let the densities be such that $m'=m\dfrac{k}{OQ}$ ....(2).

Multiplying equations (1) and (2) together, we see that the potential at P' of m' is equal to that at P of m, after multiplication by a quantity k/OP' which is independent of the position of Q.

Let any number of particles of given masses m_1, m_2, &c. be placed at different points Q_1, Q_2, &c., and let the corresponding masses m_1', m_2', &c., be placed at the inverse points Q_1', Q_2', &c. Then since an equation similar to (2) holds for each pair of masses, we have by addition

$$\left(\begin{array}{c}\text{Potential at } P' \\ \text{of the inverse system}\end{array}\right)=\left(\begin{array}{c}\text{Potential at } P \\ \text{of the given system}\end{array}\right)\frac{k}{OP'} \quad \text{.........(3)}$$

which may be compendiously written $V'=V\dfrac{k}{OP'}$.

* The Method of Inversion is due to Sir W. Thomson, now Lord Kelvin. In a letter addressed to M. Liouville and published in *Liouville's Journal*, 1845, a short history and a brief account of some of its applications are given. This letter may also be found in the *Reprint of papers on Electrostatics and Magnetism.*

169. If the given masses m_1, m_2, &c. are arranged so as to form an arc, surface or solid, the inverse masses will also be arranged in the same way. It will therefore be necessary to discover some rule by which we can compare the density at any point of the given system with that at the corresponding point of the inverse system.

Using the same figure as before but changing the meaning of P, let PQ now represent any elementary arc of the locus of Q, then $P'Q'$ represents the corresponding inverse arc. If the locus of Q is a curve, we infer from the similarity of the triangles POQ, $P'OQ'$ that the lengths of the elementary arcs $P'Q'$, PQ are in the ratio OQ'/OP, i.e. OQ'/OQ ultimately. Hence by (2) the ratio of the line densities of the arcs $P'Q'$, PQ is equal to k/OQ'.

If the locus of Q is a surface, the elementary areas $P'Q'$, PQ are in the ratio of the squares of the homologous sides, i.e. as OQ'^2 to OQ^2. Hence by (2) the ratio of the surface densities at Q' and Q is equal to $(k/OQ')^3$.

If Q travel over all points of space enclosed by a surface, the elementary volumes at Q', Q are ultimately in the ratio $OQ'^2 d\omega . d(OQ')$ to $OQ^2 d\omega . d(OQ)$. Since $OQ . OQ' = k^2$, this ratio is equal to OQ'^3/OQ^3. Hence by (2) the ratio of the densities at Q' and Q is equal to $(k/OQ')^5$.

Summing these results, we see that

$$\left(\begin{array}{c}\text{density at } Q' \\ \text{of the inverse system}\end{array}\right) = \left(\begin{array}{c}\text{density at } Q \\ \text{of the given system}\end{array}\right) \cdot \left(\frac{k}{OQ'}\right)^{2d-1} \quad \ldots(4),$$

where d represents the dimensions of the system, i.e. $d=1$, 2, or 3 according as the system is an arc, a surface or a volume. When the system is a point, $d=0$; the equation (4) then agrees with (2) and gives the relation between m and m'.

170. *The mass of any portion of the inverse body is equal to the potential at the centre of inversion of the corresponding portion of the primitive body multiplied by the radius k of inversion.* By Art. 168, we have $m' = mk/OQ$, i.e. m' is equal to the potential of m at O, multiplied by k. The theorem being true for each element of mass is necessarily true for any finite portion of the body.

171. Ex. If the law of force be the inverse nth power of the distance, the potential of a particle m takes the form $\frac{1}{n-1}\frac{m}{r^{n-1}}$. Prove that the equations corresponding to (2), (3), and (4) become

$$m' = m\left(\frac{k}{OQ'}\right)^{1-n}, \qquad \rho' = \rho\left(\frac{k}{OQ'}\right)^{2d+1-n}, \qquad V' = V\left(\frac{k}{OP'}\right)^{n-1}.$$

When the law of force is the inverse distance $n=1$, and the potential of the attracting mass takes a different form. In this case the quantity here called V becomes $\Sigma m/(n-1)$, and is therefore proportional to *the mass of the body*. The theorems therefore of inversion, though they no longer apply to the attractions of bodies, will still enable us to find their masses when their densities vary as some power of the distance from a point. See *Quarterly J.*, 1857.

172. Some Geometrical properties. It is convenient to notice that if the points P, Q invert into P', Q', then $\dfrac{P'Q'}{PQ}=\dfrac{k^2}{OP\,.\,OQ}$, where PQ, $P'Q'$ are the linear distances between P, Q and P', Q' respectively. For example the ratio $\dfrac{PQ\,.\,RS}{QR\,.\,SP}$ is unaltered by inversion; because each letter occurs the same number of times in the numerator and denominator.

173. *To find the inverse of a sphere.* Let Q describe a sphere whose centre is C, and let $OQ\,.\,OQ'=k^2$. Let OQQ' cut the primitive sphere in R, then since $OQ\,.\,OR$ is constant, it follows that OQ'/OR is constant. The locus of Q' is therefore similar to that of R, that is, *the inverse is a sphere and O is a centre of similitude.*

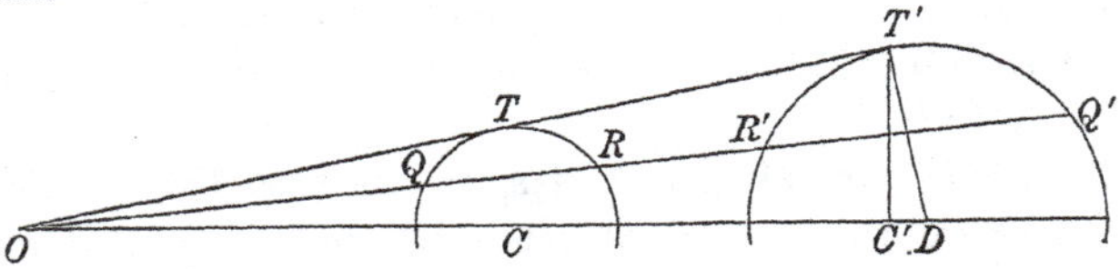

The centre D of the inverse sphere lies in OC produced, and by the properties of similar figures, is at such a distance from O that OD/OC is equal to the constant ratio OQ'/OR. The centre C of the primitive sphere does not invert into the centre D of the inverse sphere, but into some point C' such that $OC\,.\,OC'=k^2$. It is easy to see, by similar triangles, that C' lies on the polar line of the centre of inversion O with regard to the inverse sphere.

A sphere inverts into a plane when the centre of inversion O is on the surface of the primitive sphere. The inverse of a plane with regard to any centre O of inversion is a sphere which passes through O.

A circle is the intersection of two spheres and in general inverts into a circle, but when the centre of inversion lies on the circle, the inverse is a straight line.

Ex. Let P, P' be two inverse points with regard to a sphere S; prove that every sphere passing through P, P' cuts S orthogonally. Conversely, if a sphere S' cuts S orthogonally and CPP' is any chord through the centre of S, then P, P' are inverse points with regard to S. See figure of Art. 86.

174. *An angle is not altered by inversion.* Let PQ, PR be *elementary arcs* of two curves which meet in P and are not necessarily in the same plane with the centre O of inversion. Let $P'Q'$, $P'R'$ be the inverse arcs, we have to prove that the angles QPR, $Q'P'R'$ are ultimately equal. Describe a sphere through the four points P, Q, R and P'; then since the products $OP\,.\,OP'$, $OQ\,.\,OQ'$ and $OR\,.\,OR'$ are equal, the sphere also passes through Q', R'. The planes $OPQP'Q'$ and $OPRP'R'$ cut the sphere in two circles whose planes intersect in OPP'. The opposite angles QPR, $Q'P'R'$ contained by the tangents to these circles are evidently equal by symmetry. It is also evident that *the planes of the angle and its inverse*, viz. QPR and $Q'P'R'$, *make equal angles with the opposite directions of OPP'.*

175. It follows at once, from the theorem, that *two given orthogonal spheres invert into orthogonal spheres.*

176. Ex. 1. The potential of a homogeneous spherical surface at a point P is $4\pi a\rho$ or $4\pi a^2\rho/CP$ according as P is inside or outside the surface, where C is the centre and a is equal to the radius. It is required to invert this theorem with regard to an external point O.

Since the product of the segments $OQ \,.\, OQ'$ is constant in a sphere, it is clear that if we take k equal to the length of the tangent OT, the sphere will be its own inverse. When only one sphere occurs in the system this choice of the value of k will simplify the process, but when there are several spheres it will be more convenient to keep the value of k indeterminate.

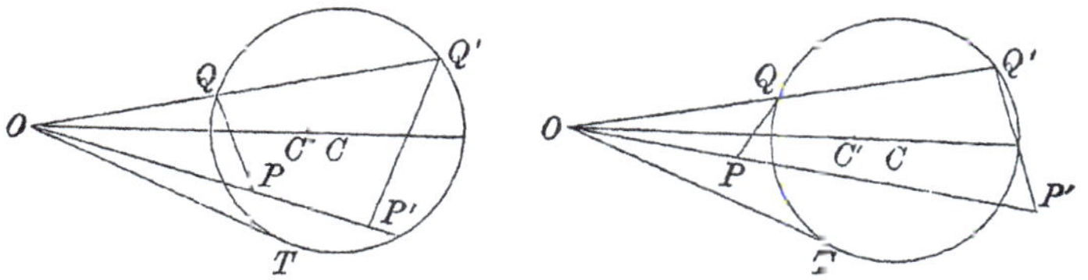

If P is within the sphere, the inverse point P' is also within the sphere. By (4) the density of the inverse sphere at Q' is equal to $\rho(k/OQ')^3$, and its potential at P is $4\pi a\rho k/OP'$.

If P is without the sphere, P' is also without. The density at Q' of the inverse system is the same as before, but the potential at P' is $\frac{4\pi a^2\rho}{CP} \cdot \frac{k}{OP'}$. Let C' be the point on the straight line OC such that C and C' are inverse points. Then by the similar triangles COP, $C'OP'$ we have $CP \,.\, OP' = OC \,.\, C'P'$. The potential at P' is therefore $\frac{4\pi a^2\rho}{OC} \cdot \frac{k}{C'P'}$.

If M' is the mass of the inverse system, the relation between M' and ρ may be easily deduced from either of these expressions for the potential. Take the first, where P' is inside the sphere, we notice that since every element of the sphere is equally distant from the centre, the potential at the centre is M'/a. Hence putting P' at the centre and comparing the two values of the potential, we have $M' = 4\pi\rho a^2 k/OC$. Take the second case, when P' is without the sphere, we notice that the potential at a very distant point must be *mass divided by distance*. By equating these two values of the potential, we arrive at the same value of M' as before. This value of M' may also be easily deduced from Art. 170.

Taking both these results, we arrive at the following inverse theorem.

Let a mass M' be distributed over a spherical surface, centre C, so that its density at any point Q' is $\rho(k/OQ')^3$, where O is an external point, and k is the length of the tangent from C. Then $\rho = M'c/4\pi a^2 k$, where $c = OC$; and the potential at any point P' is $M'\frac{c}{a}\frac{1}{OP'}$ or $\frac{M'}{C'P'}$, according as P' lies within or without the sphere. The points C' and C are inverse points with regard to O, and it is easy to see that C' lies on the polar line of O.

The potential of this heterogeneous spherical stratum at all external points is the same as if its whole mass M' were collected at C', and at all internal points is the same as if a mass $M'c/a$ were collected at O.

It follows from Art. 136 that the centre of gravity of the heterogeneous stratum is at C' and that every straight line through C' is a principal axis.

Ex. 2. If the density of a spherical surface vary as the inverse cube of its distance from an internal point O, find its potential at any point.

If the centre of inversion O is inside the primitive sphere we can still make the sphere its own inverse by drawing OQ' from O in the direction opposite to OQ, and taking k^2 equal to the product of the segments of all chords through O. With these changes we may show that the potential at all external points is the same as if its whole mass M' were collected at O, and at all internal points is the same as if the mass $M'c/a$ were collected at C'.

Ex. 3. The potential of a homogeneous solid sphere at an external point P is $\frac{4}{3}\pi\rho a^3/CP$, where C is the centre and a the radius. Invert this theorem with regard to an external point O.

The result is that the potential at an external point of a heterogeneous sphere, whose density at any point Q' is $\rho(k/OQ')^5$, is the same as if its whole mass M' were collected into a fixed point C'. This point C' is the inverse of the centre with regard to O and is also the centre of gravity of the sphere. The constant ρ may be found from the relation $M'c=\frac{4}{3}\pi\rho a^3 k$, where $c=OC$, and k is the length of the tangent from O.

Ex. 4. A heterogeneous spherical shell is bounded by eccentric spheres whose radii are a, b, and its density at any point Q is m/OQ^5, where m is a constant and O a given external point. Show that its potential at any internal point P is

$$\frac{2}{3}\pi m\left[3\left(\frac{a^2}{f^4}-\frac{b^2}{g^4}\right)\frac{1}{OP}-\left(\frac{AP^2}{OA^2}-\frac{BP^2}{OB^2}\right)\frac{1}{OP^3}\right],$$

where A and B are the points where the polar planes of O intersect the diameters drawn through O, and f, g are the tangents from O.

Ex. 5. An infinitely thin layer of matter is placed on the surface of elasticity $e^2r^4=a^2x^2+b^2y^2+c^2z^2$, so that the surface density at any point distant r from the centre varies as p/r^5, where p is the perpendicular from the origin on the tangent plane. Show that the potential at any external point is the same as if the whole mass were collected at its centre of gravity.

177. If S is a level surface of any attracting points, the inverse of S is not in general a level surface of the inverse of the attracting points, because the ratio of the potentials (being given by $V'=Vk/r'$) is not constant. But *if S is a level surface of zero potential, the inverse of S is also a level surface of zero potential of the inverse attracting points.*

178. Let P, P' be inverse points with regard to a sphere S. If Q be any point on the surface, the ratio $PQ/P'Q$ is constant by the similar triangles OPQ, $OP'Q$, (Art. 86). Let this ratio be α/β, then $\alpha/PQ-\beta/P'Q=0$, that is *the sphere is a level surface of zero potential of two particles placed at P, P'*, whose masses are measured by α and $-\beta$.

179. Two points P, P' are inverse to each other with regard to a sphere S. Let the inverse of this system, taken with regard to a new origin O, be the points Q, Q' and the sphere S'. Then the points Q, Q' are inverse points with regard to the sphere S'.

By putting particles of proper masses at P, P', the sphere can be made a level surface of zero potential. The inverse of S with regard to the new origin O is therefore also a level surface of zero potential of the inverse masses at Q, Q'. Hence Q, Q' are inverse points with regard to S'.

A purely geometrical proof of this theorem is given in Lachlan's *Modern Geometry*.

180. If a particle of finite mass m is at the centre of inversion O, the inverse is a distribution of matter at infinitely great distances from O. The theory of inversion gives the potential of the whole inverse system including the infinitely distant matter. If we wish to remove the latter from the field under consideration we must subtract its potential. Now by equation (3) of Art. 168 its potential at any point P' is $V'=V\dfrac{k}{OP'}=\dfrac{mk}{OP\,.\,OP'}=\dfrac{m}{k}$. We may therefore disregard this infinitely distant matter if we subtract from the potential of the inverse body as given by the theory, the constant m/k.

If the mass at O merely forms part of a stratum passing through O, the mass actually at O is zero and the constant to be subtracted is also zero.

181. Inversion from a line. Instead of inverting the attracting system with regard to a point O we may invert it with regard to some straight line Oz. Let a point Q move in any manner, and let QN be a perpendicular on the axis Oz. If on NQ we take a point Q' so that $NQ\,.\,NQ'=k^2$, where k is a given constant, then Q' is the inverse of Q with regard to the axis of z.

With this definition it is clear that any cylindrical surface with its generators parallel to Oz inverts into another cylindrical surface also having its generators parallel to that axis. This method of inversion will therefore help us to deduce the potential of one cylindrical surface or solid from that of its inverse. We shall suppose that the density of the cylindrical body is uniform along any generating line but varies from one generator to another.

182. If an infinite rod is parallel to the axis of z, its attraction at any point P on the plane of xy is known to be $2m/QP$, where Q is the intersection of the rod with the plane of xy and m is the line density. The potential of such a rod at P is therefore $V=C-2m\log QP$, where C is some constant, Art. 50. Let us invert this rod with regard to the axis of z into a parallel rod, and P into another point P'. Supposing the inverse rod to have the same line density as the primitive rod, its potential at P' is $V'=C-2m\log Q'P'$. But by Art. 168 $P'Q'=PQ\dfrac{OP'}{OQ}$. Hence

$$V'+2m\log OP'=V+2m\log OQ \quad\text{.......................}(1).$$

Let there be a system of rods intersecting the plane of xy in the points Q_1, Q_2, &c., and let the inverse rods intersect the same plane in Q_1', Q_2', &c. Let m_1, m_2, &c. be the line densities of the several pairs. Then for each pair we have an equation similar to (1); adding all these together we find

(Potential at P' of inverse system)

– (Potential at P' of the whole mass collected at the axis)

= (Potential at P of given system) – (Potential at O of given system).

183. If the primitive system of rods intersect the plane of xy in an arc or an area, the inverse system will also be arranged in the same way. To compare the densities we observe that *the masses of the given system and the inverse are the same but differently distributed.* If the locus of Q is an arc, the ratio of the elementary arcs at Q', Q is equal to OQ'/OQ, and the ratio of the line densities is therefore equal to OQ/OQ', i.e. $(k/OQ')^2$. If the locus of Q is an area, the ratio of the surface densities is equal to $(k/OQ')^4$.

We should notice that m is the mass per unit of length of a rod. Hence when the attracting rods form a cylindrical surface whose surface density is ρ, we have $m = \rho ds$, where ds is an element of arc of the section of the cylinder by a plane perpendicular to the axis. For example, in the case of a right circular cylinder of radius a we have $\Sigma m = 2\pi a\rho$. If the rods form a cylindrical volume of density ρ, we have $m = \rho dA$, where dA is an element of area of the curve of section.

Ex. 1. *A heterogeneous stratum is placed on a right circular cylinder, the density being uniform along any generator. It is required to compare the potentials at an internal and an external inverse point.* If we invert the system with regard to the axis and the radius k of inversion be the radius of the cylinder, the stratum inverts into itself. If P, P' be the internal and external points, V, V' the potentials, we have by Art. 182 $V' - (C' - 2\Sigma m \log OP') = V - V_0$. Collecting all the constant terms into one, we have $V' - V = A - 2\Sigma m \log OP'$. The corresponding proposition for a sphere is given in Art. 86.

Ex. 2. Invert the following theorem with regard to an eccentric internal straight line. The potential of a homogeneous right circular cylindrical surface at any internal point is constant and equal to that along the axis.

The resulting theorem is as follows. If matter be distributed in a thin stratum over a right circular cylinder so that the surface density at any point Q' is proportional to the inverse square of the distance of Q' from an internal straight line OZ parallel to the generators, the potential at any external point is the same as if the whole mass were evenly distributed over the straight line OZ.

184. Extended theory. Let $Q_1, Q_2, \ldots Q_n$, be n points arranged at equal distances on the circumference of a circle of radius ρ. Taking the centre O as origin, let the polar coordinates of these points be (ρ, ϕ), $(\rho, \phi + \alpha)$, $(\rho, \phi + 2\alpha)$ &c., where $n\alpha = 2\pi$. Let P be any point and let (r, θ) be its coordinates. By De Moivre's property of the circle we have

$$r^{2n} - 2r^n\rho^n \cos n(\theta - \phi) + \rho^{2n} = PQ_1^2 \,.\, PQ_2^2 \ldots PQ_n^2 \ldots\ldots\ldots\ldots\ldots (1).$$

Let us now take two other points Q', P' whose coordinates (ρ', ϕ') and (r', θ') are such that $\rho' = c(\rho/c)^n$, $\phi' = n\phi$; $r' = c(r/c)^n$, $\theta' = n\theta$, where c is any constant. It immediately follows that the left side of (1) is equal to $c^{2(n-1)} \,.\, (P'Q')^2$. Taking the logarithm of both sides, we find

$$\log P'Q' + (n-1)\log c = \log PQ_1 + \log PQ_2 + \text{\&c.} + \log PQ_n \ldots\ldots\ldots\ldots (2).$$

Let us now suppose that two infinite thin rods, each of uniform line density m, are placed perpendicularly to the plane of the circle at P and P' respectively. It follows at once from equation (2) that the potential of the second rod at Q' differs by a constant from the sum of the potentials of the first rod at the points Q_1, Q_2, &c.

In the same way, by properly placing pairs of corresponding rods we may build up two corresponding cylindrical bodies, which have the property that the potential of the second body at Q' differs by a constant from the sum of the potentials of the first at $Q_1 \ldots Q_n$.

We may express this result in the form of a theorem. *An infinitely long cylindrical body has its density uniform along any generating line and attracts according to the law of nature. The body, being referred to cylindrical coordinates with the axis of z parallel to the generators, is transformed into another cylindrical body by moving each cylindrical element* (r, θ) *into the position* (r', θ'), *where* $r' = c\,(r/c)^n$, $\theta' = n\theta$, *without altering the mass of element. If the potentials of the original body at the n points* (ρ, ϕ), $(\rho, \phi+\alpha)$, $(\rho, \phi+2\alpha)$ *&c. be* V_1, V_2, V_3 *&c. then the potential of the transformed body at* (ρ', ϕ'), *where* $\rho' = c\,(\rho/c)^n$, $\phi' = n\phi$, *differs by a constant from the sum* $V_1 + V_2 + \text{\&c.} + V_n$.

If one be a continuous cylindrical solid, the other body may be made also continuous by altering the areas of the sections of the transformed elements, keeping the mass unchanged. Since the elementary areas at P, P' are respectively $r\,d\theta\,dr$ and $r'd\theta'dr'$ we easily see that the volume densities at P, P' must be in the ratio of $(nr')^2$ to r^2.

If one body be a continuous surface, the other may be made also a continuous surface. Since the masses on the corresponding arcs ds, ds' are equal, *the surface densities* σ, σ', *must be such that* $\sigma ds = \sigma' ds'$. This ratio may be put into other forms. Let ψ, ψ' be the angles these arcs make with their respective radii vectores, then since $r' = r^n/c^{n-1}$, $\theta' = n\theta$,

$$\tan\psi' = r'\frac{d\theta'}{dr'} = r\frac{d\theta}{dr} = \tan\psi.$$

It appears that *the radial angle* ψ *is unaltered by the transformation*. Since $\sin\psi = r\,d\theta/ds$, $\sin\psi' = r'd\theta'/ds'$, we see that $ds/ds' = r/nr'$, and therefore $r\sigma = nr'\sigma'$.

Since the coordinates of the corresponding points of the two figures are connected by the relations $r' = r^n/c^{n-1}$, $\theta' = n\theta$, it is clear that when θ' has increased from 0 to 2π, θ has varied from 0 to $2\pi/n$, and thus an arc only and not a closed curve is obtained. If P' travel n times round its curve, the curve traced out by P will consist of n equal and similar arcs, fitting together and forming a closed curve. Since $\alpha = 2\pi/n$, it is also evident that these n arcs are similarly placed with regard to the n points Q_1, Q_2, &c., and that therefore the potential of the whole closed curve at each of these points is the same.

The potential therefore at Q' of the n coincident cylindrical strata generated by the rod P' in n revolutions (which of course is n times that of the cylinder taken once) is equal to n times the potential of the complete cylindrical stratum generated by the rod P at any one of the points Q_1, Q_2, &c. It follows that *the potentials of the two closed cylinders (each taken once) are equal at the corresponding points* Q_1 *and* Q'. If one stratum (like an electrical stratum) is equipotential throughout all space on one side of the surface, the other is also equipotential on the corresponding side.

Ex. Thin layers of attracting matter are placed on the cylinders

$$A\,(x^4 + y^4) + 2Bx^2y^2 = 1,$$

$$Ax^6 + 3\,(3B - 2A)\,x^4y^2 + 3\,(3A - 2B)\,x^2y^4 + By^6 = 1;$$

if the surface densities are proportional respectively to r^2p and r^4p, where r is the distance from the axis and p is the perpendicular on the tangent plane, prove that the potentials are constant at all internal points.

If a thin stratum is placed on the cylinder $Lx'^2+My'^2=1$ whose surface density is $\sigma'=\kappa p'$, the potential inside is constant, Art. 72. Transform this theorem by writing $r'=r^2/c$, $\theta'=2\theta$. The elliptic cylinder becomes the first of the two cylinders in the example. The surface density σ follows at once from $\sigma r=n\sigma' r'$, if we remember that $p/r=p'/r'$ at corresponding points.

Circular rings and anchor rings.

185. When the potential of a thin heterogeneous circular ring for any law of attraction is known at all points in its plane within the circle, the potential at every point of space may be deduced by inversion.

Let the plane of the circle be the plane of xy, the centre O being the origin. Let the plane of xz contain the point R at which the potential is required and let it cut the circle in A', A. In the figure the attracting circle is represented by *the dotted line.* Describe a circle through the points A, A' and R, then

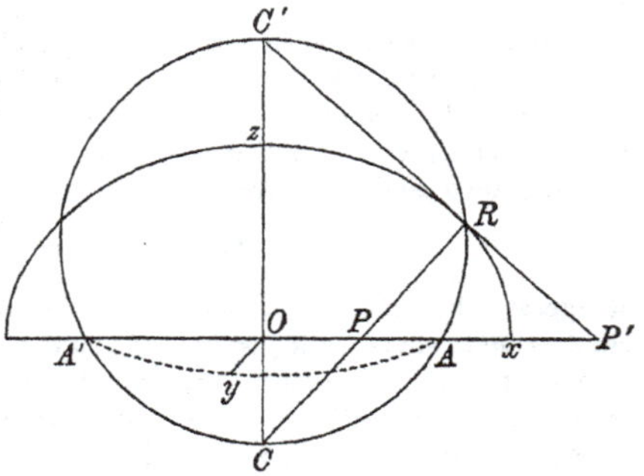

$$CP \cdot CR = CO \cdot CC' = CA^2.$$

The points P and R are therefore inverse with regard to C. If then V, V'' are the potentials of the ring at P and R, when the law of force is the inverse κth power, we have $V''=V\left(\frac{c}{r''}\right)^{\kappa-1}$ where $c=CA$ and $r''=CR$.

When the law of attraction is the inverse distance, the potential takes a logarithmic form, Art. 43. Let m be the mass of a particle of the ring situated at Q. Its potentials at P and R are $C-m\log QP$ and $C-m\log QR$. But since the triangles QCP, QCR are similar (Art. 168) $QP/QR=c/CR$. If then V and V'' are the potentials of the whole ring at P and R, we have $V''=V+M\log\frac{c}{r''}$, where M is the mass of the heterogeneous ring.

186. *The geometrical relations between the positions of P and R* are most easily obtained by describing an ellipse or hyperbola whose foci are A, A' and which passes through R.

Since the angles ARC, $A'RC$ stand on equal arcs, these angles are equal and *RPC is therefore a normal to the ellipse.* We thus have $r = e^2x$ where $r = OP$, x is the distance of R from the axis Oz of the ring, and e is the eccentricity. We also have CA/CR or $c/r'' = e$. If ρ, ρ' are the least and greatest distances of R from the ring, the focal distances $AR = \rho$, $A'R = \rho'$. Hence, a being the radius of the ring, $ex = \frac{1}{2}(\rho' - \rho)$ and $a/e = \frac{1}{2}(\rho' + \rho)$.
The semi-major axis of the ellipse is $a' = a/e$.

187. The result may be stated as follows. *Let the potential at an internal point P in the plane of the ring be $V = f(r)$. Then, if the law of force is the inverse κth power of the distance, the potential at R is $V'' = e^{\kappa-1}f(e^2x)$. If the law of force is the inverse distance $V'' = V + M \log e$. We may use any of the preceding geometrical results to express e in terms of the coordinates of R.*

The points R and P' are inverse points with regard to a sphere whose centre is C' and radius $C'A$. These may be used to deduce by the same rule the potential at R from that at the external point P'. Instead of the ellipse we then use the hyperbola which has its foci at A', A and which passes through R.

188. Ex. Prove that the component forces at R along the tangents to the ellipse and hyperbola are $e^{\kappa+1}F \sin RPA$ and $e'^{\kappa+1}F' \sin RP'A$, where F, F' are the forces at P and P' resolved in the directions OP, OP' respectively; and e, e' are the eccentricities of the ellipse and hyperbola. Prove also that $ee' = a/x$, and that P, P' are inverse points with regard to the ring.

189. Ex. The potential of a uniform circular ring, when the law of force is the inverse distance, is known to be constant at all points within the circle and in its plane, Art. 56. The potential at any point R of space is therefore $V'' = C + M \log e$. It follows that the level surfaces are oblate spheroids having the circle for a focal conic.

Prove that the resultant force at R takes either of the forms

$$F = M \, . \, n/\rho\rho' = M \, . \, (1 - e^2)/n,$$

where ρ, ρ' are the focal distances and n is the length of the normal RP.

If the ring is heterogeneous, let its law of density be given as described in Art. 58. The potential at any point R of space is then $V'' = \Sigma E_n \frac{\pi a}{n}\left(\frac{e^2x}{a}\right)^n + C$ (except when $n=0$), where a is the radius of the ring.

190. *To find the potential of a thin uniform circular ring of line density m, the law of attraction being the inverse κth power of the distance.*

First, let the point P at which the potential is required be in the plane of the circle. Taking the figure of Art. 55, let $OA = a$, $OP = r$, we then have

$$a \sin\phi = r \sin\theta, \quad u^2 - 2ur\cos\theta + r^2 - a^2 = 0 \quad\ldots\ldots(1),$$

where $\theta = OPR$, $u = PR$. When P is outside the circle both the values of u given by the quadratic are positive and represent geometrically the distances PQ, PR; these we distinguish as u_1, u_2. When P is inside the circle the geometrical distances are $-u_1$ and u_2.

The elementary masses at Q, R being $mu_1\,d\theta/\cos\phi$ and $mu_2\,d\theta/\cos\phi$, the potential V of the whole ring at an external point is

$$V = \frac{m}{\kappa-1}\int\frac{d\theta}{\cos\phi}S_{\kappa-2}, \quad S_{\kappa-2} = \frac{1}{u_1^{\kappa-2}} + \frac{1}{u_2^{\kappa-2}} \quad\ldots\ldots\ldots(2);$$

when P is inside we write $-u_1$ for u_1; we notice that when κ is an even integer, the same formula represents the potential whether P is internal or external. The value of $S_{\kappa-2}$ may be deduced from the quadratic, thus $S_0 = 2$, $S_1 = 2r\cos\theta/(r^2 - a^2)$.

The limits of the integral are different according as P is outside or inside the circle. When P is outside, ϕ varies from 0 to $\frac{1}{2}\pi$ and $\sin\theta$ from 0 to a/r, the final result being doubled. When P is inside, θ varies from 0 to $\frac{1}{2}\pi$ and $\sin\phi$ from 0 to r/a, the final result being doubled. To simplify the limits we express the integral V in terms of ϕ or θ according as P is outside or inside.

Representing these potentials by V_κ' and V_κ, we have after using (1)

$$V_\kappa' = \frac{2m}{\kappa-1}\int\frac{S_{\kappa-2}\,a\,d\phi}{(r^2 - a^2\sin^2\phi)^{\frac{1}{2}}}, \quad V_\kappa = \frac{2m}{\kappa-1}\int\frac{S_{\kappa-2}\,a\,d\theta}{(a^2 - r^2\sin^2\theta)^{\frac{1}{2}}},$$

the limits in both integrals being 0 to $\frac{1}{2}\pi$. When the law of force is the inverse square,

$$V_2' = \int\frac{4ma\,d\phi}{(r^2 - a^2\sin^2\phi)^{\frac{1}{2}}}, \quad V_2 = \int\frac{4ma\,d\theta}{(a^2 - r^2\sin^2\theta)^{\frac{1}{2}}}.$$

When the law of force is the inverse cube we find for *an external point* (using (1))

$$V_3' = \int\frac{ma\,d\phi}{r\cos\theta}\,\frac{2r\cos\theta}{r^2 - a^2} = \frac{m\pi a}{r^2 - a^2}.$$

The potential at an internal point may be deduced from the general expression for V_κ, if we remember to write $-u_1$ for u_1. It follows however at once from the expression for V_3' by using the rule of inversion (Art. 171). We write a^2/r_1 for r and multiply by $(a/r_1)^{\kappa-1}$, where $r_1 = OP$ and $\kappa = 3$. We thus find $V_3 = \dfrac{m\pi a}{a^2 - r_1^2}$.

The attraction of the ring at any external point P' in the plane of the ring is the sum of resolved attractions of the elements at Q and R. In this way we find $X' = \dfrac{M}{\pi r}\int S_{\kappa-1}d\phi$, where the limits are 0 to $\frac{1}{2}\pi$.

The potentials for these two laws of force being known, the corresponding potentials for any other inverse law may be deduced from the theorem (Art. 97)

$$V_{\kappa+2} = \frac{1}{\kappa+1}\frac{1}{a^2 - r^2}\left\{2r\frac{dV_\kappa}{dr} + (\kappa - 1) V_\kappa\right\}.$$

The potentials at points in the plane of the ring being known *the potential at any point R of space may be found by the rule of Art.* 187. If (x, z) be the coordinates of R, we write $e^2 x$ for r and multiply by $e^{\kappa-1}$.

For example, when the law of force is the inverse square, the potential at R is $\displaystyle\int\frac{8mad\theta}{\{(\rho'+\rho)^2 - (\rho'-\rho)^2\sin^2\theta\}^{\frac{1}{2}}}$, the limits being 0 to $\frac{1}{2}\pi$.

Instead of using the angles $OPR = \theta$, $ORP = \phi$, we may use the third angle $POR = \psi$ of the triangle OPR, or the angle χ subtended by OR at A, so that

$$\pi - \psi = 2\chi.$$

Supposing *the law of force to be that of the inverse square of the distance*, the potential at P is then

$$V_2 = \int\frac{mds}{PR} = \int\frac{2mad\psi}{(a^2 + r^2 - 2ar\cos\psi)^{\frac{1}{2}}} = \int\frac{4mad\chi}{\{(a+r)^2 - 4ar\sin^2\chi\}^{\frac{1}{2}}},$$

where the limits are $\psi = 0$ to π, and $\chi = 0$ to $\frac{1}{2}\pi$, and ds is an element of arc. These results hold whether P is internal or external, provided it is in the plane of the circle.

191. Ex. 1. Investigate Plana's theorem that the attraction of a uniform circular ring at an external point in the plane of the ring is

$$\frac{2M}{\pi}\frac{1}{r^2 - a^2}\int\left(1 - \frac{a^2}{r^2}\sin^2\phi\right)^{\frac{1}{2}}d\phi,$$

the limits being 0 to $\frac{1}{2}\pi$. [Turin Memoirs, 1820.]

Ex. 2. Prove that the potential of a thin uniform attracting ring at an internal point P in its plane very close to the circumference is ultimately $2m\log 8a/\xi$, where ξ is the distance of P from the ring.

We take the general expression for the potential given as an elliptic integral in Art. 190 and put $r/a = k$, where k is to be ultimately put equal to unity. We have

$$\frac{V_2}{4m} = \int\frac{d\theta}{(1 - k^2\sin^2\theta)^{\frac{1}{2}}} = \int d\theta\left\{\left(\frac{1 - k\sin\theta}{1 + k\sin\theta}\right)^{\frac{1}{2}} + \frac{k\sin\theta}{(1 - k^2\sin^2\theta)^{\frac{1}{2}}}\right\} \quad \ldots\ldots(1).$$

The last integral can be found, and is equal to $-\log\{(1-k^2)^{\frac{1}{2}}/(1+k)\}$. The other integral presents no singularities and we may put $k=1$ before integration; its value is then $\log 2$.

This value of the potential agrees with that given by H. Poincaré in his *Théorie du Potentiel Newtonien*, 1899, p. 132. The use of the first integral on the right-hand side of (1) to find the elliptic integral K when $k=1$ was suggested in a College examination paper, Dec. 1896.

A plane drawn through the axis of the ring will cut the level surfaces in a series of curves. By using the theorem $V''=eV$ of Art. 187 we may prove that these are circles in the immediate neighbourhood of the ring.

Ex. 3. Prove that the level surfaces of a thin circular ring, when the law of attraction is the inverse cube, are given by $\rho\rho'=\mu^2$, where ρ, ρ' are the greatest and least distances of any point R from the ring, and the constant μ is given by $2\mu^2=M/V''$.

192. Anchor rings. An anchor ring is generated by the revolution of a circle of radius a about an axis Oz in the plane of the circle, the centre C describing a circle of radius c. *A thin homogeneous layer is placed on its surface.* Prove that the potential of the layer at any point P of the axis is

$$V=\frac{2M}{\pi R^2}\int\sqrt{(R^2-a^2\sin^2\phi)}\,d\phi,\quad \phi=0 \text{ to } \frac{\pi}{2},$$

where $R=CP$, and M is the whole superficial mass. If m be the surface density, $M=2\pi a\,.\,2\pi c\,.\,m$ by Guldin's theorem.

Let QQ' be an arc of the generating circle; let PQ make an angle ϕ with the outward normal CQ to the anchor ring.

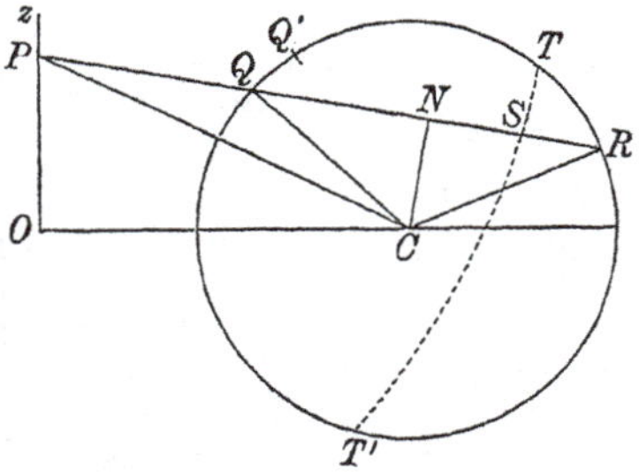

Let the angles $CPQ=\theta$, $CPO=\beta$, and $PQ=\rho$. Since the arc $QQ'=\rho d\theta\sec\phi$, the potential at P of the annulus generated by the revolution of QQ' about Oz is

$$V=m\int\sec\phi d\theta\,.\,2\pi\rho\sin(\theta+\beta)$$
$$=2\pi m\int\sec\phi d\theta\rho(\cos\theta\sin\beta+\sin\theta\cos\beta).$$

Since the integration extends over the whole circumference of the generating circle, the last term is zero. Also

$$\sin\theta=\sin\phi\,(a/R),$$
$$\therefore\ \cos\theta d\theta=\cos\phi d\phi\,(a/R).$$
$$\therefore\ V=2\pi m\sin\beta\,(a/R)\int\rho d\phi.$$

The limits are 0 to π if we double the result. Produce PQ to cut the circle again in R and let $PR=\rho'$. Then

$$V=4\pi m\,\frac{ac}{R^2}\int(\rho+\rho')\,d\phi,\quad \phi=0 \text{ to } \tfrac{1}{2}\pi.$$

Since $\rho+\rho'=2\,.\,PN=2\sqrt{(R^2-a^2\sin^2\phi)}$ this reduces to the result given above without difficulty.

193. Ex. 1. The potential of *a solid homogeneous anchor ring* at any point P of the axis may be expressed in either of the following forms

$$V=\frac{4M}{\pi R^2}\int\cos^2\phi\sqrt{(R^2-a^2\sin^2\phi)}\,d\phi=\frac{2M}{\pi}\int\frac{\sin^2\psi d\psi}{\sqrt{(R^2+a^2-2aR\cos\psi)}}.$$

The limits for ϕ are 0 to $\frac{1}{2}\pi$, and for ψ, 0 to π. If μ be the density the whole mass $M=\pi a^2\,.\,2\pi c\,.\,\mu$.

Let r, θ be the polar coordinates of an element of area at S of the generating circle referred to P as origin and PC as axis of x. The potential is then

$$V = \iint \frac{rd\theta dr}{r} \cdot 2\pi\mu r \sin(\theta+\beta) = 2\pi\mu \sin\beta \iint rd\theta dr \cos\theta,$$

since, as in the last example, one term of the integral is zero. *Let us integrate this first with regard to r.* Then using the geometrical relations connecting θ, ϕ, ρ, ρ' given in the last article, we find

$$\frac{V}{2\pi\mu \sin\beta} = \int \cos\theta d\theta \, \frac{\rho'^2 - \rho^2}{2} = 2\int \sqrt{\left(1 - \frac{a^2}{R^2}\sin^2\phi\right)} \, a^2 \cos^2\phi d\phi.$$

The limits of $\sin\theta$ are $-a/R$ to a/R and those of ϕ are $-\frac{1}{2}\pi$ to $\frac{1}{2}\pi$.

Let us integrate the double integral first with regard to θ. Let the circle whose centre is P and radius $r = PS$ cut the generating circle in T, T'. Let the angles $CPT = \theta_1$, $PCT = \psi$. Then

$$\sin\theta_1 = \sin\psi\,(a/r), \quad r^2 = R^2 + a^2 - 2aR\cos\psi.$$

We now find $$\frac{V}{2\pi\mu \sin\beta} = \int rdr \cdot 2\sin\theta_1 = \int aR \sin\psi d\psi \cdot 2\sin\psi \frac{a}{r},$$

the limits of ψ are evidently 0 to π. This is equivalent to the second expression given above. The second expression for V agrees with that given by Dyson, *Phil. Trans.* 1893, p. 55.

Ex. 2. Express the potential of a solid anchor ring in elliptic functions. Let $X^2 = 1 - k^2\sin^2\phi$, $k = a/R$, then

$$I = \int \cos^2\phi X d\phi = \int X \cos\phi d \sin\phi$$
$$= \int \sin^2\phi X d\phi + k^2 \int \sin^2\phi \cos^2\phi \, (d\phi/X)$$

by integrating by parts. The integrated part is zero at each limit.

$$\therefore \; I = \int X d\phi - I + \int (1 - X^2) \cos^2\phi \, (d\phi/X)$$
$$= \int X d\phi - 2I + \int (1 - \sin^2\phi)\,(d\phi/X).$$

Substituting for $\sin^2\phi$ its value in terms of X, we find

$$3I = \left(1 + \frac{R^2}{a^2}\right)\int X d\phi + \left(1 - \frac{R^2}{a^2}\right)\int \frac{d\phi}{X},$$

where the limits throughout are $\phi = 0$ to $\frac{1}{2}\pi$.

Attraction of Ellipsoids.

194. For the sake of brevity we shall adopt in this section two new terms taken from Thomson and Tait's *Natural Philosophy*.

A *homoeoid* is a shell bounded by two surfaces similar and similarly situated with regard to each other. In what follows we shall somewhat restrict this definition and use the term only when the shell is bounded by concentric ellipsoids.

A *focaloid* is a shell bounded by two confocal ellipsoids.

Thomson and Tait restrict these terms to infinitely thin shells, but it will be convenient for us to use them in a more general sense, distinguishing the shells as thick or thin according as the thickness is finite or infinitely small.

A shell bounded by two similar and similarly situated surfaces has been called *a homothetic shell* by Chasles in the *Jour. Pol.*, Tome xv., 1837. This is a convenient term when the surfaces are either not concentric or not ellipsoids.

195. Let (a, b, c) be the semi-axes of the internal surface of a thin ellipsoidal shell, $(a + da$, &c.) those of the external surface. Let OPQ be any radius vector drawn from the common centre O cutting the ellipsoids in P and Q, let $OP = r$. Let p be the perpendicular from O on the tangent plane at P, $p + dp$ the perpendicular on a *parallel* tangent plane to the outer ellipsoid. Then dp is equal to the thickness at P.

When the thin shell is a homoeoid we have by the properties of similar figures $$\frac{da}{a} = \frac{db}{b} = \frac{dc}{c} = \frac{dp}{p} = \frac{dr}{r} = dk.$$
Since the volume of a solid ellipsoid is $\frac{4}{3}\pi abc$, we find by differentiation that the volume v of the shell is $v = 4\pi abcdk$. Two thin homoeoids are said to be confocal when their inner boundaries are confocal conicoids.

When the shell is a focaloid, we have $a'^2 = a^2 + \lambda$, $b'^2 = b^2 + \lambda$, &c., where (a', b', c') are the semi-axes of the external surface. These give for a thin shell $ada = bdb = cdc = pdp = \frac{1}{2}d\lambda$. The volume v of the shell may be shown by differentiation to be
$$v = \frac{4\pi}{3}\frac{b^2c^2 + c^2a^2 + a^2b^2}{abc}\frac{d\lambda}{2}.$$

If we regard either shell as a thin stratum placed on an ellipsoidal surface the mass on any elementary area dA is $\rho dp \,.\, dA$ where ρ is the density. The surface density is therefore ρdp and it varies directly or inversely as p according as the stratum is the limit of a homoeoid or a focaloid.

196. Thick homoeoid, internal point. *To find the potential of a thick homogeneous homoeoid at an internal point.*

It has been shown in Art. 68 that the attraction of such a shell at all internal points is zero. The potential is therefore constant throughout the interior, and it will be sufficient to find the potential at the centre.

Taking polar coordinates with the centre as origin, the mass of any element is $\rho r^2 dr d\omega$, where ρ is the density of the element. The potential V of the whole solid at the centre is therefore $V = \rho\int\int r dr d\omega$. If r_1, r_2 be the radii vectores of the two surfaces of the shell, we have $V = \frac{1}{2}\rho\int r_2^2 d\omega - \frac{1}{2}\rho\int r_1^2 d\omega.$

The determination of the potential at the centre of a thick shell, bounded by any concentric ellipsoids, depends therefore on the evaluation of the integral $\int r^2 d\omega$ taken over the superficies of an ellipsoidal surface.

When the shell is a homoeoid these surfaces are similar. Let (a, b, c), (ma, mb, mc) be the semi-axes of the external and internal surfaces. We then find $V = \frac{1}{2}\rho(1 - m^2)\int r^2 d\omega$, where r is the radius vector of the external boundary.

When the shell is a thin homoeoid m is nearly equal to unity. The surface density is $\rho dk \,.\, p$ and the potential is $\rho dk \int r^2 d\omega$ where $dk = 1 - m$. When the surface density is μp the potential is $\mu \int r^2 d\omega$ and the whole mass is $4\pi abc\mu$.

It easily follows that the potential of a thin homoeoid is two-thirds of the potential at the centre of a solid homogeneous ellipsoid of equal mass and having the same external boundary.

197. To find the integral $\int r^2 d\omega$ we write $d\omega = \sin\theta d\theta d\phi$. Substituting for r^2 its value found from the equation to the ellipsoid, we have

$$\int r^2 d\omega = \iint \frac{\sin\theta d\theta d\phi}{\dfrac{\cos^2\theta}{c^2} + \sin^2\theta\left(\dfrac{\cos^2\phi}{a^2} + \dfrac{\sin^2\phi}{b^2}\right)},$$

where the integration extends over the whole surface of the ellipsoid. Taking only an octant, the limits are $\theta = 0$ to $\theta = \frac{1}{2}\pi$, $\phi = 0$ to $\phi = \frac{1}{2}\pi$. The order of integration is immaterial.

Let us integrate first with regard to ϕ. Dividing both numerator and denominator by $\cos^2\phi$, we find

$$\tfrac{1}{8}\int r^2 d\omega = \iint \frac{\sin\theta d\theta d\tan\phi}{\dfrac{\cos^2\theta}{c^2} + \dfrac{\sin^2\theta}{a^2} + \left(\dfrac{\cos^2\theta}{c^2} + \dfrac{\sin^2\theta}{b^2}\right)\tan^2\phi}.$$

By obvious processes in the integral calculus

$$\int_0^\infty \frac{dt}{A + Bt^2} = \left[\frac{1}{\sqrt{(AB)}}\tan^{-1} t\sqrt{\frac{B}{A}}\right]_0^\infty = \frac{\pi}{2\sqrt{(AB)}}.$$

It therefore follows that

$$\tfrac{1}{8}\int r^2 d\omega = \frac{\pi}{2}\cdot\int \frac{\sin\theta d\theta}{\sqrt{\left(\dfrac{\cos^2\theta}{c^2} + \dfrac{\sin^2\theta}{a^2}\right)}\sqrt{\left(\dfrac{\cos^2\theta}{c^2} + \dfrac{\sin^2\theta}{b^2}\right)}}.$$

To interpret this expression, let us produce the radius vector OP or r to cut the tangent plane drawn at the extremity C of the axis of z. Let R be the point of intersection and let $CR = u$, then

$u = c\tan\theta$. Since the limits of θ are 0 and $\frac{1}{2}\pi$, those of u are 0 and ∞. Substituting, we find

$$\int r^2 d\omega = 2\pi abc \int_0^\infty \frac{du^2}{(a^2+u^2)^{\frac{1}{2}}(b^2+u^2)^{\frac{1}{2}}(c^2+u^2)^{\frac{1}{2}}},$$

where *the integration on the left side extends over the whole surface of the ellipsoid.*

198. If we write $I = \int_0^\infty \frac{du}{(a^2+u)^{\frac{1}{2}}(b^2+u)^{\frac{1}{2}}(c^2+u)^{\frac{1}{2}}}$, we find that the potential of a thick homoeoid is

$$V = \rho(1-m^2)\pi abc\,.\,I = \tfrac{3}{4}MI\frac{1-m^2}{1-m^3}.$$

The potential of a thin homoeoid is $V = \frac{1}{2}M\,.\,I$, where, in each case, M is the mass of the attracting body.

It follows that *the integral I may also be defined as the ratio of the internal potential of a thin homoeoid to half its mass.* If the homoeoid represent an electrical stratum $2/I$ is the capacity.

199. Since the first integration in Art. 197 has been made with regard to ϕ it is evident that we may introduce any function of θ as a factor without disturbing the argument. We therefore have

$$\int r^2\cos^2\theta\, d\omega = 2\pi abc\int_0^\infty \frac{\cos^2\theta\, du^2}{(a^2+u^2)^{\frac{1}{2}}(b^2+u^2)^{\frac{1}{2}}(c^2+u^2)^{\frac{1}{2}}}.$$

Since $u = c\tan\theta$ and $z = r\cos\theta$, this gives $\int z^2 d\omega = -2\pi abc\,.\,c\frac{dI}{dc}$,

$$\therefore\ \int x^2 d\omega = -2\pi abc\,.\,a\frac{dI}{da}, \qquad \int y^2 d\omega = -2\pi abc\,.\,b\frac{dI}{db},$$

where the integrations extend over the whole surface of the ellipsoid.

The polar equation of the ellipsoid is

$$\frac{1}{r^2} = \frac{l^2}{a^2} + \frac{m^2}{b^2} + \frac{n^2}{c^2}, \qquad \therefore \frac{1}{r^3}\frac{dr}{dc} = \frac{n^2}{c^3}.$$

Differentiating $\int r^2 d\omega = 2\pi abcI$ with regard to c, we have $\int r^4 n^2 d\omega = \pi abc^3\frac{d(Ic)}{dc}$, with similar expressions for $\int r^4 l^2 d\omega$ and $\int r^4 m^2 d\omega$.

200. Ex. 1. Prove $\int \lambda^{2f}\mu^{2g}\nu^{2h} d\omega = \frac{4\pi}{2n+1}\frac{L(f)L(g)L(h)}{L(n)}$, where (λ, μ, ν) are the direction cosines of a radius vector, $n = f+g+h$, and $L(f)$ stands for the quotient of all the natural numbers up to $2f$ by the product of the same numbers up to f, both included. A short proof is given in the author's *Rigid Dynamics*, Vol. I. Art. 9.

Ex. 2. Show that the integral I, and therefore the potential V, may be expressed as an elliptic integral. Thus $I = \frac{2}{\sqrt{(c^2-b^2)}}\int\frac{d\psi}{\sqrt{(1-\lambda\sin^2\psi)}}$, where $\lambda = \frac{c^2 - a^2}{c^2 - b^2}$ and the limits are $\psi = 0$ to $\sin\psi = \sqrt{\frac{c^2 - b^2}{c^2}}$. The integral is real if the axis of c is the longest of the three.

To prove this we revert to the value of I (Art. 197) after the integration with regard to ϕ has been performed. Putting $\cos\theta = v$, the integral takes a known form. This is reduced to the standard form given in the example by putting

$$(c^2 - b^2)\, v^2 = c^2 \sin^2\psi.$$

Ex. 3. Show that $\left(\frac{d}{da^2} + \frac{d}{db^2} + \frac{d}{dc^2}\right) I = -\frac{1}{abc}$,

$$\left(a^2 \frac{d}{da^2} + b^2 \frac{d}{db^2} + c^2 \frac{d}{dc^2} + \tfrac{1}{2}\right) I = 0, \qquad 2\,(a^2 - b^2)\, \frac{d^2 I}{da^2 db^2} = \frac{dI}{da^2} - \frac{dI}{db^2}.$$

Let $Q^2 = (a^2 + u)(b^2 + u)(c^2 + u)$, then $I = \int du/Q$;

$$\therefore\ \frac{dI}{da^2} = -\tfrac{1}{2} \int \frac{du}{Q\,(a^2 + u)}, \qquad \frac{2}{Q}\frac{dQ}{du} = \frac{1}{a^2 + u} + \frac{1}{b^2 + u} + \frac{1}{c^2 + u}.$$

The results follow by simple substitution.

By writing $b = ma$, $c = na$, $u = va^2$, we see that *I is a homogeneous function of a, b, c of -1 dimensions.* The second result then follows from Euler's theorems on homogeneous functions.

Ex. 4. If $\int r^{2m} d\omega = abc R_m$, prove that

$$\left\{a^2 \frac{d}{da^2} + b^2 \frac{d}{db^2} + c^2 \frac{d}{dc^2} + (\tfrac{3}{2} - m)\right\} R_m = 0, \qquad \int r^{2m} z^2 d\omega = \frac{abc^3}{m} \left\{\tfrac{1}{2} R_m + c^2 \frac{dR_m}{dc^2}\right\},$$

$$\left\{a^4 \frac{d}{da^2} + b^4 \frac{d}{db^2} + c^4 \frac{d}{dc^2} + \frac{a^2 + b^2 + c^2}{2}\right\} R_m = m R_{m+1}.$$

The first result follows from Euler's theorem on homogeneous functions. By differentiating $\int r^{2m} d\omega = abc R_m$ with regard to c^2 we obtain $\int r^{2m+2} n^2 d\omega$ as in Art. 199. The two other results follow easily.

Ex. 5. Instead of the standard integral represented by I we may use the integral

$$J = \int_0^\infty \frac{abc\, du}{u\,(a^2 + u)^{\frac{1}{2}}\,(b^2 + u)^{\frac{1}{2}}\,(c^2 + u)^{\frac{1}{2}}}.$$

We then have $\frac{dJ}{da} = -\frac{bc}{a}\frac{dI}{da}, \quad \frac{dJ}{db} = -\frac{ca}{b}\frac{dI}{db}$, &c.

If we write α, β, γ for the *reciprocals* of a^2, b^2, c^2 we easily find

$$\int z^2 d\omega = -4\pi \frac{dJ}{d\gamma}, \qquad \int r^2 d\omega = -4\pi \left(\frac{d}{d\alpha} + \frac{d}{d\beta} + \frac{d}{d\gamma}\right) J = 2\pi abc I,$$

$$\int x^{2i} y^{2j} z^{2k} d\omega = \frac{4\pi\,(-1)^{i+j+k}}{\Gamma\,(i + j + k)} \left(\frac{d}{d\alpha}\right)^i \left(\frac{d}{d\beta}\right)^j \left(\frac{d}{d\gamma}\right)^k J,$$

where the integrations extend over the surface of the ellipsoid.

Differentiating $\int r^2 n^2 d\omega = -4\pi dJ/d\gamma$, i times with regard to α, j times with regard to β and $k - 1$ times with regard to γ we arrive at the last result.

Ex. 6. If $f(l^2, m^2, n^2)$ be a homogeneous function of l^2, m^2, n^2 of s dimensions, prove that

$$\int r^q f(l^2, m^2, n^2)\, d\omega = Nf\left(\frac{d}{d\alpha}\ \frac{d}{d\beta}\ \frac{d}{d\gamma}\right) \int_0^\infty \frac{v^{p - \frac{1}{2}} dv}{Q},$$

where $N = \frac{2\pi \,.\, 2^p}{1 \,.\, 3 \,.\, 5 \ldots (2p - 1)} \cdot \frac{(-1)^s}{1 \,.\, 2 \,.\, 3 \ldots (s - p)}$ and $q = 2\,(s + 1 - p)$.

Prove also that $\int \frac{f(x, y, z)\, d\omega}{r^{p-1}} = Nf\left(\frac{d}{d\alpha}\ \frac{d}{d\beta}\ \frac{d}{d\gamma}\right) \int_0^\infty \frac{v^{p - \frac{1}{2}} dv}{Q}$.

201. Theorems on thin homoeoids. *The potential at any internal point of a thin homoeoid being known it is required to find the potential at any external point.*

Let two ellipsoids have for their semi-axes (a, b, c), (a', b', c'); points on these are said to correspond when their coordinates are connected by the relations

$$\frac{x}{a}=\frac{x'}{a'},\quad \frac{y}{b}=\frac{y'}{b'},\quad \frac{z}{c}=\frac{z'}{c'} \quad\text{.................(1).}$$

Let $d\sigma$, $d\sigma'$ be two triangular elements of area at P, P' such that the corners are corresponding points; let p, p' be the perpendiculars from the centre O on the tangent planes. The volumes of the tetrahedra whose bases are $d\sigma$, $d\sigma'$ and common vertex O are respectively $\frac{1}{3}pd\sigma$ and $\frac{1}{3}p'd\sigma'$. The first of these volumes is expressed by one sixth of the determinant in the margin, where the several rows express the coordinates of the corners. The second volume is expressed in the same way with accented letters to represent the corresponding points on the second ellipsoid. It is evident from the relations (1) that these determinants are in the ratio $abc : a'b'c'$. We therefore infer that the elements of surface of the two ellipsoids are connected by

$$\begin{vmatrix} x & y & z \\ x_1 & y_1 & z_1 \\ x_2 & y_2 & z_2 \end{vmatrix}$$

the equation
$$\frac{pd\sigma}{p'd\sigma'}=\frac{abc}{a'b'c'} \quad\text{.........................(2).}$$

Since any elementary areas at P and P' may be subdivided into triangles, it is evident that this relation holds for elementary areas $d\sigma, d\sigma'$ of any shape, provided only their boundaries are formed by corresponding points.

Since the thickness of a thin homoeoid is represented by kp, it follows that *the volumes of corresponding elements of two thin homoeoids are in a constant ratio.* Adding these elementary volumes together, it is easily seen that *this constant ratio is equal to that of the whole volumes of the two shells.* If the shells are of such thicknesses that their whole volumes are equal, then the volumes of all corresponding elements are equal. See Vol. I. Art. 428.

202. We shall now require the following geometrical theorem:—*the distance between two points one on each of two confocal ellipsoids is equal to the distance between their corresponding points.* A proof may be found in Smith's *Solid Geometry*, Art. 166. This theorem is usually called Ivory's theorem after its discoverer, who also applied it to determine the potential of an ellipsoid at an external point.

Let P, P' be two corresponding points, one on each of two confocal thin homoeoids of equal volume; let also Q, Q' be any two corresponding elementary volumes each equal to dv. Let the equal distances PQ', $P'Q$ be represented by R. If $f'(R)$ represent the law of attraction, the potentials at P and P' of these elementary *volumes* are each $f(R)\,dv$. Integrating over the whole surfaces of the shells, we see that *the potential of the inner thin homoeoid at the external point P' is equal to that of the outer thin homoeoid at the corresponding internal point P, provided the densities are equal at corresponding points**.

Thus when the potentials of thin homoeoids at all internal points are known, their potentials at all external points are also known.

203. It is evident that *the potentials of these shells are equal whatever be the law of attraction* provided the potential is a function of the distance only.

The potentials are also equal if the shells are heterogeneous, and the density at any point is a function of $(x/a,\ y/b,\ z/c)$. In this case it is evident that the densities of the shells are equal at corresponding points. The equality of the potentials is also true *when the shells are incomplete,* provided only the existing parts "correspond" to each other.

204. The theorem may also be used (though not so simply) to compare the potentials even when the density is any function of the coordinates. It will be convenient to express this result in an analytical form.

Let the density ρ of a thin homoeoid (semi-axes a, b, c) be $f(x, y, z)$, and let ϖ be the volume of the shell. It is required to find its potential at any external point (ξ', η', ζ'). Let a confocal ellipsoid be described passing through the point (ξ', η', ζ') so that

* Chasles in his *Nouvelle solution du problème de l'attraction d'un ellipsoïde hétérogène sur un point extérieur*, Liouville, vol. v. 1840, shows that thin confocal homoeoids have potentials at corresponding points proportional to their masses, but considers only the case in which they are homogeneous. Knowing that the potential of the outer at an internal point is constant, he deduces several theorems on the attractions of the inner shell at external points. He finds the attraction of a solid heterogeneous ellipsoid by dividing it into thin elementary homoeoids, the strata of equal density being the elementary homoeoids. The case in which the homoeoid is heterogeneous is not discussed.

its semi-axes a', b', c' are given by $a'^2 - a^2 = \lambda$, $b'^2 - b^2 = \lambda$, $c'^2 - c^2 = \lambda$, where λ is a root of the equation

$$\frac{\xi'^2}{a^2+\lambda} + \frac{\eta'^2}{b^2+\lambda} + \frac{\zeta'^2}{c^2+\lambda} = 1 \quad \ldots\ldots\ldots\ldots\ldots(1).$$

Let this ellipsoid be the inner boundary of a second thin homoeoid whose volume is equal to that of the former. Let its density at any point (x', y', z') be $\rho' = f(ax/a', by/b', cz/c')$. The potential of this second homoeoid at the internal point $(a\xi'/a', b\eta'/b', c\zeta'/c')$ is equal to the potential required.

We shall in general take a^2, b^2, c^2 to be in descending order of magnitude. If λ is either positive, or negative and numerically greater than c^2, the surface (1) is an ellipsoid. If λ is negative and numerically greater than c^2 the surface is one of the hyperboloids or is imaginary. The root of the cubic to be chosen must therefore be the algebraically greatest root. Since the attracted point is external to the ellipsoid $a^2 + \lambda$ is necessarily greater than a^2, the greatest root is therefore positive.

205. Taking the case in which the two thin homoeoids are homogeneous, the potential of the outer has been proved constant for all internal points, Art. 68. It immediately follows that the potential of the inner is the same at all external points which lie on the same confocal. We therefore infer that *the level surfaces of any thin homogeneous homoeoid are confocal ellipsoids.*

It follows from this proposition that the direction of the attraction of a thin homoeoid at any external point P' is normal to the confocal ellipsoid which passes through that point.

It is proved in treatises on solid geometry that this normal is also the axis of the cone which has its vertex at P' and envelopes the ellipsoid.

This result was given by Poisson (*Mém. de l'Institut*, 1835). There is an elementary demonstration by Steiner in *Crelle's Journal*, vol. XII.

206. Since two thin confocal homoeoids have the same level surfaces, their potentials can be made equal over any level surface enclosing both by properly adjusting their masses. It immediately follows that their potentials are also equal throughout all external space, Art. 130. Since the potentials of finite bodies vanish at infinity in the ratio of their masses, it is evident that the masses of the two homoeoids must be equal. We have therefore the following theorem, *the potentials, and therefore also the resolved* *attractions, of two confocal thin homoeoids of equal masses are equal throughout all space external to both.*

207. Lines of force. The lines of force of a homogeneous thin homoeoid are the orthogonal trajectories of all the confocal ellipsoids. Let (a', b', c'), (a'', b'', c''), (a''', b''', c''') be the semi-axes of the confocal ellipsoid and hyperboloids which pass through any external point (ξ', η', ζ'). Then by a theorem in solid geometry

$$\xi' = \frac{a'a''a'''}{\sqrt{(a^2-b^2)(a^2-c^2)}},\quad \eta' = \frac{b'b''b'''}{\sqrt{(b^2-a^2)(b^2-c^2)}},\quad \zeta' = \frac{c'c''c'''}{\sqrt{(c^2-a^2)(c^2-b^2)}},$$

see Salmon's *Solid Geometry*, Art. 160. Since these conicoids intersect at right angles, the curve of intersection of the two hyperboloids is an orthogonal trajectory of all the confocal ellipsoids. The required trajectories are therefore found by regarding (a'', b'', c'') and (a''', b''', c''') as constants. It follows that ξ'/a', η'/b', ζ'/c' are constant for the same orthogonal. Thus it appears that *any line of force of a homogeneous thin homoeoid intersects all the confocal ellipsoids in corresponding points.*

208. Thin homoeoid, external point. *To find the potential and the attraction of a homogeneous thin homoeoid at an external point P'.* The potential V' of the given homoeoid at P' has been proved equal to that of a confocal homoeoid of equal mass having P' just outside (Art. 206). This again is equal to the potential at a point just inside (Art. 145). It follows from Art. 198 that the potential at P' is

$$V' = \tfrac{1}{2}MI' = \tfrac{1}{2}M\int_0^\infty \frac{du'}{(a'^2+u')^{\frac{1}{2}}(b'^2+u')^{\frac{1}{2}}(c'^2+u')^{\frac{1}{2}}},$$

where M is the mass of the homoeoid and (a', b', c') the semi-axes of the confocal which passes through P'.

This integral may be put into another form which contains the semi-axes a, b, c of the given homoeoid instead of those a', b', c' of the confocal. Putting $a'^2 = a^2+\lambda$, $b'^2 = b^2+\lambda$, $c'^2 = c^2+\lambda$ and $u' = u-\lambda$, we have

$$V' = \frac{M}{2}\int_\lambda^\infty \frac{du}{(a^2+u)^{\frac{1}{2}}(b^2+u)^{\frac{1}{2}}(c^2+u)^{\frac{1}{2}}},$$

where λ is defined in Art. 204.

209. *To deduce the resultant attraction,* we notice that the level surfaces of the given homoeoid are confocal quadrics. The resultant force F therefore acts parallel to the perpendicular p'

drawn from the centre O on the tangent plane at P'. Hence $F = dV'/dp'$, and

$$F = \frac{dV'}{d\lambda}\frac{d\lambda}{dp'} = -\frac{M}{2}\frac{2p'}{(a^2+\lambda)^{\frac{1}{2}}(b^2+\lambda)^{\frac{1}{2}}(c^2+\lambda)^{\frac{1}{2}}} = -\frac{Mp'}{a'b'c'},$$

since, by Art. 195, $d\lambda = 2p'dp'$.

The expression for the attraction F may be obtained independently. The attraction of the homoeoid at P' is equal to that of a confocal homoeoid having P' just outside (Art. 206) and therefore, by Art. 71, $F = -\dfrac{Mp'}{a'b'c'}$*.

Again, assuming this value of F, the potential V' follows at once by integration.

210. Ex. 1. If an attracting body has an ellipsoid enclosing the whole attracting mass for one of its level surfaces, prove that all the external level surfaces are confocal ellipsoids. See Art. 130.

Ex. 2. The attractions of a given thin homoeoid on two corresponding elementary areas taken on any two confocal ellipsoids are equal. [Chasles, *Journal Polytechnique*, 1837, Tome xv.]

Ex. 3. The attraction of a thin homoeoid at any point situated on its external surface is proportional to the thickness of the shell at that point. [Chasles.]

Ex. 4. A thin prolate spheroidal shell of mass M is divided into two portions by a diametral plane perpendicular to its axis. Prove that the pressure per unit of length on the line of separation, due to the mutual attraction of the parts, is

$$\frac{M^2}{8\pi b}\frac{\log a - \log b}{a^2 - b^2}. \qquad \text{[Math. Tripos.]}$$

The resultant force on any element of the shell is half the force just outside, Arts. 68 and 143. If $d\sigma$ be an elementary area, l the cosine of the angle p makes with the axis, the resultant pressure is $\frac{1}{2}\iint Fl d\sigma dp$, where F has the value given in Art. 209. Putting $d\sigma = 2\pi y ds$, $dp/p = da/a$, the integration can be effected.

Ex. 5. The mutual potential of a thin homoeoid (mass M, semi-axes a, b, c) and any internal mass M' is $\frac{1}{2}MM'I$, where I is the integral defined in Art. 198.

The mutual potential of the same homoeoid and any external mass M' placed as a stratum on a confocal quadric (semi-axes a', b', c') is $\frac{1}{2}MM'I'$, where I' is the same integral with a', b', c' written for a, b, c. [See Arts. 61, 208.]

211. Solid homogeneous ellipsoid. *To find the potential at an internal point P whose coordinates are (ξ, η, ζ).*

Describe a double cone with vertex P cutting the surface in two opposite elementary areas Q_1, Q_2. If R_1, R_2 are the distances PQ_1, PQ_2, the potential of the double cone at P is $\frac{1}{2}\rho\int(R_1^2 + R_2^2)\,d\omega$ (Art. 196). It is evident that if we integrate this expression all round the point P every element of volume of

* This expression for the resultant force is given by Chasles in the *Journal Polytechnique*, 1837, Tome xv. See also the *Quarterly Journal*, 1867.

the ellipsoid will be taken twice over; we must therefore halve the result. The potential at P is then

$$V = \tfrac{1}{4}\rho \int (R_1^2 + R_2^2)\, d\omega \quad\text{.................(1).}$$

The distances R_1, R_2 are the roots of the quadratic

$$\left(\frac{\xi - lR}{a}\right)^2 + \left(\frac{\eta + mR}{b}\right)^2 + \left(\frac{\zeta + nR}{c}\right)^2 = 1 \quad\text{........(2),}$$

where (l, m, n) are the direction cosines of Q_2PQ_1. This quadratic may be written shortly $\dfrac{R^2}{r^2} + 2FR - E = 0$ .... (3),

where r is the semi-diameter parallel to Q_2PQ_1. Hence

$$R_1^2 + R_2^2 = 4F^2r^4 + 2Er^2,$$

$$\therefore\ V = \frac{\rho}{2}\int \left\{2\left(\frac{l\xi}{a^2} + \frac{m\eta}{b^2} + \frac{n\zeta}{c^2}\right)^2 r^4 + \left(1 - \frac{\xi^2}{a^2} - \frac{\eta^2}{b^2} - \frac{\zeta^2}{c^2}\right) r^2\right\} d\omega.$$

It is obvious that the term containing the product lm disappears on integration, for the elements corresponding to (l, m) and $(l, -m)$ destroy each other. The terms containing mn and nl are also zero.

Hence $$V = \frac{\rho}{2}\int \left\{\left(\frac{2l^2r^4}{a^4} - \frac{r^2}{a^2}\right)\xi^2 + \&c. + \&c. + r^2\right\} d\omega \quad\text{......(4).}$$

We have $$\int r^2 d\omega = 2\pi abcI \quad\text{.........................(5).}$$

By differentiating (5) with regard to a, as in Art. 199, we have

$$\int 2r\frac{l^2r^3}{a^3}\, d\omega = 2\pi bc\frac{d(Ia)}{da} = 2\pi abc\left(\frac{dI}{da} - \frac{I}{a}\right) \quad\text{.......(6).}$$

After substituting from (5) and (6) in (4), we find

$$\frac{V}{\pi\rho abc} = I + \frac{dI}{ada}\xi^2 + \frac{dI}{bdb}\eta^2 + \frac{dI}{cdc}\zeta^2 \quad\text{...........(7).}$$

212. If we substitute $I = \int du/Q$ this becomes

$$\frac{V}{\pi\rho abc} = \int_0^\infty \frac{du}{Q}\left\{1 - \frac{\xi^2}{a^2+u} - \frac{\eta^2}{b^2+u} - \frac{\zeta^2}{c^2+u}\right\} \quad\text{......(8),}$$

where $Q^2 = (a^2+u)(b^2+u)(c^2+u)$. These two important expressions are often written, for brevity, in the form

$$V = \tfrac{1}{2}\rho\{D - A\xi^2 - B\eta^2 - C\zeta^2\} \quad\text{..............(9).}$$

Here $$\frac{D}{2\pi abc} = I = \int\frac{du}{Q},\quad \frac{A}{2\pi abc} = -\frac{dI}{ada} = \int_0^\infty \frac{du}{(a^2-u)Q} \quad\text{...(10),}$$

with similar values for B and C.

The component forces at any internal point P are then

$$X = -A\rho\xi,\qquad Y = -B\rho\eta,\qquad Z = -C\rho\zeta \quad\text{......(11).}$$

213. By putting ξ, η, ζ equal to zero, we see that $\frac{1}{2}\rho D$ is the potential at the centre of the solid ellipsoid. We also notice that by Art. 199

$$A = \int \frac{x^2}{a^2} d\omega, \quad B = \int \frac{y^2}{b^2} d\omega, \quad C = \int \frac{z^2}{c^2} d\omega \text{......(12)},$$

where the integrations extend over the whole surface of the body,

$$\left.\begin{array}{l} \therefore\ A + B + C = 4\pi, \\ Aa^2 + Bb^2 + Cc^2 = \int r^2 d\omega = 2\pi abcI = D \end{array}\right\} \text{.......(13)}.$$

The first of the results (13) follows also from Poisson's theorem since $d^2V/dx^2 = -A\rho$ &c. The second may also be deduced from (4); for the sum of the coefficients of ξ^2, η^2, ζ^2 after multiplication by a^2, b^2, c^2 is evidently $-r^2$.

Since $\frac{1}{r^2} = \frac{l^2}{a^2} + \frac{m^2}{b^2} + \frac{n^2}{c^2}$, we see by substitution either in (4) or (12) that *the constants A, B, C are functions of the ratios of the axes* and are therefore the same for all similar ellipsoids.

214. The four integrals A, B, C, D have here been expressed in terms of the integral I and its differential coefficients with regard to a, b, c. Other standard integrals might also have been taken. Thus we might use the integral called J in Art. 200, Ex. 5. We might also express the components X, Y, Z in terms of any one of the four integrals A, B, C, D. We deduce from the third part of Ex. 3, Art. 200,

$$\frac{Bb^2 - Aa^2}{b^2 - a^2} = b\frac{dA}{db}, \quad \frac{Cc^2 - Aa^2}{c^2 - a^2} = c\frac{dA}{dc} \text{........................(14)}.$$

These relations enable us to deduce the formulæ for X, Y, Z given by Laplace in the *Mécanique Céleste*, vol. II. p. 12.

215. Ex. Prove that the three numerical constants A, B, C lie between v/a^3 and v/c^3 where v is the volume and a, c are the greatest and least axes of the ellipsoid. Prove also that D lies between $4\pi a^2$ and $4\pi c^2$.

To prove the first theorem we notice that the integral (10) is decreased by writing a for b and c; the integration can then be effected. A superior limit is found by writing c for a and b. The second theorem follows from the equations (13) Art. 213 by eliminating first A and then C.

216. *To find the level surfaces inside the attracting ellipsoid.* These surfaces are given by $A\xi^2 + B\eta^2 + C\zeta^2 = K$, where K is a constant. Since A, B, C are necessarily positive, the level surfaces are similar and similarly situated ellipsoids.

To trace their forms, we must consider the magnitudes of the coefficients A, B, C. We have (Art. 212)

$$A = 2\pi abc \int_0^\infty \frac{du}{Q(a^2+u)}, \quad B = 2\pi abc \int_0^\infty \frac{du}{Q(b^2+u)},$$

$$\therefore -\frac{A-B}{a^2-b^2} = \int_0^\infty \frac{2\pi abc \,.\, du}{Q(a^2+u)(b^2+u)}, \quad \frac{Aa^2 - Bb^2}{a^2-b^2} = \int_0^\infty \frac{2\pi abc \,.\, udu}{Q(a^2+u)(b^2+u)}.$$

Both these integrals are essentially positive. It follows that when a, b, c are in descending order of magnitude, both $1/A$, $1/B$, $1/C$ and Ac^2, Bb^2, Cc^2 are also in descending order.

A level surface so far resembles the attracting ellipsoid that both quadrics have their longest, their shortest and their mean axes respectively in the same directions. The axes of a level quadric are $(K/A)^{\frac{1}{2}}$, $(K/B)^{\frac{1}{2}}$, $(K/C)^{\frac{1}{2}}$. If c be the least axis of the attracting ellipsoid and $K=Cc^2$, the level quadric touches the ellipsoid at the extremities of the least axis, while the other two axes are less than the corresponding axes of the ellipsoid. The level quadric therefore lies wholly within the ellipsoid, for if not it would cut the ellipsoid in two curves one on each side of the plane of xy and also touch it at the extremities of the axis of z. This of course is impossible. *Any level quadric therefore lies wholly within the attracting ellipsoid or intersects its surface according as K is less or greater than Cc^2.*

217. Ex. 1. Prove that the level quadrics are more spherical than the bounding surface of the attracting ellipsoid.

The eccentricities of the sections of the two quadrics by the plane of xy are respectively given by $e'^2=1-A/B$ and $e^2=1-b^2/a^2$. It follows immediately that e'^2-e^2 is negative.

Ex. 2. If a concentric ellipsoidal cavity be cut out of a solid homogeneous sphere, show that within the cavity the equipotential surfaces are given by

$$(2A-B-C)\,x^2+(2B-C-A)\,y^2+(2C-A-B)\,z^2=\text{constant},$$

where A, B, C are constants depending on the *shape* of the cavity.

[St John's Coll. 1887.]

218. Other laws of force. The potential of a *solid homogeneous ellipsoid* at an internal point P *when the law of force is the inverse κth power of the distance* may be found by the method used in Art. 211.

By describing a double cone with the vertex at P as before, we find that the potential is

$$V=\tfrac{1}{2}\frac{-1}{(\kappa-1)(\kappa-4)}\int S\,d\omega+C,\quad S=R_1^{4-\kappa}+(-R_2)^{4-\kappa}.$$

When κ is even, S is a symmetrical function of the roots of the quadric (2) of Art. 211. The double integrals take forms similar to that in equation (4) and may be reduced to single integrals by differentiations of $\int r^2 d\omega=2\pi abc I$.

We notice that when $\kappa>4$ the expression S is an integral rational function of the direction cosines (l, m, n) and the final integrals can be evaluated without difficulty (Art. 200, Ex. 1). *The potential for these laws of force can therefore be found in finite terms free from all signs of integration.*

When the law of force is the inverse fourth power of the distance we have for the potential at an internal point ξ, η, ζ, $V_4=\frac{2}{3}\pi\rho\log E+C$, $E=1-\dfrac{\xi^2}{a^2}-\dfrac{\eta^2}{b^2}-\dfrac{\zeta^2}{c^2}$.

From this result the potential for the inverse sixth, &c., powers may be deduced free from integrals by using Jellett's theorem (Art. 96).

The component attractions at an external point may be deduced by Ivory's theorem (to be presently proved). Thence by integration the potential for the inverse fourth power of the distance is found to be

$$V_4'=-\frac{2\pi}{3}\int_\infty^\lambda \frac{abc}{a'b'c'}\frac{d\lambda}{\lambda}\ \text{(Art. 204)}.$$

The potential of a thin homogeneous homoeoid may be found in a similar manner, but it may also be deduced from that of a solid ellipsoid by taking the total differential with regard to a, b, c on the supposition that the ratios $a:b:c$

are unaltered. See Arts. 195 and 92. The potentials at an internal and external point are for the law of the inverse fourth power,

$$V_4 = \tfrac{4}{3}\pi\rho\frac{da}{a}\cdot\frac{1}{E}+C,\quad V_4' = \tfrac{4}{3}\pi\rho\frac{da}{a}\cdot\frac{abc}{a'b'c'}\Big/\lambda\left(\frac{x^2}{a'^4}+\frac{y^2}{b'^4}+\frac{z^2}{c'^4}\right).$$

219. Spheroids. *To find the potential and attraction of the solid spheroid whose semi-axes are a, a, c at an internal point.*

To find the constants A, C we use the equations

$$2A + C = 4\pi,\quad 2Aa^2 + Cc^2 = \int r^2 d\omega = \iint \frac{\sin\theta\, d\theta\, d\phi}{\cos^2\theta/c^2 + \sin^2\theta/a^2}.$$

The limits of integration are $\theta = 0$ to π and $\phi = 0$ to 2π. If we put $\cos\theta = z$, the second equation becomes

$$2Aa^2 + Cc^2 = -2\pi a^2 c^2 \int \frac{dz}{c^2 + (a^2 - c^2)z^2},$$

where the limits are $z = 1$ to $z = -1$.

If the spheroid is oblate, a is greater than c, and

$$D = 2Aa^2 + Cc^2 = \frac{4\pi a^2 c}{\sqrt{(a^2 - c^2)}}\tan^{-1}\frac{\sqrt{(a^2 - c^2)}}{c} \quad\text{.......(1)}.$$

If the spheroid is prolate, a is less than c, and

$$D = 2Aa^2 + Cc^2 = \frac{4\pi a^2 c}{\sqrt{(c^2 - a^2)}}\log\frac{c + \sqrt{(c^2 - a^2)}}{a} \quad\text{......(2)}.$$

We also have $2\pi abcI = \int r^2 d\omega = D$. Thus the values of A and C may be found either by solving these equations or by using the formulæ $A = -2\pi abc dI/ada$ &c. The potential at any internal point is then $V = \frac{1}{2}\rho\{D - A(\xi^2 + \eta^2) - C\zeta^2\}$. We notice that $\tan^{-1}\sqrt{(a^2 - c^2)}/c$ in an oblate spheroid *is equal to the angle subtended at the extremity of the axis of revolution by the distance between the centre and either focus.*

220. Ex. 1. The earth being regarded as an oblate homogeneous spheroid the ratio $c/a = 1 - \epsilon$ where ϵ is the ellipticity. Since the value of ϵ is 1/300 nearly, it is generally sufficient to retain only the first powers. Prove that

$$A = \frac{4\pi}{3}\left(1 - \frac{2}{5}\epsilon\right),\quad C = \frac{4\pi}{3}\left(1 + \frac{4}{5}\epsilon\right),\quad D = 4\pi a^2\left(1 - \frac{2}{3}\epsilon\right).$$

[We have $C = \int \frac{z^2}{c^2} d\omega = \frac{a^2}{c^2}\iint \frac{\cos^2\theta\sin\theta\, d\theta\, d\phi}{1 + 2\epsilon\cos^2\theta}$. Expand the subject of integration in powers of ϵ.]

Ex. 2. Show that an attracting homogeneous oblate spheroid of eccentricity $\frac{1}{2}$, in the centre of which there acts a repulsive force μr, will have its own surface for one of its level surfaces if $3\mu = 8\pi\rho(5\pi\sqrt{3} - 27)$. [Coll. Ex. 1888.]

221. Nearly spherical ellipsoids. Ex. 1. The axes of an ellipsoid are so nearly equal that the square of the difference can be neglected. Prove that

$$A = \tfrac{4}{3}\pi\left\{1 + \tfrac{2}{5}\frac{b + c - 2a}{a}\right\}.$$

Put $b/a=1-\beta$, $c/a=1-\gamma$; then since A is a function of b/a and c/a we have $A=\frac{4}{3}\pi(L+M\beta+N\gamma)$ nearly where L, M, N are independent of the axes. Now A cannot be affected by interchanging b and c, hence $M=N$. Also when $\beta=0$ and $\gamma=\epsilon$ the ellipsoid becomes a spheroid and the expression for A must become identical with that found in Art. 220. Hence $L=1$, $N=-\frac{2}{5}$. This proof is commonly ascribed to D'Alembert.

Ex. 2. Prove that to a second approximation the constants of the attraction (Art. 212) are

$$\begin{aligned}
A&=\tfrac{4}{3}\pi\left\{1-\tfrac{2}{5}(\beta+\gamma)-\tfrac{1}{35}(9\beta^2-8\beta\gamma+9\gamma^2)+\ldots\right\},\\
B&=\tfrac{4}{3}\pi\left\{1-\tfrac{2}{5}(-2\beta+\gamma)-\tfrac{1}{35}(-18\beta^2+4\beta\gamma+9\gamma^2+\ldots\right\},\\
C&=\tfrac{4}{3}\pi\left\{1-\tfrac{2}{5}(\beta-2\gamma)-\tfrac{1}{35}(9\beta^2+4\beta\gamma-18\gamma^2)-\ldots\right\},\\
D&=4\pi a^2\left\{1-\tfrac{2}{3}(\beta+\gamma)-\tfrac{1}{15}(3\beta^2-8\beta\gamma+3\gamma^2)+\ldots\right\},\\
I&=\frac{2}{a}\left\{1+\tfrac{1}{3}(\beta+\gamma)+\tfrac{1}{15}(2\beta^2+3\beta\gamma+2\gamma^2)+\ldots\right\},
\end{aligned}$$

when $b/a=1-\beta$ and $c/a=1-\gamma$. We notice that since $A+B+C=4\pi$ for all values of β and γ, the sum of the coefficients of any power in the three first expansions must be zero.

222. Ivory's theorem. *To find the attraction of a solid homogeneous ellipsoid at an external point P' whose coordinates are ξ', η', ζ'.*

Let R be the distance of any element QQ' of the ellipsoid from P', and let ϕ be the angle this distance makes with the axis of x. Thus $R=QP'$, $\phi=P'QQ'$. If $f'(R)$ be the law of attraction, the x component of the attraction of this element at P' is

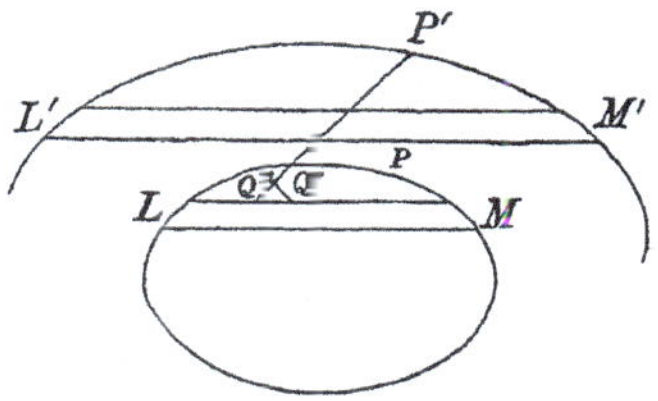

$$\rho\,dx\,dy\,dz\,f'(R)\cos\phi.$$

Draw $Q'n$ perpendicular to $P'Q$, then

$$\cos\phi=\frac{Qn}{QQ'}=-\frac{dR}{dx}.$$

The x attraction at the element at P', measured positively in the positive direction of x, is therefore $\rho\,dy\,dz\,f'(R)\,dR$. Let LM be a column having its length LM parallel to the axis of x and the elementary area $dydz$ for base. Integrating with regard to R we find that the x component of its attraction at P' is

$$\rho\,dy\,dz\int f'(R)\,dR=\rho\,dy\,dz\,\{f(P'M)-f(P'L)\}.$$

Let us now describe an ellipsoid through P' confocal to the external surface of the attracting solid. Let a', b', c' be the semi-axes of this new ellipsoid. If L', M', P be points corresponding to L, M, P', the column $L'M'$ will have for its base the elementary area $dy'dz'$, where $y'/b'=y/b$ and $z'/c'=z/c$. The coordinates

ξ, η, ζ of P are known in terms of those of P' by similar relations; see Art. 201. The attraction of the column $L'M'$ at P is

$$\rho\, dy'dz' \{f(PM') - f(PL')\}.$$

By Ivory's theorem, $P'M = PM'$, $P'L = PL'$, Art. 202; the x attractions of the columns LM, $L'M'$ are therefore in the ratio of the areas $dydz$, $dy'dz'$ of their bases, i.e. the x attractions are in the constant ratio bc to $b'c'$.

If we fill one ellipsoid with columns like LM, the other ellipsoid is filled by the corresponding columns, and the x attractions of the corresponding columns are in the same ratio. We therefore infer that

$$\frac{x \text{ att}^{\text{n}} \text{ of inner ellip}^{\text{d}} \text{ at } P'}{x \text{ att}^{\text{n}} \text{ of outer ellip}^{\text{d}} \text{ at } P} = \frac{bc}{b'c'}.$$

Similar theorems apply to the y and z components of the attractions of the two ellipsoids.

This theorem was enunciated and proved by Ivory in the *Phil. Trans.* for 1809. We ought perhaps to speak of it as Ivory's demonstration of Laplace's Theorem. But Ivory's own proof is not now exactly followed, as further simplifications have been introduced. The extension of the theorem to any law of force is due to Poisson, *Bulletin...la Société Philomathique* 1812, 1813.

223. *When the law of attraction is the inverse square*, the axial components of the attraction of the outer ellipsoid at the internal point P or (ξ, η, ζ) are

$$X = -A'\rho\xi, \quad Y = -B'\rho\eta, \quad Z = -C'\rho\zeta.$$

The axial components of the inner ellipsoid at the external point P' or (ξ', η', ζ') are therefore given by

$$X' = -A'\rho\xi\frac{bc}{b'c'} = -A'\rho\frac{abc}{a'b'c'}\xi', \; Y' = -B'\rho\frac{abc}{a'b'c'}\eta', \; Z' = -C'\rho\frac{abc}{a'b'c'}\zeta'.$$

Here a', b', c' are the semi-axes of the confocal drawn through the attracted point P', and A', B', C' are the same functions of the ratios of the axes a', b', c' that A, B, C in Art. 213 are of the ratios of a, b, c.

224. From these values of X', Y', Z' we may at once deduce a theorem often called Maclaurin's theorem. If we compare the attractions at the same point of two different ellipsoids bounded by confocals, we notice that a', b', c' are the same for each, so that each of the components X', Y', Z' is proportional to abc, i.e. to the product of the axes. *The attractions* therefore *at the same external point of different homogeneous ellipsoidal bodies bounded by confocals are the same in direction and their magnitudes are* *proportional to their masses. The law of attraction is that of the inverse square of the distance.*

Let V, V' be the potentials of the two ellipsoids at any external point P', then since the component attractions are proportional to the masses M, M', the ratios V/M and V'/M' can differ only by a quantity which is independent of the coordinates of P'. Since both potentials are zero at an infinite distance this constant must also be zero.

Hence *the potentials of two confocal solid homogeneous ellipsoids at any point external to both are proportional to their masses.* Since a focaloid is the difference of two confocal ellipsoids, it follows that *the potentials of thick focaloids* are also proportional to their masses.

225. *To find the potential V' of a solid homogeneous ellipsoid at an external point P' whose coordinates are ξ', η', ζ'*.*

Through the external point P' describe an ellipsoid confocal with the given ellipsoid. If the matter composing the given ellipsoid be made to fill the confocal (by changing the density from ρ to ρ') the attraction, and the potential, are unaltered at all external points. Let a', b', c' be the semi-axes of the confocal ellipsoid, then $\rho' a'b'c' = \rho abc$.

Since the point P' is on the surface of the confocal ellipsoid the potential is the same as that found in Art. 212 for an internal point. We therefore have by (9)

$$V' = \tfrac{1}{2}\rho \frac{abc}{a'b'c'} \{D' - A'\xi'^2 - B'\eta'^2 - C'\zeta'^2\},$$

where A', B', C', D' are the same functions of a', b', c' that A, B, C, D are of a, b, c. The potential may also be written in either of the two other forms given in Art. 211.

$$\therefore \frac{V'}{\pi\rho abc} = I' + \frac{dI'}{a'da'}\xi'^2 + \frac{dI'}{b'db'}\eta'^2 + \frac{dI'}{c'dc'}\zeta'^2$$

$$= \int_0^\infty \frac{du'}{Q'}\left\{1 - \frac{\xi'^2}{a'^2+u'} - \frac{\eta'^2}{b'^2+u'} - \frac{\zeta'^2}{c'^2+u'}\right\},$$

where $$Q'^2 = (a'^2+u')(b'^2+u')(c'^2+u').$$

* The expressions for the potentials of a homogeneous ellipsoid at an external and internal point were given by Rodrigues as early as 1815 (*Correspondance sur l'Ecole Royale Polytechnique*, vol. III.). An analysis of his method is given by Cayley in the *Quarterly Journal*, vol. II. 1858. There is a memoir by Poisson on the attraction of a homogeneous ellipsoid (*Memoires de l'Institut de France*, 1835) in which he gives a history. He finds the component attractions of the ellipsoid.

226. If we put $a'^2 = a^2 + \lambda$, $b'^2 = b^2 + \lambda$, $c'^2 = c^2 + \lambda$, and $u' = u - \lambda$, the last expression becomes

$$\frac{V'}{\pi\rho abc} = \int_\lambda^\infty \frac{du}{Q}\left\{1 - \frac{\xi'^2}{a^2+u} - \frac{\eta'^2}{b^2+u} - \frac{\zeta'^2}{c^2+u}\right\},$$

where $Q^2 = (a^2+u)(b^2+u)(c^2+u)$ and λ is to be found from the cubic of Art. 204.

To deduce the component forces X', Y', Z' we differentiate V' with regard to ξ', η', ζ' respectively. Since the lower limit λ is a function of the coordinates we shall here require the value of $dV'/d\lambda$. But the subject of integration is zero when we write λ for u, hence $dV'/d\lambda = 0$. We therefore have at once

$$\frac{X'}{\pi\rho abc} = \int_\lambda^\infty \frac{du}{Q}\frac{-2\xi'}{a^2+u} = -2\xi' \int_0^\infty \frac{du'}{(a'^2+u')\,Q'} = -\frac{A'\xi'}{\pi a'b'c'},$$

by Art. 212. This agrees with the result in Art. 223.

227. Ex. Let p, q, r be the lengths of the axial intercepts of any external level surface of a solid homogeneous ellipsoid $(a > b > c)$. Prove (1) that p is greater than q but less than qa/b, (2) that $p^2 - a^2$ is greater than $q^2 - a^2$ but less than $q^2 - b^2$.

Putting $V' = \pi\rho abcK$, the intercepts are given by

$$\int_{p^2-a^2}^\infty \left(1 - \frac{p^2}{a^2+u}\right)\frac{du}{Q} = \int_{q^2-b^2}^\infty \left(1 - \frac{q^2}{b^2+u}\right)\frac{du}{Q} = \&c. = K.$$

See Art. 226. If the inequalities to be proved were reversed it may be shown that these equations could not be true.

228. Some special cases. Ex. 1. The attraction of a thin homoeoid at any external point is the same as that of a thin disc bounded by its elliptic focal conic and having the surface density at any point P *inversely proportional to* $\left(1 - \frac{x^2}{a^2} - \frac{y^2}{b^2}\right)^{\frac{1}{2}}$, where (x, y) are the coordinates of the point P and $2a$, $2b$ the axes of the focal conic. Prove also that the level surfaces of the disc are confocal quadrics.

This follows from the theorem in Art. 206, since the disc may be regarded as a confocal homoeoid in which the axis c is evanescent. To find its law of density we notice that the mass on any elementary area $dxdy$ is $2\rho \frac{dz}{dc} dxdydc$. Now $\frac{dz}{dc} = \frac{c}{z}$ because $z^2 = c^2 - \frac{c^2}{a^2}x^2 - \frac{c^2}{b^2}y^2$, and, the surfaces of the disc being similar, c/a and c/b are constants. The masses being made equal, the result follows.

Ex. 2. The attraction of a solid ellipsoid at any external point is the same as that of a thin disc, of equal mass, bounded by its elliptic focal conic, axes $2a$, $2b$, and having its density at any point *directly proportional to* $\left(1 - \frac{x^2}{a^2} - \frac{y^2}{b^2}\right)^{\frac{1}{2}}$. Use Maclaurin's theorem, Art. 224.

Ex. 3. The attraction of a thin prolate spheroidal homoeoid at any external point is the same as that of a thin homogeneous straight rod joining the foci.

This result may be deduced from that given in Art. 224, but it follows more easily from Art. 131. The thin shell and the straight line have the same level surfaces (viz. confocal conicoids) and masses, hence their attractions are also the same.

Ex. 4. The attraction of a solid prolate spheroid at any external point is the same as that of a straight rod joining the foci, and having its line density at any point P proportional to $SP \cdot PH$.

Ex. 5. If V_s be the potential of a thin focaloid at an internal point P, prove

$$V_s = \frac{\delta v}{v} V - \pi\rho\lambda\left(1 - \frac{\xi^2}{a^2} - \frac{\eta^2}{b^2} - \frac{\zeta^2}{c^2}\right),$$

where v is the volume enclosed by the shell, δv that of the shell itself, V is the potential at the same point of the enclosed volume supposed to be of the same density as the shell itself, and λ is the difference of the squares of the semi-axes of the two boundaries of the shell. See Art. 195.

For a solid ellipsoid we have $\frac{V}{\pi\rho abc} = I + 2\frac{dI}{da^2}\xi^2 + \&c.$, as in Art. 211. To deduce the potential of a thin focaloid we find δV on the supposition that a^2, b^2, c^2 are each increased by the same quantity λ. This is evidently effected by performing on both sides of the equation, as it stands, the operation $\delta = \lambda\left(\frac{d}{dc^2} + \frac{d}{db^2} + \frac{d}{dc^2}\right)$. The result follows at once from Ex. 3, Art. 200.

Ex. 6. Show that the potential of a thin focaloid at an external point is $\frac{\delta v}{v} V$.

229. Mutual attraction. Ex. 1. A homogeneous ellipsoid attracts a body M according to the law of the inverse square; prove that if M be a spherical or cubical portion of the mass of the ellipsoid itself, the resultant attraction will be the same as if the mass M were collected at its centre of gravity. Prove also that if M be a segment of a thin exterior confocal ellipsoidal shell, and if its principal axes at its centre of gravity be parallel to the axes of the ellipsoid, the attraction of the ellipsoid on it will reduce to a single force through its centre of gravity.

[Math. Tripos.]

Ex. 2. A solid homogeneous ellipsoid is divided into two parts by a plane perpendicular to an axis. Prove that the mutual attraction of the parts for varying positions of the plane varies as the square of the area of section.

[May Exam. 1881.]

Ex. 3. Show that any plane divides a solid homogeneous ellipsoid into two parts such that the attraction between them reduces to a single force.

[Em. Coll. 1891.]

230. Elliptic coordinates. We may express the potential of an attracting ellipsoid at any internal or external point P in terms of its elliptic coordinates by using a geometrical theorem usually ascribed to Chasles.

Let a', a'', a''', be the semi-major axes of the three confocal quadrics which pass through the point ξ, η, ζ; let A, B, C be the semi-axes of any arbitrary confocal, then

$$(A^2 - a'^2)(A^2 - a''^2)(A^2 - a'''^2) = A^2B^2C^2\left\{1 - \frac{\xi^2}{A^2} - \frac{\eta^2}{B^2} - \frac{\zeta^2}{C^2}\right\}.$$

To apply this theorem, we put $A^2 = a^2 + u$, $B^2 = b^2 + u$, $C^2 = c^2 + u$ and substitute in the formulæ already found for the potential in Arts. 212 and 225. We thus find that the potential of a solid homogeneous ellipsoid at the point a', a'', a''', is given by

$$\frac{V}{\pi\rho abc} = \int \frac{du}{Q^3}(a^2 + u - a'^2)(a^2 + u - a''^2)(a^2 + u - a'''^2).$$

At an external point the limits are $u = a'^2 - a^2$ to $u = \infty$; at an internal point $u = 0$ to $u = \infty$.

231. Linear and quadratic layers. Ex. 1. If a thin layer of attracting matter, distributed over the surface of an ellipsoid, be such that the surface density ρ at any point (x, y, z) is $p(Lx+My+Nz)$, where p is the perpendicular on the tangent plane, prove (1) that *the axial components of the attraction at any internal point are constant* and respectively equal to La^2A, Mb^2B, Nc^2C, where A, B, C, have the meaning given in Art. 212 and (2) that *the potential is a linear function of the coordinates.*

To prove this we refer to Art. 92. Since the component attractions of a homogeneous ellipsoid at an internal point are $A\rho\xi$, $B\rho\eta$, $C\rho\zeta$, the potential of a thin superficial layer of surface density $\rho\cos\phi$ is $A\rho\xi$. Since $\cos\phi = px/a^2$, the potential of a layer of surface density pLx is $La^2A\xi$. The x component of attraction is therefore La^2A, while the y and z components are zero. It is evident from the symmetry of the law of density that the mass is zero. The potential is $La^2A\xi + Mb^2B\eta + Nc^2C\zeta$.

This example has an electrical meaning. An uncharged ellipsoid is placed in a field of uniform force, the direction cosines of the constant force being proportional to the arbitrary quantities La^2A, Mb^2B, Nc^2C. Since the resultant force due to the electricity and to the field must be zero at all internal points, the electrical density must be represented by $-\rho$. The result shows that the ratio ρ/p is a linear function of the coordinates.

In the same way we enquire in the next example what must be the field of force that ρ/p may be a quadratic function of the coordinates.

If the ellipsoid is charged with a quantity E of electricity, this quantity is to be so distributed over the surface that its attraction at any internal point is zero (Art. 68). The additional electrical surface density is therefore κp, where κ is such that the whole quantity is equal to E. By Art. 71 or 195 this gives $\kappa = E/4\pi abc$.

Ex. 2. If a thin layer of attracting matter, distributed on the surface of an ellipsoid, be such that the surface density at any point (x, y, z) is $pf(x, y, z)$, where f is a homogeneous quadratic function of (x, y, z), prove (1) that the potential at any internal point is also a quadratic function of the coordinates of that point together with a constant, and (2) that the axial components of the attraction at any internal point are linear functions of the coordinates of that point.

Let us regard the layer as occupying the space between two concentric ellipsoids having their axes nearly coincident in direction. The second ellipsoid is derived from the first by small rotations $\delta\theta$, $\delta\phi$, $\delta\psi$ round the axes and a change of the axes a, b, c into $a+\delta a$ &c. By choosing $\delta\theta$, $\delta\phi$, $\delta\psi$ and δa &c. properly, this thin layer may be made to represent the given quadratic distribution over the surface.

Consider first the rotation $\delta\psi$. The component displacements of a point Q are $\delta x = -y\delta\psi$, $\delta y = x\delta\psi$, $\delta z = 0$; the direction cosines of the normal at Q are $\lambda = px/a^2$ &c. The thickness of the layer is the sum of the projections on the normal. Omitting the factor $\delta\psi$, the surface density becomes $p\dfrac{a^2-b^2}{a^2b^2}xy$, and the potential of the shell at an internal point

$$\frac{dV}{d\psi} = x\frac{dV}{dy} - y\frac{dV}{dx} = (A-B)\,xy.$$

When the surface density is pxy/ab the potential becomes $2\pi abc\displaystyle\int\frac{ab\,.\,xy\,.\,du}{(a^2+u)(b^2+u)Q}$, by substituting for A, B their values, Art. 212. The upper limit is ∞ and the lower limit is zero or λ (Art. 226) according as the point is internal or external. See Art. 93, Ex. 2.

Consider next the change of a into $a+\delta a$. Let r be the radius vector measured from the centre. The thickness of the shell at Q, being the projection of δr on the normal, is $p\,\delta r/r$. But, since $1/r^2=l^2/a^2+\&c.$, where l, m, n are the direction cosines of r, we have $\delta r/r=x^2\delta a/a^3$. After omitting the factor $\delta a/a$ *the surface density of the shell becomes* $\rho x^2/a^2$ and its potential at P is $a\,dV/da$, where V is the potential of the solid ellipsoid at the same point. After substituting for V the potential becomes $2\pi abc\int\frac{du}{(a^2+u)\,Q}\left\{\frac{1}{2}u\left(1-\frac{\xi^2}{a^2+u}-\frac{\eta^2}{b^2+u}-\frac{\zeta^2}{c^2+u}\right)-\frac{a^2\xi^2}{a^2+u}\right\}$, where the limits are 0 to ∞ or λ to ∞ according as the point P is inside or outside the shell.

232. Elliptic cylinders. *To find the attraction at an internal point of a solid homogeneous cylinder whose cross section is an ellipse and whose length is infinite in both directions.*

The axial components of this attraction may be immediately deduced from those of an ellipsoid by making one of the axes infinite. Let us make $c=\infty$, so that the infinite cylinder stands on an ellipse whose axes are along the axes of x and y. The axial components of the attraction at any internal point (ξ, η, ζ) are
$X=-A\rho\xi$, $Y=-B\rho\eta$, $Z=0$, where $A=\int\frac{x^2}{a^2}d\omega$ and $B=\int\frac{y^2}{b^2}d\omega$.

Since in a cylinder (x, y) may be regarded as the coordinates of any point on the elliptic section, we have obviously

$$A+B=4\pi, \qquad Aa^2+Bb^2=\int r'^2 d\omega,$$

where r' is the radius vector of the cross section in the plane of xy. Putting for $d\omega$ its usual polar value $\sin\theta\, d\theta\, d\phi$ we have

$$\int r'^2 d\omega=\int\sin\theta\, d\theta\,.\int r'^2 d\phi,$$

where the limits are $\theta=0$ to π and $\phi=0$ to 2π. The first integral is obviously equal to 2 and the second integral is twice the area of the ellipse, i.e. $2\pi ab$. We thus have $Aa^2+Bb^2=4\pi ab$. The axial components are therefore

$$X=-4\pi\rho\frac{ab}{a+b}\frac{\xi}{a}, \qquad Y=-4\pi\rho\frac{ab}{a+b}\frac{\eta}{b}.$$

233. The potential also may be deduced from that for an ellipsoid, Art. 212. After substituting the values of A, B, and putting $C=0$, we find

$$V=\tfrac{1}{2}\rho D-2\pi\rho\frac{ab}{a+b}\left(\frac{\xi^2}{a}+\frac{\eta^2}{b}\right)\ldots\ldots\ldots\ldots\ldots(1).$$

Since the potential of the cylinder is infinite at points on its axis, Art. 50, it is evident that D is an infinite constant which may be omitted when the axes a, b are not varied. This expression for V may also be obtained by integrating the expressions $dV/d\xi = X,\ dV/d\eta = Y.$

The level surfaces inside the attracting cylinder are similar and similarly situated concentric cylinders. Considering a cross section we deduce from (1) that the longest axis of a level surface is in the direction of the longest axis of the attracting cylinder. See Art. 216.

234. Ex. If in a spheroid the axis of revolution c is very great, the spheroid becomes a cylinder. Prove that C and Cc are ultimately zero, while Cc^2 is infinite, Art. 219.

235. *To find the attraction at an external point of a solid homogeneous elliptic cylinder.*

The attraction at an external point may be deduced from that at an internal point by an application of either Ivory's or Maclaurin's theorem. Let $a'b'c'$ be the semi-axes of a confocal through the attracted point P'. Then $a'^2 - a^2 = b'^2 - b^2 = c'^2 - c^2$. Since $a'^2 - a^2$ is finite it follows that when c and c' are both infinite their ratio is unity. Since the components of attraction of the given cylinder and the confocal are proportional to their masses, we have (as in Art. 224)

$$X' = -A'\rho\frac{abc}{a'b'c'}\xi' = -4\pi\rho\frac{b'}{a'+b'}\frac{ab}{a'b'}\xi',$$

by substituting for A' its value found in Art. 232.

In this way we find that the axial components X', Y', Z' of the attraction of a solid cylinder at an external point (ξ', η', ζ') are

$$X' = -4\pi\rho\frac{ab}{a'+b'}\frac{\xi'}{a'},\qquad Y' = -4\pi\rho\frac{ab}{a'+b'}\frac{\eta'}{b'},\qquad Z' = 0,$$

where (a', b') are the semi-axes of a cross section of a confocal cylinder drawn through the attracted point.

236. Ex. 1. Show that the resultant attraction of an infinite cylinder is the same in magnitude at all *internal* points situated on a coaxial cylinder *similar and similarly situated to the boundary*. Show also that the direction of the attraction at any point on the surface of such a cylinder is parallel to the eccentric line of that point.

Ex. 2. Show that the resultant attraction of an infinite cylinder is the same in magnitude at all *external* points situated on a cylinder *confocal with the boundary*. Show also that its direction at any point on a confocal is parallel to the eccentric line of that point.

Ex. 3. If a thin stratum of attracting matter distributed on the surface of an infinite elliptic cylinder be such that the surface density ρ at any point (x, y, z) is $p\left(L\frac{x}{a} + M\frac{y}{b} + N\right)$, prove that the axial components of the attraction at an internal point (ξ, η, ζ) are $X = L\frac{4\pi ab}{a+b}$, $Y = M\frac{4\pi ab}{a+b}$, $Z = 0$, where the coordinate axes are the principal diameters of a cross section and the axis of the cylinder.

This result has an electrical meaning. If the electrical density on the surface of the elliptic cylinder be represented by $-\rho$, the electricity will be in equilibrium when the system is placed in a field of uniform force whose components are X, Y, Z; see also Art. 231, Ex. 1.

Ex. 4. If the surface density ρ of a thin stratum of attracting matter placed on the surface of an infinite elliptic cylinder be given by $\rho=p\left(L\frac{x^2}{a^2}+M\frac{xy}{ab}+N\frac{y^2}{b^2}\right)$, prove that the x component of the attraction at any internal point (ξ, η) is $X=\frac{4\pi ab}{(a+b)^2}\{(L-N)\xi+M\eta\}$, with a similar expression for the y component.

Ex. 5. Show that the potential at an internal point of an infinite cylindrical mass bounded by two coaxial cylinders is infinite. Art. 50.

Ex. 6. The components of the attraction of a right elliptic cylinder whose section is $(x/a)^2+(y/b)^2=1$, and whose ends are any two planes perpendicular to the axis, at an external point ξ', η', ζ', are X', Y', Z'. A confocal cylinder having the same ends is described through ξ', η', ζ', and attracts an internal point ξ, η, ζ, with components X, Y, Z. Show that if $\xi/a=\xi'/a'$, $\eta/b=\eta'/b'$, $\zeta=\zeta'$, then $X'/X=b'/b'$, $Y'/Y=a/a'$. [Math. T. 1879.]

237. *To find the potential at an external point of an elliptic cylinder* we use Maclaurin's theorem.

Let V, V' be the potentials of two confocal cylinders whose semi-axes are respectively a, b and a', b'. Since their component attractions at all external points are proportional to their masses, we must have

$$V=\frac{ab}{a'b'}V'+E,$$

where the constant E is independent of the coordinates ξ', η' of the attracted point but may be a function of the axes of either cylinder.

Let the external cylinder (a', b') pass through the attracted point P', then by Art. 233

$$\frac{V'}{2\pi\rho}=-\frac{a'b'}{a'+b'}\left(\frac{\xi'^2}{a'}-\frac{\eta'^2}{b'}\right)+\frac{D}{4\pi},\quad \therefore \frac{V}{2\pi\rho}=-\frac{ab}{a'+b'}\left(\frac{\xi'^2}{a'}+\frac{\eta'^2}{b'}\right)+E',$$

where D and E' are independent of ξ', η' but are functions of a', b'. Let $2f$ be the distance between the foci of the given elliptic cylinder, then $a'^2-b'^2=f^2$.

To find E', we place the attracted point on the axis of x, then $\xi'=a'$ and $\eta'=0$. By Art. 235 we have

$$\frac{dV}{da'}=X=-4\pi\rho\frac{ab}{a'+b'}=-4\pi\rho\frac{ab}{f^2}(a'-b');$$

after substituting $b'=\sqrt{(a'^2-f^2)}$, we find by an easy integration

$$V=-2\pi\rho\frac{ab}{f^2}\{a'^2-a'\sqrt{(a'^2-f^2)}+f^2\log(a'-b')\}+G,$$

where the constant G of integration is independent of a', b'.

Comparing the two values of V we see that $E' = -ab\log(a'+b')$.

Hence $$V = -2\pi\rho\frac{ab}{a'+b'}\left(\frac{\xi'^2}{a'}+\frac{\eta'^2}{b'}\right) - 2\pi\rho ab\log(a'+b') + G.$$

238. The four variables ξ', η', a', b' in the expression for V are connected by the relations $\xi'^2/a'^2+\eta'^2/b'^2=1$ and $a'^2-b'^2=f^2$. It is often convenient to reduce these to two independent coordinates. Let

$$\xi'=a'\cos\theta,\ \eta'=b'\sin\theta;\ a'=\tfrac{1}{2}f(e^{\phi}+e^{-\phi}),\ b'=\tfrac{1}{2}f(e^{\phi}-e^{-\phi}).$$

The value of ϕ determines the particular confocal elliptic cylinder on which the attracted particle P lies, and θ (being the eccentric angle) determines the position of P on that cylinder. Substituting we find*

$$V = -\pi\rho ab\,(e^{-2\phi}\cos 2\theta + 2\phi) + G,$$

where some constant terms, functions of a, b, have been included in the infinite constant G.

239. Heterogeneous ellipsoid, similar strata. *To find the potential of an ellipsoidal shell whose strata of equal density are similar to, and concentric with, a given ellipsoid.*

Let a, b, c be the semi-axes of the given ellipsoid; ma, mb, mc, $(m+dm)a$ &c. those of the inner and outer boundaries of an elementary homoeoid. If (x, y, z) be any point on this homoeoid, the value of m is given by $$\frac{x^2}{a^2}+\frac{y^2}{b^2}+\frac{z^2}{c^2}=m^2 \quad\text{.................. (1).}$$

Let the density at any point (xyz) of this homoeoid be $\rho = f(m^2)$. The mass of the element is therefore $f(m^2)\,.\,4\pi abc m^2 dm$.

The potential of this homoeoid at any point P is (by Art. 208)

$$2\pi abc m^2 dm f(m^2)\int_{\lambda'}^{\infty}\frac{dv}{(m^2a^2+v)^{\frac{1}{2}}(m^2b^2+v)^{\frac{1}{2}}(m^2c^2+v)^{\frac{1}{2}}} \quad \text{...(2),}$$

where the lower limit λ' is 0 when the point P is internal and is the greatest root of the cubic

$$\frac{\xi^2}{m^2a^2+\lambda'}+\frac{\eta^2}{m^2b^2+\lambda'}+\frac{\zeta^2}{m^2c^2+\lambda'}=1 \quad\text{............(3),}$$

when P is external (Art. 204). The potential of the heterogeneous shell may be obtained by integrating (2). To simplify the integration we put $v=m^2u$, $\lambda'=m^2\mu$. The potential V of the shell is then given by $$\frac{V}{\pi abc}=\int 2m\,dm f(m^2)\int_{\mu}^{\infty}\frac{du}{Q} \quad\text{.................(4),}$$

where $Q^2=(a^2+u)(b^2+u)(c^2+u)$, and μ is zero, or the greatest root of $$\frac{\xi^2}{a^2+\mu}+\frac{\eta^2}{b^2+\mu}+\frac{\zeta^2}{c^2+\mu}=m^2 \quad\text{.................(5),}$$

according as P is internal or external. The limits for m depend on the internal and external boundaries of the shell.

* See a paper by Prof. Lamb in the *Messenger of Mathematics*, 1878.

First, *let us find the potential V_1 at an external point of the mass enclosed by the ellipsoid defined by $m=n$.* To effect this we change the order of integration in (4).

To make this process clear we trace the curve AB whose ordinate u is given as a function of the abscissa m by the equation

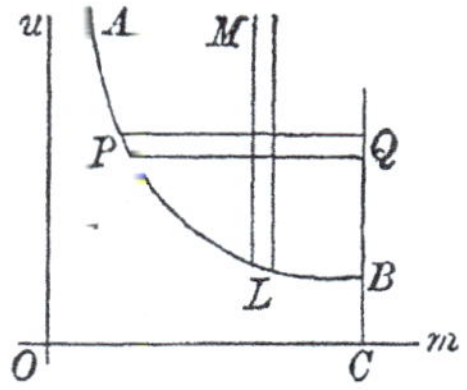

$$\frac{\xi^2}{a^2+u}+\frac{\eta^2}{b^2+u}+\frac{\zeta^2}{c^2+u}=m^2 \quad \text{............(6)},$$

say $\phi(u)=m^2$. When $m=0$, u is infinite, and when $m=n$, u has the value ϵ given by

$$\frac{\xi^2}{a^2+\epsilon}+\frac{\eta^2}{b^2+\epsilon}+\frac{\zeta^2}{c^2+\epsilon}=n^2 \quad \text{............(7)}.$$

In the order of integration indicated in (4) we integrate along a column LM from $u=\mu$ to $u=\infty$ and then sum the columns from $m=0$ to $m=OC=n$. In the reversed order we integrate first along a row PQ from $m^2=\phi(u)$ to $m^2=n^2$ and then sum the rows from $u=CB=\epsilon$ to $u=\infty$.

The equation (4) then becomes

$$\frac{V_1}{\pi abc}=\int_\epsilon^\infty \frac{du}{Q}\int_{m^2}^{n^2} f(m^2)\,dm^2=\int_\epsilon^\infty \frac{du}{Q}\{f_1(n^2)-f_1(m^2)\} \ldots (8),$$

where $Q^2=(a^2+u)(b^2+u)(c^2+u)$, m^2 is given as a function of u by (6) and ϵ by (7). This formula gives the potential of the mass enclosed by the ellipsoid defined in equation (1) by putting $m=n$. If the attracted point P is on the surface of this ellipsoid, we have, by (7), $\epsilon=0$. *If the potential of the whole mass enclosed by the ellipsoid x^2/a^2+ &c. $=1$ be required, we have $n=1$ and $\epsilon=\lambda$,* where λ is defined in Art. 204.

240. Secondly, *let us find the potential V_2 at an internal point of the mass between the ellipsoids defined by $m=n$ and $m=n'$.* The limits for the integral (4) are now $u=0$ to ∞ and $m=n$ to n'. These are constants and the order can be immediately reversed. The potential is therefore given by

$$\frac{V_2}{\pi abc}=\{f_1(n'^2)-f_1(n^2)\}\int_0^\infty \frac{du}{Q} \quad \text{..............(9)}.$$

241. Lastly, *let us find the potential V_3 at a point P situated in the substance of a solid ellipsoid bounded by the surface $m=n'$.* Let the point P be situated on the ellipsoid defined by $m=n$. The potentials at P of the two portions of the solid separated by this ellipsoid are given by the values V_1, V_2 found above. We find V_3 by adding (8) and (9) together.

$$\therefore \frac{V_3}{\pi abc}=\int_0^\infty \frac{du}{Q}\{f_1(n'^2)-f_1(m^2)\} \quad \text{..........(10)},$$

where $Q^2 = (a^2+u)(b^2+u)(c^2+u)$, and m^2 is given by (6) as a function of u.

One result may be briefly stated, as follows. *The potential V at P of the solid ellipsoid $x^2/a^2 + \&c. = 1$ is given by*

$$\frac{V}{\pi abc} = \int \frac{du}{Q} \{f_1(1) - f_1(m^2)\} \quad \text{..............(11)},$$

where the limits are 0 to ∞ when P is internal and λ to ∞ when P is external. The value of λ is given in Art. 204, and

$$m^2 = \frac{\xi^2}{a^2+u} + \frac{\eta^2}{b^2+u} + \frac{\zeta^2}{c^2+u} \quad \text{..............(12)}.$$

242. *The component forces X, Y, Z due to the attraction of the solid ellipsoid* may be found by differentiating the expression for V just found. Since both m^2 and λ are functions of the coordinates we must find $dV/d\lambda$ and dV/dm^2. When $u=\lambda$, m^2 becomes unity and the subject of integration vanishes. Hence $dV/d\lambda$ is zero,

$$\therefore \frac{X}{\pi abc} = \int_\lambda^\infty - f(m^2) \frac{dm^2}{d\xi} \frac{du}{Q} = - \int_\lambda^\infty \frac{2\rho\xi}{a^2+u} \frac{du}{Q}.$$

The corresponding values of Y, Z follow at once. For an internal point $\lambda = 0$.

243. The expression (11) for the potential of a solid ellipsoid may be put into another form in which the limits are constant by putting $u = v + \lambda$, $a^2 + \lambda = a'^2$ &c. Writing the formula at length we have at an external point

$$\frac{V}{\pi abc} = \int_0^\infty \left\{ f_1(1) - f_1\left(\frac{\xi^2}{a'^2+v} + \frac{\eta^2}{b'^2+v} + \frac{\zeta^2}{c'^2+v}\right)\right\} \frac{dv}{(a'^2+v)^{\frac{1}{2}}(b'^2+v)^{\frac{1}{2}}(c'^2+v)^{\frac{1}{2}}}.$$

The axes of the attracting ellipsoid have disappeared from the right-hand side and are replaced by the axes a', b', c' of the confocal which passes through the attracted point.

244. Ex. The density at any internal point T of an ellipsoid is $k.OR/OT$, where OR is the semidiameter which passes through T and k is a constant. Prove that the integrations to find X, Y, Z can be effected in finite terms.

Prove also that the axial components are the same, at all internal points on any given radius vector.

The last result is proved by noticing that before integration each component is a homogeneous function of ξ, η, ζ of zero dimensions. It should also be remarked that though the density at the centre is infinite the components X, Y, Z are finite. Poisson, *Connaissance &c.*, 1837.

245. If we write $f(m^2) = A(1-m^2)^n$, the density ρ at (x, y, z) of the solid ellipsoid, and the potential V at (ξ, η, ζ), become

$$\rho = A\left(1 - \frac{x^2}{a^2} - \frac{y^2}{b^2} - \frac{z^2}{c^2}\right)^n, \quad V = A\frac{\pi abc}{n+1}\int\left\{1 - \frac{\xi^2}{a^2+u} - \frac{\eta^2}{b^2+u} - \frac{\zeta^2}{c^2+u}\right\}^{n+1}\frac{du}{Q},$$

where the limits are λ to ∞ for an external point and 0 to ∞ for an internal point, Art. 204.

Since the density is zero at the surface of the ellipsoid, it follows from Art. 92 that we may differentiate each of these expressions, so that if the density is $\rho' = d\rho/dx$, the potential is $V' = dV/d\xi$. This enables us to find the potential of a heterogeneous ellipsoid when the density is x, xy, x^2, &c.

Let $$E = 1 - \frac{x^2}{a^2} - \frac{y^2}{b^2} - \frac{z^2}{c^2}, \quad R = 1 - \frac{\xi^2}{a^2+u} - \frac{\eta^2}{b^2+u} - \frac{\zeta^2}{c^2+u},$$

then $$x = -\frac{a^2}{2}\frac{dE}{dx}, \qquad xy = \frac{a^2b^2}{8}\frac{d^2E^2}{dx\,dy}, \qquad x^2 = \frac{a^4}{8}\frac{d^2E^2}{dx^2} + \frac{a^2}{2}E.$$

The potentials for these three laws of density are

$$\rho = Ax, \qquad V = \pi a^3 bcA\xi \int \frac{R\,du}{(a^2+u)\,Q},$$

$$\rho = Axy, \qquad V = \pi a^3 b^3 cA\xi\eta \int \frac{R\,du}{(a^2+u)\,(b^2+u)\,Q},$$

$$\rho = Ax^2, \qquad V = \pi a^3 bcA \int \left\{\tfrac{1}{4}uR^2 + \frac{a^2\xi^2 R}{a^2+u}\right\} \frac{du}{(a^2+u)\,Q}.$$

The proper forms for the three following laws of density may be found by differentiating E^3. We then have

$$\rho = Ax^2y, \qquad V = \pi a^3 b^3 cA\eta \int \left\{\tfrac{1}{4}uR^2 + \frac{a^2\xi^2 R}{a^2+u}\right\} \frac{du}{(a^2+u)\,(b^2+u)\,Q},$$

$$\rho = Axyz, \qquad V = \pi a^3 b^3 c^3 A\xi\eta\zeta \int \frac{R\,du}{Q^3},$$

$$\rho = Ax^3, \qquad V = \pi a^5 bcA\xi \int \left\{\tfrac{3}{4}uR^2 + \frac{a^2\xi^2 R}{a^2+u}\right\} \frac{du}{(a^2+u)^2\,Q}.$$

The case in which $\rho = Ax^f y^g z^h$ is considered by Ferrers in the *Quarterly Journal*, 1877, vol. XIV.

It will presently be proved that the potentials of a homoeoid, whose surface density σ is numerically equal to $p\rho$ (where p is the perpendicular on the tangent plane), may be deduced from that of the solid ellipsoid by differentiating with regard to R and doubling the result, Art. 249.

The potentials for a homoeoid are therefore

$$\sigma = Apx, \qquad V = 2\pi a^3 bcA\xi \int \frac{du}{(a^2+u)\,Q},$$

$$\sigma = Apxy, \qquad V = 2\pi a^3 b^3 cA\xi\eta \int \frac{du}{(a^2+u)\,(b^2+u)\,Q},$$

$$\sigma = Apx^2, \qquad V = 2\pi a^3 bcA \int \left\{\frac{a^2\xi^2}{a^2+u} + \tfrac{1}{2}uR\right\} \frac{du}{(a^2+u)\,Q}.$$

The limits are 0 to ∞ for an internal point and λ to ∞ for an external point. See Art. 204.

These agree with the results obtained in Art. 231 by an elementary method.

246. We may use *the method of Art.* 211 *to find the potential at any internal point* P, (ξ, η, ζ), *of a heterogeneous ellipsoid whose density at any point* Q *is* $\phi(x, y, z)$.

We describe as before a double cone with its vertex at P cutting the ellipsoid in two elementary areas Q_1, Q_2. The distances $PQ_1 = R_1$, $PQ_2 = -R_2$ are given by the quadratic (2) of Art. 211. Let $PQ = R$.

The density ρ at any point Q of the cone is

$$\rho=\phi(\xi+lR,\ \eta+mR,\ \zeta+nR)$$
$$=\phi+R\,\delta\phi+\ldots+R^i\delta^i\phi/L(i)+\ldots \quad\ldots\ldots(1),$$

where $\delta=ld/d\xi+md/d\eta+nd/d\zeta$ and $L(i)=1\,.\,2\ldots i$.

The potential at P of the positive side of the double cone is $\Sigma\int R^{i+1}dR\,d\omega\,\delta^i\phi/L(i)$, where the limits are $R=0$ to $R=R_1$ and Σ implies summation for all existing values of i. To find the potential of the negative side of the double cone we let R be the distance of one of its elements Q' taken positively. The density at Q' is found by writing $-R$ for R in the series for ρ. The potential at P of the negative side of the cone is therefore $\Sigma\int R^{i+1}dR(-1)^i d\omega\,\delta^i\phi/L(i)$, and the limits are $R=0$ to $R=-R_2$.

Taking the two conical elements together we find for the potential of the ellipsoid

$$V=\iint d\omega\,\Sigma\,\frac{S_{i+2}}{i+2}\frac{\delta^i\phi}{L(i)},\qquad S_{i+2}=R_1^{i+2}+R_2^{i+2}\ \ldots\ldots(2).$$

As we shall integrate all round the point P, every element is taken twice over, *we must* therefore *halve the value of V thus found.*

The quantity S_i is a symmetrical function of the roots of the quadratic (2) of Art. 211. We therefore have $S_{i+2}+2Fr^2S_{i+1}-Er^2S_i=0$ ……(3).
The initial terms are $S_0=2$, $S_1=-2Fr^2$. Assume that S_i and S_{i+1} contain only terms of the form $H_\kappa r^{\kappa+i}$ and $H_\kappa' r^{\kappa+i+1}$, where H_κ represents a homogeneous function of (l, m, n) of κ dimensions. It follows that S_{i+2} contains terms of the same forms, viz. $H_\kappa r^{\kappa+i+2}$, $H'_{\kappa+1}r^{\kappa+i+3}$. Again $\delta^i\phi$ is a function of l, m, n of i dimensions, hence $S_{i+2}\delta^i\phi$ is of the form $H_{\kappa+i}r^{\kappa+i+2}$. *The determination of V is therefore reduced to the integration of expressions in which the index of r exceeds by 2 the sum of the indices of l, m, n.*

The terms of (2) which contain any odd exponents of l, m, n give zero after integration, as in Art. 211. Omitting these it is clear that every term of V is of the form $\xi^{f_1}\eta^{g_1}\zeta^{h_1}$, where
$$u=\iint\frac{l^{2f}m^{2g}n^{2h}d\omega}{(\alpha l^2+\beta m^2+\gamma n^2)^{f+g+h+1}},$$
and α, β, γ are the reciprocals of a^2, b^2, c^2. Now by Art. 197

$$\int\frac{d\omega}{\alpha l^2+\beta m^2+\gamma n^2}=2\pi abc\int\frac{du}{Q}=2\pi\int\frac{v^{-\frac{1}{2}}dv}{Q_1},$$

where $Q^2=(a^2+u)(b^2+u)(c^2+u)$, $Q_1{}^2=(\alpha+v)(\beta+v)(\gamma+v)$ and $v=1/u$ and the limits for u and v are 0 to ∞. Differentiating this, we find

$$u=\frac{(-)^{f+g+h}}{L(f+g+h)}\left(\frac{d}{d\alpha}\right)^f\left(\frac{d}{d\beta}\right)^g\left(\frac{d}{d\gamma}\right)^h 2\pi\int\frac{v^{-\frac{1}{2}}dv}{Q_1}$$
$$=\frac{[1\,.\,3\,.\,5\ldots(2f-1)][1\,.\,3\ldots(2g-1)][1\,.\,3\ldots(2h-1)]}{2^{f+g+h}L(f+g+h)}\int\frac{2\pi v^{-\frac{1}{2}}dv}{Q_1(\alpha+v)^f(\beta+v)^g(\gamma+v)^h},$$

where the limits are $v=0$ to ∞.

The remaining integrations cannot be effected. The potential has thus been found expanded in powers of the coordinates ξ, η, ζ of the attracted point, with single integrals with regard to v for coefficients. The several powers of ξ, η, ζ may be collected together, once for all. We then arrive at the formula given in Art. 247. The algebraic process of collection is however tedious when the density ρ contains high powers of x, y, z. It is given at length in the *Phil. Trans.* 1895, vol. CLXXXVI.

Ex. 1. The density of a solid ellipsoid at any point Q is a homogeneous function of i dimensions of the coordinates of Q. Prove that the potential of the ellipsoid at an internal point P is the sum of a series of homogeneous functions of the coordinates of P of the dimensions $i+2$, i, $i-2$, &c.

[This may be deduced from the equation (3) by noticing that S_0 and S_1 are respectively of zero and one dimension. Thence by induction the dimensions of S_i in terms of the coordinates can be found.]

Ex. 2. The density of a homoeoid is Ax and the law of force is the inverse fourth power of the distance. Prove that the potential at an internal point is $\frac{2}{3}\pi A \frac{da}{a} \cdot \frac{2x}{E}$.

247. Two theorems. Let the density of a solid heterogeneous ellipsoid (when $\kappa > 0$) be

$$\rho = A\left(1 - \frac{x^2}{a^2} - \frac{y^2}{b^2} - \frac{z^2}{c^2}\right)^{\kappa-1} \phi\left(\frac{x}{a}, \frac{y}{b}, \frac{z}{c}\right).$$

Let
$$R = 1 - \frac{\xi^2}{a^2+u} - \frac{\eta^2}{b^2+u} - \frac{\zeta^2}{c^2+u},$$

$$D = \frac{a^2+u}{a^2}\frac{d^2}{d\xi^2} + \frac{b^2+u}{b^2}\frac{d^2}{d\eta^2} + \frac{c^2+u}{c^2}\frac{d^2}{d\zeta^2},$$

$$M_\kappa = \frac{R^\kappa}{\kappa} + \frac{R^{\kappa+1}}{\kappa(\kappa+1)} \cdot \frac{uD}{L(1).2^2} + \ldots + \frac{R^{\kappa+n}}{\kappa(\kappa+1)\ldots(\kappa+n)} \cdot \frac{u^n D^n}{L(n)2^{2n}} + \&c.,$$

$Q^2 = (a^2+u)(b^2+u)(c^2+u)$ and $L(n) = 1.2.3 \ldots n$.

The potential of the ellipsoid at any point (ξ, η, ζ) is

$$V = \pi abcA \int \frac{du}{Q} M_\kappa \phi\left(\frac{a\xi}{a^2+u}, \frac{b\eta}{b^2+u}, \frac{c\zeta}{c^2+u}\right) \ldots\ldots(\text{I}),$$

where the limits are $u = 0$ to ∞ for an internal point, and $u = \lambda$ to ∞ for an external point, Art. 204.

Let the surface density of a thin homoeoid be

$$\sigma = Ap\psi\left(\frac{x}{a}, \frac{y}{b}, \frac{z}{c}\right).$$

$$M' = \frac{dM_1}{dR} = 1 + \frac{RuD}{1.2^2} + \ldots + \frac{R^n u^n D^n}{[L(n)]^2 2^{2n}} + \&c.$$

The potential at any point (ξ, η, ζ) is

$$V = 2\pi abcA \int \frac{du}{Q} M'\psi\left(\frac{a\xi}{a^2+u}, \frac{b\eta}{b^2+u}, \frac{c\zeta}{c^2+u}\right) \ldots\ldots(\text{II}),$$

where the limits are 0 to ∞ or λ to ∞, according as the attracted point is inside or outside.

The advantages of these formulæ* are (1) that the only differentiations to be performed are those on the expression for

* These formulæ were first given in this form by Dyson in the *Quarterly Journal*, 1891, vol. xxv. By computing the potentials of a homoeoid for several laws of density he discovered by induction the forms assumed by the potential when the density is $Ax^f y^g z^h$. Assuming the potentials to be known he deduces the attracting body by reasoning similar to that given in Art. 164. He gives the necessary differentiations at length.

the density, and (2) that most of the terms containing ξ, η, ζ have been collected together and expressed in powers of the function R.

The component attractions at any point may be deduced from these potentials by differentiation with regard to ξ, η, ζ. When the attracted point is internal the limits are absolute constants and we merely differentiate the subject of integration. When the point is external, the lower limit λ is a function of ξ, η, ζ (Art. 204) which makes $R=0$ when $u=\lambda$. Hence (as in Art. 226) we may treat λ as a constant during the differentiation, except in the first term of M'.

248. To prove the two theorems* in Art. 247 we shall adopt the method described in Art. 164. We assume as given the two expressions for the potential of a homoeoid at an internal and external point, and we shall deduce the attracting body.

Let the potentials at an internal and external point be distinguished as V and V'. Then since $\lambda=0$ at the surface of the ellipsoid we have $V=V'$ at all points of the separating ellipsoid. It is also evident by inspection that V' is zero at infinitely distant points.

The expressions for V, V' are found by actual differentiation to satisfy Laplace's equation, Art. 95. As these differentiations with regard to ξ, η, ζ, present no peculiar difficulties but lead to long algebraical processes they will not be reproduced here. We infer however from the result that the attracting matter resides solely on the separating ellipsoid, Art. 164.

Let σ be the surface density of the separating stratum. If dn

* The potentials of heterogeneous ellipsoids and shells have been investigated in several ways. First there is Green's paper, *Camb. Phil. Soc.* 1833, where the law of attraction is the inverse κth and the density $E^n f(x, y, z)$ where $E=1-x^2/a^2-$&c. Green uses Cartesian coordinates, but a solution by means of Lamé's functions is given in Ferrers' *Spherical Harmonics*. The values of X, Y, Z given in Art. 242 are due to Poisson, *Connaissance des Temps*, 1837 (published three years earlier). He begins with the formulæ for the component attractions of a homogeneous ellipsoid which he had obtained in 1835 (*Mém. de l'Institut de France*). Cayley gives a formula for the potential of certain heterogeneous ellipsoids in his memoir on Prepotentials, *Phil. Trans.* 1875, which is really an extension of the theorem of Art. 239 to the case in which the force varies as the inverse κth power of the distance. In the *Quarterly Journal*, 1877, Ferrers applies these results to determine the potential of a solid ellipsoid whose density is $x^f y^g z^h$, see Art. 245. He also first discovered the rule to find the potential of a shell by differentiating with regard to R. His proof is different from that in Art. 249. In vol. xxv. of the *Quarterly Journal* Dyson gave the important formulæ mentioned in Art. 247. There are other valuable papers by W. D. Niven, *Phil. Trans.* 1879, and by Hobson, *London Math. Soc.* 1893. There is also a paper by the author on these potentials, *Phil. Trans.* 1895. A second memoir is given by Hobson in the *Lond. Math. Soc.* 1896. There is also a paper by G. Prasad in the *Messenger of Math.* 1900. Most of these give the applications to discs and laminæ and assume that the law of force is the inverse κth power.

be an element of the normal to the ellipsoid drawn outwards we have (since V, V' differ only in the limit λ)

$$-4\pi\sigma = \frac{dV'}{dn} - \frac{dV}{dn} = \frac{dV'}{d\lambda}\frac{d\lambda}{dn} \quad\text{..............(1).}$$

Now $\lambda = 0$ at every point of the ellipsoid and $\xi = x$, &c., hence

$$\frac{dV'}{d\lambda} = -2\pi abc \frac{A}{abc}\psi\left(\frac{x}{a}, \frac{y}{b}, \frac{z}{c}\right).$$

Again λ and p are given by the equations

$$\frac{x^2}{a^2+\lambda} + \&c. = 1, \quad \frac{x^2}{a^4} + \&c. = \frac{1}{p^2}.$$

Differentiate the former and put $\lambda = 0$, we have

$$\frac{1}{p^2}\frac{d\lambda}{dn} = \frac{2x}{a^2}\frac{dx}{dn} + \&c. = \frac{2}{p},$$

since the direction cosines dx/dn, &c., of the normal are px/a^2, &c. Substitute these values in (1) and we find that $\sigma = Ap\psi(x/a, \&c.)$. This is therefore the density of the stratum which produces the potentials V, V'.

249. The potential of *a solid ellipsoid whose density is* $\rho = Ax^fy^gz^h$ *being known* we can immediately deduce that of *a thin homoeoid having the same law of density.* We write $a = ma_1$, $b = mb_1$, $c = mc_1$, $u = m^2u_1$ and then differentiate with regard to m. The thickness dp of the homoeoid thus obtained is given by $dp/p = d(ma_1)/mc_1 = dm/m$, Art. 195. The surface density $\sigma = \rho\, dp = \rho p\, dm/m$. After the differentiation has been effected it is convenient to put $m = 1$, so that a_1, b_1, c_1, u_1 become again a, b, c, u. We may also omit the factor dm and regard the homoeoid as a layer of finite density $\sigma = \rho p$. It is supposed that A is independent of the axes a, b, c.

The potential of the solid ellipsoid becomes after these changes have been made

(since $\kappa = 1$) $\quad V = \pi a_1 b_1 c_1 A \int_{\lambda_1}^{\infty} \frac{du_1}{Q_1}(m^2M_1)\,\xi^f\eta^g\zeta^h \frac{a_1^{2f}b_1^{2g}c_1^{2h}}{(a_1^2+u_1)^f(b_1^2+u_1)^g(c_1^2+u_1)^h}.$

The operator D is unaltered, $\lambda = \lambda_1 m^2$, and

$$m^2R^{n+1}u^n = \left(m^2 - \frac{\xi^2}{a_1^2+u_1} - \&c.\right)^{n+1} u_1^n = S^{n+1}u_1^n,$$

where S represents the quantity in brackets. Since $R = 0$ when $u = \lambda$ and therefore $S = 0$ when $u_1 = \lambda_1$, we may as before treat λ_1 as constant when differentiating with regard to m. We now see that m enters into the expression for V only implicitly through S. The differential coefficient of V is therefore $2m\,dV/dS$. When $m = 1$, $S = R$ and exactly replaces R in the formula for V, hence $dV/dm = 2dV/dR$. *The potential of the homoeoid is therefore found by differentiating that of the solid ellipsoid with regard to R and doubling the result.*

250. The potential of *a heterogeneous homoeoid whose surface density is* $\sigma = Apx^fy^gz^h$ being known, that of *a solid ellipsoid whose density is*

$$\rho = (1 - x^2/a^2 - \&c.)^{\kappa-1}x^fy^gz^h$$

can be deduced by integration. Let a, b, c be the semi-axes of the ellipsoid whose potential is required, ma, &c., $(m+dm)a$, &c. those of an elementary homoeoid as described in Art. 239. If ρ' be the density of the homoeoid, its surface density is $\rho' p\, dm/m$. The potential of this element may be found by writing ma, &c., m^2u, $m^2\lambda_1$ for a, &c., u, λ in the general expression (II.) of Art. 247, and multiplying the density by $a^f b^g c^h\, dm/m$. The potential is thus seen to be

$$2\pi abcA \int_{\lambda_1}^{\infty} \frac{du}{Q} (m\, dm\, M') \left(\frac{a^2\xi}{a^2+u}\right)^f \left(\frac{b^2\eta}{b^2+u}\right)^g \left(\frac{c^2\zeta}{c^2+u}\right)^h,$$

where $$M' = \Sigma \left[m^2 - \frac{\xi^2}{a^2+u} - \&c.\right]^n \frac{(m^2u)^n D^n}{m^{2n}(Ln)^2 2^{2n}}.$$

We shall write $m^2 - 1 + R$ for the quantity in square brackets, this being its value when expressed in terms of R.

To find the potential of the ellipsoid we multiply by $(1-m^2)^{\kappa-1}$ (see equation (1) of Art. 239) and integrate from $m=0$ to $m=1$. The potential of the solid heterogeneous ellipsoid at an external point is therefore

$$V' = \Sigma \int_0^1 dm^2 \int_{\lambda_1}^{\infty} du\, (1-m^2)^{\kappa-1} (m^2-1+R)^n F(u),$$

where $F(u)$ contains all the factors which are independent of m, thus

$$F(u) = \frac{\pi abc}{Q} \frac{Au^n}{(Ln)^2 2^{2n}} D^n \left(\frac{a^2\xi}{a^2+u}\right)^f \left(\frac{b^2\eta}{b^2+u}\right)^g \left(\frac{c^2\zeta}{c^2+u}\right)^h.$$

Since the attracted point is external to the ellipsoid, λ_1 is not zero and it is necessary to change the order of integration by the process described in Art. 239. The $\phi(u)$ of that article is here called $1-R$ and since $n=1$, we have $\epsilon = \lambda$. The new limits are at once seen to be $m^2 = 1-R$ to 1 and $u=\lambda$ to ∞. The potential V is now

$$V' = \int_{\lambda}^{\infty} du \int_0^1 dm^2 (1-m^2)^{\kappa-1} (m^2-1+R)^n F(u).$$

The integration with regard to m^2 is effected by putting $1-m^2 = Rv$ so that $dm^2 = -R\,dv$ and the limits become $v=1$ to 0. We then find by using Euler's gamma function,

$$V' = \Sigma \int_{\lambda}^{\infty} du R^{n+\kappa} \frac{\Gamma(\kappa)\Gamma(n+1)}{\Gamma(n+\kappa+1)} F(u).$$

Substitute for $F(u)$ its value given above and this at once reduces to the expression for V' given in Art. 247.

251. *To find the potential at an external point of an elliptic disc whose surface density at any point x, y, z is $\sigma = \left(1 - \frac{x^2}{a^2} - \frac{y^2}{b^2}\right)^n$ where n is not necessarily integral.*

We regard the disc as the limit of a solid ellipsoid whose axis of c is zero. By Art. 245, corresponding values of the density ρ and potential V of the ellipsoid are

$$\rho = A\left(1 - \frac{x^2}{a^2} - \frac{y^2}{b^2} - \frac{z^2}{c^2}\right)^{\kappa-1}, \qquad V = A\frac{\pi abc}{\kappa} \int \left\{1 - \frac{\xi^2}{a^2+u} - \&c.\right\}^{\kappa} \frac{du}{Q},$$

where the limits are λ to ∞.

The mass enclosed by a prism standing on the base $dxdy$ and extending *both ways to the surface of the ellipsoid* is

$$2dxdy \int_0^{pc} (p^2c^2 - z^2)^{\kappa-1} \frac{A}{c^{2\kappa-2}} dz,$$

where $p^2 = 1 - x^2/a^2 - y^2/b^2$. Put $z = pc \sin\theta$ and this when κ is positive reduces to $dxdy\left(1-\frac{x^2}{a^2}-\frac{y^2}{b^2}\right)^{\kappa-\frac{1}{2}}\frac{\Gamma(\kappa)\Gamma(\frac{1}{2})}{\Gamma(\kappa+\frac{1}{2})}Ac.$

We now put $\kappa = n + \frac{1}{2}$ and $Ac = \Gamma(\kappa+\frac{1}{2})/\Gamma(\kappa)\Gamma(\frac{1}{2})$, and the mass of the prism is $\sigma dxdy$. Since $\Gamma(\frac{1}{2}) = \sqrt{\pi}$, the potential at an external point of the heterogeneous elliptic disc is

$$V' = \frac{\Gamma(n+1)\Gamma(\frac{1}{2})}{\Gamma(n+\frac{3}{2})}\int_\lambda^\infty \frac{ab\,du}{Q}\left(1-\frac{\xi^2}{a^2+u}-\frac{\eta^2}{b^2+u}-\frac{\zeta^2}{u}\right)^{n+\frac{1}{2}},$$

where $Q^2 = (a^2+u)(b^2+u)u$, $n > -\frac{1}{2}$, and λ is defined in Art. 204.

To find the potential of a homogeneous elliptic disc at any external point, we put $n = 0$. *The potential is therefore*

$$V' = \int_\lambda^\infty \frac{2ab\,du}{Q}\left(1-\frac{\xi^2}{a^2+u}-\frac{\eta^2}{b^2+u}-\frac{\zeta^2}{u}\right)^{\frac{1}{2}}.$$

By using Chasles' geometrical theorem (Art. 230) we may express the potential of the disc in elliptic coordinates. We find at once for a homogeneous disc

$$V' = \int \frac{2ab\,du}{Q^2}\left[(a^2+u-a'^2)(a^2+u-a''^2)(a^2+u-a'''^2)\right]^{\frac{1}{2}}$$

where the limits are $a'^2 - a^2$ to ∞.

252. Ex. 1. The surface density of an elliptic disc is

$$\sigma = (1 - x^2/a^2 - y^2/b^2)^{\kappa-\frac{1}{2}}\phi(x, y),$$

where κ is positive. Prove that the potential at an external point is

$$V' = \pi ab\frac{\Gamma(\kappa+\frac{1}{2})}{\Gamma(\kappa)\Gamma(\frac{1}{2})}\int\frac{du}{Q}M_\kappa\phi\left(\frac{a^2\xi}{a^2+u}, \frac{b^2\eta}{b^2+u}\right),$$

where $Q^2 = (a^2+u)(b^2+u)u$, the limits are λ to ∞, M_κ has the same meaning as in Art. 247 and $R = 1 - \frac{\xi^2}{a^2+u} - \frac{\eta^2}{b^2+u} - \frac{\zeta^2}{u}$, $D = \frac{a^2+u}{a^2}\frac{d^2}{d\xi^2} + \frac{b^2+u}{b^2}\frac{d^2}{d\eta^2}$.

[Proceed as in Art. 251, using Art. 247.]

Ex. 2. The line density of an elliptic ring is $\rho' = p\phi(x, y)$. Prove that its potential at $(\xi\eta\zeta)$ may be deduced from that of the elliptic disc in Ex. 1 by putting $\kappa = \frac{1}{2}$, differentiating V' with regard to R and doubling the result.

Ex. 3. The density of a solid elliptic cylinder is $\rho = A(1 - x^2/a^2 - y^2/b^2)^{\kappa-1}.x^f y^g z^h$. Prove that the potential at an external point is

$$V = \pi ab A\int\frac{du}{Q_1}.M_\kappa\left(\frac{a^2\xi}{a^2+u}\right)^f\left(\frac{b^2\eta}{b^2+u}\right)^g\zeta^h$$

where $Q_1{}^2 = (a^2+u)(b^2+u)$, and the limits are λ to ∞. If the attracted point is internal the limits are 0 to ∞.

The potentials of an elliptic cylindrical shell follow by the rule of differentiation.

[Put $\phi = (ax)^f(by)^g(cz)^h$ and $c = \infty$ in the formula of Art. 247.]

253. Confocal level surfaces. Ex. Let the law of force be the inverse κth power. Prove that the *level surfaces of an elliptic disc whose surface density is* $\sigma = \left(1-\frac{x^2}{a^2}-\frac{y^2}{b^2}\right)^n$, *where* $n = \frac{1}{2}(\kappa-3)$, *are confocal quadrics of the disc.* Conversely prove that if the level surfaces are confocal quadrics the surface density is that given by the expression for σ.

Thence find the potential for any level surface by placing the attracted point on the axis of the disc. The proofs may be found in the *Phil. Trans.* 1895.

254. The component attraction at P of a uniform elliptic disc in a direction perpendicular to its plane may be found by using Playfair's theorem (Art. 27). We describe a cone whose vertex is P and base the elliptic area. The normal attraction of the disc is equal to the solid angle of the cone multiplied by the surface density of the thin disc.

When *the attracted point lies on the hyperbolic focal conic of the attracting ellipse, the cone is known to be a right cone and the solid angle may be found by elementary solid geometry.*

If in this last case the distance of P from the plane be ζ, and the major axis of the confocal ellipsoid through P be a', we have

$$\frac{Z}{2\pi}=1-\left(\frac{a'^2-a^2}{a'^2-h^2}\right)^{\frac{1}{2}}=1-\frac{a\zeta}{(a^2\zeta^2+b^4)^{\frac{1}{2}}},$$

$$\frac{X}{2\pi}=\frac{b}{h}\left\{\frac{a'}{(a'^2-h^2)^{\frac{1}{2}}}-1\right\}=\frac{ab}{h}\left\{\left(\frac{\zeta^2+b^2}{a^2\zeta^2+b^4}\right)^{\frac{1}{2}}-\frac{1}{a}\right\},$$

$$\frac{V}{2\pi}=\frac{b}{a}\left\{a'-(a'^2-a^2)^{\frac{1}{2}}\right\}=(\zeta^2+b^2)^{\frac{1}{2}}-\zeta,$$

where $2h$ is the distance between the foci of the attracting disc, and the surface density is unity.

Ex. 1. Prove that the solid angle of the cone $Ax^2+By^2-Cz^2=0$ is $\int\frac{2u^{\frac{1}{2}}du}{Q}$, where $Q^2=(C-u)(A+u)(B+u)$. The limits are $u=0$ to C, where A, B, C are positive quantities.

Ex. 2. Prove that the solid angle of the cone

$$Ax^2+By^2+Cz^2+2Dyz+2Ezx+2Fxy=0$$

is $\int\frac{2\sqrt{u}\,du}{\sqrt{(-\Delta)}}$, where Δ is the determinant in the margin. The limits are $u=0$ to that root of the equation $\Delta=0$ which has a different sign to the other roots.

$$\begin{vmatrix} A+u, & F, & E \\ F, & B+u, & D \\ E, & D, & C+u \end{vmatrix}$$

Potentials of rectilinear figures.

255. Potential of a lamina. *To find the potential at any point P of a plane lamina of unit surface density.*

Let PN be the perpendicular from P on the plane. Let the plane of the lamina be the plane of xy, N the origin and NP the axis of z. Let $NP=\zeta$. Let (r, θ) be the polar coordinates of a point on the plane of xy.

If QQ' be any elementary arc of the curvilinear boundary, the potential of the triangular area NQQ' is $\int\frac{r\,d\theta\,dr}{(r^2+\zeta^2)^{\frac{1}{2}}}$, where the limits of integration are $r=0$ and $r=r$. If $R=PQ$, this reduces to $(R-\zeta)\,d\theta$.

Integrating this again for all the elements of the boundary, we see that *the potential V' at P of the area of any closed plane curve* is $\int(R-\zeta)\,d\theta$. In this expression the limits are determined by making the point Q (whose coordinates are r, θ) travel completely round the curve in the positive direction, the elementary angle $d\theta$ having its proper sign according as the radial angle θ is increasing or decreasing when Q passes over each element of the perimeter.

When the perpendicular PN falls within the lamina, the limits of θ are 0 and 2π, the expression for the potential is then $\int R d\theta - 2\pi\zeta$. When the perpendicular falls outside the lamina the upper and lower limits of θ are the same, so that $\int \zeta d\theta = 0$ and the expression for the potential is simply $\int R d\theta$.

256. We may put the expression just found for the potential into another form which is sometimes more useful.

If $r d\theta dr$ is any element of the area of the triangle NQQ', u its distance from P and ϕ the angle u makes with the normal to the plane, the solid angle $d\omega$ subtended at P by the triangle is

$$d\omega = \int \frac{r\, d\theta\, dr}{u^2} \cos\phi = \int \frac{\zeta\, du\, d\theta}{u^2} = \left(1 - \frac{\zeta}{R}\right) d\theta,$$

the limits of u being ζ and R.

The potential of the triangular area NQQ' at P is, by Art. 255, equal to

$$\frac{R^2 d\theta}{R} - \zeta d\theta = \frac{r^2 d\theta}{R} - \zeta d\theta \left(1 - \frac{\zeta}{R}\right) = \frac{r^2 d\theta}{R} - \zeta d\omega.$$

In fig. 1, the perpendicular PN falls within the attracting area. We then find, by integrating all round the perimeter of the area, that the potential at P is

$$V' = \int \frac{r^2 d\theta}{R} - \zeta\omega,$$

where ω is now the solid angle subtended at P by the area.

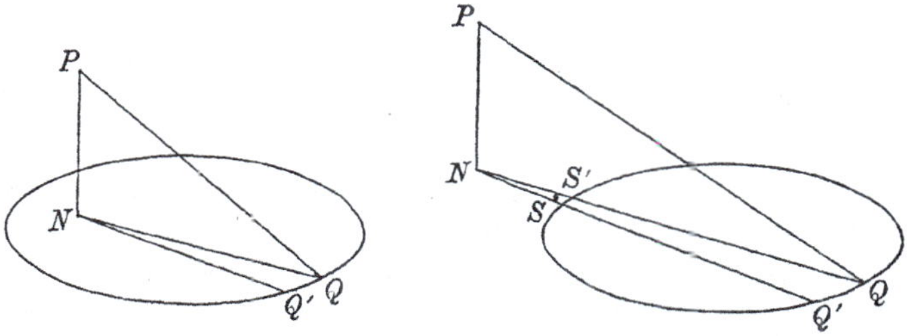

Fig. 1. Fig. 2.

In fig. 2, the perpendicular PN falls without the area. In this case we must subtract from the potential of NQQ' that of NSS'. Since $d\theta$ is positive for QQ' and negative for $S'S$ when a point travels round the curve in the positive direction, the form of the result is unaltered.

Let ds be the length of any elementary arc QQ' of the perimeter, p the perpendicular from N on the tangent at Q. Then since $r^2 d\theta = p\,ds$, the potential at P of the area takes the form $V' = \int \frac{p\,ds}{R} - \zeta\omega$, where the integration extends all round the perimeter, and ω is the solid angle subtended by the lamina at P.

Ex. If the law of force be the inverse fifth power of the distance, show that the potential of a plane lamina of unit density at a point P is $\frac{1}{8\zeta^2}\int \frac{p\,ds}{R^2}$, where the integration extends all round the perimeter and the letters have the same meaning as in Art. 255.

257. When the lamina is bounded by rectilinear sides, p is constant for each side and may therefore be brought outside the integral sign. The integral $\int ds/R$ is then the potential of that side at P. We therefore have the following theorem.

If V' be the potential at any point P, of the area contained by any plane rectilinear figure regarded as of unit surface density; V_1, V_2, &c. the potentials at the same point of its sides each regarded as of unit line density, ω the solid angle subtended at P by the area, then $V' = -\zeta\omega + p_1V_1 + p_2V_2 + \&c.$, where ζ is the length of the perpendicular PN on the area, and p_1, p_2, &c. are the perpendiculars from N on the sides taken with their proper signs.

The signs of the perpendiculars are determined by the following rule. If the point Q travel round the perimeter in the direction of the motion of the hands of a watch, the perpendicular p is positive or negative according as the origin N lies on the right or left-hand side of the tangent at Q.

258. Potential of a solid. *If V'' be the potential at any point P of a solid, of unit density, and bounded by plane rectilinear faces; V_1', V_2', &c. the potentials at the same point of its faces each regarded as of unit surface density, then*

$$2V'' = \zeta_1 V_1' + \zeta_2 V_2' + \dots,$$

where ζ_1, ζ_2, &c. are the perpendiculars from P on the faces taken with their proper signs.

Describe an elementary cone whose vertex is P and whose base is any element of area of the boundary of the solid. Let $d\omega$ be its solid angle. The volume of an element of the cone being $r^2 d\omega dr$, the potential of the cone at P is

$$\int \frac{r^2 d\omega dr}{r} = \tfrac{1}{2} r^2 d\omega = \tfrac{1}{2}\frac{p d\sigma}{r},$$

where r is now the radius vector drawn from P to the elementary area $d\sigma$ and p is the perpendicular from P on the tangent plane. The potential of the whole solid body at P is therefore $\frac{1}{2}\int \frac{p d\sigma}{r}$.

When the boundaries of the solid are planes, p is constant for each plane and $\int p d\sigma / r$ is the potential of that plane face at P. We have at once $V'' = \frac{1}{2}\Sigma pV'$.

259. The solid angle subtended at any point P by any triangle ABC is the area of the unit sphere enclosed by the planes PAB, PBC, PCA. This area is the same as that of the spherical triangle traced on the sphere by these planes, and a finite expression for its value is given in books on spherical trigonometry. Since any polygonal area can be divided into triangles it follows that the solid angle subtended at P by any rectilinear figure can always be found. The result may be complicated but it involves no integrations which cannot be effected.

It immediately follows from Arts. 257, 258 that *the potentials of all rectilinear figures and the potentials of all solids bounded by plane rectilinear faces can be found. Thus the three integrals which express the components of the attraction of a rectilinear lamina or solid can be found in finite terms.*

260. Components of Attraction. Some simple expressions may be found for the components of the attraction of the lamina. We know by Playfair's theorem, that the component along the perpendicular PN on the lamina is equal to the solid angle subtended at P by the lamina, see Art. 27.

We may obtain an expression for *the resolved part of the attraction along a straight line drawn in the plane.* If this straight line be called the axis of x and the boundary of the lamina be a closed curve in the plane of xy, the x component of the attraction is $X' = \int \frac{dy}{R}$, where R is the distance of an element of the boundary from the attracted point P.

Divide the lamina into elementary rectangles having their lengths parallel to the axis of x, and let the breadth of each be dy. If AB be any one of these (regarded as of unit surface density), its x attraction on P in the direction AB is $\left(\frac{1}{PA} - \frac{1}{PB}\right) dy$, see Art. 11. The attraction of the whole lamina is therefore $\int dy/R$, where R stands for either PA or PB, and dy is taken positive or negative according as the ordinate y is increasing or decreasing when a point Q travelling round the curve passes A or B.

261. *A solid body of unit density is bounded by plane faces: it is required to find the resolved part of its attraction at a given point P in a given direction Px.*

Whatever the form of the solid may be, its component of attraction in the direction Px is $X'' = \int \frac{d\sigma \cos\phi}{R}$, where $d\sigma$ is an element of the surface, ϕ the angle the normal at $d\sigma$ makes with the given direction Px and R is the distance of $d\sigma$ from P.

When the solid is bounded by plane faces, $\cos\phi$ is the same for all the elements of the same face. It may therefore be brought outside the integral sign. Since the integral $\int d\sigma/R$ is obviously the potential at P of the face, we have at once

$$X'' = V_1' \cos\phi_1 + V_2' \cos\phi_2 + \ldots\ldots = \Sigma V \cos\phi,$$

where V_1', V_2', &c. are the potentials at P of the plane faces regarded as of unit surface density, and ϕ_1, ϕ_2, &c. are the angles the normals measured inwards make with the direction in which X is measured.

262. Ex. 1. If $\alpha, \beta, \gamma, \delta$ be the quadriplanar coordinates of a point P referred to the faces of a tetrahedron, show that the potential of the solid contained by the tetrahedron regarded as of unit density is $\frac{1}{2}(V_1\alpha + V_2\beta + V_3\gamma + V_4\delta)$ where V_1, V_2, V_3, V_4 are the potentials at the same point of the several faces regarded as of unit surface density.

Ex. 2. Show that the solid angle ω subtended at any point P by a triangular area ABC is given by

$$\left(\tfrac{3}{2} v \operatorname{cosec} \frac{\omega}{2}\right)^2 = \frac{(q+r)^2 - a^2}{4} \cdot \frac{(r+p)^2 - b^2}{4} \cdot \frac{(p+q)^2 - c^2}{4},$$

where v is the volume of the tetrahedron $ABCP$ and p, q, r are the distances of P from the angular points of the triangle.

Ex. 3. The triangle OBC is right-angled at B, and at O a straight line OP is drawn perpendicular to its plane. If the triangle be of unit surface density, prove that its attractions at P resolved parallel to OP, OB, and BC respectively are

$$\tan^{-1} \frac{b}{ac} (a^2 + b^2 + c^2)^{\frac{1}{2}} - \tan^{-1} \frac{b}{c}$$

$$\frac{c}{(b^2 + c^2)^{\frac{1}{2}}} \log \frac{(b^2 + c^2)^{\frac{1}{2}} + (a^2 + b^2 + c^2)^{\frac{1}{2}}}{a} - \log \frac{c + (a^2 + b^2 - c^2)^{\frac{1}{2}}}{(a^2 + b^2)^{\frac{1}{2}}}$$

$$\log \frac{b + (a^2 + b^2)^{\frac{1}{2}}}{a} - \frac{b}{(b^2 + c^2)^{\frac{1}{2}}} \log \frac{(b^2 + c^2)^{\frac{1}{2}} + (a^2 + b^2 + c^2)^{\frac{1}{2}}}{a},$$

where $a = OP$, $b = OB$, $c = BC$. Since any rectilinear figure in the plane of xy may be divided into right-angled triangles having a common corner O by dropping perpendiculars from O on the sides and joining O to the corners these results give the three resolved attractions of any plane rectilinear figure. [Knight's problem. Todhunter's *History*, p. 474.]

Laplace's Functions and Spherical Harmonics.

263. In many parts of the theory of Attractions, the integrations are shortened and made more comprehensive by the use of Laplace's functions. In other parts the necessary processes could not be effected without their help. There are several treatises on these functions from which the reader may acquire a knowledge of this important branch of Pure Mathematics. The propositions however which are wanted in Attractions are not very numerous and these books contain much more than is here required. At the same time the subject of Attractions is generally approached by the student at a period of his course when he has not yet reached the proper study of these functions. For these reasons it seems proper to make a preliminary statement of a few elementary theorems which the reader acquainted with Laplace's functions may pass over.

264. Expansion of the inverse distance. Let P, P' be two points, one of which will afterwards be taken as a point of the attracting mass and the other as the point at which the attraction is required. Let (x, y, z), (x', y', z') be their Cartesian coordinates referred to any rectangular axes, (r, θ, ϕ), (r', θ', ϕ') their corresponding polar coordinates. Let R be the distance between the points and let $p = \cos POP'$. We therefore have

$$\frac{1}{R} = \frac{1}{\sqrt{\{(x-x')^2+(y-y')^2+(z-z')^2\}}} = \frac{1}{\sqrt{\{r^2-2rr'p+r'^2\}}} \quad \ldots(1).$$

It will be found convenient to expand $1/R$ in a convergent series of ascending powers of either r/r' or r'/r. Supposing first $r < r'$, we write $h = r/r'$. We then have by the binomial theorem

$$(1-2ph+h^2)^{-\frac{1}{2}} = 1 + \tfrac{1}{2}(2ph-h^2) + \tfrac{3}{8}(2ph-h^2)^2 + \ldots$$

Expanding these terms and writing P_1, P_2, &c. for the coefficients of the several powers of h

$$(1-2ph+h^2)^{-\frac{1}{2}} = 1 + P_1h + P_2h^2 + \ldots \quad \ldots\ldots\ldots(2).$$

The terms containing h^n are evidently the first in $(2ph-h^2)^n$, the second in $(2ph-h^2)^{n-1}$, and so on. It is therefore clear that *P_n is a rational integral function of p, whose highest power is p^n and whose powers descend two at a time, the terms being alternately positive and negative.* Thus P_n is of the form

$$P_n = A_np^n + A_{n-2}p^{n-2} + \ldots \quad \ldots\ldots\ldots\ldots\ldots\ldots(3),$$

where A_n, A_{n-2}, &c. are constants.

These constants are easily found when n is a small integer by the use of the binomial theorem in the manner shown above, thus

$$P_1 = p, \quad P_2 = \tfrac{1}{2}(3p^2-1), \quad P_3 = \tfrac{1}{2}(5p^3-3p),$$
$$P_4 = \tfrac{1}{8}(35p^4 - 30p^2 + 3), \text{ \&c.}$$

265. The function P_n is usually called a *Legendre's function* of the nth order. It is sometimes written in the form $P_n(p)$ when it is desired to call attention to the independent variable. Regarding one of the two radii vectores OP, OP' as a fixed axis and the other as capable of moving into all positions round the origin, P_n is a function of the inclination of the latter to the fixed axis. *The fixed radius vector is called the axis of reference of the function* or more shortly the axis of the function.

266. If $(\alpha', \beta', \gamma')$ are the direction cosines of OP', we have by projecting OP on OP' $\quad pr = \alpha'x + \beta'y + \gamma'z$,

$$\therefore P_n r^n = A_n(\alpha'x+\beta'y+\gamma'z)^n + A_{n-2}(\alpha'x+\beta'y+\gamma'z)^{n-2}(x^2+y^2+z^2)+\ldots.$$

Regarding OP' as fixed in space and OP as moving about O we see that $P_n r^n$ *is a homogeneous rational and integral function of the coordinates of* P.

The quantity $1/R$, regarded as a function of the variables (x, y, z), is known to satisfy Laplace's equation, Art. 95. Since this is true whatever (x', y', z') may be, provided they are fixed, it follows that the coefficient of every power of $1/r'$ in the expansion

$$\frac{1}{R} = \frac{1}{r'} + \frac{P_1 r}{r'^2} + \frac{P_2 r^2}{r'^3} + \ldots \ldots\ldots\ldots\ldots (4)$$

satisfies Laplace's equation.

267. *Any homogeneous function of (x, y, z) which satisfies Laplace's equation* is called a *spherical harmonic function.* Its degree may be any positive or negative integer, it may be fractional or imaginary.

When the function is such that it may be written in the form $r^n f(\theta)$ where θ is the inclination of the radius vector to a fixed straight line, it is called a *zonal* spherical harmonic. We therefore see that $P_n r^n$ is a zonal spherical harmonic of the nth order.

268. The expansion (4) has been made in powers of r/r' on the supposition that r is less than r'. If the contrary be the case we must make the expansion in powers of r'/r in order that the series may be convergent. We then have

$$\frac{1}{R} = \frac{1}{r} + \frac{P_1 r'}{r^2} + \frac{P_2 r'^2}{r^3} + \ldots \ldots\ldots\ldots\ldots (5).$$

It follows in the same way that the coefficient of r'^n, viz. $P_n r^{-(n+1)}$, is a homogeneous function of the $-(n+1)$th order which satisfies Laplace's equation. Thus both $P_n r^n$ and $P_n r^{-(n+1)}$ are zonal harmonics of different orders.

269. It is useful to notice that the values of P_n when $p = \pm 1$ and $p = 0$ follow at once from the series (2). Thus when $p = \pm 1$, P_n is the coefficient of h^n in the expansion of $(1 \mp h)^{-1}$. When $p = 1$, $P_n = 1$ and when $p = -1$, $P_n = +1$ or -1 according as n is even or odd. Both cases are included in the statement that $P_n = p^n$ when $p = \pm 1$.

It follows that the sum of the coefficients of the several powers of p in the expansion of P_n is unity.

After differentiating the series (2) κ times we find

$$\mu (1 - 2ph + h^2)^{-\frac{2\kappa+1}{2}} = \frac{d^\kappa P_\kappa}{dp^\kappa} + \frac{d^\kappa P_{\kappa+1}}{dp^\kappa} h + \&c. + \frac{d^\kappa P_n}{dp^\kappa} h^{n-\kappa} + \&c.,$$

where $\mu = 1 . 3 . 5 ... (2\kappa - 1)$. It follows that when $p = \pm 1$, $\dfrac{d^\kappa P_n}{dp^\kappa} = \dfrac{L(n+\kappa)}{L(n-\kappa) L(\kappa)} \dfrac{p^{n-\kappa}}{2^\kappa}$.

The value when $p = 0$ is somewhat more complicated.

270. Any integral rational function of p of the nth degree, say $F(p)$, can be expanded in a series of the form

$$F(p) = B_n P_n + B_{n-1} P_{n-1} + \dots + B_0 P_0.$$

Since the highest power of P_n is $A_n p^n$, we can, by properly choosing the constant B_n, make $F(p) - B_n P_n = F_2(p)$ contain p^{n-1} as the highest power. Choosing again the constant B_{n-1} properly we can make $F_2(p) - B_{n-1} P_{n-1}$ contain p^{n-2} as the highest power and so on until we arrive at zero. In this way, we find

$$p^2 = \tfrac{1}{3}(2P_2 + P_0),\ p^3 = \tfrac{1}{5}(2P_3 + 3P_1),\ p^4 = \tfrac{1}{35}(8P_4 + 20P_2 + 7P_0),\ \&c.$$

It follows from Art. 269 that the sum of the coefficients of the functions P_n in any one of these expressions is unity.

271. *To prove that* $\displaystyle P_n = \frac{1}{2^n} \frac{1}{\lfloor n} \frac{d^n}{dp^n} (p^2 - 1)^n.$

Let $u = p + \frac{1}{2}(u^2 - 1) h$, then by Lagrange's theorem

$$u = \Sigma \frac{h^n}{2^n \lfloor n} \frac{d^{n-1}}{dp^{n-1}} (p^2 - 1)^n \text{ .... (A).}$$

By solving the quadratic and differentiating we find

$$\frac{du}{dp} = \mp (1 - 2ph + h^2)^{-\frac{1}{2}} \text{ .... (B).}$$

The coefficient of h^n in the expansion (B) is by definition P_n. By differentiating (A) and comparing the two expansions the theorem follows at once. The positive sign in (B) must be taken, because when $n=1$, $P_n=p$. This expression for P_n is due to *Rodrigues*.

272. COR. Since all the roots of $(p^2-1)^n=0$ are real and lie between ± 1 inclusively, those of $\frac{d}{dp}(p^2-1)^n=0$ are also real and lie between those of $(p^2-1)^n=0$. By continuing this process we see that *all the roots of $P_n=0$ are real and lie between* ± 1.

273. The two following equations are important

$$\frac{d}{dp}\left\{(1-p^2)\frac{dP_n}{dp}\right\}+n(n+1)P_n=0 \ldots\ldots\ldots\ldots(1),$$

$$(n+1)P_{n+1}-(2n+1)pP_n+nP_{n-1}=0\ldots\ldots\ldots(2).$$

The first is usually called *the differential equation* and the second *the equation of differences* and sometimes the scale of relation.

To prove these we notice that, if $u=\Sigma P_n h^n$, the left-hand sides of the equations are the coefficients of h^n in the expressions

$$\frac{d}{dp}\left\{(1-p^2)\frac{du}{dp}\right\}+h\frac{d^2}{dh^2}(uh), \quad (1-2ph+h^2)\frac{du}{dh}+(h-p)u.$$

By substituting $u=(1-2ph+h^2)^{-\frac{1}{2}}$ these expressions are found to be zero.

The following theorems are also useful

$$c+\int P_n dp=\frac{P_{n+1}-P_{n-1}}{2n+1}\ldots\ldots(3), \qquad \frac{dP_n}{dp}=(n+1)\frac{P_{n+1}-pP_n}{p^2-1}\ldots\ldots(4).$$

These may be proved by substituting in

$$2h^2\frac{du}{dh}+uh=(1-h^2)\frac{du}{dp}, \qquad (1-p^2)\frac{du}{dp}+(1-ph)\frac{du}{dh}-pu=0.$$

274. *The equation $P_n=0$ has no equal roots*, for if P_n and dP_n/dp were zero simultaneously it would follow from the differential equation (Art. 273) that either $p=\pm 1$ or $d^2P_n/dp^2=0$. The first alternative is impossible since these values of p make $P_n=\pm 1$. Differentiating again we prove in the same way that $d^3P_n/dp^3=0$ and so on. But this would make $d^nP_n/dp^n=0$ which it is not, for by Art. 264 it is equal to $A_n \lfloor n$.

275. *The roots of $P_n=0$ lie between those of $P_{n+1}=0$.* Let $\alpha_1, \alpha_2, \ldots \alpha_n$ be the n roots of $P_n=0$ in increasing order of magnitude. Then dP_n/dp is alternately + and − when we give these values to p; it has the same sign as P_n when $p>\alpha_n$ and is therefore positive when $p=\alpha_n$. But by (4) of Art. 273 P_{n+1} and dP_n/dp have opposite signs when $P_n=0$ and $p<1$. Hence P_{n+1} is alternately − and + when we put $p=\alpha_1$, &c. α_n, and is negative when $p=\alpha_n$. Again P_{n+1} is positive when $p=1$ (being in fact unity), hence one root of $P_{n+1}=0$ is $>\alpha_n$, $n-1$ roots lie between those of $P_n=0$, and the $(n+1)$th root must be $<\alpha_1$.

276. The reader is recommended to trace the polar curve $r=a+bP_n$ for the values $n=1, 2, 3,$ &c. where $P_1, P_2,$ &c. have the values given in Art. 264 and

$p=\cos\theta$. The constant b should be regarded as much smaller than a. The two theorems of Arts. 274 and 275 will be found useful in tracing the relations which exist between the several functions.

277. It is important to notice that the function P_n is not numerically greater than unity for any value of p less than unity. For the proof of this we have here no room.

Supposing h to be less than unity, the series

$$(1-2ph+h^2)^{\frac{1}{2}}=P_0+P_1h+P_2h^2+\dots$$

is convergent even when we replace every coefficient by its greatest numerical value and make every term positive. *The series is therefore absolutely convergent when both p and h are less than unity.*

278. *To prove that* $\int_{-1}^{+1} f(p)\,P_n dp = 0$, *where* $f(p)$ *is any integral rational function of* p *of less than* n *dimensions. It follows from this that when* m *and* n *are unequal (so that one is less than the other)* $\int_{-1}^{+1} P_m P_n dp = 0$.

By a theorem in the integral calculus we have

$$\int u\,dv = uv - u'v_1 + u''v_{11} - \dots + (-1)^n \int u^n dv_n \dots\dots\dots(1),$$

where accents denote differentiations and suffixes denote integrations. Let Q be finite between the limits and let

$$v=\frac{d^{n-1}Q^n}{dp^{n-1}},\quad v_1=\frac{d^{n-2}Q^n}{dp^{n-2}},\quad v_{11}=\frac{d^{n-3}Q^n}{dp^{n-3}},\ \&c.$$

Each of the terms $v, v_1 \dots v_{n-1}$ contains the factor Q at least once and therefore vanishes when p is put equal to any root of $Q=0$. If we also put $u=f(p)$ the series terminates before we arrive at the final integral. It follows that the integral (1) is zero when the limits are any two unequal roots of $Q=0$. Let $Q=p^2-1$, the integral is then zero when the limits are $p=\pm 1$. See Art. 271.

279. If $f(p)$ is of n or higher dimensions, the only term on the right-hand side of (1) (Art. 278) which is not zero is the final integral. *This is also true if* $f(p)=p^\kappa$ (where κ is a positive quantity = or $>n$) *and the limits are* $p=0$ *to* $p=1$. In this case all the terms up to $u^\kappa v_\kappa$ are zero because the first factor vanishes when $p=0$ and the second when $p=1$. The final integral is made one of the standard forms in the integral calculus by putting $p=\cos\theta$ and its value can be written down. As these integrals are not required here, it is sufficient to state the result in the form

$$\int_0^1 p^\kappa P_n dp = \frac{\kappa+2-n}{\kappa+1+n}\int_0^1 p^\kappa P_{n-2} dp.$$

This result is also true when the limits are -1 to $+1$ and κ is integral. For if $\kappa+n$ is even each side is then doubled and if odd each side becomes zero.

Ex. Prove $\int p^\kappa P_n dp = \frac{\kappa}{\kappa+n+1}\int p^{\kappa-1}P_{n-1}dp$ where the limits are 0 to 1, and κ is a positive quantity greater than or equal to n.

280. *To find* $\int P_m P_n dp$ *between any limits.* The functions P_m, P_n satisfy

$$m(m+1)P_m = -\frac{d}{dp}\left\{(1-p^2)\frac{dP_m}{dp}\right\}, \qquad n(n+1)P_n = -\frac{d}{dp}\left\{(1-p^2)\frac{dP_n}{dp}\right\}.$$

Multiply the first by P_n and the second by P_m and integrate each product by parts. We then obtain by subtraction

$$(m-n)(m+n+1)\int P_m P_n dp = (1-p^2)\left\{P_m\frac{dP_n}{dp} - P_n\frac{dP_m}{dp}\right\},$$

where the right-hand side is to be taken between the given limits.

It immediately follows that where the limits are -1 to $+1$ the integral is zero. When m is even and n odd, we deduce from Art. 269

$$\int_0^1 P_m P_n dp = \frac{n(-1)^{\frac{1}{2}(n+m-1)}}{(n-m)(n+m+1)} \cdot \frac{1.3.5\ldots(m-1)}{2.4.6\ldots m} \cdot \frac{1.3.5\ldots(n-2)}{2.4.6\ldots(n-1)}.$$

When m and n are both even or both odd, the integral is half that of the same integral with the limits ± 1 and is therefore zero.

Since $P_0 = 1$ we find $m(m+1)\int_0^1 P_m dp$ is equal to the value of dP_m/dp when $p=0$. Also $\int_{-1}^{+1} P_m dp = 0$.

281. *To prove that* $\displaystyle\int_{-1}^{+1} P_n^2 dp = \frac{2}{2n+1}$.

This important result may be deduced from Art. 278 by putting $f(p) = P_n$, but the following method is of more general application. We multiply the equation of differences, viz.

$$nP_n - (2n-1)pP_{n-1} + (n-1)P_{n-2} = 0,$$

by P_n and integrate between the limits $p = \pm 1$. We then have

$$n\int P_n^2 dp - (2n-1)\int pP_nP_{n-1}dp = 0.$$

In the same way, if we multiply by P_{n-2} and integrate between the same limits, we find

$$-(2n-1)\int pP_{n-1}P_{n-2}dp + (n-1)\int P^2_{n-2}dp = 0.$$

We now write $n+1$ for n in the last equation and eliminate $\int pP_nP_{n-1}dp$. We thus arrive at

$$(2n+1)\int P_n^2 dp = (2n-1)\int P^2_{n-1}dp,$$

provided n is not zero. By continued reduction we find that each of these is equal to $\int P_0^2 dp = 2$. The result follows at once.

282. Ex. 1. Prove $\displaystyle\int_{-1}^{+1}\left(\frac{dP_n}{dp}\right)^2 dp = n(n+1)$, $\displaystyle\int_{-1}^{+1}\left(\frac{dP_n}{d\theta}\right)^2 dp = \frac{2n(n+1)}{2n+1}$, where $p = \cos\theta$.

To prove the first, integrate by parts and notice that since $d^2P_n/dp^2 = P''$ is of lower dimensions than P_n, $\int P_nP''dp = 0$. To prove the second, write $d\theta = -dp/\surd(1-p^2)$, integrate by parts and use the differential equation.

Ex. 2. Prove that $\displaystyle\int_{-1}^{+1}\frac{dP_m}{dp}\frac{dP_n}{dp}dp = m(m+1)$ if $n > m$, and $m+n$ is even. It is evidently zero if $m+n$ is odd.

Ex. 3. Let $(1-2ph+h^2)^{-\frac{1}{2}(\kappa-1)}=\Sigma Q_n h^n$, $\quad \phi(p)=(1-p^2)^{\frac{1}{2}(\kappa-2)}$.

Prove
$$\frac{d}{dp}\left\{(1-p^2)\,\phi(p)\,\frac{dQ_n}{dp}\right\}+n(n+\kappa-1)\,\phi(p)\,Q_n=0,$$
$$(n+1)\,Q_{n+1}-p\,(2n+\kappa-1)\,Q_n+(n+\kappa-2)\,Q_{n-1}=0,$$
$$\int\phi(p)\,Q_m Q_n\,dp=0, \qquad \int\phi(p)\,Q_n{}^2\,dp=N\int\phi(p)\,Q^2_{n-1}\,dp,$$
where $N=(2n+\kappa-3)(n+\kappa-2)/n\,(2n+\kappa-1)$ and the limits are -1 to $+1$.

Ex. 4. Prove that $\int_{-1}^{+1}(1-p^2)^\kappa \frac{d^\kappa P_m}{dp^\kappa}\frac{d^\kappa P_n}{dp^\kappa}=0$, if m and n are unequal. [It follows at once from Ex. 3 by using Art. 269.]

283. Potential of a body*. To apply these expansions to find the potential of a body, we regard (x', y', z') as the coordinates of any particle m of the attracting mass. We now multiply $1/R$ by m and sum or integrate the result for all the attracting particles. At some points of the body we may have $r'>r$, at others $r>r'$; we may therefore have to use both the expansions in Arts. 266 and 268 each for the appropriate portion of the attracting mass. In this way we find

$$V=\Sigma\frac{m}{R}=Y_0+Y_1r+Y_2r^2+\ldots+\frac{Z_0}{r}+\frac{Z_1}{r^2}+\frac{Z_2}{r^3}\ldots\ldots(6),$$

where
$$Y_n=\Sigma\frac{mP_n}{r'^{n+1}} \text{ and } Z_n=\Sigma mr'^n P_n.$$

These summations cannot be effected until the form and law of density of the heterogeneous body are known. We notice however that both Y_n and Z_n are the sums of a number of Legendre's functions with coefficients and axes depending on the given structure and shape of the body. Regarded as a function of (x, y, z) both Y_nr^n and Z_nr^n are integral rational spherical harmonics. When therefore we use Cartesian coordinates we write the series in the form

$$V=S_0+S_1+S_2+\ldots+\frac{T_0}{r}+\frac{T_1}{r^3}+\frac{T_2}{r^5}+\ldots$$

where S_n, T_n are spherical harmonic functions of x, y, z of n dimensions.

284. Laplace's equations. In this way we have been led to an expansion of V in powers of r which must hold for all attracting masses. Let this be written $V=\Sigma Y_n r^n$, where n may be either a positive or a negative integer. Substituting this

* This expression for the potential V is given by Sir G. Stokes in his memoir on the Variation of Gravity, &c. *Camb. Trans.* 1849. He obtains the expression by solving Laplace's equation.

series for V in Laplace's equation as expressed in polar coordinates (Art. 108) and equating the coefficient of r^n to zero, we have

$$\frac{d}{d\mu}\left\{(1-\mu^2)\frac{dY_n}{d\mu}\right\}+\frac{1}{1-\mu^2}\frac{d^2Y_n}{d\phi^2}+n(n-1)Y_n=0 \;\ldots\; (7),$$

where $\mu=\cos\theta$.

The corresponding equation for Y_m is found by writing m for n. If we choose m so that $m(m+1)=n(n+1)$ we have $m=n$ or $m=-(n+1)$. It follows that there are two powers of r, and only two, viz. r^n and $r^{-(n+1)}$, such that their coefficients in the series (6), viz. Y_n and Z_n, satisfy the differential equation (7). It appears therefore that Y_n and Z_n are both solutions of the differential equation (7) and differ only in the arbitrary functions or constants which occur in the solution.

Any function of two independent angular coordinates (such as the direction angles θ, ϕ of the radius vector) which satisfies equation (7) is called a *Laplace's function*. Thus Y_n is a Laplace's function of the order n. The corresponding function $Y_n r^n$ when expressed in terms of (x, y, z) satisfies Laplace's equation and is a spherical harmonic, Art. 267. A Laplace's function when expressed as a function of the Cartesian coordinates of the point at which the radius vector intersects some given sphere with its centre at the origin is called a *spherical surface harmonic*.

285. If θ', ϕ' be the direction angles of a fixed radius vector OP' and $\cos POP'=p$, we have $p=\cos\theta\cos\theta'+\sin\theta\sin\theta'\cos(\phi-\phi')$.

The Legendre's function P_n is therefore a symmetrical function of θ, ϕ and θ', ϕ'. Regarded as a function of θ, ϕ, we see, by comparing the series (4) and (5) of Arts. 266, 268 with (6) of Art. 283, that P_n is a special case of Y_n. It follows that P_n must also satisfy Laplace's equation (7).

If the axis of the function P_n, i.e. OP', be taken as the axis of reference, we have $\mu=p$ and $dP_n/d\varphi=0$. The differential equation then becomes

$$\frac{d}{dp}\left\{(1-p^2)\frac{dP_n}{dp}\right\}+n(n+1)P_n=0 \;\ldots\ldots\ldots\ldots\; (8).$$

The general solution of the differential equation (8) has two arbitrary constants. To find the general solution when a partial solution has been found we use a rule given in the theory of differential equations (see Forsyth's *Diff. Eq.* Art. 58). The general solution is thus found to be

$$AP_n+BP_n\int\frac{dp}{P_n^2(p^2-1)},$$

where A and B are the two arbitrary constants. Since P_n is an integral rational function of p we may by using partial fractions effect this integration. The process is rather long and the results will not be required. It will be sufficient to notice that the part of the solution derived from the integral is not an integral rational function of p. It follows that the only integral rational solution is AP_n.

In the same way the general solution of the equation of differences

$$(n+2)\,u_{n+2}-(2n+3)\,pu_{n+1}+(n+1)\,u_n=0$$

is $u_n=AP_n+BQ_n$ where

$$Q_n=\frac{2n-1}{n}P_{n-1}+\frac{2n-5}{n-1}\frac{P_{n-3}}{3}+\text{\&c.}+\frac{2n-4r+3}{n-r+1}\frac{P_{n-2r+1}}{2r-1}+\text{\&c.}$$

Both these partial solutions are integral rational functions of p. This result is easily verified by substitution: if we remember that the equation is satisfied by $u_n=P_n$, we find that the coefficient of every P_n is zero.

286. We have seen in Art. 283 that the potential of any body can be expanded in a series of spherical harmonics of integral orders. In this expansion Y_nr^n and Z_nr^n are both integral and rational functions of x, y, z of a positive integral order. Changing to polar coordinates we find that Y_n is an integral function of $\cos\theta$, $\sin\theta\cos\phi$, $\sin\theta\sin\phi$. Expanding the powers of $\sin\phi$, $\cos\phi$ in multiple angles, we have

$$Y_n=A_0+(A_1\cos\phi+B_1\sin\phi)+(A_2\cos2\phi+B_2\sin2\phi)+\ldots+(A_n\cos n\phi+B_n\sin n\phi)\ldots(9),$$

where $A_0, A_1\ldots A_n, B_1\ldots B_n$ are all integral and rational functions of $\sin\theta$ and $\cos\theta$.

Substituting this value of Y_n in (7), we see that both A_k and B_k satisfy

$$\frac{d}{d\mu}\left\{(1-\mu^2)\frac{dA_k}{d\mu}\right\}+n(n+1)A_k=\frac{k^2}{1-\mu^2}A_k\ \ldots\ldots\ldots\ldots\ldots\ldots (10),$$

where $\mu=\cos\theta$.

Since the equation (10) reduces to the form (8) when $k=0$, we have $A_0=a_0P_n(\mu)$, where a_0 is an arbitrary constant.

The values of A_1, B_1 &c. will not be required; it will therefore be sufficient to mention that their values found from equation (10) are

$$A_k=a_k(\sin\theta)^k\frac{d^kP_n(\mu)}{d\mu^k},\qquad B_k=b_k(\sin\theta)^k\frac{d^kP_n(\mu)}{d\mu^k},$$

where a_k and b_k are arbitrary constants.

The function $(a_k\cos k\phi+b_k\sin k\phi)(\sin\theta)^k\dfrac{d^kP_n(\mu)}{d\mu^k}$ is called *a tesseral surface harmonic* of degree n and order k. In the particular case in which $k=n$, the function is called *a sectorial surface harmonic* of degree n.

287. The case in which $Y_n=P_n(p)$ is sometimes useful in the theory of attractions. Since p is the cosine of the angle between the directions $(\theta,\ \phi)$, $(\theta',\ \phi')$, P_n is a symmetrical function of $(\theta,\ \phi)$, $(\theta',\ \phi')$. We therefore have

$$P_n(p)=a_0P_nP_n'+\Sigma a_k(\sin\theta\sin\theta')^k\frac{d^kP_n}{d\mu^k}\frac{d^kP_n'}{d\mu'^k}\cos k(\phi-\phi'),$$

where $P_n=P_n(\mu)$, $P_n'=P_n(\mu')$, $\mu=\cos\theta$, $\mu'=\cos\theta'$ and Σ implies summation from $k=1$ to n. By putting $\theta=0$, $\theta'=0$ we see that $a_0=1$. In a similar way by putting $\theta=\frac{1}{2}\pi$, $\theta'=\frac{1}{2}\pi$ we deduce that $a_k=2\dfrac{L(n-k)}{L(n+k)}$. When $k=0$, we take half this value.

288. Three theorems. The great utility of Laplace's functions depends on three theorems. To these we now turn our attention.

Theorem I. *If Y_m, Y_n be two Laplace's functions of different orders then $\int Y_mY_nd\omega=0$, where $d\omega$ is an elementary solid angle and the integration extends over the whole surface of the unit sphere.*

The following is Kelvin's proof. Put $V = Y_m r^m$, $V' = Y_n r^n$ and apply Green's theorem (Art. 150) to the surface of a sphere of radius a, whose centre is at the origin, then

$$\int V \frac{dV'}{dr} d\sigma = \int V' \frac{dV}{dr} d\sigma.$$

Substitute for V, V' and we have

$$a^{m+n-1} n \int Y_m Y_n d\omega = a^{m+n+1} m \int Y_m Y_n d\omega;$$

hence unless m and n are equal, $\int Y_m Y_n d\omega = 0$.

When m and n are positive these values of V and V' are both finite throughout the sphere. If however m, or n, is negative it is necessary to integrate over the two surfaces of a spherical shell, to avoid the infinity at the centre. If a and b be the radii we then have $(a^{m+n+1} - b^{m+n+1}) n \int Y_m Y_n d\omega = (a^{m+n+1} - b^{m+n+1}) m \int Y_m Y_n d\omega$.

It follows that $\int Y_m Y_n d\omega = 0$ unless $m = n$ or $m + n + 1 = 0$. We have also $\int Y_m P_n d\omega = 0$ and since $P_0 = 1$, $\int Y_m d\omega = 0$, where the integration extends over the whole unit sphere.

289. Theorem II. *Let Y_n be a Laplace's function of the angular coordinates (θ, ϕ) and P_n a Legendre's function of the same coordinates having (θ', ϕ') for its axis. Let both these be of the same order, viz. n, then $\int Y_n P_n d\omega = \frac{4\pi}{2n+1} Y_n'$, where the integration extends over the whole unit sphere, and Y_n' is the value of Y_n when (θ', ϕ') have been substituted for (θ, ϕ).*

To find the value of $\int Y_n P_n d\omega$, let us take as the axis of z, the axis of P_n, (Art. 265) so that $P_n = P_n(\mu)$, where $\mu = \cos\theta$. Also $d\omega = \sin\theta d\theta d\phi$ becomes $-d\mu d\phi$. The limits of integration are $\mu = 1$ to -1, $\phi = 0$ to 2π.

Taking the value of Y_n given in Art. 286, viz.

$$Y_n = a_0 P_n(\mu) + \Sigma (A_k \cos k\phi + B_k \sin k\phi),$$

we notice that $\int \cos k\phi d\phi = 0$ and $\int \sin k\phi d\phi = 0$ when the limits of ϕ are 0 to 2π. Hence

$$\int Y_n P_n d\omega = -a_0 \iint P_n^2 d\mu d\phi = a_0 . 2\pi . \frac{2}{2n+1}.$$

It remains to find the value of a_0. Referring to equation (10) of Art. 286, we see $A_k = 0$ and $B_k = 0$ when $\mu = 1$ except when $k = 0$. Also $P_n(\mu) = 1$ when $\mu = 1$. Thus a_0 is the value of Y_n at the point where the positive direction of the axis of z cuts the unit sphere. Since the axis of P_n has been taken as the axis of z

it follows that a_0 is the value of Y_n at the positive extremity or pole of the axis of P_n, and this value has been represented in the enunciation by Y_n'.

290. **Theorem III.** *Any function of the two angular coordinates of the radius vector can be expanded in a series of Laplace's functions, and the expansion can be made in only one way.*

For a discussion of this important theorem we must refer the reader to the treatises on these functions. It will suffice here if we consider how we may practically use the theorem in those simpler cases which generally occur in the theory of attraction.

Let us first suppose that the given function is an integral rational function of the direction cosines of the radius vector, i.e. of $\sin\theta\cos\phi$, $\sin\theta\sin\phi$, and $\cos\theta$. On transforming to Cartesian coordinates and multiplying each term by the proper power of r the function becomes an integral rational function of x, y, z, which we can arrange in a series of homogeneous functions. Taking any one of these, say $f_n(x, y, z)$, we shall show how it may be expanded in a series of spherical harmonics combined with powers of r. Thence (if it be necessary) we deduce the expansion in Laplace's functions by giving r any constant value.

Subtract from f_n the expression $(x^2+y^2+z^2)f_{n-2}$, where f_{n-2} is an arbitrary integral and rational function of (x, y, z) of the $(n-2)$th degree, viz.

$$f_{n-2}=A_0x^{n-2}+A_1x^{n-3}y+B_1x^{n-3}z+\dots.$$

Substituting $V=f_n-(x^2+y^2+z^2)f_{n-2}$ in ∇^2V, there results a homogeneous function of (x, y, z) of the $(n-2)$th degree, which therefore contains as many terms as there are ways of making homogeneous products of x, y, z of that degree. But f_{n-2} is an arbitrary homogeneous function of the same degree and contains an equal number of terms. There are therefore just enough arbitrary constants A_0, A_1, B_1 &c. to enable us to make the coefficients of every term in ∇^2V equal to zero. Assuming that the linear equations thus formed to find A_0, A_1 &c. are not inconsistent with each other, the expression $f_n-(x^2+y^2+z^2)f_{n-2}=S_n$ satisfies Laplace's equation and is therefore a spherical harmonic.

Repeating this process with the function f_{n-2}, we have

$$f_{n-2}-(x^2+y^2+z^2)f_{n-4}=S_{n-2},$$

and so on. We finally end with a constant or an expression of the first degree according as n is an even or odd integer.

Writing r^2 for $x^2+y^2+z^2$ we have $f_n=S_n+r^2S_{n-2}+r^4S_{n-4}+\dots$, where S_n, S_{n-2} &c. are all spherical harmonics. It should be noticed that this equality is a mere algebraical transformation, and involves no assumptions as to the meaning of the letters.

If we now regard r as the radius of the unit sphere or any suitable sphere, S_n, S_{n-2} &c. become Laplace's functions, and the required expansion has been made.

When the function does not contain powers of x, y, z above the cube, this process will be unnecessary, for the arrangement in harmonics can then be generally performed at sight.

291. When the Cartesian equivalent of the given function is not an integral rational function of the coordinates, an expansion in a finite number of terms cannot be obtained. We then proceed in another way. Assume that the expansion can be effected in a convergent series, say $f(\theta, \phi)=Y_0+Y_1+Y_2+\dots$, where Y_n is a Laplace's

function of the nth order. Let P_n be the Legendre's function having (θ', ϕ') for its axis, so that P_n is a symmetrical function of (θ, ϕ) and (θ', ϕ'); Art. 285. Multiply both sides of the equation by P_n and integrate over the whole surface of the unit sphere; then by Art. 289
$$\iint f(\theta, \phi) P_n d\mu d\phi = \frac{4\pi}{2n+1} Y_n',$$
where Y_n' is the value of Y_n when (θ', ϕ') have been written for (θ, ϕ). When the integration on the left-hand side has been effected, the result will be a known function of θ', ϕ' only. Since θ', ϕ' are arbitrary we can replace them by θ, ϕ and thus the form of Y_n has been found.

Laplace's expansion is an extension to two independent variables of Fourier's expansion of a function of one variable in a series of sines and cosines of its multiples, and like that theorem is subject to limitations. The process of expansion given above is not in any way a proof, it is to be regarded as merely a convenient method of applying Laplace's theorem to special cases. It fails to give the limitations and must be used with caution when the function to be expanded is not single-valued.

292. Ex. 1. What are the conditions that

(1) $ax+by+cz$, (2) $Ax^2+By^2+Cz^2+2Dyz+2Ezx+2Fxy$

may be spherical harmonics? The first is always so, the second when $A+B+C=0$.

Ex. 2. Expand $\sin^3\theta\cos^3\phi$ in Legendre's functions.

This is the same as p^3 if the axis of x be taken as the axis of reference. Now $P_3=\frac{1}{2}(5p^3-3p)$, hence $p^3-\frac{2}{5}P_3=\frac{3}{5}p$. The result is $p^3=\frac{2}{5}P_3+\frac{3}{5}P_1$.

Ex. 3. Expand $\sin^2\theta\sin\phi\cos\phi+\cos^3\theta$ in Laplace's functions.

The result is $Y_1+Y_2+Y_3$, where $Y_1=\frac{3}{5}\cos\theta$, $Y_2=\sin^2\theta\sin\phi\cos\phi$, $Y_3=\frac{1}{5}(5\cos^3\theta-3\cos\theta)$.

Ex. 4. Expand $\log(1+\operatorname{cosec}\frac{1}{2}\theta)$ in Legendre's functions. [Coll. Ex.]

The result is $P_0+\frac{1}{2}P_1+\frac{1}{3}P_2+\frac{1}{4}P_3+\dots$.

Ex. 5. Prove by successive induction or otherwise the equalities
$$P_0^2+3P_1^2+\&c.+(2n+1)P_n^2=(n+1)^2P_n^2-(p^2-1)\left(\frac{dP_n}{dp}\right)^2,$$
$$\left(\frac{dP_0}{dp}\right)^2+3\left(\frac{dP_1}{dp}\right)^2+\&c.+(2n+1)\left(\frac{dP_n}{dp}\right)^2=\tfrac{1}{3}\left\{(n+2)^2\left(\frac{dP_n}{dp}\right)^2-(p^2-1)\left(\frac{d^2P_n}{dp^2}\right)^2\right\}.$$

Ex. 6. If $\frac{dP_n}{dp}=\Sigma a_rP_r$ and $\frac{d^2P_n}{dp^2}=\Sigma b_sP_s$, prove that
$$a_r=2r+1 \text{ and } b_s=\tfrac{1}{2}(2s+1)(n-s)(n+1+s).$$
Multiply the series by P_r and P_s respectively and integrate by parts between the limits ±1. The expansion of the mth differential coefficient of P_n is investigated in the *Proceedings of the London Math. Soc.* 1894.

Ex. 7. If $p^\kappa=a_\kappa P_\kappa+\dots+a_nP_n+\dots$ prove that
$$\frac{a_{n-2}}{a_n}=\frac{2n-3}{2n+1}\,\frac{\kappa+n+1}{\kappa-n+2},\qquad a_n=\frac{(2n+1)\,\underline{|\kappa}}{2.4.6\dots(\kappa-n).1.3.5\dots(\kappa+n+1)}.$$

293. Ex. 1. The polar equation of a nearly spherical surface is
$$r=a\{1+\beta(Y_0+Y_1+\dots)\},$$
where β is a small quantity whose square can be neglected. Prove the following results,

(1) The volume is $\frac{4}{3}\pi a^3(1+3\beta Y_0)$ and the surface is $4\pi a^2(1+2\beta Y_0)$.

(2) If $rY_1=Ax+By-Cz$, the coordinates of the centre of gravity of the volume

are $\bar{x}=\beta Aa$, $\bar{y}=\beta Ba$, $\bar{z}=\beta Ca$. The centre of gravity of the surface coincides with that of the volume.

(3) If $r^2Y_2 = Ax^2 + By^2 + Cz^2 + 2Dyz + 2Ezx + 2Fxy$, the moment of inertia about the axis of z is $\frac{4\pi a^3}{3}\frac{2a^2}{5}(1-\beta C+5\beta Y_0)$, and the product of inertia about the axes of x, y is $\frac{4\pi a^3}{3}\cdot\frac{2a^2}{5}\beta F$.

It follows from this example that when the origin is placed at the centre of gravity of the volume the term Y_1 is absent from the equation. When the constant a is so chosen that it is equal to the radius of the sphere of equal volume, the term Y_0 is absent.

To obtain any of these results, we proceed as follows. Let M be the volume, $P_1=\cos\theta$, &c., then $M\bar{z}=\iint r^2drd\omega\,.\,z=\int\frac{1}{4}r^4d\omega P_1$. Substitute for r, expand and use Art. 289. The result is $\frac{4}{3}\pi a^4\beta Y_1'$, where Y_1' is the value of Y_1 at the extremity of the axis of z and in the small terms this is C. Similarly the moment of inertia is $\iint r^2drd\omega\,.\,r^2\sin^2\theta=\int\frac{1}{5}r^5d\omega\,.\,\frac{2}{3}(1-P_2)$. We then proceed as above.

Ex. 2. The polar equation of a nearly spherical surface is $r=a(1+\beta P_n)$ where β is a small quantity whose powers *above the second* may be neglected. Prove that the area of the surface exceeds the area of a sphere of radius a by $2\pi a^2\beta^2\,.\,\frac{n^2+n+2}{2n+1}$, except when $n=0$. [Math. T.]

Ex. 3. Prove that the surfaces $r=a(1+\beta Y_1)$, $r=a\{1+\beta(Y_0+Y_1+Y_2)\}$, where the square of β can be neglected, are respectively a sphere and a conicoid. The coordinates of the centre are the same as those of the centre of gravity already found.

294. Attraction of a spherical stratum. *A thin heterogeneous stratum of attracting matter is placed on a sphere of radius a. It is required to find its potential at any internal or external point.*

Let ρ be the surface density at any point Q of the sphere, $d\sigma$ an element of area at Q; θ, ϕ the polar coordinates of Q, then $d\sigma=\sin\theta d\theta d\phi$. Let P be the point at which the attraction is required, and let the coordinates of P be (r', θ', ϕ').

If R be the distance between the points Q and P, the potential of the whole stratum at P is $V=\int\rho d\sigma/R$. Let p be the cosine of the angle between the positive directions of the radii vectores OQ and OP, then $R^2=a^2+r'^2-2apr'$.

If the point P is inside the sphere, r' is less than a, and we may expand $1/R$ in a convergent series of ascending powers of r'/a. If the point attracted is outside the sphere, we must expand in powers of a/r'. Since R is a symmetrical function of a and r we have

$$\frac{1}{R}=\frac{1}{a}\left\{P_0+P_1\frac{r'}{a}+P_2\left(\frac{r'}{a}\right)^2+\dots\right\}$$

or

$$=\frac{1}{r'}\left\{P_0+P_1\frac{a}{r'}+P_2\left(\frac{a}{r'}\right)^2+\dots\right\}.$$

The surface density ρ is a given function of the coordinates of Q; let it be expanded in a series of Laplace's functions or surface harmonics, thus $\rho = Y_0 + Y_1 + Y_2 + \ldots\ldots\ldots$

Substituting these values of ρ and $1/R$ in the expression for V, we have by the theorems I. and II. in Arts. 288, 289,

$$V = 4\pi a \left\{Y_0' + \frac{1}{3} Y_1' \frac{r'}{a} + \frac{1}{5} Y_2' \left(\frac{r'}{a}\right)^2 + \ldots \frac{1}{2n+1} Y_n' \left(\frac{r'}{a}\right)^n + \ldots\right\},$$

$$V' = \frac{4\pi a^2}{r'} \left\{Y_0' + \frac{1}{3} Y_1' \frac{a}{r'} + \frac{1}{5} Y_2' \left(\frac{a}{r'}\right)^2 + \ldots \frac{1}{2n+1} Y_n' \left(\frac{a}{r'}\right)^n + \ldots\right\},$$

according as r' is less or greater than a. *The first of these two expansions gives the potential at any internal point, the second at any external point.*

If Y_n is expressed as a function of the angular coordinates (θ, ϕ) of Q, then as already explained (Art. 289) Y_n' is the value of Y_n when the polar coordinates θ', ϕ' of the attracted point P have been written for (θ, ϕ). If however Y_n is expressed as a homogeneous function of the Cartesian coordinates (x, y, z) of Q, then Y_n' is obtained from Y_n by writing the Cartesian coordinates of P for (x, y, z) and multiplying the result by $(a/r')^n$.

We notice that by Art. 86, *the potentials at two inverse points* are connected by the equation $V' = Va/r'$. It follows that either of the series in the brackets must change into the other when we write a^2/r' for r'.

295. Ex. 1. The surface density at any point Q of a sphere is a quadratic function of the Cartesian coordinates of Q. Find the potential at any point whose coordinates are (x', y', z').

Let the surface density ρ be given by $\rho = Ax^2 + By^2 + Cz^2 - 2Dyz + 2Ezx + 2Fxy$. Let us represent this function by $f(x, y, z)$.

As this function would be a spherical harmonic if $A + B + C = 0$, we make the necessary expansion in surface harmonics by subtracting and adding $G(x^2 + y^2 + z^2)$, where $3G = A + B + C$. We therefore have $\rho = Y_0 + Y_2$, where

$$Y_0 = Ga^2, \quad Y_2 = f(x, y, z) - G(x^2 + y^2 + z^2).$$

The required potential at the point P is therefore

$$V = 4\pi a \left\{Y_0' + \tfrac{1}{5} Y_2' \left(\frac{r'}{a}\right)^2\right\}, \quad \text{or } V' = \frac{4\pi a^2}{r'} \left\{Y_0' + \tfrac{1}{5} Y_2' \left(\frac{a}{r'}\right)^2\right\},$$

according as P is inside or outside the sphere. Here $Y_2' = \{f(x', y', z') - Gr'^2\} \left(\frac{a}{r'}\right)^2$, and $Y_0' = Ga^2$. Substituting these values for Y_0' and Y_2' in the formulæ for V and V' the required potentials have been found.

Ex. 2. The surface density at any point of a sphere is $\rho = mxy$: show that its potential at any point (x', y', z') is $\dfrac{4\pi am}{5} x'y'$ or $\dfrac{4\pi am}{5} x'y' \left(\dfrac{a}{r'}\right)^5$, according as the point is within or without the sphere.

Ex. 3. The surface density at any point of a sphere is $mxyz$: show that the potential at an internal point is $\frac{4}{35}\pi am\,x'y'z'$.

Ex. 4. Matter of mass M is distributed on a spherical surface whose centre is at O and radius a, so that its density at any point is proportional to the square of its distance from a point C outside the sphere where $OC=b$; prove that the potential at an external point P distant r from the centre is $M\left\{\frac{1}{r}-\frac{2a^2b}{3(a^2+b^2)}\frac{x}{r^3}\right\}$, where $x=r\cos POC$. [Caius Coll. 1897.]

Ex. 5. If the surface density at any point Q be an integral rational function of the Cartesian coordinates of Q of a degree not higher than the nth, prove that the potential at any internal point P is an integral rational function of the Cartesian coordinates of P also of a degree not higher than the nth.

296. Attraction of a solid sphere. *To find the potential of a solid heterogeneous shell bounded by concentric spheres when the density ρ at any point is a homogeneous function of the coordinates of the kth degree.*

Let the density ρ be expanded in a series of the form

$$\rho = r^k\{Y_0+Y_1+Y_2+\dots\dots\},$$

where Y_n is a Laplace's function of the angular coordinates. The potentials of an elementary shell whose radii are r and $r+dr$ at an internal and external point respectively are

$$dV = 4\pi r^{k+1}\,dr\,\Sigma\frac{Y_n'}{2n+1}\left(\frac{r'}{r}\right)^n,\quad dV' = \frac{4\pi r^{k+2}dr}{r'}\,\Sigma\frac{Y_n'}{2n+1}\left(\frac{r}{r'}\right)^n.$$

The potentials of the solid sphere are found by integrating these expansions between the limits a and b, where a, b are the internal and external radii of the given shell.

Ex. 1. The density of a shell bounded by concentric spheres of radii a and b is given by $\rho=mxy$. Show that the potential at an internal point is $\frac{2}{5}m\pi(b^2-a^2)\,x'y'$.

Ex. 2. The density of a solid sphere of radius a is given by $\rho=mxyz$. Show that its potential at an external point is $\frac{4}{63}\pi ma^9\,x'y'z'/r'^7$.

297. Nearly spherical bodies. *The strata of equal density of a solid are nearly spherical and both its internal and external boundaries are surfaces of equal density. Find to a first approximation its potential at an internal and an external point*.*

Let any surface of equal density be $r=a+af(\theta,\phi,a)$, where a is a constant and f *a function whose square can be neglected.*

* The formulæ here given are those used by Laplace to find the potential of the earth regarded as a stratified heterogeneous body, *Méc. Céleste*, vol. II. p. 44. When the strata are not so nearly spherical that the square of $f(\theta,\phi)$ can be neglected the algebraical processes become very complicated. For these the reader is referred to memoirs by Poisson in the *Connaissance des Temps* for 1829 and 1831.

The quantity a is the parameter of the strata, i.e. by its variation we pass from one stratum to another. Let the internal and external boundaries be defined by $a = a_0$ and $a = a_1$. Let the density of any stratum be $\rho = F(a)$.

Let the equation of the stratum be expanded in a series of Laplace's functions, viz. $r = a(1 + \Sigma Y_n)$ ....(1).
The solid bounded by this surface may be regarded as a sphere of radius a, together with a stratum of surface density $a\Sigma Y_n$ placed on its external boundary.

The potentials of this solid, regarded *as homogeneous and of unit density,* at an internal and an external point are respectively

$$U = 4\pi\left\{\frac{a^2}{2} - \frac{r'^2}{6} + \Sigma\frac{Y_n'}{2n+1}\frac{r'^n}{a^{n-2}}\right\} \quad \text{............(2)},$$

$$U' = \frac{4\pi}{r'}\left\{\frac{a^3}{3} + \Sigma\frac{Y_n'}{2n+1}\frac{a^{n+3}}{r'^n}\right\} \quad \text{................(3)}.$$

If we differentiate each of these with regard to a, we obtain the potentials of a stratum of unit density bounded by the surfaces whose parameters are a and $a + da$. The actual density of the stratum is $\rho = F(a)$; if then we multiply the differential coefficients by ρ and integrate between the limits $a = a_0$ and $a = a_1$, the required potentials at an internal and external point are found to be

$$V = 4\pi\int\rho\left\{a + \frac{d}{da}\Sigma\frac{Y_n'}{2n+1}\frac{r'^n}{a^{n-2}}\right\}da \quad \text{.........(4)},$$

$$V' = \frac{4\pi}{r'}\int\rho\left\{a^2 + \frac{d}{da}\Sigma\frac{Y_n'}{2n+1}\frac{a^{n+3}}{r'^n}\right\}da \quad \text{.........(5)},$$

the limits of the integrals being a_0 and a_1.

We may also find the potential at any point of the solid defined by the value $a = a'$ of the parameter. In this case the point is external to the strata between a_0 and a' and internal to those between a' and a_1. The required potential V'' is therefore the sum of the two expressions for V and V', the first between the limits a_0 and a' and the second between a' and a_1. The result is

$$V'' = 4\pi\int_{a'}^{a_1}\rho\left\{a + \Sigma\frac{r'^n}{2n+1}\frac{d}{da}\left(\frac{Y_n'}{a^{n-2}}\right)\right\}da$$
$$+ \frac{4\pi}{r'}\int_{a_0}^{a'}\rho\left\{a^2 + \Sigma\frac{1}{r'^n}\frac{1}{2n+1}\frac{d}{da}(Y_n' a^{n+3})\right\}da \quad \text{...(6)},$$

where Σ implies summation for all the values of n which occur in the equation (1), r', θ', ϕ' are the coordinates of the attracted point P, Y_n' is a known function of θ', ϕ', a', and ρ is a function of a.

After the integration has been effected, the potential V'' is expressed as a function of r', θ', ϕ', and a'. In the terms which contain the small factor Y_n' we may put $a'=r'$. In the first term of the second line where there is no small factor, we use the equation $r'=a'(1+\Sigma Y_n')$.

To obtain the component attractions at P it is necessary to differentiate the potential with regard to the coordinates of P. *If no substitution has been made for a'* we must remember that a' is a function of r', θ', ϕ'. We shall however immediately prove that *the partial differential coefficient* $dV''/da'=0$, so that *the first differential coefficients* of V'' with respect to r', θ', ϕ' may be correctly found by treating a' as a constant.

We have by differentiating (6)

$$\frac{dV''}{da'}=4\pi\rho\left\{-a'-\Sigma\frac{r'^n}{2n+1}\frac{d}{da'}\left(\frac{Y_n'}{a'^{n-2}}\right)+\frac{a'^2}{r'}+\Sigma\frac{1}{r'^{n+1}}\frac{1}{2n+1}\frac{d}{da'}(Y_n'a'^{n+3})\right\}.$$

We now put $a'^2/r'=a'(1-Y_n')$ and in the remaining terms $r'=a'$. It is then easily seen that the terms independent of Y_n' cancel, while the coefficients of both Y_n' and dY'/da' are zero. There are some remarks of Poisson on this point in the memoir already referred to.

Another proof. The change of a' into $a'+da'$ transfers an element from one integral of (6) to the other and this is equivalent to moving the stratum bounded by the surfaces a' and $a'+da'$ from one side of the point P to the other. But this change does not alter the potential of *that stratum* at a point on its surface, (Art. 145), that is $dV''/da'=0$. The potential at P is therefore only altered by the direct change of the coordinates of P.

298. Ex. There is some reason to suppose that the strata of the earth are elliptical and that the density decreases from the centre to the surface. Assuming then that $r=a(1+Y_2)$ and that $\rho=ga^m$, where m is greater than -2, prove that the potential at any internal point is

$$4\pi g\left\{\frac{a^{2+m}}{2+m}+\frac{\mathrm{a}^{3+m}-a^{3+m}}{3+m}\frac{1}{r'}+\frac{Y_2'}{r'^3}\frac{\mathrm{a}^{5+m}-a^{5+m}}{5+m}\right\},$$

where a is the value of a at the boundary, and $r'=a(1+Y_2')$.

299. *Let the potential be given at every point of the surfaces of two concentric spheres, radii a and b, there being no attracting matter between the spheres. Find the potential throughout the intervening space.*

The potentials, being given functions of θ, ϕ when $r=a$ and $r=b$, may be expanded in one way only in a series of surface harmonics, Art. 290. Let these expansions be respectively $V=\Sigma S_n$ and $V'=\Sigma S_n'$, where S_n and S_n' are known functions of θ, ϕ. The general expression for the potential is

$$V=\Sigma(Y_nr^n+Z_n/r^{n+1}).$$

The conditions of the question are satisfied if we take

$$Y_na^n+Z_n/a^{n+1}=S_n,\quad Y_nb^n+Z_n/b^{n+1}=S_n'.$$

Thus Y_n and Z_n are found. We know by Art. 133 that there is but one value of V which satisfies the given conditions.

If the inner sphere (radius a) include all the attracting matter we may put $b=\infty$, and then $Y_n=0$. The potential V takes the form $V=\Sigma S_n(a/r)^{n+1}$ and has only the inverse powers of r.

If all the attracting matter is outside the sphere $r=b$ we may put $a=0$. We then have $Z_n=0$ and the potential has only the direct powers given by $V=\Sigma S_n(r/a)^n$.

300. Solid of revolution. *To find the potential of a solid of revolution at any point P not occupied by matter.*

Let the axis of the solid be taken as the axis of z with any suitable origin. We have then by Art. 283,

$$V=Y_0+\frac{Z_0}{r}+Y_1r+\frac{Z_1}{r^2}+\dots \quad \dots\dots\dots\dots(1).$$

Since the attracting body is symmetrical about the axis of z it is evident that V cannot be a function of the angular coordinate ϕ. Hence by Art. 286, $Y_0=c_0P_0$, $Z_0=c_0'P_0$, $Y_1=c_1P_1$ &c., where c_0, c_0' &c. are as yet undetermined constants. To find these we put the attracted point on the axis; we then have $P_0=1$, $P_1=1$, &c. The equation (1) thus becomes

$$V=c_0+\frac{c_0'}{r}+c_1r+\frac{c_1'}{r^2}+\dots \quad \dots\dots\dots\dots(2).$$

Suppose then we know the potential of the solid at all points of its axis in a convergent series, then (2) is a known series, and therefore the coefficients c_0, c_0', &c. are also known. The series (1) for the potential at P then becomes

$$V=\left(c_0+\frac{c_0'}{r}\right)P_0+\left(c_1r+\frac{c_1'}{r^2}\right)P_1+\dots \quad \dots\dots\dots(3).$$

Thus the potential has been found.

In this way we arrive at a theorem of Legendre, viz. *if the attraction of a solid of revolution is known for every external point which is on the prolongation of its axis, it is known for every external point.* See Todhunter's *History*, Arts. 782, 791.

301. It may happen that the expansion (2) giving the potential at points on the axis takes different forms at different points. Thus when r is less than some quantity a there may be only positive powers of r, and when r is greater than a there may be only negative powers. Again, if the solid of revolution have a cavity extending to the axis, (2) may assume one form within the cavity and another outside the solid.

If the solid have a ring-like hollow symmetrically placed about the axis of revolution but not extending to it, it is clear that a point P situated in this hollow has no corresponding point Q on the axis from which the potential may be derived. In such a case the values of some of the constants c_0, c_1, &c. may be determined when we know the values of V along some line passing through the cavity and making an angle $\theta = \alpha$ with the axis. It should however be noticed that one of Legendre's functions may vanish when $\theta = \alpha$ and *the unknown constant which accompanies that function would remain undetermined.* Since each Legendre's function is unity when $\theta = 0$ this does not occur when the values of the potential along the axis are given.

302. By integration $\int_0^\pi \frac{d\psi}{a + b\cos\psi} = \frac{\pi}{\sqrt{(a^2 - b^2)}}$. We write $a = 1 - hp$, $b = h\sqrt{(p^2 - 1)}$ and expand both sides in powers of h. Since only the first power of h occurs in the denominator on the left-hand side, the general term is easily found. Comparing the coefficients of h^n we have

$$\frac{1}{\pi}\int_0^\pi \{p \pm \sqrt{(p^2 - 1)}\cos\psi\}^n d\psi = P_n \text{ ....(4).}$$

This formula is given by Laplace, *Mécanique Céleste,* Tome v., page 40.

Since p is less than unity, this integral appears to be imaginary. If however we expand the nth power, the integrals of the odd powers of $\cos\psi$ will vanish between the limits, and a real expression for P_n will remain. We may therefore take either of the signs before the radical. There is another integral which may be deduced from (1), viz.

$$P_n = \frac{1}{\pi}\int_0^\pi \frac{d\psi}{(p \pm \sqrt{p^2 - 1}\cos\psi)^{n+1}} \text{ ....(5).}$$

Suppose that for any portion of the axis the potential is given by $V = f(r)$, where $f(r)$ is such an expansion as (2) Art. 300 with either positive or negative powers of r or both. Substituting for P_n in (3), the integral (4) in the terms with positive powers of r, and the integral (5) in those with negative powers, we have

$$V = \frac{1}{\pi}\int_0^\pi f(rp \pm r\sqrt{p^2 - 1}\cos\psi)\, d\psi \text{...........(6).}$$

Thus *when the potential is known along the axis in the form*

$V=f(r)$, the potential at other points is known in the form of the definite integral (6).

Other forms for P_n and therefore for V may be obtained by other substitutions. For example if we begin with $\int_0^\pi \frac{d\psi}{a+b\cos 2\psi + c\sin 2\psi} = \frac{\pi}{\sqrt{(a^2-b^2-c^2)}}$ and put $a=1-ph$, $b=ph$, $c=\mp h\sqrt{(-1)}$ we find

$$P_n = \frac{2^n}{\pi}\int_0^\pi (\sin\psi)^n \{p\sin\psi \pm \sqrt{(-1)}\cos\psi\}^{-n} d\psi.$$

This result is due to Catalan, *Bulletin de Soc. Math. de France*, 1883, vol. XVI., p. 129.

303. Ex. 1. *To find the potential of a uniform circular ring of infinitely small section at any point not on the axis.*

Let the origin be the centre of the ring and let the axis of the ring be the axis of z. Let a be the radius of the ring, M its mass.

The potential at any point Q on the axis distant r from the origin is evidently $M/\sqrt{a^2+r^2}$. We shall expand this in powers of r/a or a/r according as r is less or greater than a. Taking the first supposition, we have

$$V=\frac{M}{a}\left\{1-\frac{1}{2}\left(\frac{r}{a}\right)^2+\frac{1.3}{2.4}\left(\frac{r}{a}\right)^4-\frac{1.3.5}{2.4.6}\left(\frac{r}{a}\right)^6+\dots\right\}.$$

When r is greater than a the expression may be deduced from that just written down by interchanging a and r.

The potential of the ring at any point P not on the axis is therefore

$$V=\frac{M}{a}\left\{1-\frac{1}{2}P_2\left(\frac{r}{a}\right)^2+\frac{1.3}{2.4}P_4\left(\frac{r}{a}\right)^4-\frac{1.3.5}{2.4.6}P_6\left(\frac{r}{a}\right)^6+\&c.\right\},$$

$$V'=\frac{M}{r}\left\{1-\frac{1}{2}P_2\left(\frac{a}{r}\right)^2+\frac{1.3}{2.4}P_4\left(\frac{a}{r}\right)^4-\frac{1.3.5}{2.4.6}P_6\left(\frac{a}{r}\right)^6+\&c.\right\},$$

according as r is less or greater than a.

Ex. 2. A solid ring is generated by the revolution of a closed curve about an axis Oz and is symmetrical about the equatorial plane. Prove that the level surfaces in the immediate neighbourhood of the intersection O of the axis with that plane are given by $2z^2-x^2-y^2=\beta$ where β is a constant.

Since the potential at a point on the axis is of the form $A+Br^2$, the result follows from Legendre's rule, Art. 300.

Ex. 3. A solid anchor ring is generated by the revolution of a circle of small radius a, the centre describing a circle of radius c. Prove that in the neighbourhood of the origin the potential at the point xyz is $V=\frac{M}{c}\left\{1-\frac{a^2}{8c^2}-\frac{2z^2-x^2-y^2}{4c^2}\right\}$.

Ex. 4. Prove that the potential V' of a homogeneous oblate spheroid of mass M at an external point P is

$$V'=\frac{M}{r}\left\{1-\frac{3.P_2}{3.5}\left(\frac{ae}{r}\right)^2+\frac{3.P_4}{5.7}\left(\frac{ae}{r}\right)^4-\&c.+\frac{(-1)^n.3.P_{2n}}{(2n+1)(2n+3)}\left(\frac{ae}{r}\right)^{2n}+\&c.\right\},$$

where r, θ are the polar coordinates of P referred to the centre and axis of revolution, and e is the *eccentricity* of the generating ellipse.

To prove this we first find the potential V' at an external point on the axis and then use Legendre's rule.

By using Laplace's rule, Art. 297, we at once deduce that *the potential of a heterogeneous spheroid* whose strata of equal density are co-axial spheroids and whose boundary is a surface of equal density is $\int \rho \frac{dV'}{da} da$, the limits being $a=0$ to a. Here a is the semi-axis major of any spheroid, $\rho=f(a)$, $e=\psi(a)$ are the corresponding density and eccentricity and $a=$a at the surface.

If this body represent the earth, we notice that e is very small and *a few terms only of the series are necessary* to find the potential even at points near the surface.

304. Clairaut's theorem. *To investigate the law according to which gravity at any point on the surface of the earth varies with the position of that point**.

Without making any hypothesis respecting the distribution of matter in the interior of the earth, we assume the principle that *the surface of the earth is a level surface of the attraction of the earth and of the centrifugal forces.* If ω be the angular velocity of the earth, the centrifugal acceleration at a distance p from the axis is $\omega^2 p$ and the potential is $\frac{1}{2}\omega^2 p^2$. At all points of the surface we have therefore $$V + \tfrac{1}{2}\omega^2 r^2 \sin^2\theta = \kappa \quad\text{.....................}\quad (1),$$ where θ is the co-latitude of the point, r the radius vector and κ a constant.

The potential V is therefore such that at all points of the surface its value is given by (1), and at all points infinitely distant $V=0$. It follows by Art. 133 that *the potential V is determinate at all points of space external to the surface.*

Let the equation of the surface of the earth be

$$r = c(1 + u_1 + u_2 + \dots) \quad\text{....................}\quad (2),$$

where u_1, u_2, &c. are Laplace's functions of the first and higher

* This famous theorem was given by Clairaut in his *Théorie de la figure de la terre*, 1743. No assumption was made about the law of density in the interior except that the strata of equal density are spheroids of small ellipticity, and that the external surface is one of equilibrium. The theorem was extended by Laplace who, assuming only that the strata are nearly spherical and the surface stratum one of equilibrium, established a connexion between the form of the surface and the variation of gravity which in the particular case of an oblate spheroid gives directly Clairaut's theorem. Stokes, without making any hypothesis respecting the state of the interior of the earth but assuming that the surface is one of equilibrium and nearly spherical, obtained Laplace's equations. *Camb. Phil. Trans.* 1849. O'Brien in his *Mathematical Tracts*, 1840, remarks that if the surface of the earth and also the law of variation of gravity are known the effects of the earth's attraction on the moon follows as a natural consequence independently of any theory except that of universal gravitation. These effects may also be deduced from MacCullagh's theorem on the potential of a body given in Art. 135. See also the author's treatise on *Rigid Dynamics*, vol. II. chap. XII.

The extension of Clairaut's theorems to include terms of the second order of small quantities was first effected by Airy, *Phil. Trans.* 1826, part III. This is also investigated by Callandreau, *Annales de l'Observatoire*, Paris, 1889. There is also a paper by G. H. Darwin in the *Monthly Notices of the Astronomical Society*, London, 1899, who gives a short summary of the works of Helmert, Callandreau, Wiechert on the terms of the second order.

orders. We shall assume as the result of observation that *the surface is so nearly spherical that all the terms after the first are small quantities.* The origin of coordinates is either on the axis or distant from it by small quantities of the first order. In the latter case the term $\omega^2 r^2 \sin^2\theta$ in (1), which already contains the small factor ω^2, is altered only by terms of the second order. The constant c is the radius of the sphere of equal volume and the term u_0 has therefore been omitted, Art. 293. The term u_1 would also be zero if the origin were taken at the centre of gravity of the volume.

The potential at all points external to the earth is given by

$$V = \frac{Y_0}{r} + \frac{Y_1}{r^2} + \dots \qquad (3),$$

where the constants in Y_0, Y_1, &c. depend on those in u_1, u_2, &c.

Since ω^2 is small, it follows from (1) that V is nearly constant over the surface of the earth. Hence when we put $r = c$, the expression (3) for V must differ from its first term only by small quantities. It follows that the functions Y_1, Y_2, &c. are small.

Using (1) and (3) we find

$$\frac{Y_0}{r} + \frac{Y_1}{r^2} + \dots + \tfrac{1}{2}\omega^2 c^2 \left(\tfrac{2}{3} + \tfrac{1}{3} - \cos^2\theta\right) = \kappa,$$

where $\sin^2\theta$ has been arranged as the sum of two Laplace's functions. This equation gives r as a function of θ, ϕ and must therefore reduce to an identity if we substitute for r from (2). In this substitution we write the value of r true to a first approximation in the term Y_0/r, but in the subsequent small terms it is sufficient to put $r = c$. We therefore have

$$\frac{Y_0}{c}(1 - u_1 - u_2 - \&c.) + \frac{Y_1}{c^2} + \frac{Y_2}{c^3} + \&c. + \tfrac{1}{2}\omega^2 c^2 \left(\tfrac{2}{3} + \tfrac{1}{3} - \cos^2\theta\right) = \kappa.$$

Equating to zero the functions of the same order, we deduce that

$$\frac{Y_0}{c} + \frac{\omega^2 c^2}{3} = 0, \quad Y_1 = cY_0 u_1, \quad Y_2 = c^2 Y_0 u_2 - \tfrac{1}{2}\omega^2 c^5 \left(\tfrac{1}{3} - \cos^2\theta\right), \ \&c.$$

$$\therefore \ V = Y_0\left(\frac{1}{r} + \frac{c u_1}{r^2} + \frac{c^2 u_2}{r^3} + \&c.\right) - \frac{\omega^2 c^5}{2r^3}\left(\tfrac{1}{3} - \cos^2\theta\right) \dots (4).$$

This formula expresses the potential of the attraction at any point of external space when the form of the surface is known. It is evident that Y_0 is here the mass of the earth.

305. The force of gravity at a point on the earth's surface is the resultant of the attraction of the earth and the centrifugal force due to the rotation. If ν be the angle between the vertical and the radius vector, $g\cos\nu$ is the component along the radius vector. Since ν is very small, we have

$$g = -\frac{d}{dr}(V + \tfrac{1}{2}\omega^2 r^2 \sin^2\theta)$$
$$= Y_0\left(\frac{1}{r^2} + \frac{2cu_1}{r^3} + \frac{3c^2u_2}{r^4} + \&c.\right) - \frac{3\omega^2c^5}{2r^4}(\tfrac{1}{3} - \cos^2\theta) - \omega^2 r \sin^2\theta,$$

after substituting for r from (2) and rejecting the squares of small quantities we find

$$g = \frac{Y_0}{c^2}(1 - 2u_1 - 2u_2 - \&c.) + \frac{Y_0}{c^2}(2u_1 + 3u_2 + \&c.)$$
$$- \tfrac{3}{2}\omega^2 c(\tfrac{1}{3} - \cos^2\theta) - \omega^2 c(\tfrac{2}{3} + \tfrac{1}{3} - \cos^2\theta).$$

Let G be the mean value of g taken over the whole surface of the earth, then (Art. 288)

$$G = \iint g \sin\theta\, d\theta\, d\phi / 4\pi = \frac{Y_0}{c^2} - \tfrac{2}{3}\omega^2 c.$$

Let m represent $\omega^2 c/G$, we then have

$$g = G\{1 - \tfrac{5}{2}m(\tfrac{1}{3} - \cos^2\theta) + u_2 + 2u_3 + 3u_4 + \&c.\} \ldots\ldots (5).$$

The law of variation of gravity is therefore found, when the form of the surface is given.

306. The surface of the earth is known to be very nearly an oblate spheroid of such small ellipticity that the difference of the polar and equatorial semi-diameters is only 1/300th part of either. We may therefore write its equation in the form

$$r = a(1 - \epsilon\cos^2\theta) \ldots\ldots\ldots\ldots\ldots\ldots\ldots (6).$$

Putting $\theta = \tfrac{1}{2}\pi$ and $\theta = 0$ in turn we see that the equatorial and polar semi-diameters are a and $a(1-\epsilon)$. In order to make a comparison between the equations (6) and (2) we write (6) in the form $r = a\{1 - \tfrac{1}{3}\epsilon + \epsilon(\tfrac{1}{3} - \cos^2\theta)\} = c\{1 + \epsilon(\tfrac{1}{3} - \cos^2\theta)\}$.

We have therefore

$$c = a(1 - \tfrac{1}{3}\epsilon), \quad u_2 = \epsilon(\tfrac{1}{3} - \cos^2\theta), \quad u_1 = 0, \quad u_3 = 0, \ \&c.$$

The expression for g therefore becomes

$$g = G\{1 - (\tfrac{5}{2}m - \epsilon)(\tfrac{1}{3} - \cos^2\theta)\} = G'\{1 + (\tfrac{5}{2}m - \epsilon)\cos^2\theta\} \ldots (7),$$

where $G' = G\{1 - \tfrac{1}{3}(\tfrac{5}{2}m - \epsilon)\}$. Putting $\theta = \tfrac{1}{2}\pi$ we see that G' represents the acceleration due to gravity at the equator.

The centrifugal force at the equator is $\omega^2 a$ and the time of rotation of the earth (viz. $2\pi/\omega$) is 24 hours. Taking a to be about 3963 miles, and mean gravity to be 32·18, we find that $\omega^2 a/G = 1/289$. Since this ratio contains the small factor ω^2, we may put $a = c$ and $G = G'$. We may therefore *define the quantity* $m = \omega^2 c/G$ *to be the ratio of the centrifugal force at the equator to equatorial gravity.*

307. *The potential of the earth at any external point* follows from equation (4). If we put E for the mass of the earth, we have $Y_0 = E$, $\omega^2 c = mG = mE/c^2$. The potential is therefore

$$V = \frac{E}{r} + (\tfrac{1}{2}m - \epsilon)\frac{Ec^2}{r^3}(\cos^2\theta - \tfrac{1}{3}) \text{.............}(8).$$

If P, Q be the polar components of the attraction at any external point, say the moon, we have

$$P = -\frac{dV}{dr} = \frac{E}{r^2} + 3(\tfrac{1}{2}m - \epsilon)\frac{Ec^2}{r^4}(\cos^2\theta - \tfrac{1}{3}),$$

$$Q = \frac{dV}{rd\theta} = \quad -2(\tfrac{1}{2}m - \epsilon)\frac{Ec^2}{r^4}\sin\theta\cos\theta.$$

308. By comparing Laplace's expressions for the potential, (4) or (8), with that given by MacCullagh (Art. 135) we may obtain some information respecting the distribution of matter in the interior of the earth. If the origin in (2) be taken at the centre of gravity of the volume, the term u_1 becomes zero. Since the term containing $1/r^2$ in the potential is then absent the origin is also at the centre of gravity of the mass (Art. 135). *The centres of gravity of the volume and mass must therefore coincide.*

Since by (8) the potential is independent of the longitude, the same must be true in the expression

$$V = \frac{E}{r} + \frac{A + B + C - 3I}{2r^3} + \&c.$$

This requires that *the axis of rotation should be a principal axis of the mass.* Again writing $B = A$, and $I = A\sin^2\theta - C\cos^2\theta$, we see that

$$\frac{C - A}{Ec^2} = \frac{2}{3}\left(\epsilon - \frac{m}{2}\right).$$

309. Clairaut's theorem to a second approximation. It is not difficult to carry the approximation to the second order of small quantities if we follow the same reasoning. We make no assumption about the law of density of the earth except that the potential is symmetrical about the axis of rotation and on each side of the plane of the equator. As a trial solution, we omit the even powers of $1/r$ and take instead of (1) and (3) of Art. 304 the equations

$V + \frac{1}{2}\omega^2 r^2 \sin^2\theta = \kappa$...............(1), $\qquad V = \frac{E}{r} + \frac{\beta P_2}{r^3} + \frac{\gamma P_4}{r^5} + \ldots$ ....(3), where E is the mass of the earth; P_2, P_4, &c. are Legendre's functions and β, γ are two constants. We shall also suppose that the surface of the earth has the form

$$r = a(1 - \epsilon\cos^2\theta - p^2\sin^2\theta\cos^2\theta) \text{.........................} (2),$$

where a is the semiaxis major and if the form be a spheroid, $p^2 = \frac{3}{2}\epsilon^2$.

If we substitute from (2) and (3) in (1) as in Art. 304 the result should be an identity. This will be found to be true if β and γ are small quantities respectively of the first and second orders, and the expression for V in (3) is restricted to the first three terms. Equating to zero the coefficients of $\cos^2\theta$ and $\cos^4\theta$, (all the higher powers having coefficients of at least the third order), we thus obtain two equations to determine β and γ.

Let m be the ratio of the centrifugal force at the equator to equatorial gravity, then

$$\omega^2 a = m\left(-\frac{dV}{dr} - \omega^2 a\right),$$

where a is to be written for r after the differentiation has been performed, and $\cos\theta$ put equal to zero.

In this way we obtain the three results

$$\frac{\beta}{a^4} = \frac{2}{3}\frac{E}{a^2}\left\{\frac{m}{2} - \epsilon - \tfrac{3}{4}m^2 - \tfrac{1}{7}m\epsilon + \tfrac{5}{7}\epsilon^2 - \tfrac{4}{7}p^2\right\} \text{....................} (4),$$

$$\frac{\gamma}{a^6} = \frac{8}{35}\frac{E}{a^2}\left\{-\tfrac{5}{2}m\epsilon + 2\epsilon^2 + p^2\right\} \text{................................} (5),$$

$$\omega^2 a = \frac{E}{a^2}m\{1 + \epsilon - \tfrac{3}{2}m\} \text{..} (6).$$

After substituting these values of β and γ in (3) we have an expression for the potential of the earth at all external points.

To find gravity g at the surface, we have

$$g^2 = \left(\frac{dV'}{dr}\right)^2 + \left(\frac{dV'}{r\,d\theta}\right)^2, \quad \therefore\ g = -\frac{dV'}{dr} + \frac{1}{2}\left(\frac{dV'}{r\,d\theta}\right)^2 \Big/ \frac{E}{a^2} \text{...........} (7),$$

where $V' = V + \frac{1}{2}\omega^2 r^2\sin^2\theta$. On substituting this value of V' we soon see that the expression for g contains terms which are constant multiples of $\cos^2\theta$ and $\cos^4\theta$. We may therefore write

$$g = G'\{1 + \lambda\cos^2\theta + \mu\sin^2\theta\cos^2\theta\} \text{...........................} (8).$$

To find the three constants G', λ, μ we notice that $g = G'$ when $\theta = \frac{1}{2}\pi$. Hence G' is the value of equatorial gravity, and may be found from (7) by putting $r = a$ and $\theta = \frac{1}{2}\pi$ after the differentiations have been performed. We observe next that $\lambda G'$ is the difference between the values of gravity at the pole and the equator and that both these may be deduced from (7). Lastly we notice that $-\mu G'$ is the coefficient of $\cos^4\theta$ in the value of g; and this may be very shortly deduced from (7). In this way we find

$$G' = \frac{E}{a^2}\left\{1 - \tfrac{3}{2}m + \epsilon - \tfrac{27}{14}m\epsilon + \tfrac{9}{4}m^2 + \tfrac{1}{7}\epsilon^2 + \tfrac{4}{7}p^2\right\},$$

$$\lambda = \tfrac{5}{2}m - \epsilon - \tfrac{17}{14}m\epsilon - \tfrac{3}{7}\epsilon^2 + \tfrac{2}{7}p^2,$$

$$\mu = \tfrac{15}{2}m\epsilon + \epsilon^2 - 3p^2.$$

The angle θ is the angle the radius vector r makes with the axis of rotation. If θ' be the angle the direction of gravity makes with the axis of rotation we have

$$\theta = \theta' + 2\epsilon\sin\theta'\cos\theta'.$$

We then find by an easy substitution

$$g = G'\{1 + \lambda\cos^2\theta' + (\mu - 4\lambda\epsilon)\sin^2\theta'\cos^2\theta'\}.$$

We may extend Clairaut's theorem to a third approximation by proceeding in the same way. We then include a fourth term $\delta P_6/r^7$ in equation (3) in which δ is a small quantity of the third order. We have also an additional term in (2). The numerical calculations are troublesome and the additional terms too small to be of any interest.

310. Figure of Saturn. *To find, to a first approximation, the effect on the figure of Saturn of the attraction of the ring.* We suppose the form of Saturn to be nearly spherical, the ring to be circular, concentric, homogeneous, of small section and situated in the plane of the planet's equator. The planet rotates with a small angular velocity. The principle of the investigation is that the surface of Saturn is a level surface of the attractions of the planet, ring and the centrifugal forces.

Let the polar equation of the surface of Saturn be

$$r = c(1 + Y_1 + Y_2 + \&c.) \quad \ldots\ldots\ldots\ldots\ldots\ldots (1).$$

Since the surface is nearly spherical, all the harmonics Y_1, Y_2, &c. are small quantities whose squares and products are to be neglected. By omitting the term Y_0, we have made c to be the radius of the sphere of equal volume. Also the mass $M = \frac{4}{3}\pi\rho c^3$ where ρ is the density. By Art. 294 the potential of Saturn at an external point is

$$V_\varepsilon = \frac{4\pi\rho c^3}{r}\left\{\frac{1}{3} + Y_1\frac{c}{3r} + Y_2\frac{c^2}{5r^2} + \&c.\right\} \quad \ldots\ldots\ldots (2).$$

We now substitute from (1) in the first term of (2) and put $r = c$ in the small terms. We thus find

$$V_s = \frac{M}{c}(1 - Y_1 - Y_2 - \&c.) + \frac{3M}{c}\left(\frac{Y_1}{3} + \frac{Y_2}{5} + \&c.\right) \ldots (3).$$

The centrifugal force at any point is $\omega^2 x$, where ω is the angular velocity of the planet and x the distance from the axis of rotation. Putting $x = r\sin\theta$, the potential of the centrifugal forces becomes

$$V_c = \frac{\omega^2 x^2}{2} = \frac{\omega^2 r^2}{3}\{1 - P_2\} \quad \ldots\ldots\ldots\ldots\ldots\ldots (4).$$

Since ω^2 is small, we put $r = c$ in this formula.

Lastly if M/n is the mass of the ring, supposed to be condensed into a circle of radius a, the potential of the ring is, by Art. 303,

$$V_r = \frac{M}{na}\left\{1 - \frac{1}{2}P_2\left(\frac{r}{a}\right)^2 + \frac{1.3}{2.4}P_4\left(\frac{r}{a}\right)^4 - \&c.\right\} \quad \ldots\ldots (5).$$

Since $1/n$ is small, we again put $r = c$ in the small terms.

We now substitute these three potentials in the equation

$$V_s + V_c + V_r = \kappa \ldots\ldots\ldots\ldots\ldots\ldots\ldots\ldots (6),$$

where κ is a constant. Since there can be but one expansion of the potential in harmonic functions, the sums of the several potentials of each order must separately vanish.

The potentials V_c and V_r contain no harmonics of an odd order; hence those in V_s must also vanish. We therefore have $Y_1=0$, $Y_3=0$, &c. After substituting for V_s, V_c, V_r and equating to zero the sums of the harmonics of the second and fourth orders, we have

$$Y_2=-\frac{5}{2}\left(\frac{c^3}{2na^3}+\frac{\omega^2}{4\pi\rho}\right)P_2, \quad Y_4=\frac{9}{16n}\left(\frac{c}{a}\right)^5 P_4.$$

The remaining terms contain higher powers of c/a. Since this fraction is nearly $\frac{1}{2}$, these terms may be disregarded in a first approximation.

Representing these results by $Y_2=-\beta P_2$ and $Y_4=\gamma P_4$, we see that a near approximation to the form of Saturn is given by

$$r=c\{1-\beta P_2(\cos\theta)+\gamma P_4(\cos\theta)\} \quad \text{...........(7)},$$

where θ is the angle the radius vector makes with the axis of rotation.

If the last term of (7) were omitted the surface would be an oblate spheroid, Art. 306. The effect of the small term γP_4 is to lengthen slightly both the polar and equatorial diameters and to shorten those in middle latitudes.

The real shape of Saturn was at one time a matter of great controversy. The first observations were made by Sir W. Herschel who found that the deviation of the figure from that of an oblate spheroid was so great that the *longest diameter* was in latitude 43° 20′. Herschel believed that this peculiarity was due to the attraction of the ring. But it was soon discovered that this opinion was not confirmed by a theoretical examination of the effect of the ring. Bessel however afterwards proved by direct measurements of several diameters that the true form was very nearly that of an oblate spheroid. Probably the discrepancy was due to an optical distortion of the planet when seen through its atmosphere. These measurements of Bessel are given in a memoir *On the dimensions and position of the ring of Saturn and those of the planet.* See a translation in the *Additions à la Connaissance des Temps* for the year 1838, page 47.

311. Ex. 1. If the free surface of equilibrium of the earth is an ellipsoid, and if ϵ is the mean ellipticity of the meridians, η the ellipticity of the equator, and l the longitude reckoned from the meridian of greatest ellipticity, and λ the latitude, prove that $g=G\{1-(\frac{5}{2}m-\epsilon)(\frac{1}{3}-\sin^2\lambda)+\frac{1}{2}\eta\cos^2\lambda\cos 2l\}$. [Math. T. 1867.]

Ex. 2. *Jacobi's ellipsoid.* An ellipsoid revolves about a principal diameter with an angular velocity which is not necessarily small. Prove that the internal level surfaces due to the attraction and the centrifugal forces are similar ellipsoids. Prove also that the resultant force at any point P on a given level surface is proportional to the length of the normal intercepted between P and the principal plane perpendicular to the axis of revolution. If the boundary of the ellipsoid is itself a level surface and the angular velocity is small, prove by comparing this result with Clairaut's formula for gravity that $\epsilon = 5m/4$.

By adding to the value of V in Art. 212 the terms due to the centrifugal forces, viz. $\omega^2 (x^2 + y^2)$, we see at once that the level surfaces are similar ellipsoids. By Art. 46, the force at any point P on a given level surface is inversely proportional to the distance dp between two neighbouring level surfaces. In our case dp is proportional to p (Art. 195) and therefore inversely proportional to the length of the normal. For points on the axis of rotation but on different level surfaces, the force is $C\rho z$, (Art. 213).

312. Ex. Let the earth be a solid heterogeneous nearly spherical nucleus completely covered by a homogeneous ocean. If the system is made to rotate, with equal angular velocities, about the principal axes at the centre of gravity of the nucleus in succession, the ocean will assume three different forms. Prove that the mean of the three radii vectores in any given direction is the same as the radius vector of the ocean when supposed to be in equilibrium on the nucleus without rotation.

Let $r = a(1 + \Sigma u_n)$, $r' = b(1 + \Sigma v_n)$ be the equations of the surfaces of the nucleus and ocean as in Art. 304. Then since the nucleus and the mass of the ocean are given, a, b and u_n are known and we have to find v_n. The potential of a homogeneous mass of fluid extending from the centre to the surface of the ocean is given in (3) of Art. 297. The potential of the excess of the nucleus above that of an equal volume of fluid, and the potential of the centrifugal forces are given in Art. 304. The sum of these three potentials is constant along the surface. By equating to zero the sum of functions of the same order, we notice that v_n is independent of ω except when $n=2$. We find that $v_2 = Z_2 + A\left(\frac{1}{3} - \cos^2\theta\right)$ where Z_2 is independent of ω, and A is a multiple of ω. Since the sum of the squares of the direction cosines of a radius vector is unity, the mean of the three values of v_2 is independent of ω.

313. Ex. Let the earth consist of a spheroidal homogeneous fluid nucleus surrounded by a consolidated crust whose external surface is also a spheroid, the two spheroids being level surfaces of the attractions and centrifugal forces. If ϵ', ϵ be the ellipticities; a', a the mean radii of the inner and outer spheroids; ρ', ρ the densities of the two substances, prove that

$$\epsilon\rho + \left(\frac{\mathrm{a}'}{\mathrm{a}}\right)^5 \epsilon'(\rho' - \rho) = \tfrac{5}{6}(2\epsilon - m)\,\Delta,$$

$$(\epsilon' - \epsilon)\,\rho + \tfrac{2}{3}\epsilon'\rho' = \tfrac{5}{6} m\Delta,$$

$$\rho' - \rho = (\Delta - \rho)\left(\frac{\mathrm{a}}{\mathrm{a}'}\right)^3,$$

where the mean density Δ is given by the last equation. The whole mass is supposed to rotate about a principal axis at the centre of gravity with a small angular velocity ω.

To obtain the first two equations we use the formulæ (2) and (3) of Art. 297 to find the potentials of the two portions of the earth. The sum of these together with that of the centrifugal forces is constant along each spheroid.

In the case of the earth $\Delta = 2\rho$, $m = 1/289$, $\epsilon = 1/300$, and $\mathrm{a} = 3958$ miles. With these numbers the Rev. S. Haughton deduced from these equations that the thickness of the crust is 768 miles. *Trans. Royal Irish Academy*, 1851, vol. XXII. dated 1855. It is remarkable that the thickness should be so great. The *first attempt* to discover the thickness of the crust was made by W. Hopkins, who estimated the minimum thickness to be not less than one-fourth or one-fifth of the earth's radius, *Phil. Trans.* 1842. Much has been written on the subject since then.

Magnetic Attractions.

314. Potentials of Magnets. Two equal particles, each of mass m, are placed at two points A, B, whose distance apart is $2a$. Any particle being placed at P one of these repels the particle at P, while the other attracts it. Such a combination may be called *a simple magnet**. See the figure of Art. 316.

It will be convenient to take repulsion as the standard case. Let the mass of the particle at A be called positive, then that at B is the negative mass. The particle at P, if of positive mass, will then be repelled by the particle at A and attracted by that at B. The ends A and B are called respectively *the positive and negative poles* of the magnet.

Since the particle at each end of a magnet repels a particle of the same sign, it is a matter of convention to call one positive and the other negative. The convention adopted in Maxwell's *Electricity* is that *when used as a compass the positive pole points north* (Art. 394). It follows that the north pole of the earth attracts the positive pole of the magnet. The south pole is therefore the positive pole of the earth.

315. The line BCA is called *the axis,* and the distance BA *the length*; the positive direction is BA. The middle point C is called *the centre.* The quantity m is called *the strength* and the product of the length by the strength, viz. $2am$ or M, is called *the magnetic moment.*

If the point P lie in the axis, the magnet is said to be *end on.* If the axis is perpendicular to the distance CP, the magnet is *broadside on.*

The strengths are so measured that the force exerted by m on m' at a distance r is mm'/r^2. As explained in Art. 5 the dimensions

* The latin treatise of W. Gilbert of Colchester, *De Magnete &c.*, 1600, (translated by F. Mottelay), 1893, is generally referred to as one of the earliest. The book discusses in general terms, and without Mathematics, the magnetic theory of the Earth. The mathematics of Magnetism was first properly discussed by Poisson, and he was soon followed by other great mathematicians. In 1849 Kelvin gave a complete theory which, without assuming any hypothetical magnetic fluid, is founded on facts generally known, see the *Reprint of papers on Electrostatics and Magnetism.* The student of Magnetism will find the treatise of J. J. Thomson of great assistance, and also that of Maxwell when more advanced in the subject.

of strength are $LF^{\frac{1}{2}}$ where L represents length and F force. The dimensions of magnetic moment are $L^2F^{\frac{1}{2}}$.

In all that is here said (unless when otherwise specified) the magnets are supposed to be used in air. The effects of the medium are not included.

316. *To find the potential of a simple magnet at any point P.* Let r be the distance of P from the middle point C of AB and let θ be the angle PCA. We notice that the angle θ is measured from the positive end towards P. We have in a field without induction

$$V = \frac{m}{AP} + \frac{-m}{BP} = \frac{m}{\sqrt{(r^2 + a^2 - 2ar\cos\theta)}} - \frac{m}{\sqrt{(r^2 + a^2 + 2ar\cos\theta)}}$$
$$= \frac{2am}{r^2}\left\{\cos\theta + P_3\left(\frac{a}{r}\right)^2 + \&c.\right\}.$$

When the length $2a$ of the magnet is small compared with the distance r, it is often a sufficient approximation to reject all but the first term of this series. Put $M = 2am$, *the potential of the magnet as given by its principal term is then* $V = \frac{M\cos\theta}{r^2}$. The order of the first term rejected is the fraction $(a/r)^2$ of the term retained. Magnets in which it is sufficient to take account of the principal term only are sometimes called *small magnets.*

Since repulsion has been taken as the standard case the component forces (Art. 41) at P in the direction CP and perpendicular to CP are respectively

$$F = -\frac{dV}{dr} = \frac{2M\cos\theta}{r^3}, \quad G = -\frac{dV}{rd\theta} = \frac{M\sin\theta}{r^3},$$

the latter being measured positively in the direction which makes θ increase. In the figure the arrow-heads indicate the directions of the forces at P due to the repulsion of A and the attraction of B; while the double arrows indicate the positive directions of the components F and G.

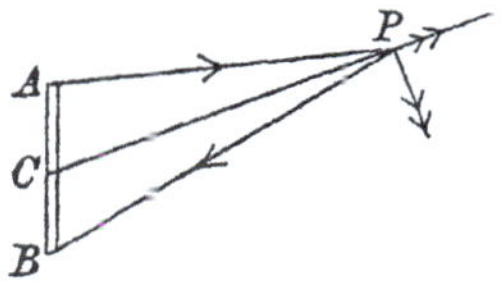

It appears from the investigation that both the potential and the force at any point P are not altered by changing the length $2a$ and the strength m provided the product $M = 2am$ is kept unchanged. *A small magnet is therefore given when we know* (1) *the position of its centre C,* (2) *the positive direction of its axis and* (3) *the magnetic moment M.*

317. Resolution of Magnets. When a small magnet of moment M' is end on to P so that $\theta = 0$, it follows from Art. 316 that the resultant force at P is directed along CP and is equal to $2M'/r^3$. When a small magnet of moment M'' is broadside on to P so that $\theta = \frac{1}{2}\pi$, the resultant force at P is perpendicular to CP and is equal to M''/r^3. If we take $M' = M\cos\theta$, $M'' = M\sin\theta$, we notice that the component forces at P due to the magnet M are the same in direction and magnitude as those due to two magnets M', M''. It therefore follows, that *the small magnet M may be resolved into two components $M\cos\theta$, $M\sin\theta$.* This rule being true for a rectangular resolution may be extended to include all cases. Hence *small magnetic moments may be compounded and resolved by the parallelogram law.*

One advantage of the resolution into components "*end on*" and "*broadside on*" is that *the direction of the force due to each component is at once evident, the direction being in every case parallel to the axis of the component magnet.* The force at P due to a magnet "end on" acts in the *positive* direction of its axis; the force due to a magnet "broadside on" acts parallel to the *negative* direction of the axis.

318. Mutual action of two small magnets. Let the two small magnets BCA, $B'C'A'$ *be in one plane* and let their moments be M, M'. Let $CC' = r$, and let r be measured positively from C to C'. Let θ, θ' be the angles the positive directions of the axes make with the positive direction of r, that is with CC' produced, and let $B'C'A' = 2a'$.

We resolve the acting magnet M into $M\cos\theta$, $M\sin\theta$. These produce forces F and G at the centre C' of the magnet $B'C'A'$ respectively where

$$F = 2M\cos\theta/r^3$$

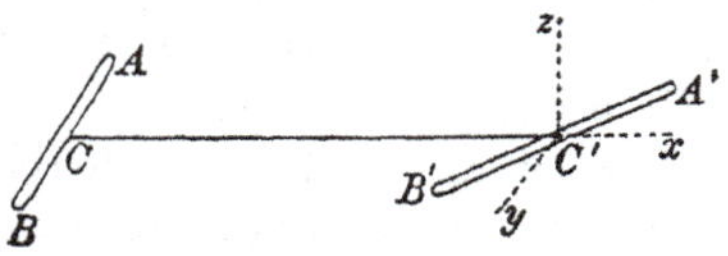

and $G = M\sin\theta/r^3$. The former acts along CC' and the latter perpendicularly to CC' in a direction *tending to increase θ.*

These may also be regarded as the forces at any point in the neighbourhood of C', provided the magnets are so small that we can reject Fa'/r and Ga'/r. We therefore apply them without alteration of magnitude or direction to the pole A' and also with *their signs reversed* to the pole B'. *The action of the one magnet on the other is therefore a couple.* See Art. 320.

To find the magnitude of the couple, we take the moment about the centre C' of the force which acts at the positive pole A' only and double the result. The couple tending to increase θ' is therefore

$$\Gamma' = -2m'a'(F\sin\theta' + G\cos\theta')$$
$$= -\frac{MM'}{r^3}(2\cos\theta\sin\theta' + \sin\theta\cos\theta').$$

319. *When the two magnets are not in one plane* we proceed in the same way. Let CC' be taken as the axis of x, and let $(\lambda\mu\nu)$, $(\lambda'\mu'\nu')$ be the direction cosines of the positive directions of the two magnetic axes. We resolve the acting magnet into $M\lambda$, $M\mu$, $M\nu$. The former being "end on" produces a force at C' which acts in the positive direction of its axis and is therefore $X = 2M\lambda/r^3$. The two others being "broadside on" produce forces which act in the negative direction of their axes and are $Y = -M\mu/r^3$ and $Z = -M\nu/r^3$. These forces are transferred to act at the positive pole A' whose coordinates are $x' = a'\lambda'$, $y' = a'\mu'$, $z' = a'\nu'$. Twice their moments about any axes having C' for origin give the couples which represent the action of one magnet on the other. The couples about the axes of x, y, z are (by Art. 257, vol. I.)

$$K_x' = \frac{MM'}{r^3}(\mu\nu' - \mu'\nu),\quad K_y' = \frac{MM'}{r^3}(2\lambda\nu' + \lambda'\nu),\quad K_z' = -\frac{MM'}{r^3}(2\lambda\mu' + \lambda'\mu).$$

To simplify the results, let the plane containing CC' and the magnetic axis $B'C'A'$ be the plane of xz. Let θ, θ' be the angles the magnetic axes make with the axis of x and let ϕ be the angle between the planes in which θ, θ' are measured. The coordinates of A' are then $x' = a'\cos\theta'$, $y' = 0$, $z' = a'\sin\theta'$. The forces $X = 2M\cos\theta/r^3$ and $Z = -M\sin\theta\cos\phi/r^3$ act in the plane xz and produce a couple

$$\Gamma' = -\frac{MM'}{r^3}\{2\cos\theta\sin\theta' + \sin\theta\cos\theta'\cos\phi\}.$$

This couple when positive tends to *increase* θ'. The force $Y = -M\sin\theta\sin\phi/r^3$ produces a couple Δ' in the plane $yC'A'$ where

$$\Delta' = \frac{MM'}{r^3}\sin\theta\sin\phi.$$

When positive this couple tends to *increase* ϕ and acts from A' to y.

When the plane xz contains the axis $B'C'A'$, $\lambda = \cos\theta$, $\mu = \sin\theta\sin\phi$, $\nu = \sin\theta\cos\phi$, $\lambda' = \cos\theta'$, $\mu' = 0$, $\nu' = \sin\theta'$. The couples $\Delta' = -K_x\sin\theta' + K_z\cos\theta'$, and $\Gamma' = -K_y$ may then be at once deduced from those of K_x', K_y', K_z'.

320. *The component forces* at the poles A', B' have been regarded as equal in magnitude but opposite in sign. To this degree of approximation the forces which tend to move the centre of gravity of the magnet $B'A'$ are zero. This means that the expressions for their magnitudes contain an additional factor r in the denominator so that *the forces vary as the inverse fourth power of the distance.*

These forces are very small and are generally neglected. We must however notice that, though the moment about C' of the forces in Art. 318 which act between the poles of the magnets is Γ', the moment Γ about C of the same forces differs from Γ' by the moments of the forces which act at the centre C'. Though these forces are very small, yet the arm r is here very great and the resulting couple is of the order $1/r^3$.

It is sufficient to indicate the method of finding these forces and to state their magnitudes. Let (x, y, z), (x', y', z') be the coordinates of the positive poles A, A' of the two magnets referred to origins C, C' respectively. The distance D between A, A' is
$$D^2 = (r + x' - x)^2 + (y' - y)^2 + (z' - z)^2.$$
The forces X, Y, Z are then
$$X' = \Sigma \frac{mm'}{D^3}(r + x' - x), \quad Y' = \Sigma \frac{mm'}{D^3}(y' - y), \quad Z' = \&c.$$
We now expand these expressions in inverse powers of r and effect the summation of each term for positive and negative values of m, m'. Finally we write $x = a\lambda$, $x' = a'\lambda'$ &c. We then find
$$X' = -\frac{3MM'}{r^4}(2\lambda\lambda' - \mu\mu' - \nu\nu'), \quad Y' = \frac{3MM'}{r^4}(\lambda\mu' + \lambda'\mu), \quad Z' = \frac{3MM'}{r^4}(\lambda\nu' + \lambda'\nu).$$

Ex. Two small magnets float horizontally on the surface of water, one along the direction of the straight line joining their centres and the other at right angles to it. Prove that the action of each magnet on the other reduces to a single force at right angles to the straight line joining the centres and meeting that line at one-third of its length from the longitudinal magnet. [Coll. Ex. 1900.]

321. Potential energy. *A small magnet, whose moment is M', is acted on by a number of given magnets; it is required to find the potential energy.* Let m' be the strength, $2a'$ the length of the small magnet $B'C'A'$, then $M' = 2a'm'$. Let V, V' be the potentials of the field at the negative and positive poles of the small magnet, then $F' = -(V' - V)/2a'$ is the component of force, due to the field, at the small magnet in the positive direction of the axis, Art. 40. The mutual potential energy is, by Art. 59,
$$W = -Vm' + V'm' = -M'F'.$$
The required potential energy is therefore found by multiplying the moment M' of the small magnet by the axial component of force F' and changing the sign.

322. *To find the potential energy of two small magnets.* We use the same notation as in Art. 319. The component forces due to the magnet M are $X = 2M\lambda/r^3$, $Y = -M\mu/r^3$, $Z = -M\nu/r^3$. The resolved part F' of these along the axis of the magnet M' is

$$F' = \frac{M}{r^3}(2\lambda\lambda' - \mu\mu' - \nu\nu').$$

The required potential energy is, by Art. 321,

$$W = -\frac{MM'}{r^3}(2\lambda\lambda' - \mu\mu' - \nu\nu') \quad\ldots\ldots\ldots\ldots\ldots (1).$$

If ψ is the angle which the positive directions of the magnetic axes make with each other

$$\cos\psi = \lambda\lambda' + \mu\mu' + \nu\nu';$$

$$\therefore\ W = -\frac{MM'}{r^3}(3\lambda\lambda' - \cos\psi) \quad\ldots\ldots\ldots\ldots\ldots (2).$$

If the magnetic axes BCA, $B'C'A'$ make angles θ, θ' with CC' and if the planes ACC', $A'CC'$ make an angle ϕ with each other we may put, as in Art. 319,

$$\lambda = \cos\theta, \quad \mu = \sin\theta\sin\phi, \quad \nu = \sin\theta\cos\phi,$$
$$\lambda' = \cos\theta', \quad \mu' = 0, \quad \nu' = \sin\theta',$$

$$\therefore\ W = -\frac{MM'}{r^3}\{2\cos\theta\cos\theta' - \sin\theta\sin\theta'\cos\phi\} \quad\ldots\ldots (3).$$

The potential energy W being known, we deduce without difficulty the couples which represent the action of the magnet M on M'. Referring to the figure of Art. 318 we see that $\Gamma' = -dW/d\theta'$ is the moment of the couple in the plane in which θ' is measured (Art. 41). The couple in the perpendicular plane (that is the plane $yC'A'$) is $\Delta' = -dW/\sin\theta' d\phi$.

Ex. 1. If the law of force be the inverse κth power of the distance, prove (1) that the potential of a small magnet at any point P is $V = M\cos\theta/r^\kappa$ and (2) that the potential energy of two small magnets is

$$W = \frac{MM'}{r^{\kappa+1}}\{\cos\psi - (\kappa+1)\cos\theta\cos\theta'\},$$

where the notation is the same as in Art. 322.

To prove the first part we proceed as in Art. 316. To obtain the second result we follow the method of Art. 322, using the rule in Art. 321.

Ex. 2. A small magnet free to move about its centre is acted on by another fixed magnet and the law of force between the poles is the inverse κth power of the distance. The magnets are placed with the axis of one along and that of the other perpendicular to the straight line joining the centres. Prove that the couple tending to produce rotation in the free magnet when the fixed magnet is "end on" is κ times that when "broadside on."

By making experiments on the magnitudes of these couples Gauss determined the value of κ and thus proved that the law of force is the inverse square. The experiments are shortly described in J. J. Thomson's *Electricity and Magnetism.*

323. *A series of particles whose masses (positive or negative) are m_1, m_2, &c. are placed in a straight line Ox at given points A_1, A_2, &c. Find the equations of the lines of force.*

Let r_1, r_2, &c. be the distances of any point P from A_1, A_2, &c.; θ_1, θ_2, &c. the angles these distances make with Ox. Let ϕ_1, ϕ_2, &c. be the angles the tangent to the line of force through P makes with the radii vectores r_1, r_2, &c.; then taking any one of these

$$\sin\phi = rd\theta/ds.$$

Since the resultant force at P acts along the line of force, we have

$$\Sigma\frac{m}{r^2}\sin\phi = 0, \qquad \therefore\ \Sigma m\frac{d\theta}{r} = 0.$$

When the points A_1, A_2, &c. lie in the axis of x,

$$r_1\sin\theta_1 = r_2\sin\theta_2 = \text{\&c.}$$

Hence

$$\Sigma m\sin\theta d\theta = 0, \quad \therefore\ \Sigma m\cos\theta = K.$$

The equations of the lines of force and the level surfaces written at length, are therefore

$$m_1\cos\theta_1 + m_2\cos\theta_2 + \text{\&c.} = K,$$
$$m_1/r_1 + m_2/r_2 + \text{\&c.} = K',$$

where K and K' are arbitrary constants.

In a magnet $m_2 = -m_1$, the lines of force and the level surfaces reduce to

$$\cos\theta - \cos\theta' = K_1, \qquad 1/r_1 - 1/r_2 = K_1'.$$

Line of force from one particle to another. When a line of force passes through one of the attracting or repelling particles, the radius vector at that particle becomes a tangent and θ is then the angle that tangent makes with the positive direction of the axis of x. Let a line of force pass between the points A_i, A_k. Then, equating the values of K at these two points, we have

$$m_1 + \text{\&c.} + m_i\cos\theta_i - m_{i+1} - \text{\&c.} = m_1 + \text{\&c.} + m_k\cos\theta_k - m_{k+1} - \text{\&c.}$$
$$\therefore\ m_i\sin^2\tfrac{1}{2}\theta_i + m_{i+1} + \text{\&c.} + m_{k-1} + m_k\cos^2\tfrac{1}{2}\theta_k = 0.$$

If all the masses have the same sign the only line of force which can pass from one particle to another is the straight line Ox on which all the particles are situated.

Line of force from a particle to an infinite distance. Let a line of force pass from the particle m_i to a point at an infinite distance in a direction which ultimately makes an angle β with the axis of x. We then have in the same way

$$m_1 + m_2 + \text{\&c.} + m_i\cos\theta_i - m_{i+1} - \text{\&c.} = (\Sigma m)\cos\beta.$$

In a magnet where $\Sigma m = 0$, no line of force can pass to an infinite distance except the one along Ox.

Parallel rods. We may obtain a corresponding theorem for a series of thin parallel attracting rods. Let the rods be cut by a perpendicular plane in the points A_1, A_2, &c. and let (r_1, θ_1), (r_2, θ_2), &c. be the polar coordinates of any point P in this plane referred to A_1, A_2, &c. as origins. If m_1, m_2, &c. are the line densities of the rods, *the lines of force and level curves in this plane are respectively* $\Sigma m\theta = K$, $\Sigma m \log r = K'$.

324. Ex. 1. Prove that the lines of force of a simple magnet BCA (not necessarily small) are symmetrical curves concave to the magnet and passing through its poles. If P be the middle point of one of the lines of force, prove that the curvature at P is three halves that of the circle BPA, and that the curvatures at B and A are zero. If BPA be an equilateral triangle prove that the line of force meets the magnet at right angles. [Math. T. 1871.]

Ex. 2. A small fixed magnet BCA acts on a small magnet $B'C'A'$ free to turn about its centre. Prove that when the free magnet is in equilibrium its axis lies in the plane ACC' and that $\tan\theta' = -\frac{1}{2}\tan\theta$.

Let the magnetic forces of the earth be represented by those of a small magnet placed at the centre with its positive pole pointing south. The north-seeking pole of the compass needle is then its positive pole. It follows that in north magnetic latitude λ, the dip D below the horizon of a small magnet free to turn about its centre of gravity (usually called a dipping needle) is found by writing $\frac{1}{2}\pi + \lambda$ for θ and $\frac{3}{2}\pi - D$ for θ'. Hence *the tangent of the dip is twice the tangent of the magnetic latitude.*

Ex. 3. A small fixed magnet BCA acts on a small magnet $B'C'A'$ free to turn about its centre in the plane ACC'. Prove that the two positions of $B'C'A'$ in which the couple Γ', tending to produce rotation, is greatest and zero are at right angles. Prove also that the maximum couple is $E(1 + 3\cos^2\theta)^{\frac{1}{2}}$ where $E = MM'/r^3$, and that when the magnet $B'C'A'$ makes an angle ϕ with its position in equilibrium the couple is proportional to $\sin\phi$.

Ex. 4. A compass needle $B'C'A'$ is free to turn about its centre C' in a horizontal plane and is acted on by a small vertical magnet whose centre C lies on the circumference of a horizontal circle having its centre in the vertical $C'Z$. Prove that, if ϕ, ϕ' be the angles the planes $ZC'C$, $ZC'A'$ make with the magnetic meridian, $\sin(\phi - \phi')/\sin\phi'$ is approximately the same for all positions of the disturbing magnet.

Ex. 5. Three small magnets are placed with their centres at the angular points of an equilateral triangle ABC and being free to move about their centres rest in the following positions. The magnet at A is parallel to BC whilst those at B and C are at right angles to AB and AC respectively. Prove that the magnetic moments are in the ratios $\sqrt{3} : 4 : 4$. [Math. T. 1880.]
(Use Art. 318.)

Ex. 6. Two small magnets of moments M, M' are fixed at two corners of an equilateral triangle with their axes bisecting the angles. A third small magnet is free to move at the other angular point. Prove that its axis makes with the bisector of the third angle an angle whose tangent is $\sqrt{3}\,(M \sim M')/7\,(M+M')$.

[Math. T. 1882.]

Ex. 7. Point charges e, $-e'$, $-e'$ are placed at O, A, B respectively which are in a straight line and $OA = OB$. Prove that, if $e > 2e'$, the greatest angle a line of force leaving O and entering A can make with OA is α, where $e \sin^2 \frac{1}{2}\alpha = e'$.

[Coll. Ex. 1900.]

[If the line of force pass from O to an infinite distance we must have $e \cos\theta < e - 2e'$; if it arrive at A, we have $e \sin^2 \frac{1}{2}\theta = e' \sin^2 \frac{1}{2}\theta'$, where θ, θ' are the angles the tangents at O, A make with OA and AO respectively. If θ is greater than the value of α given above, the line cannot go to A; if less it cannot go to an infinite distance. See Art. 323.]

325. To determine by experiment the numerical values of (1) the horizontal force H due to the earth's magnetism and (2) the magnetic moment M of a given magnet. There are several ways of effecting this, but in general two experiments have to be made, one to determine the ratio H/M and the other the product HM. The two following examples will explain the process without details. A minute account of the methods of conducting these and other experiments for the same purpose is given in Maxwell's *Electricity*, vol. II. chap. VII. The quantity H represents the horizontal component of force on a unit pole and is directed towards magnetic north.

Ex. 1. A small compass needle free to turn round its centre in a horizontal plane is acted on by a fixed magnet of moment M whose length is perpendicular to the magnetic meridian and whose centre is in the horizontal plane. If the deviation of the compass needle from the magnetic meridian be ϕ, prove that $\tan\phi = 2M/Hr^3$. This determines M/H when ϕ has been observed. It also gives the value of M or H when the other is known.

Ex. 2. A magnet of moment M is suspended by two fine threads of length l from two points D, E of a horizontal bar. The strings are attached to two points D', E' of the magnet which are equally distant from the centre. The magnet being acted on by the earth's horizontal force assumes a position of equilibrium. Let the bar be turned round a vertical axis until the magnet, when again in equilibrium, is perpendicular to the magnetic meridian. In this position let the bar make an angle θ with the magnet. Prove that $(l^2 - 4b^2 \sin^2 \frac{1}{2}\theta)^{\frac{1}{2}} = Wb^2 \sin\theta / HM$, where W is the weight of the magnet and $2b$ the length of either DE or $D'E'$. This experiment determines the product HM.

326. Potential of a magnetic body. We have hitherto supposed that the attracting and repelling particles of a magnet were situated at two definite points of the axis, called the poles. But there are no such ideal magnets in nature. When a real magnet is broken into pieces the fragments continue to exhibit polarity. We must therefore suppose that the magnetism (whatever that may be) is distributed throughout the body. We shall here assume as a working hypothesis that each element of volume of a magnetic body acts on an external magnetic element as if it were occupied by a small simple magnet whose strength and length are indefinitely small. Let m and $2a$ be the strength and length of the small magnet which occupies the element dv of volume, and let $M = 2am$ be its moment. *The moment per unit of volume is* $2am/dv$. *Representing this ratio by* I, we have the relation $Idv = 2am = M$. The positive direction of the axis of this ideal magnet represents the positive direction of magnetisation of the body at the element dv, and the intensity of the magnetisation is measured by I. *The potential of any element of a magnetic body at a point* P *which is at a finite distance* r *from the element is* $\frac{Idv}{r^2}\cos\theta$ *where* θ *is the angle which the distance* r *makes with the positive direction of magnetisation.*

327. Elementary rule. The potential $Idv \cos\theta/r^2$ is the same as the repulsion of the element dv, supposed to be of density I, when resolved in the direction of magnetisation. It immediately follows that *when the direction of magnetisation is uniform throughout the body the potential at a point* P *is the same as the repulsion at* P *of that body, supposed to be of density* I, *when resolved in the direction of magnetisation.* If the intensity I is not also uniform, the body is supposed to be heterogeneous. This simple rule frequently enables us to write down the potential of a magnetic body.

328. Magnetic rod. The potential of a thin uniformly magnetised rod AB of volume v and length l at any external point P is

$$\frac{Iv}{l}\left(\frac{1}{AP} - \frac{1}{BP}\right) \text{ or } \frac{Iv}{lp}(\sin\beta - \sin\alpha),$$

by Arts. 10, 11, according as the direction of magnetisation is along, or perpendicular, to the length. In the former case we see that *the magnetic rod acts as if it were a simple magnet of equal length whose strength is* Iv/l.

This result may also be arrived at by *à priori* reasoning. The effect of the elementary magnet in any element dv of volume is not altered if its length is increased (without changing the moment Idv) so that the magnet occupies the full length of each element. The positive and negative ends of the successive magnets then destroy each other, leaving a positive element of magnetism at one end of the rod and a negative element at the other.

It follows from Art. 27 that the potential at P of *a thin circular disc*, of volume v, area A, uniformly magnetised perpendicularly to its plane is $Iv\omega/A$ where ω is the solid angle subtended by the disc at P.

329. **Magnetic sphere.** Since the attraction or repulsion of a homogeneous solid sphere of volume v and unit density is v/r^2, it follows immediately that the potential at P of the same sphere when uniformly magnetised is $Iv\cos\theta/r^2$, where r is the distance of P from the centre and θ the angle r makes with the direction of magnetisation. *The potential of a uniformly magnetised solid sphere is therefore the same as that of a small concentric simple magnet*, (called the equivalent magnet), *whose moment is* $M = Iv$ *and whose axis is in the direction of magnetisation.*

When equivalent magnets can be determined for two bodies we can at once deduce from Art. 322 *their potential energy.* In this way we see that the mutual potential energy of two spheres uniformly magnetised in different directions is

$$\frac{II'vv'}{r^3}(\cos\psi - 2\cos\theta\cos\theta'),$$

where r is the distance between the centres and ψ, θ, θ' have the same meaning as in the Art. just referred to.

330. **Magnetic ellipsoid.** The potential of an ellipsoid uniformly magnetised in a given direction can be obtained at once by using the rule. The component repulsions of a homogeneous ellipsoid at an internal point are IAx, IBy, ICz. By resolving these in the direction of magnetisation (l, m, n) we find that the magnetic potential at an internal point (ξ, η, ζ) is

$$V = I(Al\xi + Bm\eta + Cn\zeta)$$

where A, B, C are the quantities defined in Art. 212. The components of magnetic force at any internal point are therefore $X = -IAl$, $Y = -IBm$, $Z = -ICn$: Art. 41. These are constant in magnitude and direction at all internal points.

At an external point, the magnetic potential is

$$V' = I\frac{abc}{a'b'c'}(A'l\xi + B'm\eta + C'n\zeta),$$

where A', B', C' and a', b', c' have the meanings defined in Art. 223.

331. *An ellipsoid is placed in a field of uniform magnetic force, it is required to find the magnetism induced in the ellipsoid.*

The theory of induced magnetism is discussed in the section on *magnetic induction.* It is enough for our present purpose to say that when certain neutral bodies are acted on by magnetic forces each element dv of volume becomes magnetised in the direction of the resultant force F which acts on that element and that the intensity $I = kF$. The constant k is called the *magnetic susceptibility*; another constant $\mu = 1 + 4\pi k$ afterwards introduced is called the *magnetic permeability.*

Let l, m, n be the direction cosines of the direction of the induced magnetisation at any point P of the ellipsoid. Let X, Y, Z be the components of force at P due to the field, X', Y', Z' those of the force due to the ellipsoid now become magnetic. The force F is the resultant of X, Y, Z and X', Y', Z'. Since the intensity I at P is given by $I = kF$, we have

$$Il = kFl = k(X + X'), \quad Im = k(Y + Y'), \quad In = k(Z + Z').$$

Let us assume as a trial solution that the ellipsoid becomes uniformly magnetised in direction and magnitude. We then have $X' = -IAl$, &c. while X, Y, Z are given constants. The equations give at once

$$Il = \frac{kX}{1 + kA}, \quad Im = \frac{kY}{1 + kB}, \quad In = \frac{kZ}{1 + kC}.$$

Since these equations give constant values for the components of magnetisation the trial solution satisfies the conditions of the problem. This therefore is one solution. If we use the constant μ instead of k, these equations become

$$Il = \frac{(\mu - 1) X}{4\pi + (\mu - 1) A}, \quad Im = \frac{(\mu - 1) Y}{4\pi + (\mu - 1) B}, \quad In = \frac{(\mu - 1) Z}{4\pi + (\mu - 1) C}.$$

332. Ex. 1. A sphere and a circular cylinder, constructed of the same kind of material, are placed in succession in a uniform magnetic field, the axis of the cylinder being perpendicular to the force. Prove that the intensities of the induced magnetisms are in the ratio $3(\mu + 1)$ to $2(\mu + 2)$. [In a sphere A, B, C are each equal to $4\pi/3$. Their values for a cylinder are given in Art. 232.]

Ex. 2. An elliptic cylinder, which has *one transverse axis very much longer than the other,* is placed in a uniform magnetic field with its infinite axis perpendicular to the direction of the force. Prove that the intensity of the induced magnetism when the transverse longest axis is in the direction of the force is approximately μ times that when the same axis is perpendicular to the force.

Ex. 3. Prove that the potential of a thin plane lamina uniformly magnetised perpendicularly to its plane at a distant point $(\xi\eta\zeta)$ is

$$V=\frac{Iv\zeta}{r^3}\left\{1-\frac{3}{2}\frac{a^2+b^2}{r^2}+\frac{15}{2}\frac{b^2\xi^2+a^2\eta^2}{r^4}+\&c.\right\},$$

where the axes of coordinates are the principal axes of inertia at the centre of gravity, va^2, vb^2 the moments of inertia about the axes of x and y, and r is the distance of the point from the origin. [To prove this we differentiate with regard to ζ MacCullagh's expression for V, Art. 135.]

333. Magnetic cylinder. Prop. 1. *The density at any point of an infinite right circular cylinder (radius a) is $\phi(x, y)$, the axis of the cylinder being the axis of z. Prove that, if $\phi(x, y)$ satisfy Laplace's equation and be of i dimensions, the potential of the cylinder at an internal point (ξ, η) is*

$$U=\pi\left\{\frac{a^2}{i}-\frac{\xi^2+\eta^2}{i+1}\right\}\phi(\xi, \eta).$$

We obtain this result by making c infinite in the first theorem of Art. 247, noticing that $Q/c=a^2+u$ when $a=b$. The potential is therefore

$$U=\pi a^2\int\frac{du}{u^2+a}\left\{\Sigma\frac{R^{1+n}u^nD^n}{L(n+1)L(n)2^{2n}}\right\}\left(\frac{a^2}{a^2+u}\right)^i\phi(\xi, \eta).$$

The operator $D=\dfrac{a^2+u}{a^2}\left(\dfrac{d^2}{d\xi^2}+\dfrac{d^2}{d\eta^2}\right)$, and ϕ satisfies Laplace's equation, hence all the terms except that given by $n=0$ are zero. The potential becomes

$$U=\pi\int\left(\frac{a^2}{a^2+u}\right)^{i+1}du\left\{1-\frac{\xi^2+\eta^2}{a^2+u}\right\}\phi(\xi, \eta).$$

At an internal point, the limits are 0 to ∞,

$$\therefore\ U=\pi\left\{\frac{a^2}{i}-\frac{\xi^2+\eta^2}{i+1}\right\}\phi(\xi, \eta).$$

At an external point, the limits are λ to ∞,

$$\therefore\ U'=\pi\left(\frac{a^2}{a^2+\lambda}\right)^i\left\{\frac{a^2}{i}-\frac{\xi^2+\eta^2}{i+1}\frac{a^2}{a^2+\lambda}\right\}\phi(\xi, \eta).$$

Prop. 2. *The x and y components of magnetisation of a right circular cylinder are $Il=df/dx$ and $Im=df/dy$, where $f(x, y)$ is a homogeneous function of x and y of i dimensions which satisfies Laplace's equation. Prove that the potential of the cylinder at an internal point is $2\pi f(\xi, \eta)$.*

The potential of a magnetic cylinder whose intensity is Il is equal to the resolved repulsion of a cylinder whose density is Il (Art. 327). The potential of the cylinder due to both components of magnetisation is therefore

$$V=-\pi\frac{d}{d\xi}\left\{\left(\frac{a^2}{i-1}-\frac{\xi^2+\eta^2}{i}\right)\frac{df}{d\xi}\right\}-\pi\frac{d}{d\eta}\left\{\left(\frac{a^2}{i-1}-\frac{\xi^2+\eta^2}{i}\right)\frac{df}{d\eta}\right\}$$

$$=\pi\left\{\frac{2\xi}{i}\frac{df}{d\xi}+\frac{2\eta}{i}\frac{df}{d\eta}\right\}=2\pi f(\xi, \eta),$$

since $df/d\xi$ and $df/d\eta$ are both of $i-1$ dimensions.

The potential at an external point is found in the same way. Since $a^2+\lambda=\xi^2+\eta^2$, the result is $V'=2\pi\left(\dfrac{a^2}{a^2+\lambda}\right)^i f(\xi, \eta)$.

Prop. 3. *A right circular cylinder is placed in a field of force whose potential is $f(\xi, \eta)$. Prove that, if $f(\xi, \eta)$ is a homogeneous function of i dimensions which satisfies Laplace's equation, the magnetic potential inside the cylinder is*

$$V_1=\frac{2}{1+\mu}f(\xi, \eta).$$

Assume as a trial solution that the x, y components of magnetisation are $L\,df/dx$ and $L\,df/dy$. The equation of condition (Art. 331) $l=k(X+X')$ becomes $L\frac{df}{d\xi}=-k\left(\frac{df}{d\xi}+2\pi L\frac{df}{d\xi}\right)$. Hence $L(1+2\pi k)=-k$. The other equation of condition leads to the same result. The potential inside the cylinder is therefore

$$V_1=f+2\pi Lf=\frac{f}{1+2\pi k}=\frac{2f}{1+\mu}.$$

The potential at a point outside the cylinder is

$$V_1'=\left\{1-\frac{\mu-1}{\mu+1}\left(\frac{a^2}{a^2+\lambda}\right)^{i}\right\}f(\xi,\eta).$$

Ex. A right circular cylinder is placed in a field of magnetic force whose potential is $A(\xi^2-\eta^2)$. Prove that the potential of the magnetic force within the cylinder is $A'(\xi^2-\eta^2)$, where $A'(1+\mu)=2A$. [Coll. Ex. 1899.]

In the same way, if the potential of the field were Axy, the magnetic potential would be $A'xy$, where A' has the same value. This result follows at once from the former because $\xi^2-\eta^2$ becomes $-2\xi'\eta'$ when the axes are turned round OZ through half a right angle.

334. *To find the mutual potential energy of two magnetic bodies.* By Art. 321 the potential energy of a magnetic body and an elementary magnet of moment M' is $-M'F'$, where F' is the component of force due to the magnetic body in the direction of the axis of the elementary magnet. If the elementary magnet represent the magnetism of an element dv' of a second magnetic body, we have $M'=I'dv'$. *The potential energy of the two bodies is therefore* $W=-\int F'I'dv'$ where the integral extends throughout the volume of the second body.

If V be the potential of one magnetic body, λ', μ', ν' the direction cosines of the direction of magnetisation at any point of the other, the expression for W takes the form

$$W=\iiint\left(\frac{dV}{dx'}\lambda'+\frac{dV}{dy'}\mu'+\frac{dV}{dz'}\nu'\right)I'dx'dy'dz'.$$

This integral is the same as that considered in Green's theorem (Art. 149), and is equivalent to

$$W=\int VI'\cos i'd\sigma'-\int V\left(\frac{dI'\lambda'}{dx'}+\frac{dI'\mu'}{dy'}+\frac{dI'\nu'}{dz'}\right)dv'.$$

If the magnetisation I' is such that its components $I'\lambda'=df/dx'$, and where f is a function which satisfies Laplace's equation, the expression for W is reduced to a surface integral.

335. Terrestrial magnetism. The phenomena of terrestrial magnetism can be roughly represented by the action of a powerful small magnet placed near the centre of the earth (Biot, *Traité de Physique,* 1816). This supposition is equivalent to treating the earth as a sphere uniformly magnetised in direction and magnitude (Art. 329). The theory altogether fails in accuracy when applied to explain the irregularities at special places. An attempt was therefore made by a Norwegian observer, Hansteen, to explain the observed facts by the action of two large magnets within the earth, both being excentric. But the results, though superior to those derived from a single magnet, were not satisfactory.

336. Gauss* investigated the potential of the magnetism at a point P on the supposition that it was distributed irregularly throughout the earth. To effect this he used a formula equivalent to that given in Art. 283, viz.

$$V = Y_0 + Y_1\frac{r}{a} + \&c. + Z_0\frac{a}{r} + Z_1\left(\frac{a}{r}\right)^2 + \&c. \quad\text{.........(1)},$$

where a is the radius and r the distance of P from the centre of the earth. If the causes of magnetism are inside the earth the second of these series alone is to be retained. When P is at a great distance from the attracting mass, this reduces to Z_0a/r. It follows that Z_0a is the attracting mass and is therefore zero. After some preliminary trials Gauss decided that it would be sufficient for a first approximation to retain only the terms up to and including $(a/r)^4$. This is to be regarded as a trial solution to be accepted or rejected after a comparison of its results with the observed facts of magnetism. With this limited value of V the theoretical components of force in three rectangular directions can be found by differentiation. Let the directions be, one parallel, a second perpendicular to the meridian, and a third vertical. Representing these components by X, Y, Z, the declination δ of the needle and the dip i are given by $(X^2+Y^2)\tan^2 i = Z^2$ and $X\tan\delta = Y$. The values of the declination, dip, and intensity were known in Gauss' time at nearly 100 places. The observations at 12 of these (properly chosen) were used to determine the 24 unknown constants which occurred in the functions Z_1 &c. Gauss then tabulated side by side the observed and computed values of the declination, dip and intensity at 91 places on the surface of the earth, so that an easy comparison could be made.

337. In general the agreement was so accurate as to leave no doubt on the fundamental correctness of the theory. The observations made since Gauss' time are also in sufficient accordance with the theory. The small discordances which remain are ascribed by Gauss partly to errors in the observations and partly to the fact that all the observations used do not correspond to the *same year*. The terms beyond the fourth order in (1) may have sensible effects and possibly other less influential causes of magnetism may exist.

* Gauss' paper is translated in *Taylor's Scientific Memoirs*, vol. II., 1841.

338. The causes of magnetism have been assumed to be inside the earth. If there are any external causes, their effects could be represented by including some terms of the first series in (1). If the causes were wholly external to the earth the potential would be represented solely by the first series in (1). The vertical force would then be $-\Sigma n Y_n/a$ instead of $\Sigma (n+1) Z_n/a$. Since the observed vertical force does closely satisfy the latter of these two very different expressions Gauss considers it proved that *only a small part* of the terrestrial magnetic force can be due to causes external to the earth. This argument does not apply to the periodical changes of the needle which have not been considered by Gauss.

339. Poisson's theorem. *To investigate a general formula for the potential of a magnetic body.* We resolve the intensity I into three components $A = I\lambda$, $B = I\mu$, $C = I\nu$. Let us find the potential due to the first of these. Let $QQ' = dx$ be an element of a column LM parallel to x (figure of Art. 222). Let $QP = r$ and let $Q'n$ be perpendicular to QP, then $Qn = -dr$ and $\cos PQQ' = -dr/dx$. The potential of the column at P is then

$$\int \frac{A\,dx\,dy\,dz}{r^2} \cos PQQ' = dy\,dz \int A \frac{d(1/r)}{dx}\,dx,$$

and the potential of the whole magnetic body at P is

$$V = \iiint \left(A\frac{d}{dx} + B\frac{d}{dy} + C\frac{d}{dz}\right)\frac{1}{r}\,dx\,dy\,dz.$$

Following the same reasoning as in Green's theorem (Art. 149) we put this into the form

$$V = \int I\cos i\,\frac{dS}{r} - \int\left(\frac{dA}{dx} + \frac{dB}{dy} + \frac{dC}{dz}\right)\frac{dv}{r},$$

where dS is an element of the boundary and dv of the volume of the magnetic body.

It follows from this equation that *the magnetic potential at P is the same as that of a quantity of matter distributed partly internally and partly superficially. The volume density ρ of the internal distribution, and the surface density σ of the superficial distribution, are*

$$\rho = -\left(\frac{dA}{dx} + \frac{dB}{dy} + \frac{dC}{dz}\right), \quad \sigma = I\cos i.$$

Here i is the angle the direction of magnetisation at any point of the surface makes with the outward normal at that point.

340. Since the total quantity of attracting matter in each elementary magnet is zero, it is evident that *the sum of the internal and superficial distributions in Poisson's theorem is also zero.*

The mass distributed over the surface S of the magnetic body, being $\int I \cos i\, dS$, is evidently the flux of the vector I across the boundary of S.

So also, if the surface S' is the boundary of any portion of the body, the mass distributed internally is equal and opposite to the flux across the surface S'. Thus $-\rho\, dx\, dy\, dz$ is equal to the outward flux across the six faces of the Cartesian element $dx\, dy\, dz$. We may therefore deduce the value of ρ for polar, cylindrical, or other orthogonal coordinates by finding as in Art. 108 the flux across the faces of the corresponding element.

If I_1, I_2, I_3 are the polar components of I in the directions in which r, θ, ϕ are measured, then (Art. 108)

$$-\rho = \frac{1}{r^2}\frac{d(I_1 r^2)}{dr} + \frac{1}{r\sin\theta}\frac{d(I_2\sin\theta)}{d\theta} + \frac{1}{r\sin\theta}\frac{dI_3}{d\phi}.$$

In cylindrical coordinates R, ϕ, z

$$-\rho = \frac{1}{R}\frac{d(I_1 R)}{dR} + \frac{1}{R}\frac{dI_2}{d\phi} + \frac{dI_3}{dz}.$$

341. Ex. A magnetic shell is bounded by spheres of radii a, b. The direction of the magnetisation at any distance r from the centre is radial and its magnitude is κr^n. Find the potential at an internal and external point.

The internal distribution is $\rho = -\dfrac{1}{r^2}\dfrac{d}{dr}(\kappa r^{n+2})$ and the superficial distribution $\sigma_1 = -\kappa a^n$ and $\sigma_2 = +\kappa b^n$ on the two boundaries. The potential of all these at an internal point (Art. 64) is

$$V = \int \rho \frac{4\pi r^2 dr}{r} + \sigma_2 \frac{4\pi b^2}{b} - \sigma_1 \frac{4\pi a^2}{a} = -4\pi \frac{\kappa\,(b^{n+1} - a^{n+1})}{n+1}.$$

The potential at an external point is zero.

342. The force of induction*. The magnetic force at a point P of space void of magnetised matter is the force on the positive pole of a magnet of unit strength, the positive pole being placed at P. To find the force at a point P situated within a magnetic body we imagine an infinitely small space round P to be removed and a positive unit pole placed at P in the cavity.

Consider the effect of this removal; the attraction of the solid distribution of Poisson which once filled the space has disappeared, and there is now a superficial distribution on the inside of the cavity. Since the attractions of similar and similarly situated bodies on the same point vary as their linear dimensions (Art. 94),

* The distinction between the magnetic force and the force of induction is due to Kelvin and is fully explained in his *Theory of Magnetism*, Arts. 479 &c. The former name is due to him, and the latter to Faraday and Maxwell. See the treatise on *Electricity and Magnetism*, Art. 428.

the solid distribution is not affected by the removal of the infinitely small quantity of matter (Art. 101). But the superficial distribution on the inside of the cavity does affect the force in a manner which depends on the form of the cavity.

Thus the resultant force at a point inside a magnetic body is made up of two components. One of these is due to (1) all external causes, (2) the whole solid distribution, (3) the superficial distribution on the external boundary. The other is due to the superficial distribution on the inside of the cavity alone. *The former component is defined to be the magnetic force at a point within the magnetic substance.*

343. Let the cavity have the form of an infinitely small cylinder whose length is $2b$ and radius a, and let the generating lines be in the direction of magnetisation. Let P be at the central point of the cylinder. The superficial surface density, being $I\cos i$, is zero along the generating lines and $\pm I$ at the two circular ends. The "outward" normal for the cavity tends *towards the point P* and therefore the surface density is $+I$ for the negative end of the cavity and $-I$ for the positive end. The repulsion of the two ends at P is $4\pi I\left\{1-\dfrac{b}{\sqrt{(a^2+b^2)}}\right\}$ by Art. 21 acting in the direction of the magnetisation of the body in the neighbourhood of P. It appears that the force depends not on the absolute dimensions of the cavity but on the ratio of the length to the breadth. Hence however small the cavity may be made, the force due to the superficial distribution on its walls will in general remain finite. If the radius a is infinitely smaller than the length b, the force due to the superficial distribution is zero. If the radius is infinitely greater than the length the force is $4\pi I$.

The actual force, due to all causes, on a positive unit pole situated at the central point P of a cylindrical cavity, whose length is in the direction of magnetisation and is infinitely greater than the breadth, is the same as the force already called *the magnetic force at P*. The actual force, due to all causes, on the pole when situated at the central point of a thin disc-like cylindrical cavity, whose plane is perpendicular to the direction of magnetisation, is called *the force of induction at P*.

By taking cavities of different forms we may contrast the two forces in other ways. Let the cavity be of a thin disc-like form the normal to the plane making an angle i with the direction of magnetisation. The distribution on the curved side is ultimately zero. The distributions $\pm I \cos i$ on the two plane faces act on P as if they were distributed on infinite planes; the repulsion at P is therefore $4\pi I \cos i$. Thus the actual force on P is the magnetic force or the force of induction according as the plane of the cavity contains or is perpendicular to the direction of magnetisation.

344. Ex. Prove that when the cavity is spherical the force at the centre due to the superficial distribution is $\frac{4}{3}\pi I$ (see Art. 93).

345. It appears from what precedes that the force of induction at P is the resultant of the magnetic force at P and a force $4\pi I$ acting at P in the direction of magnetisation of the body in the neighbourhood of P.

Let $A = I\lambda$, $B = I\mu$, $C = I\nu$ be the Cartesian components of the vector I; X, Y, Z and X_1, Y_1, Z_1 the components of the magnetic force and the force of induction. Let V be the potential of the whole magnetic body at any internal point P, as given by Poisson's theorem, Art. 339. Then

$$X = -dV/dx, \qquad X_1 = X + 4\pi A,$$
$$Y = -dV/dy, \qquad Y_1 = Y + 4\pi B,$$
$$Z = -dV/dz, \qquad Z_1 = Z + 4\pi C.$$

346. Bodies not uniformly magnetised. When the magnetism of a body is not uniform, either in direction or intensity, it becomes necessary to choose special forms for the elements.

The magnetic lines are curves such that the *direction of magnetisation* at any point is a tangent to the curve at that point. In a line of force the direction of *the force* is a tangent, Art. 47. If we draw a magnetic line through every point of a closed curve we construct a tube which is called *a magnetic tube*. When the section of the tube is very small it is sometimes called *a filament*. By analysing a magnetic body into elementary tubes or filaments we may often find its magnetic potential at any external point P with great ease.

347. Solenoids. Let $d\sigma$ be the area of a section of a magnetic filament at any point Q, ds an element of length measured in the direction of magnetisation and I the magnetic intensity. Using the same notation as before (Art. 316) we notice that θ is the angle in front of the radius vector QP and that therefore $\cos\theta = -dr/ds$. Hence since $dv = d\sigma . ds$ the potential of the filament at P is
$$V = \int \frac{Idv}{r^2}\cos\theta = -\int Id\sigma \frac{dr}{r^2}.$$

When the magnetism of the body is so distributed that $Id\sigma = d\mu$ is constant for each magnetic filament, the body is called a solenoid. The integration can then be effected at sight. If R, S be the intersections of the filament with the surface of the body, RS being the positive direction of magnetisation, the potential at any external point P is
$$V = \frac{d\mu}{SP} - \frac{d\mu}{RP}.$$

The potential of a solenoidal filament is independent of its form and depends solely on the magnetism $Id\sigma$ of a cross section and on the positions of its extremities. A closed solenoid exerts no action on any external magnet.

348. The potential of a solenoid, or *of any portion of a solenoid,* may be found by summing up the potentials of the filaments which compose the body. Let any filament intersect the boundary in an area dS and let the direction of magnetisation make an angle i with the outward normal. Then since
$$Id\sigma = IdS \cos i,$$
the potential of the body is the same as that of a thin superficial stratum on the boundary, and this stratum is the same as that given in Poisson's theorem (Art. 339).

349. Since this must be also true for every element of volume of the body, it follows that the solid distribution of Poisson must be zero. We have then
$$-\rho = \frac{dA}{dx} + \frac{dB}{dy} + \frac{dC}{dz} = 0,$$
where A, B, C are the components of I at the point x, y, z. *This is a necessary condition that the magnetism is solenoidal.*

To prove that this condition is sufficient. By Poisson's theorem the potential of every portion of a body is equivalent to that of a surface distribution $I \cos \theta$ and a volume distribution ρ. Let this portion be an *arbitrary length l* of a magnetic filament. The potential of the filament is
$$V = \int Id\sigma\, d\left(\frac{1}{r}\right) = \frac{I_1 d\sigma_1}{r_1} - \frac{I_0 d\sigma_0}{r_0} - \int \frac{1}{r} \frac{d(Id\sigma)}{ds} ds$$
where the suffixes refer to the ends of the filament. The potential of the filament is therefore the same as that of a surface distribution $I \cos \theta$ and a volume distribution $\rho' = \frac{1}{d\sigma} \frac{d(Id\sigma)}{ds}$. Since the surface distribution of the arbitrary filament is the same as that given by Poisson, the density ρ' of the volume distribution must also be the same as ρ. Hence when $\rho = 0$ we must have $Id\sigma$ constant for any filament.

350. Lamellar shells. If the magnetic lines can be cut orthogonally by a system of surfaces we can conveniently analyse the body into elementary shells. Let the equation of these surfaces be $f(x, y, z) = c$. Consider the shell bounded by the surfaces c and $c + dc$. Let $d\sigma$ be an element of area at any point Q of the first surface, t the thickness of the shell; the volume of the corresponding element of the shell is then $dv = td\sigma$. Let t be measured in the direction of magnetisation and let I be the intensity. The magnetism of the element dv is equivalent to that of a small magnet whose moment is Idv and whose axis is normal to the surface. The potential at any point P is $Idv \cos\theta/r^2$, where r and θ have the same meanings as in Art. 347.

Let $d\omega$ be the solid angle subtended by $d\sigma$ at P, then $d\sigma \cos\theta/r^2 = d\omega$ (Art. 26). The potential of the shell at P is therefore

$$V = \int I dv \cos\theta/r^2 = \int It d\omega.$$

Here the sign of $d\omega$ follows that of $\cos\theta$. Let that side of $d\sigma$ be called the positive side to which the direction of magnetisation points. Since θ is the angle QP makes with that axis, *the solid angle $d\omega$ is positive or negative according as P lies on the positive or negative side of the elementary area $d\sigma$.*

Let P travel from a position P_1 close to $d\sigma$ on its positive side to a position P_2 also close to $d\sigma$ on its negative side, the journey being made outside the elementary area. When P crosses the tangent plane to $d\sigma$ at some external point, the solid angle subtended at P changes from positive to negative. The solid angles at P_1 and P_2 are 2π and -2π, hence if we suppose P to travel from P_2 to P_1 through the element of area the solid angle is increased by 4π.

351. *The product It is called the strength of the elementary shell at Q.* When the shell is such that the strength is everywhere the same the shell is called a simple magnetic shell and *the distribution of magnetism is said to be lamellar.* If the strength varies from point to point, the shell is called a complex magnetic shell and the distribution of magnetism is said to be complex lamellar. Let $It = \phi$.

352. When the distribution of magnetism is lamellar the potential takes a simple form. Putting ω for the whole solid angle subtended at P by any portion of the elementary shell, we find that the potential at P of that portion is

$$V = \int It d\omega = It\omega.$$

It follows from this result that *if two thin lamellar shells have the same rim and the positive sides are turned the same way, the potentials and therefore the forces at any point P are equal each to each.* The dimensions of ϕ are those of potential.

Let a thin lamellar shell enclose a space. The potential at P of any portion has been shown to be $It\omega$. Let this portion increase and finally cover the shell. If P be inside the empty space the solid angle ω subtended at P increases and is finally 4π. If P be outside, the angle ω will presently begin to decrease and will be finally zero. It follows that *the potential of a closed lamellar shell at an internal point is $4\pi It$, at an external point the potential is zero.*

353. If a thin lamellar shell is in the presence of a number of magnets, the mutual potential energy of any element dv and the field is $-FIdv$ by Art. 321, where F is the axial component of force at the element dv. Since $Idv = \phi d\sigma$ and ϕ is constant when the distribution of magnetism is lamellar, *the mutual potential energy of the whole shell and the field is $-\phi\int F d\sigma$.* The integral $\int F d\sigma$ represents the flux of the force due to the field entering the negative side of the shell.

354. *To determine the conditions that the distribution of magnetism is lamellar.*

Let λ, μ, ν, expressed as functions of x, y, z, be the direction cosines of the tangent at any point R of a magnetic line. The analytical condition that the magnetic lines can be cut orthogonally by some system of surfaces is that $\lambda dx + \mu dy + \nu dz$ can be made a perfect differential of some function $f(x, y, z)$ by multiplication by a factor, and the orthogonal surfaces are then $f(x, y, z) = c$. Let ρ be one of these factors, the three equations

$$\rho\lambda = \frac{df}{dx}, \qquad \rho\mu = \frac{df}{dy}, \qquad \rho\nu = \frac{df}{dz} \quad \text{..........................(1)}$$

must then be satisfied by simultaneous values of ρ and f. If A, B, C be any quantities *proportional* to λ, μ, ν, say $A = m\lambda$ &c., we find, by eliminating ρ/m and f,

$$A\left(\frac{dB}{dz} - \frac{dC}{dy}\right) + B\left(\frac{dC}{dx} - \frac{dA}{dz}\right) + C\left(\frac{dA}{dy} - \frac{dB}{dx}\right) = 0 \quad \text{..............(2)}$$

Let a shell be formed by the two surfaces $f = c$, $f = c + dc$. Let x, y, z; $x + dx$, &c. be the coordinates of two adjacent points R, S, one on each surface. The thickness t of the shell at R is the sum of the projections of dx, dy, dz on the normal at R, hence $\rho t = \rho(\lambda dx + \mu dy + \nu dz) = df$.

The product ρt is therefore constant and equal to dc for the shell. Now two quantities (say ρt and It) cannot both be constant for the same shell, unless I bear a constant ratio to ρ. Thus It will be constant only if $I = P\rho$, where P is a function of x, y, z which is constant all over the surface $f = c$. Hence $P = F(f)$, and it is evident that $\rho' = P\rho$ is another factor which also makes $\lambda dx +$ &c. a perfect differential, viz. the differential $F(f)\, df$. It is therefore necessary that I should be equal to some one of the values of ρ which satisfy the conditions (1).

Let the magnetism of the body be given by the components A, B, C of the vector I expressed as known functions of x, y, z. *The necessary and sufficient conditions that the distribution of magnetism should be lamellar are that*

$$A = \frac{df}{dx}, \qquad B = \frac{df}{dy}, \qquad C = \frac{df}{dz},$$

where f is an arbitrary function of x, y, z. The condition (2) that the magnetic lines can be cut orthogonally by some system of surfaces is satisfied by these values of A, B, C. The function f is called *the potential of magnetisation.* It must be distinguished from the magnetic potential V.

355. Ex. Each element Q of a thin spherical shell is magnetised along the direction OQ, where O is a given point on the surface, with an intensity I which varies as the distance OQ. Prove that the potential at any external point P is proportional to $\cos\theta/r^2$, where r is the distance of P from the centre C and θ is the angle r makes with OC.

Resolve the magnetism at Q into the two directions CQ and OC. Taking the former alone, the distribution is lamellar and the external potential is zero. Taking the latter, the distribution is uniform and the potential is known.

356. *To find the magnetic force exerted by a lamellar shell of strength ϕ on a unit pole at P.*

Describe a cone whose vertex is P and whose generators pass through the rim or margin AQQ' of the shell, and let this cone be cut by a sphere, whose centre is P and whose radius c is very great, in the spherical segment BRR'. We replace the given shell by another shell with the same rim, but having for its surface the spherical segment and that portion of the cone which lies between the rim and the segment (Art. 352).

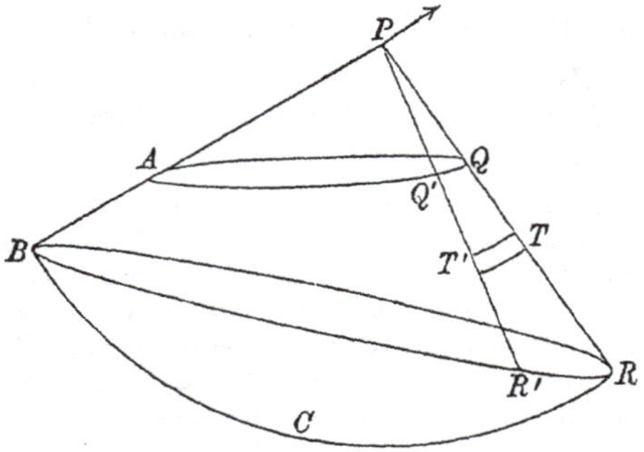

The small magnet equivalent to the magnetism at TT' on any elementary area dS of the cone is "broadside on" to P and the force exerted at P is therefore $\phi\, dS/r^3$, where $r = PT$. When P is on the positive side of the given shell the positive pole of the small magnet at T is directed *inwards* towards the given shell and the force at P tends outwards in the direction indicated by the arrow (Art. 317). Let the angle $QPQ' = d\psi$, then $dS = r\, d\psi\, dr$. The force at P due to the magnetism on the strip $QQ'RR'$ becomes by integration $\left(\frac{1}{r} - \frac{1}{c}\right)\phi\, d\psi$, where r is now PQ. This reduces to its first term when c is very great.

The equivalent magnets which represent the magnetism on the spherical segment BRR' are "end on" to P. The force on P due to any elementary area dS is $2\phi\, dS/c^3$. Since $dS = c^2 d\omega$ (where $d\omega$ is the solid angle subtended at P) this force is zero when c is very great.

To find the force at P due to a thin lamellar shell we divide the rim into elements. The force due to the element QQ' is equal to $\phi\, d\psi/r$ and also to $\phi\, dA/r^3$, where $r = PQ$, $QPQ' = d\psi$ and dA is twice the area QPQ'. This force acts at P perpendicularly to the plane QPQ'. The resultant of these elementary forces is the force on a unit pole at P.

357. We notice that the magnitude of the force due to the elementary arc QQ' and its direction relatively to the plane PQQ' are not changed by rotating that plane about QQ' as axis. The side of the plane to which the force tends (when not already obvious) is therefore easily found. If P is brought by the rotation from the positive to the negative side of the shell, the force on a positive unit pole at P acts in the direction of motion.

When the rim is a plane curve and the point P lies in that plane, the force at P is normal to the plane and equal to $\phi \int d\psi/r$.

358. *To find the Cartesian components of the force at P due to a lamellar shell.* Let (ξ, η, ζ) be the coordinates of P, (x, y, z) those of a point Q on the rim. Let (λ, μ, ν) be the direction cosines of the normal to the plane PQQ'. The z component of force is therefore $\nu\phi\, dA/r^3$. By projecting the area dA on the plane of xy we have
$$(x-\xi)\, dy - (y-\eta)\, dx = \nu\, dA.$$

Hence $Z = \phi \int \frac{(x-\xi)\, dy - (y-\eta)\, dx}{r^3}$ and similarly

$$X = \phi \int \frac{(y-\eta)\, dz - (z-\zeta)\, dy}{r^3}, \quad Y = \phi \int \frac{(z-\zeta)\, dx - (x-\xi)\, dz}{r^3}.$$

The integration in each case is conducted round the rim in the positive direction. The left-handed system of coordinates being used (vol. I. Arts. 97 and 272); the positive direction is clockwise as seen by an observer with his feet on the shell and his head on the positive side.

We may put these expressions for the forces into another form. Let

$$F = \int \frac{dx}{r}, \quad G = \int \frac{dy}{r}, \quad H = \int \frac{dz}{r}.$$

Then since $r^2 = (x-\xi)^2 + \&c.$, we have $\frac{d}{d\xi}\left(\frac{1}{r}\right) = \frac{x-\xi}{r^3}$, &c.,

$$\therefore\ Z = \phi \int \left\{\frac{d}{d\xi}\left(\frac{1}{r}\right) dy - \frac{d}{d\eta}\left(\frac{1}{r}\right) dx\right\} = \phi\left(\frac{dG}{d\xi} - \frac{dF}{d\eta}\right).$$

Similarly $$X = \phi\left\{\frac{dH}{d\eta} - \frac{dG}{d\zeta}\right\}, \quad Y = \phi\left\{\frac{dF}{d\zeta} - \frac{dH}{d\xi}\right\}.$$

359. *To find the potential of a lamellar magnetic body.* Let the components of I be $A = df/dx$, &c., and let r be the distance of a point P from any element of volume dv. The potential at P is (by Art. 339)

$$V = \int\left(A\frac{d}{dx} + B\frac{d}{dy} + C\frac{d}{dz}\right)\frac{dv}{r} = \int\left(\frac{df}{dx}\frac{d\,1/r}{dx} + \frac{df}{dy}\frac{d\,1/r}{dy} + \frac{df}{dz}\frac{d\,1/r}{dz}\right) dv$$
$$= \int f \frac{d\,1/r}{dn}\, dS - \int f \nabla^2 \frac{1}{r}\, dv$$

by Green's theorem. Now $\frac{d\,1/r}{dn} = -\frac{1}{r^2}\frac{dr}{dn} = \frac{1}{r^2}\cos\theta$, where θ is the angle the distance r makes with the outward normal (Art. 347). Also $\nabla^2 1/r = 0$ or -4π according as P is without or within the substance of the body. Hence

$$V = \int f \frac{\cos\theta\, dS}{r^2} + 4\pi(f) \quad \text{.......................(1)},$$

where (f) is the value of f at the point P.

Let the surface integral $\int f \frac{\cos\theta\, dS}{r^2}$ be represented by Ω.

Then $$V = \Omega + 4\pi(f) \quad \text{.......................(2)}.$$

The components of magnetic force at an internal point P are given by $X = -dV/dx$ &c., and the components of magnetisation by $A = d(f)/dx$ &c. It follows at once from Art. 345 that *the components of magnetic induction are related to* Ω *by the equations* $X_1 = -d\Omega/dx$ &c.

Let Ω_1, Ω_2 be the values of Ω at points P_1, P_2 respectively just outside and just inside the surface S. Now $\cos\theta\, dS/r^2$ is the solid angle subtended by dS at P and hence, by the same reasoning as in Art. 142, $\Omega_1 - \Omega_2 = (f)(\omega_1 - \omega_2)$, where ω_1, ω_2 are the solid angles subtended at P_1, P_2 by the elementary area dS. The difference of these solid angles is 4π by Art. 350. We therefore have $\Omega_1 - \Omega_2 = 4\pi(f)$. If V_1, V_2 are the values of V at the same points, P_1 and P_2, we have by (2) $V_1 = \Omega_1$ and $V_2 = \Omega_2 + 4\pi(f)$. Hence $V_1 = V_2$. The magnetic potential is therefore continuous when the point P enters the substance of the magnetic body, but the potential Ω is not continuous.

360. *The mutual potential energy of two thin lamellar shells may be found by the formula* $W = -\phi\phi' \iint \frac{\cos\epsilon}{r}\, ds\, ds'$, *which involves only integrations round the two rims.* Here ds, ds' are elements of the two rims, ϵ is the angle their positive directions make with each other, and ϕ, ϕ' are the strengths of the shells.

Let us first find the mutual potential energy of the shell whose strength is ϕ and a portion of the other shell which is so small that we may regard its rim as a plane curve. Let this plane be taken as the plane of xy and let (ξ', η') be the coordinates of any element $d\sigma'$ of the area. The potential energy is by Art. 321

$$= -\int ZI'dv' = -\int ZI't'd\sigma' = -\phi\phi' \iint \left(\frac{dG}{d\xi'} - \frac{dF}{d\eta'}\right) d\xi' d\eta'.$$

By the application of Green's method to plane curves (Art. 149) this surface integral is replaced by an integration round the rim which in our case gives $\int (F d\xi' + G\, d\eta')$. We now substitute for F and G their values from Art. 358 and remember that $d\zeta' = 0$. The expression for the potential energy is then

$$-\phi\phi' \iint \frac{dx\, d\xi' + dy\, d\eta' + dz\, d\zeta'}{r} = -\phi\phi' \iint \frac{\cos\epsilon}{r}\, ds\, ds'.$$

To find the energy when the second shell is of finite size we integrate the expression just found. Let two adjacent elements touch along the arc AB. When integrating round these two elements, we pass over the arc AB in opposite directions and therefore for this arc the angle ϵ (being the angle the direction of integration makes with an arc ds of the first shell) has supplementary values in the two elements. The sum of the integrals for both elements may therefore be found by integrating round both as if they were one, omitting the common arc AB. The same reasoning applies to all adjacent elements, hence the total energy for two shells of finite size may be found by integrating round their perimeters.

361. The theory of thin lamellar shells derives additional importance from its connection with electric currents. According to Ampère's theory the forces on a magnetic pole due to a closed electric circuit are the same as those of a thin lamellar magnetic shell, of proper strength, whose rim is the closed circuit.

The direction of these currents may be usefully remembered thus: if the earth's magnetism were due to currents round the axis, their general direction would follow the sun, that is, would be from east to west. Now the south pole of the earth has been taken as its positive pole (Art. 314). Hence the currents equivalent to a small magnet flow round it anti-clockwise when viewed by a person with his head at the positive and feet at the negative pole.

362. Examples. Ex. 1. Prove that the force due to a thin lamellar shell, whose rim is a circle of radius a, (1) at the centre is $Z_0 = 2\pi\phi/a$ and (2) at a point

P situated in its plane at a small distance ξ from the centre is $Z=Z_0(1+3\xi^2/4c^2)$. [Use the formula $Z=\phi\int d\psi/r$.]

Ex. 2. Prove that the potential of a thin lamellar shell whose rim is a circle of radius a at a point P distant ζ from the plane and at a small distance ξ from the axis is

$$V=2\pi\phi\left(1-\frac{\zeta}{v}-\frac{3}{4}\frac{a^2\xi^2\zeta}{v^5}+\&c.\right),$$

where $v^2=a^2+\zeta^2$. Prove also that the component forces at P are

$$X=2\pi\phi\frac{3a^2\xi\zeta}{2v^5}+\&c.,\quad Z=2\pi\phi\frac{a^2}{v^3}\left\{1+\frac{3}{4}\frac{\xi^2(a^2-4\zeta^2)}{v^4}+\&c.\right\}.$$

It is evident from the symmetry of the figure on each side of the plane $\eta\zeta$ that the potential can contain only even powers of ξ. The expression for V is therefore correct up to the cubes of the small quantity ξ.

Ex. 3. Prove that the mutual potential energy W of a thin circular lamellar shell and a small magnet whose centre C is on the axis of the circle is

$$W=-\frac{2\pi\phi a^2M}{l^3}\left\{\sin\theta+\frac{4h^2-a^2}{4l^4}\rho^2\sin\theta\,(2\sin^2\theta-3\cos^2\theta)+\&c.\right\},$$

where a is the radius of the shell, h, l the distances of C from the centre and rim of the shell respectively, ρ the half-length of the magnet, θ the inclination of its axis to the plane of the shell, and M its moment.

If the magnet C be placed on the axis of the shell at *a distance from its plane equal to half the radius*, the terms of the order ρ^2 are zero. It follows that *the action of the shell on the small magnet C is equal to that of another small magnet whose moment is $M'=\pi a^2\phi/5\sqrt{5}$ placed end on at the centre of the shell.* A similar theorem for electric currents is given by M. Gaugain in the *Comptes Rendus*, 1853.

We may obtain this result by expanding the expression for the potential found in Ex. 2 in powers of ρ, then twice the sum of the odd powers after multiplication by the strength of the small magnet is the energy required.

Ex. 4. If the law of force be the inverse κth power of the distance, prove that the mutual potential energy W of two thin lamellar shells of elementary areas is

$$W=-\left(\frac{\kappa}{2}\right)^2 R^{\frac{(\kappa-2)(\kappa+1)}{\kappa}}\phi\phi'\iint\frac{\cos\epsilon}{r^{2/\kappa}}\,ds\,ds',$$

where R is the distance between two points one in each area. It appears that we cannot extend the theorem to shells of finite area unless the law of force be either the inverse square or the direct distance.

Electrical Attractions.

363. Introductory statement*. When certain bodies are electrified one evidence of the presence of electricity is the

* Many theorems in Attractions are important because they are used in the theory of Electricity and would seem almost purposeless without some notice of these applications. To enable these to be properly understood it is necessary to give a brief introductory account of the principles to be afterwards assumed. In this statement only so much of these principles is given as is required in what follows. For example, the experimental proofs of the laws of electricity are not described. These may be found fully discussed in Everett's edition of Deschanel's *Natural Philosophy*, 1901, in J. J. Thomson's *Treatise on Electricity*, and in that of Maxwell.

attraction exerted on other electrified bodies. For the purpose of illustrating the theory of attractions we shall here replace the electricity by that system of repelling particles which exerts an equal force at every external point.

364. If m, m' be the masses of two particles we assume here that the force of repulsion exerted by one on the other is mm'/r^2 (Art. 1). But the masses are not restricted to be positive, two particles whose masses are of opposite signs attract each other. The electricity is called positive or negative according as it is represented by particles of positive or negative mass. Since like particles repel each other the fundamental formulæ connecting potential and force are $X = -dV/dx$ &c. (Art. 41).

It is obvious from this definition that either kind of electricity may be called positive, provided the other is called negative. A convention is therefore necessary. Let a piece of glass and a piece of resin be chosen which exhibit no signs of electricity. Let these be rubbed together and separated. Each body is now found to attract the other. The electrification on the glass is called positive, that on the resin negative. If a body electrified in any manner repels the glass and attracts the resin, it is positively electrified. If it attract the glass and repel the resin, its electricity is negative.

365. When a particle A, say positively electrified, is brought into the presence of certain bodies it is found that electricity is immediately developed in them. Some positive electricity is repelled and driven to the parts of the body most remote from A and some negative electricity is attracted to the nearer parts. If a second and then a third particle be made to act on the body more positive and negative electricity are developed and separated as before. In all these cases the arrangement of the electricity when in equilibrium is altered by the approach of a new particle. This result is interpreted to mean that *when electricity is in equilibrium* the force acting on each element of the volume of the body is zero. If it were not zero, more electricity would be developed and the existing arrangement would not be in equilibrium. It follows that *the electric potential due to its own charge and to the external electrical particles is constant throughout the body.*

366. We have here supposed that the electricity is able to move without constraint from one element of volume to another. A body which permits this transference is called *a perfect conductor.* There are also bodies in which the electricity cannot travel through the volume but is forced to remain in the place where it has been developed. These are called *perfect non-conductors.* There are in nature no perfect conductors and no perfect non-conductors, but in some bodies the developed electricity travels so easily and in others with such difficulty that they are usually distinguished as conductors and non-conductors. Metals, fluids and living bodies are conductors, while dense dry gases, glass, silk are non-conductors.

367. If we represent the electricity by repelling points we must be able to apply Poisson's theorem to a body which is without constraints, Art. 105. We then have $4\pi\rho = -\nabla^2 V$, where ρdv is the excess of the positive over the negative electricity in the elementary volume dv. Since X, Y, Z are zero in equilibrium, Art. 365, this equation gives $\rho = 0$. *The element therefore contains equal quantities of positive and negative electricity.*

This holds at every internal point but not at the boundary of the solid, for here the surface constraint comes into play. The conductor is supposed to be surrounded by a non-conducting medium through which the electricity cannot pass. This medium by its pressure constrains the electricity to remain in the conductor. *There may therefore be an indefinitely thin layer of attracting or repelling particles on the boundary.*

When equal quantities of positive and negative electricity occupy an element, that element is said to be *neutral.* It exerts no force at any external point. When there is an excess of either kind in any element, that excess is said to be *free.* *In a conductor the free electricity resides on the surface.*

368. The potential of the electricity is the same as that of an indefinitely thin layer of repelling matter placed on the surface of each electrified conductor. *We measure the amount of the electricity at any point by the surface density of this equivalent layer.* The whole quantity of electricity is measured by the mass of the layer.

369. *Surface density.* Since the potential is constant throughout the interior of a conductor, the theorem of Green proved in

Art. 142 takes a simpler form. Let ρ be the surface density at any point P, let X be the normal force in the non-conducting medium at a point infinitely close to P and *let the positive direction of X be from the conductor into the non-conductor*, then $4\pi\rho = X$. If dn be an element of the normal *drawn outwards*, V the potential due to all causes, this equation may be written in the form $4\pi\rho = -dV/dn$.

The repulsive force on the electricity which covers any element $d\sigma$ of the surface is $\frac{1}{2}X\rho\, d\sigma$, (Art. 143). This may be written in either of the equivalent forms $2\pi\rho^2 d\sigma$ or $(X^2/8\pi)\, d\sigma$. Since this is always positive, *the direction of the force is necessarily outwards.*

370. If an electrified conductor is joined by a wire of conducting material to the earth its potential must become the same as that of the earth (Art. 365). At the same time the potential of so large a body as the earth is not affected by any transference of electricity to or from it. Supposing the earth to be in its ordinary neutral state, the potential then becomes zero. When a body is thus joined to the earth, it is said to be *uninsulated.*

Electrified bodies are in general supposed to be insulated by air, unless otherwise stated. When the density of the air is diminished its resistance to the escape of the electricity also decreases. When the pressure is still further reduced its resistance increases again. A vacuum, that is to say, that which remains in a vessel after we have removed everything which we can remove from it, is a strong insulator. Maxwell's *Electricity &c.* Art. 51.

371. *Capacity.* If one conductor is insulated, while all the other conductors in the field are kept at zero potential by being put into communication with the earth, and if the conductor, when charged with a quantity E of electricity, has a potential V, *the ratio of E to V is called the capacity of the conductor.*

In this definition the capacity is supposed to be independent of the special nature of the non-conducting medium which surrounds and separates the conductors. But the medium is itself acted on and reacts on the conductors. To take account of these actions it is necessary to introduce into the definition a new factor called *the Specific Inductive Capacity* of the medium. The effect of this factor is a subject for separate discussion. In what follows (until otherwise stated) it will be supposed that the medium is such that the effects of the induction on it can be disregarded. This is the case if the medium is air or generally any dry gas. In these media the specific inductive capacity is nearly equal to unity.

Ex. Let a sphere of radius a be at a great distance from all other bodies and let it be charged with a quantity E of electricity. The potential is E/a, (Art. 64). The capacity of an insulated sphere is therefore equal to its radius.

372. The electrical problem. Let a conductor have a charge M of electricity and be acted on by an external charge M'. Then in equilibrium a mass $M+\mu$ of free positive electricity and a mass $-\mu$ of negative electricity (where μ is one of the unknown quantities to be found) will be arranged on the surface of the conductor in such a manner that the sum of the potentials of M', $M+\mu$ and $-\mu$ is constant throughout the interior. The electrical problem is to find the superficial density at every point. Conversely, if the electricity be thus arranged it will be in equilibrium. First, we notice that the component forces at every internal point are zero. Next, the tangential component just inside the surface is zero and therefore by Art. 166 the tangential force at any point of the stratum is zero. The resultant force on any superficial element of the electricity is therefore normal and by Art. 369 acts outwards. It is therefore balanced by the reaction at the boundary, (Art. 367).

373. Ex. 1. The potentials of an electrical system at the corners of a small tetrahedron are V_1, V_2, V_3, V_4; prove that the potential at the point which is the centre of mass of particles m_1, m_2, m_3, m_4 placed at the corners is $\Sigma mV/\Sigma m$.

This follows at once from Taylor's theorem. [Trin. Coll. 1897.]

Ex. 2. An insulated conductor of finite volume is charged; prove that *the electrical layer completely covers the conductor.*

If there were any finite area on the surface unoccupied by electricity, the potential must also be constant throughout all external space which can be reached without passing in the immediate neighbourhood of repelling matter, Art. 117. Hence $X=0$ both on the inside and on the outside and the surface density would be everywhere zero.

Ex. 3. A conductor is charged by repeated contacts with a plate which after each contact is re-charged so that it always carries the same charge E. Prove that, if e is the charge of the conductor after the first operation, the ultimate charge is $Ee/(E-e)$. [Coll. Ex.]

When the plate touches the conductor the whole charge on both is divided between the two bodies, so that their potentials become equal. If the whole charge were increased in any ratio the potentials would be increased in the same ratio and equilibrium would still exist. It follows that just after each contact the quantities of electricity on the plate and the conductor are in a ratio $\beta : 1$ which only depends on their forms.

Let x_n be the quantity on the conductor after n contacts. After the next contact, $x_{n+1}-x_n$ is taken from the plate. Hence the ratio of $E-x_{n+1}+x_n$ to x_{n+1} is $\beta : 1$. After the first operation the quantities on the plate and conductor are $E-e$ and e, and this ratio also is $\beta : 1$. To find the ultimate ratio (when n is very great) we put $x_{n+1}=x_n$. We then find x_{n+1} by eliminating β.

Ex. 4. A soap bubble is electrified; prove that the difference between the pressures of the air inside and outside is $2T/r-2\pi\rho^2$, where T is the surface tension, r the radius and ρ the surface density of the electricity.

374. Two spheres joined by a wire. Two small spheres of radii a, b, placed at a considerable distance from each other, are joined by a thin wire and the system is insulated from all other attracting bodies. The spheres and the wire are made of some conducting material. A charge E of electricity is given to the system, determine approximately how it is distributed over the several bodies.

Let l be the distance between the centres A, B of the spheres; x, y the quantities of electricity on their surfaces, z the quantity on the wire. The electricity is so distributed that the potential is constant throughout the interior and therefore the potentials at A, B are equal. Since the centre A is equally distant from every point of that sphere, the potential at A of the electricity on its surface is x/a. Since A is very distant from every point of the other sphere, the potential at A of the electricity on the sphere B is *very nearly* y/l. Neglecting the electricity on the wire for the moment we have the two equations

$$\frac{x}{a}+\frac{y}{l}=\frac{x}{l}+\frac{y}{b},\quad x+y=E\ldots\ldots\ldots\ldots\ldots\ldots(1);$$

when the radii are very small compared with their distance we can reject the terms with l in the denominator. *The electricity is then distributed over the surfaces of the two spheres nearly in the ratio of the radii.*

We shall now prove that *the quantity of electricity on the cylindrical wire may be neglected if the radius c is sufficiently small.* Let D be the average surface density on the wire, then $z=2\pi clD$. The potential of a cylinder of length l and of uniform surface density at the middle point of its axis is

$$V=\int\frac{2\pi cDd\xi}{\sqrt{(\xi^2+c^2)}}=4\pi cD\log\left(\frac{l}{c}\right)=\frac{2z}{l}\log\frac{l}{c},$$

very nearly, since c/l is very small. Since V is nearly equal to x/a or y/b it is clear that z can be made as small as we please by using a wire sufficiently thin compared with its length.

375. Ex. 1. A conducting sphere, of radius a, is joined to the earth by a fine wire and is acted on by an electrical point Q at a distance b from the centre of the sphere. Prove that the electricity on the sphere is $-Qa/b$. Prove also that the mutual attraction between the sphere and the point approximately varies inversely as the cube of the distance and as the square of the charge.

Ex. 2. Two conducting spheres (radii a, b) are joined by a long thin conducting wire, and the total charge is zero. A cloud charged with a quantity E of electricity passes much nearer to one sphere than the other, but at a considerable distance from both. Prove that the transference of electricity from one sphere to the other is nearly $abE/(a+b)\,l'$, where l' is the distance of the cloud from the nearest sphere.

Ex. 3. Two conducting spheres of capacities c, c' are at a great distance l from each other and are connected by a long fine wire. Prove that the capacity of the conductor so formed is approximately $c+c'-2cc'/l$. [Coll. Ex. 1900.]

376. An ellipsoidal Conductor. Let a solid ellipsoid be charged with a quantity M of electricity and be not acted on by any external forces. We know by Art. 68 that the stratum enclosed between the given ellipsoid and a similar and similarly situated concentric ellipsoid exerts no attraction at any internal point. We therefore infer that the infinitely thin stratum of electricity will be in equilibrium, when distributed on the given ellipsoid so that the surface density ρ at any point P is proportional to the thickness of the thin shell. By Art. 71 the surface density at P is $\rho=\dfrac{Mp}{4\pi abc}$, where p is the perpendicular from the centre on the tangent plane at P.

377. Let points on two ellipsoids be said to "correspond" when their coordinates referred to the axes are in the ratio of the parallel axes, thus $x/a=x'/c'$ &c. Let two curves be drawn, one on each ellipsoid, such that the points on one correspond to those on the other and let the spaces enclosed be called corresponding spaces. *The quantities of electricity on corresponding spaces are in the ratio of the whole charges given to the ellipsoids.*

This theorem follows at once from the proof in Art. 201 if we regard each electrical stratum as a thin homoeoid. It may also be proved by the method of projecting one ellipsoid into the other as explained in vol. I. Art. 428.

Ex. 1. *Prove that the quantity of electricity on the portion of the ellipsoid bounded by any two parallel planes is the same fraction of the whole electricity that the portion of the diametral line between the planes is of the whole diameter.*

If a portion of an indefinitely thin shell formed by two concentric spheres be cut off by any two parallel planes we know that the intercepted volume is proportional to the distance between the planes, (vol. I. Art. 450). Projecting the spherical shell into an ellipsoidal one, the plane sections project into planes and the theorem enunciated follows at once.

Ex. 2. Two planes drawn through any diameter POP' of the ellipsoid intersect the diametral plane of POP' in OR, OR', and D is the diameter parallel to the chord RR'. Prove that the electricity E on the lune included by these two planes is given by $E=(M/\pi)\sin^{-1}(RR'/D)$ where M is the whole electricity on the ellipsoid.

378. The constant potential inside the electrified ellipsoid can be found only as a definite integral, (Art. 197). When the ellipsoid is one of revolution so that $a=b$, the integration can be effected without difficulty.

Let the axis of z be the axis of revolution. The quantity of electricity between the planes z and $z+dz$ is $dz \,.\, M/2c$, Art. 377. The potential at the centre is therefore

$$V=\frac{M}{2c}\int\frac{dz}{r}, \quad r^2=a^2+\frac{c^2-a^2}{c^2}z^2.$$

Effecting the integrations we find

$$V=\frac{M}{\sqrt{(a^2-c^2)}}\sin^{-1}\frac{\sqrt{(a^2-c^2)}}{a}, \quad V=\frac{M}{\sqrt{(c^2-a^2)}}\log\frac{c+\sqrt{(c^2-a^2)}}{a},$$

according as the spheroid is oblate $(a>c)$ or prolate $(c>a)$. The internal potential of the prolate spheroid is found more easily by taking V to be the potential at the focus, for then $r=c+ez$.

The potential of the oblate spheroid is also $V=M\phi/f$, where f is half the distance between the foci of the generating ellipse and ϕ is half the angle subtended by $2f$ at the extremity of the axis of revolution.

379. We know by Art. 205 that all the external level surfaces of the ellipsoidal conductor are confocal ellipsoids. If P be any point situated on the confocal whose semi-axes are a', b', c' the potential at P is $\frac{1}{2}M\int\frac{d\lambda}{\{(a^2+\lambda)(b^2+\lambda)(c^2+\lambda)\}^{\frac{1}{2}}}$, where the limits are $\lambda=a'^2-a^2$ to ∞, (Art. 208).

380. The external surface of a conductor charged with a given quantity of electricity is not acted on by any external body. Prove that *the electricity at every point has the same sign.*

Let V_0 be the potential and first let this be positive. If there be any point P on the surface at which there is negative electricity, dV/dn must there be positive because the surface density m is given by $4\pi m=-dV/dn$ (Art. 369). Hence the potential increases outwards along the line of force at P. But this is impossible since the potential at every point between the surface of the conductor and the surface of a sphere of infinite radius must lie between V_0 and zero, (Art. 116, Ex. 2). A similar proof applies if V_0 is negative.

381. A conductor, charged with a given quantity of electricity, is acted on by given forces. Prove that *there is but one arrangement of the electricity which could be in equilibrium.*

If possible let there be two distributions of the electricity, either of which could be in equilibrium, though the potentials are not necessarily the same. By subtracting one of these from the other, as in Art. 129, we obtain a distribution of electricity in equilibrium in which the external forces are absent and the total mass is zero. This distribution must have both positive and negative electricity on the surface, but this has just been proved to be impossible.

382. Elliptic Disc. To deduce the distribution of electricity on an insulated elliptic conducting disc from that on an ellipsoid we put $c=0$. Then

$$\left(\frac{c}{p}\right)^2 = c^2\left(\frac{x^2}{a^4}+\frac{y^2}{b^4}\right)+\frac{z^2}{c^2}=\frac{z^2}{c^2}.$$

The surface density ρ at any point (x, y) of the disc is then given by

$$\rho=\frac{Mp}{4\pi abc}=\frac{M}{4\pi ab}\frac{1}{\sqrt{(1-x^2/a^2-y^2/b^2)}},$$

where M is the *whole charge* on the disc. This value of ρ gives the surface density on *either side*.

The internal potential due to the electricity on both sides is

$$V=\frac{M}{4\pi ab}\iint\frac{r\,d\theta\,dr}{r}\cdot\frac{2}{\sqrt{(1-r^2q^2)}},\quad q^2=\frac{\cos^2\theta}{a^2}+\frac{\sin^2\theta}{b^2},$$

where the limits are $r=0$ to $1/q$ and $\theta=0$ to 2π. Effecting the integration with regard to r and writing $\theta=\frac{1}{2}\pi-\phi$, we find

$$V=\frac{M}{a}\int_0^{\frac{1}{2}\pi}\frac{d\phi}{\sqrt{(1-e^2\sin^2\phi)}}.$$

For a circular disc we have, if ρ be the surface density at any distance r from the centre,

$$\rho=\frac{M}{4\pi a}\frac{1}{\sqrt{(a^2-r^2)}},\text{ and } V=\int\frac{2\pi r\,dr\,2\rho}{r}=\frac{M\pi}{2a}.$$

The capacity M/V is therefore $2a/\pi$.

383. The quantity of electricity on any elementary area $dx\,dy$ of the disc is $\frac{M}{4\pi ab}\frac{dx\,dy}{\sqrt{(1-x^2/a^2-y^2/b^2)}}$. If then two elliptic discs (say an ellipse and its auxiliary circle) have equal charges, the quantities of electricity on corresponding elements are also equal, for in these elements $x/a=x'/a'$ &c., $dx/a=dx'/a'$ &c. *The quantities on any corresponding finite areas are therefore also equal.* This result follows at once from Art. 377.

384. *The potential V' of the elliptic conductor at an external point P* follows from the result stated for an ellipsoid in Art. 379. Let the confocal on which P is situated cut the axis of the disc in C, C' and let its semi-axes be a', b', c'. We find after putting $c=0$, $\lambda=\mu^2$ and $\mu=b\tan\theta$,

$$V'=M\int\frac{d\mu}{\{(a^2+\mu^2)(b^2+\mu^2)\}^{\frac{1}{2}}}=\frac{M}{a}\int\frac{d\theta}{\{1-e^2\sin^2\theta\}^{\frac{1}{2}}},$$

where the limits are $\mu = c'$ to $\mu = \infty$ and $\theta = \frac{1}{2}\pi - \phi$ to $\theta = \frac{1}{2}\pi$ where ϕ is half the angle subtended by the minor axis of the disc at C or C'.

For a circular disc the potential at an external point is $M\phi/a$ where ϕ is half the angle subtended at C by any diameter of the disc.

385. Ex. 1. A chord is drawn on an elliptic insulated disc, prove that the quantities of electricity on each side of the chord are in the ratio of the segments of the conjugate diameter, (Art. 377).

Ex. 2. Prove that the quantity of electricity on an elliptic sector bounded by the semi-diameters CP, CP' is $M(\phi' - \phi)/2\pi$, where ϕ, ϕ' are the eccentric angles of P, P' and M is the whole quantity of electricity (Art. 383).

Ex. 3. A similar coaxial ellipse whose semi-axes are na, nb is described on the electrified disc. Prove that the quantity of electricity between this ellipse and the rim of the disc is $M\sqrt{(1-n^2)}$.

Ex. 4. Prove that the *line density of a thin electrified insulated rod is constant.*

Regard the rod as an evanescent ellipsoid in which a and b are zero and c finite. The line density ρ' is such that $\rho' dz$ represents the mass between two planes whose abscissæ are z and $z + dz$. This we know is $Mdz/2c$, Art. 377.

386. Conductor with a cavity. A body is bounded by two surfaces S_1, S_2 which do not intersect. It is charged with a given quantity M of electricity and is acted on by a given system of electrical points (mass M_1) situated within S_1.

Let x be the quantity of electricity on S_1, $M - x$ that on S_2. These are so distributed that the sum of the potentials of M_1, x and $M - x$ is constant throughout the space between S_1, S_2. Describe a surface σ between S_1, S_2 and enclosing S_1, then by Gauss' theorem (Art. 106) $4\pi(M_1 + x)$ is equal to the flux of force across σ, and this is zero. Hence $x = -M_1$; the charge on S_1 is therefore $-M_1$ and that on S_2 is $M + M_1$.

387. If the charge $M = -M_1$, the total quantity on S_2 is zero. It immediately follows from the argument in Art. 380 that the charge *on each element* is also zero. For by that article, the electricity on every element of S_2 must have the same sign, which is impossible when the whole quantity is zero. The whole of the free electricity is therefore concentrated on S_1.

The sum of the potentials of the system M_1 and of the electricity $x = -M_1$ is constant throughout the space between S_1 and S_2 and therefore throughout all space which can be reached without passing in the immediate neighbourhood of attracting matter, Art. 117. It is therefore constant throughout all space external to S_1 and is the same as that at an infinitely distant point. The sum of the potentials of M_1 and x is therefore zero.

It appears that the form of the surface S_2 may be changed without in any way disturbing the electricity on the surface S_1.

388. Let a *solid conductor* whose boundary is the surface S_2 be acted on by a given external system of electrical points (mass M_2) and let the charge given to the conductor be y. The condition of equilibrium is that the sum of the potentials of M_2 and y is constant throughout the interior. Since this condition is not affected by removing any portion of the inside matter the equilibrium of the electricity on S_2 will not be disturbed when the surface S_1 is made the internal boundary.

389. If we now superimpose these two conductors, we have a conductor bounded by the surfaces S_1, S_2 with a given electrical mass M_1 inside S_1 and a mass M_2 outside S_2. Let the charge given to the conductor be M.

There will now be a charge $x = -M_1$ on S_1 so arranged that the sum of the potentials of M_1 and x is zero at all points external to S_1. There will be charge $y = M + M_1$ on S_2 so arranged that the sum of the potentials of M_2 and y is *constant* at all points internal to S_2. The condition that the equilibrium should remain undisturbed is that the sum of the four potentials should be constant at all points between S_1, S_2 and this condition is evidently satisfied.

We observe that the distributions of electricity on the two boundaries are independent of each other.

390. Screens. We notice how completely the electricity on the surface S_1 screens the repelling masses M_1 from observation by an external spectator. If the masses forming the system M_1 be moved about in any way within the cavity the electricity on S_1 rearranges itself continually so that in equilibrium the resultant force at every external point is zero.

In the same way if the external masses M_2 be moved about, the electricity on the surface S_2 is rearranged and the motion is imperceptible to an observer within the cavity.

391. We shall now prove that *there is but one possible distribution in equilibrium* on the two surfaces S_1, S_2 when the charge M and the electrical masses M_1, M_2 are given.

If possible let there be another arrangement. Then subtracting one of these from the other, we have an arrangement of electricity in equilibrium, in which there are no internal or external masses, while there are charges x and y on S_1 and S_2, whose sum $x+y$ is zero.

The sum of the potentials of x and y is constant over S_1 and therefore over a surface just inside S_1 (Art. 145). This surface contains no attracting matter and therefore the sum of the potentials of x and y is constant throughout all space within S_1. The forces X, X' just inside and just outside S_1 are therefore zero. It follows that the density of the electricity at every point of S_1 is zero (Art. 142).

Consider next the surface S_2. The total charge on it is zero. Hence, as in Art. 381, the charge on each element is zero.

The two possible distributions must therefore have been the same.

392. Ex. 1. A solid sphere, radius a, is concentric with a spherical shell, radii b and c, both being perfect conductors, and charged with quantities E, E' of electricity. To find the potential of the sphere.

The quantities of electricity on the three surfaces whose radii are a, b, c are respectively E, $-E$, and $E'+E$ (Art. 389). By Art. 64, the potentials at any points inside the substance of the sphere and shell are respectively

$$V=\frac{E}{a}-\frac{E}{b}+\frac{E'+E}{c}, \qquad V'=\frac{E'+E}{c}.$$

If the shell is joined to earth, or if the radius c is very great, in either case $V'=0$ and the capacity of the sphere, viz. E/V, becomes $ab/(b-a)$. If a and b are also very nearly equal to each other, the capacity is very great. Supposing the potential of the sphere to be finite, the charge on the sphere and the equal opposite charge on the inner surface of the shell are very large.

When two conductors, insulated from each other, are placed very near each other the system is called *a condenser*.

Ex. 2. Three insulated conductors A, B, C, are in the form of thin concentric spherical shells of radii a, b, c; and are so charged that their potentials are V_1, V_2, V_3. Prove that the charge on the intermediate shell B is

$$\frac{b}{(a-b)(b-c)}\cdot\{a(b-c)V_1+b(c-a)V_2+c(a-b)V_3\}. \qquad \text{[Coll. Ex. 1897.]}$$

Ex. 3. A condenser is formed of two concentric spherical thin conducting sheets, the radius of the inner being b, that of the outer a. A small hole exists in the outer sheet through which an insulated wire passes connecting the inner sheet with a third conductor, of capacity c, at a large distance r from the condenser. The outer sheet of the condenser is put to earth, and the charge on the two connected conductors is E. Prove that approximately the force on the third conductor is $ac^2E^2\Big/\left(\frac{ab}{a-b}+c\right)^2 r^3$. [Trin. Coll. 1897.]

393. Green's Method. The law of distribution of a given quantity of electricity on a given surface under the influence of given forces cannot always be discovered. *Two methods of finding a solution in certain limited cases are in general use.* The *method of inversion*, by which when one problem has been solved the solution of another can be deduced, has been explained in Arts. 168 &c. We shall now proceed to *Green's Method*, by which the law of distribution of a certain quantity of electricity can be found when the boundary of the conductor is a level surface of a known system of repelling bodies. This method has been already discussed in Art. 156. Before however proceeding to its application we shall give an elementary proof in small type which more fully illustrates the principles of attraction.

394. Let M_1, M_2 be two given systems of attracting or repelling particles and let S be a surface enclosing M_1 only within its finite space. If the masses M_2 were removed and S made *the inner boundary of a conductor*, a quantity of electricity, equal to $-M_1$, would be found on the surface S and its potential together with that of the system M_1 would be zero throughout all space external to S (Art. 387). Let this distribution of electricity on S be called E_1.

Let us now change the sign of every element of the electricity E_1 and constrain it to remain, otherwise unaltered, on the surface S. Let this new distribution be called E_2. The *quantity* of the electricity E_2 is then *equal to* $+M_1$ and *the potential of* E_2 *is the same as that of* M_1 *throughout all space external to S.*

Let us next suppose that S is a level surface of M_1 and M_2 and let the potential at any point of S be V_s. It must therefore also be a level surface of E_2 and M_2. The sum of the potentials of E_2 and M_2 is therefore constant and equal to V_s at all points of a surface just inside S. Since no particle of either E_2 or M_2 lies within this surface, *the sum of the potentials of* E_2 *and* M_2 *is constant and equal to* V_s *throughout the interior of S* (Art. 115).

If S be made *the external boundary of a conductor* and the system M_1 removed, the distribution of electricity E_2 would be in equilibrium under the influence of its own attraction and that of M_2 (Art. 372). We also know that no other distribution of the same quantity M_1 of electricity will be in equilibrium (Art. 381).

Briefly, E_1, when acted on by M_1, is in equilibrium if S is the inner surface of a conductor, and E_2, when acted on by M_2, is in equilibrium if S is the outer surface. Also E_1 and E_2 differ only in sign.

The surface density ρ_2 *at any point P of the stratum* E_2 when placed on the external surface of a conductor follows at once from Green's theorem, (Art. 142). By that theorem $4\pi\rho_2 = X$ where X is the sum of the normal forces due to M_2 and E_2 at a point just outside the substance of the conductor. But the normal force due to E_2 has been proved equal to that of M_1. Hence X is the sum of the normal forces at P, due to M_1 and M_2, measured positively from the conductor towards the non-conductor.

The density ρ_1 at any point P of the stratum E_1 when placed on the inner surface of a conductor has the sign opposite to ρ_2. Since the non-conductor is now on the opposite side of S_1 the density ρ_1 is given by the same rule, $4\pi\rho_1 = X$, where X is the sum of the normal forces due to M_1 and M_2 measured towards the non-conductor.

395. We arrive at the following rules.

1. Let S be any closed portion of a level surface (potential V_0) of a given electrical system, M_1 being inside and M_2 outside. We may remove either the mass M_2 and regard S as the internal boundary of a solid conductor acted on by the internal mass M_1, or

we may remove M_1 and regard S as the external boundary of a conductor acted on by the external mass M_2. In either case the density ρ at P is given by the rule $4\pi\rho = X$, where X is the normal force at P due to both M_1 and M_2 measured positively towards the non-conductor.

2. The quantity of electricity on S is $-M_1$ or $+M_1$ according as S is an internal or external boundary.

3. The potential of the electrical stratum at any point R on the side opposite to M_1 is *numerically* equal to that of M_1. It follows that the stratum and M_1 have not only the same mass, but also the same centre of gravity, (Art. 136). Their principal axes at the centre of gravity also coincide in direction and the difference of their moments of inertia about every straight line is the same.

4. The potential of the stratum at any point R' on the same side as M_1 is equal to $V_s - V_2$ where V_2 is the potential of M_2 at R', when S is an external boundary. It follows that when S is an internal boundary, the potential at R' is given by the same expression with its sign changed.

396. If the quantity Q of electricity which covers any given portion σ of the surface S is required we use Gauss' theorem, (Art. 106). We describe a surface closely enveloping the given portion of S both inside and out, then $4\pi Q = \int F d\sigma$. Just *inside the conductor* S the force F is zero and all *this portion of the integral may be omitted.* The required quantity Q is therefore given by the above integral, where F is the normal (or resultant) force due to the given system M_1, M_2 measured towards the non-conductor at the element $d\sigma$ of the given area, and the integral extends over that area.

When the systems M_1, M_2 consist of isolated particles of masses m_1, m_2 &c. the integral can be put into a more convenient form. For any one particle m we have, as in Art. 106, $Fd\sigma = md\omega$ where $d\omega$ is the solid angle subtended at m by $d\sigma$. This elementary solid angle is to be estimated as positive when a repulsive force emanating from m traverses $d\sigma$ outwards into the non-conducting medium. Adding up the corresponding portions for all the particles, we see that *the required quantity Q of electricity is given by* $4\pi Q = \Sigma m\omega$.

Image. When an imaginary system of points, if properly placed on one side of a surface, would produce at points on the other side of that surface the same attraction or repulsion which the actual electricity on the surface produces that system of points is called an image. Thus, when the surface S is the external surface of a conductor, M_1 is the image of M_2, because the attraction which it exerts at points on the other side of S is the same as that due to the electrification on S when acted on by M_2.

397. Electricity on a sphere. *To find the distribution of electricity on a sphere of conducting matter when acted on by an electrical point.*

Let A be the centre of the sphere, B any external point, BD a tangent and DC a perpendicular on AB. Let the distances of A, B, C from D be respectively a, b, c, so that a is the radius. Let the distances of B and C from the centre A be f and f'.

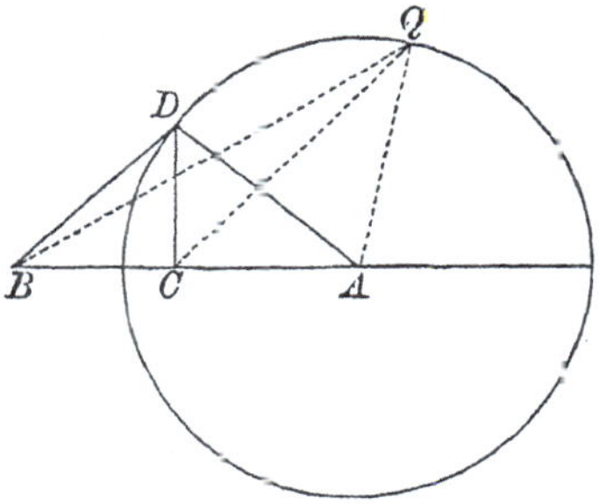

Since $a^2 = ff'$, the points B and C are inverse with regard to the sphere. If Q be any point on the sphere, the ratio CQ/BQ is constant and therefore (by putting Q at D) equal to c/b (Art. 86). We also have, by the similar triangles BDC, BAD, $c/b = a/f$.

If then we place at B and C respectively quantities of electricity $E = \mu b$, $E' = -\mu c$*, where μ is any constant, we have at any point Q of the surface

$$\frac{E}{BQ} + \frac{E'}{CQ} = 0 \quad\text{.......................(1)},$$

i.e. *the sphere is a level surface of zero potential.* We may therefore at once apply Green's principles (Art. 395).

398. If the conducting mass is to be a *solid sphere surrounded by a non-conducting medium,* we remove the electrical point C and distribute the mass E' over the surface so that the surface density ρ at any point Q is given by $4\pi\rho = F$, where F is the normal force at Q due to E and E' tending towards the side of the sphere on which the non-conducting medium lies. The

* Since $CQ^2/BQ^2 = a^2/f^2 = f'/f$, it follows that the squares of the mass particles which, placed at the inverse points B, C, make the sphere a level surface of zero potential are proportional to the distances of those points from the centre. *The centre of gravity of masses proportional to* $1/E^2$, $-1/E'^2$, *placed at* B, C *respectively, is at the centre* A *of the sphere.* This result enables us to apply any of the theorems relating to the centre of gravity given in vol. I.

stratum thus formed is in equilibrium under its own repulsion and the action of the electricity E at the external electrical point B.

The potential of the stratum at any point R outside the sphere is equal to that of the electrical point E' and is therefore E'/CR. At a point R' inside the sphere, the potential of the stratum is equal and opposite to that of the external electrical point E, because the sum of these two potentials is zero at all points inside the conductor.

399. *If the sphere is the boundary of a cavity in the conductor,* we remove the electrical point B. The surface density ρ, when the electricity is acted on by the electrical point $E' = -Ea/f$ situated at C, is the same as that just found for a solid sphere, but with the sign changed (Art. 395).

The potential of the stratum at any point within the sphere is therefore equal to that of E at the same point. The potential of the stratum at a point outside the sphere is equal and opposite to that of E'.

Another proof. Let us surround the system by a sphere of infinite radius with its centre near C. The point B is now included within the boundary formed by the given sphere and the infinite sphere. We remove the point B and spread its electricity over the double-sheeted boundary. By Art. 386 the quantity on the given sphere is equal and opposite to that of C and is therefore $-E'$; the quantity on the infinite sphere is therefore $E+E'$. Since this is a finite quantity spread on a sphere of infinite radius, both its potential and attraction at points near B or C are zero. This electricity may therefore be removed from the field.

400. We may at once *deduce either of the results given in Arts.* 398, 399 *from the other by an easy inversion with regard to the centre A of the sphere,* the radius being the radius of inversion (Art. 86).

When the sphere is *an outer boundary* the potential of a charge E at B together with that of a surface distribution ρ on the sphere is zero throughout the interior. When we invert this system with regard to the centre, the distribution on the sphere is unaltered while the charge at B is moved to C and becomes Ea/f (Art. 168). It follows, from Art. 177, that the potential of the same distribution on the *surface of a spherical hollow* together with that of a charge Ea/f at C is zero at all points outside the sphere. This distribution is therefore in equilibrium when placed on the inner surface of a conductor and acted on by the charge at C. This last result is the same as that obtained in Art. 399 except that the signs of both E' and ρ have been changed.

401. *To find the surface density ρ at any point Q in terms of the distance of Q from either electrical point.*

The outward normal force F due to the repulsions of the points at B and C is the resultant of E/BQ^2 and E'/CQ^2, see figure of Art. 397. Hence resolving perpendicularly to CQ, we have

$$\frac{E}{BQ^2}\sin BQC = -F \, . \sin AQC.$$

But $BQ \sin BQC : AQ \sin AQC$ is equal to the ratio of the perpendiculars from B and A on CQ and is therefore equal to $BC : CA$.

$$\therefore\ F = -\frac{E}{BQ^3}\frac{BC \, . \, a}{CA}.$$

By similar triangles we have $BC/CA = b^2/a^2$, hence (Art. 369)

$$4\pi\rho a = aF = -\frac{Eb^2}{BQ^3} \quad\ldots\ldots\ldots\ldots\ldots\ldots (1).$$

In the same way we find by resolving perpendicularly to BQ,

$$4\pi\rho a = aF = \frac{E'c^2}{CQ^3} \quad\ldots\ldots\ldots\ldots\ldots\ldots (2).$$

Either of these results may be deduced from the other by using the known relations $E/b + E'/c = 0$, $b/BQ = c/CQ$. Art. 397.

If the sphere is the boundary of a solid conductor, F is to be measured outwards from the sphere into the non-conductor, and these expressions give the density at any point. *If the sphere is the boundary of a cavity*, the force F must be taken positively inwards and *the signs on the right-hand sides of* (1) *and* (2) *must be changed.*

In both cases let the point (B or C) at which the acting charge is situated be called O and let the charge (E or E') be called E_1. If k^2 be the product of the segments of a chord drawn through O, the surface density ρ at any point Q on the sphere is given by $4\pi\rho a = -\dfrac{E_1 k^2}{OQ^3}$.

402. In the case of a solid conducting sphere we may superimpose a uniform stratum of any surface density ρ_0. This addition changes the potential to V_0, where $V_0 = 4\pi\rho_0 a^2/a$. If ρ' be the resulting surface density at any point Q, we have

$$4\pi\rho' = \frac{V_0}{a} - \frac{1}{a}\frac{Eb^2}{BQ^3} = \frac{V_0}{a} + \frac{1}{a}\frac{E'c^2}{CQ^3} \quad\ldots\ldots\ldots\ldots (3).$$

The quantities of electricity on the sphere due to the two strata respectively are $V_0 a$ and E', and the total quantity is $E'' = V_0 a + E'$ where $E'/a = -E/f$. *The potential at any external point R is the sum of the potentials of two electrical points, one of mass E' placed at C, the other of mass $V_0 a = E'' - E'$ placed at the centre A of the sphere.*

403. *The surface density may also be easily found by the method of inversion**. Proceeding as in Art. 176 we invert the theorem "the potential of a thin spherical stratum of density ρ_0, radius a, at an internal point P is $V_0 = 4\pi\rho_0 a^2/a$." Let k^2 be the product of chords through the centre O of inversion, so that the sphere inverts into itself. We immediately arrive at the theorem "the potential at a point P' of a thin spherical stratum of density $\rho_0' = \rho_0 (k/OQ)^3$ is $V_0' = V_0 (k/OP')$," where P' lies on the side of the sphere opposite to O.

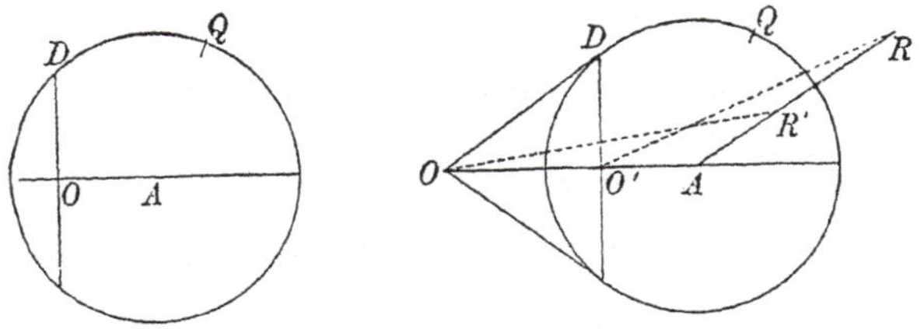

Since the expression just found for V_0' is clearly the potential at P' of a particle $V_0 k$ placed at O, it follows that "the sum of the potentials at P' of the electrical stratum and of the particle $(-V_0 k)$ placed at O is equal to zero." Let the arbitrary density ρ_0 be such that $-V_0 k = E$, then *the sum of the potentials at P' of a stratum whose density* $\rho = \dfrac{V_0}{4\pi a}\left(\dfrac{k}{OQ}\right)^3$, *and of the particle E placed at O, is zero, and is therefore constant throughout the space on the side of the sphere opposite to O.*

If the sphere is to be a solid sphere of conducting matter we place O outside, say at B in the figure of Art. 397. *If the sphere is to be a cavity in a conducting medium* we place O inside, say at C in the same figure. *In either case the density of the stratum at any point Q when acted on by an electrical point of mass E at O is given by* $4\pi\rho = \dfrac{-Ek^2}{a\,.\,OQ^3}$, *and the conducting matter is at zero potential.*

404. In the case of a solid conducting sphere the potential of this stratum alone at any point R' within the sphere (being equal and opposite to that of E) is $-E/OR'$. Placing R' at the centre, we see that the quantity E' of electricity on the sphere is given by $E'/a = -E/f$ where $f = OA$. We also find the potential V at any external point R by inverting the stratum with regard to its centre A as in Art. 399. The stratum is unaltered by this inversion. Its potential at R is therefore $V = \left(-\dfrac{E}{OR'}\right)\dfrac{a}{AR}$. If O' be a point such that $AO\,.\,AO' = a^2$, the triangles OAR', $O'AR$ are similar and $OR'\,.\,AR = O'R\,.\,OA$. The potential of the stratum at any point R is therefore $E'/O'R$ and is equal to that of a particle of mass E' placed at O'.

405. *Lines of force and level surfaces.* To simplify the discussion of these curves, let us consider the case in which the sphere is at potential zero. We may

* The first determination of the law of distribution of electricity on a sphere when influenced by an electric point was made by Poisson whose method required the use of Laplace's functions. Sir W. Thomson discovered the powerful method of inversion and used it to obtain an easy geometrical solution of this problem. He also expressed the surface density in a much simpler form than that given by Poisson. See Section v. of the *Reprint of his papers on Electrostatics and Magnetism*. The first application of Green's theorem to this problem is to be found in Maxwell's *Treatise on Electricity*, &c.

then represent the attractions by two electrical points E, E' situated at B, C; in our case $E' = -Ea/f$ is negative. We shall put $a/f = n$ for brevity, and we notice that $n < 1$. See figure on page 206.

The equations of the curves have been found in Art. 323. If (r, θ), (r', θ') are the polar coordinates of a point P referred to B and C as origins, BCA being the axis of reference, the lines of force and the level curves are given by

$$E \cos\theta + E' \cos\theta' = K, \qquad E/r + E'/r' = K'.$$

It is clear that the lines of force emanating from one electrical point must either pass to the other or proceed to an infinite distance, (Art. 114)

When a line of force proceeds from B to an infinite distance we equate the values of K at B and at infinity. Since the radius vector at the origin B coincides with the tangent and the angle θ' is there equal to π we have

$$E \cos\theta_0 - E' = (E + E') \cos\beta,$$

where θ_0 and β are the angles the tangent at B and the asymptote make with the axis of reference BCA. Since $\cos\beta$ must lie between ± 1, we see that $\cos\theta_0$ must lie between -1 and $1 - 2n$.

When a line of force proceeds from C to an infinite distance we have

$$E + E' \cos\theta_0' = (E + E') \cos\beta',$$

hence $\cos\theta_0'$ must lie between 1 and $(2 - n)/n$. Since the latter fraction is greater than unity, no line of force can pass from C to an infinite distance, except that which coincides with the straight line BCA*.

When a line of force proceeds from B to C we have

$$E \cos\theta_1 - E' = E + E' \cos\theta_1',$$

where θ_1, θ_1' are the angles the tangents at B, C make with BCA. As $\cos\theta_1$ decreases from unity to $1 - n$, the sign of $\cos\theta_1'$ is negative and the lines of force *arrive at C on the side nearest B.* When $\cos\theta_1 = 1 - n$ the line of force at C is perpendicular to BCA. When $\cos\theta_1$ lies between $1 - n$ and $1 - 2n$ the sign of $\cos\theta_1'$ is positive and the lines of force *enter C on the side furthest from B.* When $\cos\theta_1 < 1 - 2n$ the line of force goes to an infinite distance.

To trace the level surfaces we proceed as in Art. 134. The level surface which passes through the point of equilibrium X governs the whole figure. This point lies in BC produced so that $CX = BX\sqrt{n}$. There is a conical point at X, the tangents making an angle $\pm\tan^{-1}\sqrt{2}$ with BCA produced (Art. 121, Ex. 2). This surface *when complete* consists of two sheets; one of these passes between B and C because its potential is less than the infinite positive potential near B and algebraically greater than the infinite negative potential near C. The other sheet cuts ACB beyond B because its potential is less than that near B and greater than that at an infinite distance. The two sheets therefore turn from X towards B and C, one enveloping C only and the other both B and C. This level surface is represented by the thick line in the left-hand figure. Its potential is $E(1 - \sqrt{n})^2/l$, where $BC = l$.

The other level surfaces fill up the enclosed spaces and surround the two sheets. A few of these are represented by the dotted lines. The level surfaces within the smaller sheet and those outside both are at potentials less than that at X, while

* Since the sphere of zero potential surrounds C (but not B), it is clear that no line of force (except CA produced) could proceed from a point on that sphere to a point at an infinite distance at which the potential is also zero (Art. 114). It is also clear that there can be no points of equilibrium except on the axis BCA.

those between the two sheets are at greater potential. The complete level surfaces whose potentials are less than that at X and greater than zero are therefore two-sheeted surfaces. The two sheets of the level surface of zero potential are a sphere inside the smaller sheet and a sphere of infinite radius.

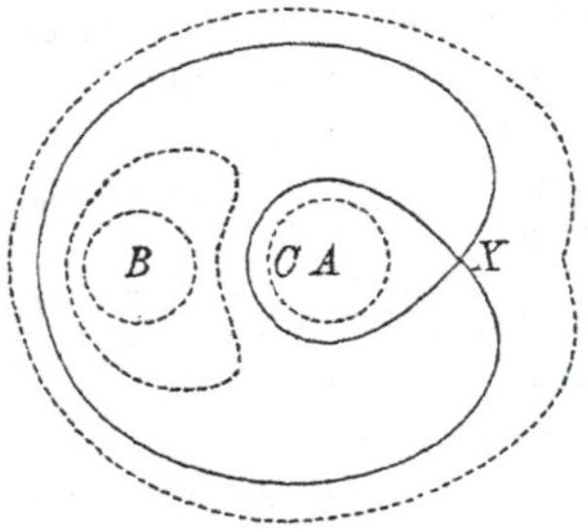

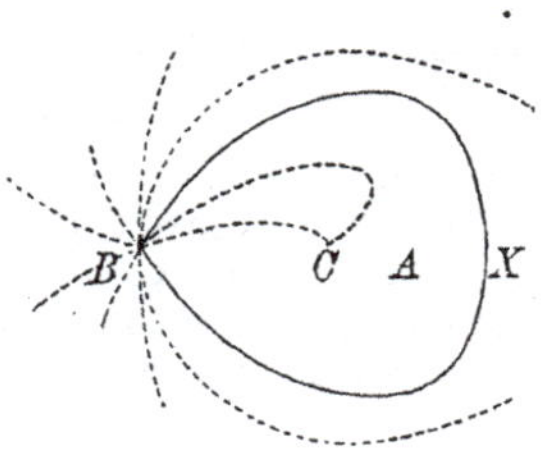

Since X lies on the axis of reference, any line of force which passes from B to X is defined by $E\cos\theta_1 - E' = E + E'$ or $\cos\theta_1 = 1 - 2n$. If the potential is to continue to decrease, this line must make a sharp turn at X, Art. 114. By doing this it could either reach C or proceed to an infinite distance in the direction BCA. The line of force from B to X is represented by the thick line in the figure on the right-hand side. The closed surface formed by all the lines of force which proceed from B to X separates the other lines of force into two systems, those inside this space pass from B to C, those outside proceed from B to an infinite distance. A few of these are represented by the dotted lines.

The figures represent the level surfaces and lines of force due to two particles placed at B and C. When C is surrounded by a spherical conductor the lines of force are cut off by the sphere, and exist only outside the sphere. The potential being constant within the conductor the level surfaces become indeterminate.

The figures are only roughly drawn. The outer sheet, for example, of the level surface which passes through X should be very much larger.

406. Ex. 1. A sphere charged with a given quantity of electricity is acted on by an external electrified point B and a tangent from B touches the sphere at D. Prove that the potential at D of the heterogeneous stratum of electricity is the same as if it were homogeneous and its density equal to that of the heterogeneous stratum at D.

Ex. 2. A conducting sphere (radius a) is at potential zero under the action of a quantity E of electricity at a point B distant f from the centre A. The sphere is cut by a plane perpendicular to the diameter BA. Prove that the quantity of electricity on the side remote from B is $\frac{E}{2}\frac{f-a}{f}\left(1-\frac{f+a}{BQ}\right)$ where Q is any point on the curve of section.

Problems of this kind may be solved in three ways: (1) by Gauss' theorem the quantity Q on any area is given by $4\pi Q = E\omega + E'\omega'$ as explained in Art. 396; (2) the heterogeneous stratum is known to be inverse of a homogeneous stratum, hence Q/k is equal to the potential at the centre B of inversion of the corresponding portion of the homogeneous stratum; (3) the result may be obtained by direct integration.

Ex. 3. Prove that the potential at any point R on the diameter BA of the electricity cut off by the plane section *as described in the last question* is

$-\frac{Eb^2}{2f}\cdot\frac{1}{BR\,.\,CR}\left(\frac{RQ}{BQ}-\frac{RH}{BH}\right)$ where H is the point of intersection of BA produced to cut the sphere.

Ex. 4. A spherical insulated conductor, charged with a given quantity E'' of electricity, is in *a uniform field of force* defined by the potential Fx. Prove that the surface density at Q is given by $4\pi\rho a^2 = E'' - 3aFx$ where x is the abscissa of Q referred to the centre. [In Art. 401 make B very distant and $E/BQ^2 = -F$.]

Ex. 5. A solid sphere being charged with a given quantity E'' of electricity is acted on by an electrical particle of mass E situated at a distance f from the centre A of the sphere. Prove that the mutual repulsion between the sphere and particle is

$$F = \frac{E''E}{f^2} + \frac{E^2a}{f^3} - \frac{E^2af}{(f^2-a^2)^2}.$$

Thence show that if the sphere be close enough to the particle, the mutual force is attractive; and if the sphere is uncharged the force is attractive at all distances. If the sphere be allowed to fall from rest towards the particle find the velocity in any position.

Ex. 6. A unit charge is brought to a point B, at a distance f from the centre of an insulated sphere, of radius a and charge E; prove that the total work done is $\frac{E}{f} - \frac{a^3}{2f^2(f^2-a^2)}$. [Coll. Ex. 1897.]

Ex. 7. Outside a spherical charged conductor there is a concentric insulated but uncharged conducting spherical shell which consists of two segments: prove that the two segments will not separate if the distance of the separating plane from the centre is $< ab/(a^2+b^2)^{\frac{1}{2}}$, where a, b are the internal and external radii of the shell. [Coll. Ex. 1897.]

Ex. 8. If a uniform circular wire charged with electricity of line density $-e$ is presented to an uninsulated sphere of radius a, the centre of which is in the line through the centre and perpendicular to the plane of the circular wire, prove that the electrical density induced at any point on the sphere, whose angular distance from the axis of the ring is θ, is

$$\frac{f^2-a^2}{\pi a}\,\frac{Eef\sin\alpha}{\{a^2-2af\cos(\theta+\alpha)+f^2\}^{\frac{1}{2}}\{a^2-2af\cos(\theta-\alpha)+f^2\}},$$

where f is the distance of any point of the ring from the centre of the sphere, α is the angle subtended at the centre by any radius of the ring, and

$$E = \int_0^{\frac{1}{2}\pi}\sqrt{(1-k^2\sin^2\phi)}\,d\phi, \qquad k^2 = \frac{4af\sin\alpha\sin\theta}{a^2-2af\cos(\theta+\alpha)+f^2}.$$

[Math. Tripos, 1879.]

The density at any point Q of the sphere due to an element of electricity $m = ef\sin\alpha\,d(2\phi)$ at a point B on the ring is given in Art. 401 and is a known multiple of m/BQ^3. To effect the integration between the limits 0 and $\frac{1}{2}\pi$ we first prove by geometry that BQ is a known multiple of $\Delta = \sqrt{(1-k^2\cos^2\phi)}$ and then use the theorem $(1-k^2)\int\Delta^{-3}\,d\phi = \int\Delta^{-1}\,d\phi$. This analytical result may be obtained by differentiating $\sin\phi\cos\phi/\Delta$ and then integrating the result between the limits 0 and $\frac{1}{2}\pi$.

407. Electricity on cylinders. We may apply either Green's theorem or the method of inversion to find the distribution of electricity on an infinite circular conducting cylinder when acted on by a thin uniformly electrified non-conducting rod placed parallel to the axis either inside or outside.

Referring to the figure of Art. 397 we see that since CQ/BQ is constant, $\log CQ - \log BQ$ is also constant for all points on the circle. Let two non-conducting thin rods infinite in both directions be placed at B and C perpendicularly to the plane of the paper; let these rods be uniformly and equally electrified but with opposite signs. The infinite cylinder whose cross section is the circle is then a level surface of the two rods (Art. 43).

If the cylinder is the boundary of a solid conductor, we remove the electrical rod C and distribute its electricity over the cylinder. The repulsion of the stratum at any external point R is the same in direction and magnitude as that of the rod C. Its magnitude is therefore $2m/CR$, where m is the line density of the rod. At any internal point R' the repulsion is equal and opposite to that of B, Art. 365.

If the cylinder is the boundary of a hollow in a conductor we remove the rod B. The distribution of the electricity on the cylinder is the same as that found for the solid cylinder but opposite in sign.

To find the surface density ρ at any point Q we follow the analysis in Art. 401. We notice that the attractions are $2m/BQ$ and $-2m/CQ$ instead of E/BQ^2 and E'/CQ^2. Making the corresponding changes in the result we find that for a solid cylinder

$$4\pi\rho a = -\frac{2mb^2}{BQ^2} = \frac{2mc^2}{CQ^2}.$$

The external rod has here the positive line density m. If the cylinder is hollow and the internal rod has a negative line density $-m$, the sign of ρ must be changed.

408. The same results follow from the method of inversion. Thus let the rod be inside the cylindrical hollow as at C. We know, by Art. 183, Ex. 2, that if the surface density at Q is proportional to $1/CQ^2$ the attraction at all external points is the same in magnitude and direction as if the attracting mass were equally distributed over the rod C. The condition of equilibrium is that the attraction due to both the surface density and the rod should be zero at all external points. This is satisfied if the surface density have a sign opposite to that of the line density of the rod.

The result for the case in which the rod is outside a solid cylinder may be deduced from this by an inversion with regard to the axis of the cylinder, see Art. 399.

409. Let the positions of the rods B, C be given; let O bisect BC, and let $BC = 2t$. Let us describe the system of co-axial circles whose radical axis is perpendicular to BC and passes through O. Let the length of the tangent drawn from O to any circle be t, then obviously B, C are the evanescent circles of the system. Let A be the centre of any circle, then

$$AB \cdot AC = (AO + OB)(AO - OC) = AO^2 - t^2 = a^2.$$

The points B and C are therefore inverse points with regard to any co-axial circle. The cross section of the cylinder may be any of these circles.

Since the line densities of the two rods are equal and opposite, it follows from Art. 323 that the lines of force are defined by $\theta_1 - \theta_2 = \kappa$ and the level curves by $r_1/r_2 = \kappa'$, where (r_1, θ_1), (r_2, θ_2) are the polar coordinates of any point P referred to B and C respectively as origins and BCA as the axis of reference. *The lines of force are therefore the circles which pass through B and C and the level curves are the co-axial circles.*

410. We may also find *the law of distribution on two circular non-intersecting cylinders (radii a, a') having their axes parallel to each other and their charges equal and opposite.*

Let A, A' be the centres of the two circles made by a perpendicular cross section of the cylinders. Then two points B, C can be found (and only two) which are inverse to each other with regard to both circles. Each cylinder is a level surface of two parallel rods passing through B and C equally electrified but with opposite signs.

Let the cylinders be solid conductors, each external to the other, and let them be separated by a non-conducting medium. We remove each rod and spread its electricity over the cylinder within which it lies, according to the law found in Art. 407. Since the attraction of one electrified cylinder (say A) at all external points is the same as that of the rod which was inside its conducting matter, the attraction of the other cylinder (A') is in equilibrium when acted on by the electrified cylinder (A). The electricity on each cylinder is therefore in equilibrium when acted on by the other.

Several arrangements of the cylinders may be made. *First*, each cylinder may be external to the other as just explained, or one cylinder may contain the other and be separated from it by the non-conducting medium. In both these cases the rods are removed and each cylinder is occupied by the electricity of the rod which was within its conducting matter. *Secondly*, the cylinders may be separate hollows in an infinite conducting medium, or one cylinder may contain the other with the conducting medium between their surfaces. In these two cases the rods are not removed; each cylinder is occupied by electricity equal in quantity but opposite in sign to that of the rod within the nearest non-conductor.

411. Ex. An infinite conducting cylinder of radius a is placed with its axis parallel to an uninsulated conducting plane and at a distance c from it. The cylinder is maintained at potential V, prove that the charge (m) per unit of length is given by $\frac{V}{2m} = \log \frac{c+\sqrt{(c^2-a^2)}}{a}$. Prove also that the surface density at any point of the cylinder is proportional to the distance from the plane. [Coll. Ex. 1880.]

Prove also that the mutual attraction between the cylinder and the plane is $m^2/(c^2-a^2)^{\frac{1}{2}}$. [Math. T. 1888.]

[Let the points A, B (through which the rods pass as described in Art. 407) be so placed that the plane bisects their distance apart at right angles. Both the plane and the cylinder are then level surfaces of the two rods.]

412. Electricity on planes. *To find the distribution of electricity on an uninsulated infinite plate when acted on by a quantity E of electricity collected into a point B at a distance h from the plate.*

Draw BM perpendicular to the plate and produce it to C so that $MC = BM$. The surface of the plate is then a level surface of zero potential of E placed at B and $-E$ at C.

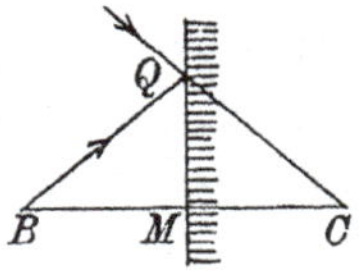

The surface of the plate may be regarded as a sphere of infinite radius enclosing conducting matter on the side C. The electricity $(-E)$ will then be in equilibrium if distributed on the surface so that $4\pi\rho$ is equal to the normal force at Q due to the electrical points *measured towards the non-conductor.* We therefore have

$$4\pi\rho = -\frac{2E}{r^2}\sin BQM = -\frac{2Eh}{r^3},$$

where $r = BQ$. The total quantity of electricity on the surface is $-E$. We obtain the same results by inverting the sphere described on BM as diameter.

413. Ex. 1. Show that half the whole electricity on the infinite plate is comprised within *any* right cone whose vertex is at the influencing point B and whose semi-angle is 60°.

Use the theorem $4\pi Q = \Sigma E\omega$, Art. 396. It also follows that all areas on the plate which subtend the same solid angle at the influencing point contain equal quantities of electricity.

Ex. 2. Prove that the quantity of electricity, on one side of any straight line X drawn on the plate, is $Q = -E\theta/\pi$, where θ is the angle a plane drawn through X

and the influencing point makes with the plate. The angle θ is measured on that side of X which makes Q and θ numerically increase together, when X is moved parallel to itself.

The solid angle ω subtended at the influencing point is here enclosed by two planes. These form a lune on the unit sphere whose area is 2θ.

Ex. 3. A spherical body with an electric charge E is at a height h above the surface of the earth, the height being large compared with the dimensions of the body. Prove that the body is attracted downwards with a force approximately equal to $E^2/4h^2$, in addition to its weight.

Prove also that its capacity is increased by the presence of the ground in the ratio $1+a/2h:1$ approximately, where a is the radius. [Coll. Ex. 1900.]

414. *The planes xOy, yOz intersecting in Oy are the boundaries of a conductor; the non-conducting medium being in the positive quadrant. The system is acted on by an electrical point at A whose coordinates are ξ, ζ. To find the distribution of electricity on the planes* (1) *when the angle xOz is a right angle, and* (2) *when that angle is π/n where n is an integer.*

(1) Let us try to find a system of electrical points such that the two planes xy, yz form part of one level surface. One of these points must be at A, all the others will be inside the conductor.

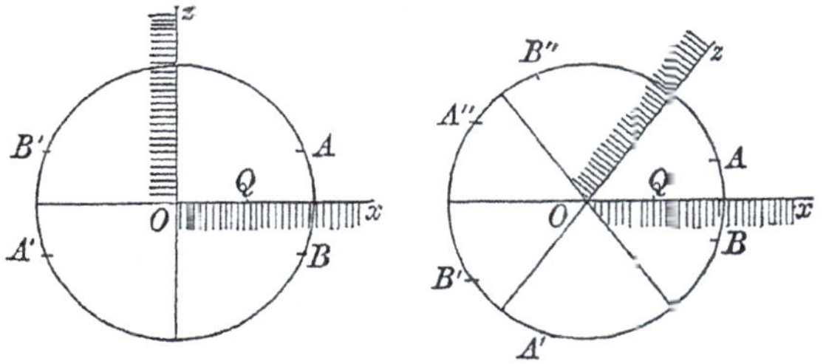

Describe a circle centre O, radius OA and let $ABA'B'$ be a rectangle. Place at A' a quantity E of electricity equal to that at the given point A and at B, B' quantities each equal to $-E$. The planes xy, yz are then evidently level surfaces of zero potential.

The surface density ρ at any point Q on the plane xy is then given (Art. 412) by $4\pi\rho=-\dfrac{2E\zeta}{AQ^3}+\dfrac{2E\zeta}{A'Q^3}$.

The quantity Q of electricity on the plane xOy is given by

$$4\pi Q=\Sigma E\omega=-4E(OAx'-OB'x')=-4E\theta,$$

where $\theta=$ angle AOB' and $B'Ax'$ is a straight line parallel to Ox. The quantity Q' on Oz is $-E\theta'/\pi$ where $\theta'=ACB$.

Ex. A straight line Y is drawn on the plane xy parallel to Oy. Prove that the quantity of electricity on the side of Y remote from O is $-E\phi/\pi$ where ϕ is the angle AYB'.

415. (2) If the angle $xOz=\pi/3$ we divide the circle into six parts by three diameters and place A, A', A''; B, B', B'' just before and just after the alternate divisions. If we suppose each A to be occupied by $+E$ and each B by $-E$, it is obvious that both the planes xy, yz are level surfaces of zero potential. In the same way we may sketch the figure for the angle $xOz=\pi/n$ and in all these cases the surface density at any point Q on either boundary can be written down by Green's rule, (Art. 395).

Ex. Prove that the quantity Q of electricity on the plane xOy is $-E(\pi-3\theta)/\pi$ where θ is the angle AOx.

416. Ex. 1. A long rod uniformly charged with electricity is placed perpendicular to a large conducting plane and with an end nearly in contact with the plane; show that if the plane be put in connexion with the earth, the density of the electricity induced on the plane will vary inversely as the distance from the rod. [Caius Coll. 1880.]

Place a similar oppositely electrified non-conducting rod on the other side of the plane. The plane is then a level surface of zero potential of the two rods and the electricity can be found by Green's method.

Ex. 2. A uniformly attracting rod is placed parallel to a large conducting plane. Prove that, if the plane is put in connexion with the earth, the density of the electricity at any point of the plane will vary inversely as the square of the distance from the rod.

Ex. 3. A conductor is bounded by the surface of a sphere, whose centre is at the origin, and by the rectangular planes xy, yz; the non-conducting medium being the portion of the positive quadrant inside the sphere. The system is acted on by an electrical point of given intensity, situated in the non-conducting medium, whose coordinates are x, y, z. Find the surface density at any point of the boundary. [Use seven other electrical points situated in the conducting medium.]

417. A simple condenser. Let a portion S of the surface of a conductor A be so near the surface of another conductor A' that the distance θ between them at any point is a very small fraction of the radii of curvature of each surface, and let β, β' be the potentials of the conductors. It is required to find, to a first approximation, the distribution of electricity on the neighbouring surfaces.

Let P, P' be two points on the conductors on the same line of force; ρ, ρ' the surface densities at these points; F, F' the forces just outside the conductors at P, P' measured in the direction PP'. Then $4\pi\rho=F$, $4\pi\rho'=-F'$. By Taylor's theorem

$$\beta'-\beta=\frac{dV}{dn}\theta+\&c.,\qquad F=-\frac{dV}{dn}.$$

As a first approximation, we have $F=F'$, and $4\pi\rho=\dfrac{\beta-\beta'}{\theta}$; similarly $4\pi\rho'=\dfrac{\beta'-\beta}{\theta}$. Hence, 4π times the surface density on either conductor is ultimately equal to the fall of the potential from that conductor to the other divided by the distance.

We notice that when the potentials β, β' are given the electrical densities on the neighbouring surfaces can be made very great by diminishing the distance θ.

If $d\sigma$ be an element of the area S, the quantity of electricity on S is $\dfrac{\beta-\beta'}{4\pi}\int\dfrac{d\sigma}{\theta}$. This is $(\beta-\beta')\,S/4\pi\theta$ when the distance θ is constant.

If the conductor A' is joined to the earth, its potential $\beta'=0$, and by the definition in Art. 371 the capacity of S becomes $S/4\pi\theta$.

To obtain a nearer approximation we take a second term in Taylor's theorem. We then have

$$\beta'-\beta=\frac{dV}{dn}\theta+\tfrac{1}{2}\frac{d^2V}{dn^2}\frac{\theta^2}{2}+\&c.$$

Here, as before, $dV/dn=-F$, and in the small additional term we write for d^2V/dn^2 its mean value, viz. $-(F'-F)/\theta$. Substituting for F and F' their values $4\pi\rho$ and $-4\pi\rho'$, we find

$$\frac{\beta'-\beta}{4\pi}=-\rho\theta+\frac{\rho'+\rho}{2}\theta \quad\text{.........(1).}$$

To obtain another equation connecting the nearly equal quantities ρ and $-\rho'$, we construct a tube of force joining P, P'. Let the areas at P, P' be $d\sigma$, $d\sigma'$, then $Fd\sigma=F'd\sigma'$, (Art. 127) and therefore $\rho d\sigma+\rho' d\sigma'=0$.........(2).

Let R, R' be the principal radii of curvature at P measured positively in the direction $P'P$. Then, as in Art. 128, Ex. 2,

$$\frac{d\sigma'}{d\sigma}=1+\left(\frac{1}{R}+\frac{1}{R'}\right)\theta \quad\text{.........(3)}$$

Solving these equations we have

$$4\pi\rho=\frac{\beta-\beta'}{\theta}\left\{1+\frac{\theta}{2}\left(\frac{1}{R}+\frac{1}{R'}\right)\right\},$$

$$4\pi\rho'=-\frac{\beta-\beta'}{\theta}\left\{1-\frac{\theta}{2}\left(\frac{1}{R}+\frac{1}{R'}\right)\right\}.$$

These two approximations were given by Green in his *Essay on Electricity and Magnetism*, pages 43, 45.

418. Ex. 1. A condenser is formed of two flat rectangular plates, each of area A, which are very near together but not quite parallel, one pair of parallel edges being at distance c and the opposite pair at distance c'. Prove that the capacity is approximately $\dfrac{A}{4\pi(c-c')}\log\dfrac{c}{c'}$.

The lower part of the condenser is fixed in a horizontal position and the other is free to turn about a horizontal axis through the centre of its under face. Show that a slight tilt which draws one pair of opposite edges together and the other apart through $1/n$th of their distance will increase the capacity approximately by the fraction $1/3n^2$ of its value. Prove that, when the upper plate is delicately balanced on its axis, charging the condenser will make its equilibrium unstable.

[St John's Coll. 1897.]

Ex. 2. A conducting plate A is inserted between two conductors B, B' terminated by plane faces parallel to those of A. Let θ, θ' be the distances of B, B' from the nearest face of A, and let S be the area of either face of A. If B, B' be maintained at equal potentials V_2 and the potential of A be V_1, prove that the ratio of the quantity Q of electricity on both faces of A to the difference $V_1 - V_2$ is $\frac{S}{4\pi}\left(\frac{1}{\theta} + \frac{1}{\theta'}\right)$.

[Coll. Ex.]

419. Cylindrical Condenser. A long straight electric cable, consisting of a conducting cylindrical core surrounded by a shell of non-conducting matter whose external surface is a co-axial cylinder, is placed in deep water. The perpendicular sections of the two cylinders are concentric circles whose external and internal radii are a', a. To find the capacity of the cable.

Let m_1, m_2 be the charges per unit of length on the outer and inner surfaces of the shell; α, β the potentials of the outer and inner conducting media.

When a non-conducting shell separates two conductors the sum of the potentials of the charges on the two surfaces of the shell is constant (and therefore zero) at all great distances. It follows from Art. 136 that the charges are equal and opposite. The proof for the special case of cylinders is nearly the same as for the general case. The potentials of the two cylinders at a point P in the external conducting medium distant r from the axis differ only by constants from $2m_1 \log r$ and $2m_2 \log r$ (Art. 56). The sum of these cannot be constant when r varies unless $m_2 = -m_1$ (Art. 365).

The potential of the inner cylinder at a point R in the non-conducting shell differs only by a constant from $2m_2 \log r$ while the potential of the outer is constant; (Arts. 55, 56). The potential at R of both cylinders is therefore

$$V = 2m_2 \log r + A$$

where A is a constant. The difference of the potentials at the two surfaces of the shell is

$$\alpha - \beta = 2m_2 (\log a' - \log a).$$

The capacity C of a length l of the core, measured by the ratio of the quantity of electricity to the difference of potentials, is

$$C = \frac{m_2 l}{\alpha - \beta} = \frac{1}{2} \frac{l}{\log a'/a}.$$

When the radii a, a' are nearly equal, the thickness of the shell is very small and the capacity is very great. Putting $x = (a' - a)/a$, the capacity becomes

$$C = \tfrac{1}{2}\frac{l}{\log(1+x)} = \tfrac{1}{2}\frac{l}{x} = \tfrac{1}{2}\frac{la}{a'-a}.$$

We may deduce this result from Art. 417. The capacity is there proved to be approximately $S/4\pi\theta$, where the area $S = 2\pi cl$ and the thickness $\theta = a' - a$.

When the axes of the cylinders bounding the shell are parallel but not coincident, we proceed in the same way. Let A, A' be the centres of a cross section of the two cylinders; a, a' the radii, $a' > a$. Let B, C be the points in which this cross section is intersected by the two rods described in Art. 407, C being inside the core and B in the water. Let m_1, $m_2 = -m_1$ be the line densities of the rods B, C respectively; r_1, r_2 the distances from B, C of any point R between the cylinders. The potential at R of the electric cylinders is (by Art. 407)

$$\begin{aligned} V &= 2m_2 \log r_2 + 2m_1 \log r_1 + A \\ &= 2m_2 \log (r_2/r_1) + A. \end{aligned}$$

When R is on the circle whose radius is a, we have $r_2/r_1 = a/f$, where f is the distance of B from the centre A. A similar result holds for the other circle. The difference of potentials at the two surfaces is therefore

$$\alpha - \beta = 2m_2 (\log a'/f' - \log a/f).$$

The capacity C' of a length l of the core is therefore

$$C' = \frac{m_2 l}{\alpha - \beta} = \tfrac{1}{2}\frac{l}{\log a'/a - \log f'/f},$$

where f, f' are the distances of the axes of the two cylinders from the external rod B. Since $f' > f$, we see that the capacity is least when the two cylinders are co-axial.

420. Nearly spherical surface. *To find to a first approximation the distribution of electricity on the surface of an insulated conductor which is nearly spherical.*

Let the given equation of the surface be expanded in a series of Laplace's functions

$$r = a\{1 + Y_1 + Y_2 + \dots\} \quad\text{.....................(1).}$$

The term Y_0 has been omitted because all constants may be included in the factor a. The terms Y_1, Y_2 &c. are so small that their squares can be neglected. Let the required distribution of electricity be $\rho = D\{1 + Z_1 + Z_2 + \dots\}$ ....(2).

If the surface were strictly spherical, the distribution of electricity would be uniform and every Z would be zero. It follows that when the surface is nearly spherical each Z is of the first order of small quantities.

Let (r, θ, ϕ) be the coordinates of any elementary area $d\sigma$ of the surface; (r', θ', ϕ') the coordinates of any internal point P. Let $d\omega = \sin\theta d\theta d\phi$. The potential V of the electricity at P is

$$V = \int \frac{\rho d\sigma}{\sqrt{(r^2 - 2rr'p + r'^2)}} = \int \frac{\rho d\sigma}{r} \Sigma P_n \left(\frac{r'}{r}\right)^n \text{.........(3)},$$

where the limits of integration are $\theta = 0$ to π, and $\phi = 0$ to 2π.

This series is convergent for all positions of P which are at a distance r' from the origin less than the least radius vector of the surface. Let ψ be the angle the radius vector r makes with the normal to the element $d\sigma$, then $r^2 d\omega = d\sigma \cos\psi$. Since ψ is a quantity whose square can be neglected, we have $r^2 d\omega = d\sigma$.

The electricity is so distributed that the potential V is constant throughout the interior, we therefore equate to zero the coefficients of the several powers of r' in the series (3). Hence

for all values of $n > 0$, $$\int \frac{\rho d\omega}{r^{n-1}} P_n = 0 \text{........................(4)}.$$

We now substitute for r and ρ their values given by (1) and (2) and reject the squares and products of the small quantities Y_1, Y_2, &c., Z_1, Z_2, &c. We then have by Art. 290

$$\int \{-(n-1) Y_n + Z_n\} P_n d\omega = 0 \text{ ....(5)}.$$

Now $\int Y_n P_n d\omega = \frac{4\pi}{2n+1} Y'_n$, $\int Z_n P_n d\omega = \frac{4\pi}{2n+1} Z'_n$, where Y'_n, Z'_n are the values of Y_n, Z_n when θ', ϕ' are written for θ, ϕ; Art. 289. We thus find $$Z'_n = Y'_n (n-1) \text{........................(6)}.$$

The conclusion is that the surface density of the electricity on the surface (1) is

$$\rho = D\{1 + Y_2 + 2Y_3 + \dots + (n-1) Y_n + \dots\}.$$

It may be noticed that the term Y_1 is absent from the expression for ρ. The reason is that the surface $r = a(1 + Y_1)$ is approximately a sphere when Y_1 is small, Art. 293, Ex. 3. The surface density is then uniform.

If E be the quantity of electricity on the surface, we have, since $\int Y_n d\omega = 0$ and the squares of Y_n are neglected,

$$E = \int \rho r^2 d\omega = 4\pi a^2 D.$$

This equation determines D when E is given. The potential at the origin is $$V = \int \rho r d\omega = 4\pi a D.$$

The capacity is therefore equal to a.

To find the potential of the stratum at an external point we make the expansion (3) in powers of r/r'. We then have

$$V' = \int \rho\, d\omega\, \Sigma P_n \frac{r^{n+2}}{r'^{n+1}}$$

$$= \Sigma D a^{n+2} \int d\omega \{1 + \&c. + Z_m + (n+2) Y_m + \&c.\} \frac{P_n}{r'^{n+1}}$$

$$= 4\pi D \Sigma \frac{Z'_n + (n+2) Y'_n}{2n+1} \frac{a^{n+2}}{r'^{n+1}}.$$

After substituting for Z'_n its value, this reduces to

$$V' = \frac{4\pi a^2 D}{r'} \left\{1 + Y_1 \frac{a}{r'} + Y_2 \left(\frac{a}{r'}\right)^2 + \&c\right\}.$$

421. Ex. 1. The surface $r = a(1 + \beta \cos^2\theta)$, where β is very small, is charged with a quantity E of electricity. Prove that the surface density is

$$(1 - \beta \sin^2\theta)\, E/4\pi a^2.$$

Ex. 2. *A nearly spherical conductor whose equation is $r = a(1 + \Sigma u_n)$ is enclosed in a nearly spherical shell, the equation of whose inner surface is $r = b(1 + \Sigma v_n)$ where u_n, v_n are Laplace's functions of (θ, ϕ). If the potentials of the solid and shell are respectively α and β, find the potential at any point P between the conductor and shell.* See Art. 392, Ex. 1.

The potential at P is, by Art. 283,

$$V = Y_0 + Y_1 r + \&c. + \frac{Z_0}{r} + \frac{Z_1}{r^2} + \&c. \quad \text{..........................(1)}.$$

If the surfaces were truly spherical, the distribution of electricity on each would be uniform and the expression for V would take the form $A - B/r$ where A, B are constants. It follows that Y_1, Y_2, &c., Z_1, Z_2, &c. are in our problem small quantities. Proceeding as in Art. 299 and rejecting the squares of small quantities

we have
$$Y_0 + Y_1 a + Y_2 a^2 + \&c. + \frac{Z_0}{a}(1 - u_1 - u_2 - \&c.) + \frac{Z_1}{a^2} - \&c. = \alpha,$$

$$Y_0 + Y_1 b + Y_2 b^2 + \&c. + \frac{Z_0}{b}(1 - v_1 - v_2 - \&c.) + \frac{Z_1}{b^2} - \&c. = \beta.$$

Equating the functions of like order, we find

$$\left.\begin{aligned} Y_0 + \frac{Z_0}{a} &= \alpha \\ Y_0 + \frac{Z_0}{b} &= \beta \end{aligned}\right\} \qquad \left.\begin{aligned} Y_n a^n - \frac{Z_0}{a} u_n + \frac{Z_n}{a^{n+1}} &= 0 \\ Y_n b^n - \frac{Z_0}{b} v_n + \frac{Z_n}{b^{n+1}} &= 0 \end{aligned}\right\}.$$

These give
$$\left.\begin{aligned} Y_0 &= \frac{b\beta - a\alpha}{b - a} \\ Z_0 &= ab\frac{\alpha - \beta}{b - a} \end{aligned}\right\} \qquad \left.\begin{aligned} Y_n &= Z_0 \frac{b^n v_n - a^n u_n}{b^{2n+1} - a^{2n+1}} \\ Z_n &= Z_0 a^n b^n \frac{b^{n+1} u_n - a^{n+1} v_n}{b^{2n+1} - a^{2n+1}} \end{aligned}\right\}.$$

Substituting these values in (1) the potential at any point P is obtained, the equations of the two surfaces being given. The surface density ρ at any point P of the internal conductor is found from $4\pi\rho = -dV/dr$, where after differentiation we put $r = a$ in the small terms and $r = a(1 + \Sigma u_n)$ in the large term. We then find

$$4\pi\rho a^2 = Z_0 [1 + \Sigma\mu \{(n+2)\, a^{2n+1} + (n-1)\, b^{2n+1}\} u_n$$
$$- \Sigma\mu (2n+1)\, a^{n+1} b^n v_n],$$

where $1/\mu = b^{2n+1} - a^{2n+1}$ and the summation Σ begins at $n=1$. The surface density ρ' at any point of the external conductor is found by interchanging (a, b) and (u, v), the sign of the first term being also changed. This problem is discussed in a nearly similar manner in Maxwell's *Electricity*.

Ex. 3. A shell is bounded internally by a nearly spherical surface whose equation is $r = b(1 + \Sigma v_n)$ and is acted on by an electrical point situated at its approximate centre. Prove that the electrical density ρ at any point P of the surface is given by $4\pi b^2 \rho = -E\{1 - \Sigma(n+2) v_n\}$, where the origin is at the electrical point, E is the quantity of electricity at that point and the summation Σ begins at $n=1$.

Ex. 4. *A nearly spherical conductor, which is also a solid of revolution with the approximate centre near the axis, is placed in a uniform field of force whose potential is Mx where the axis of x is the axis of the solid conductor. Find the law of distribution of electricity on the surface when the charge is given.*

The surface being one of revolution about the axis of reference and also nearly spherical, its equation referred to an origin on the axis can be expressed in the form

$$r = a(1 + A_1 P_1 + A_2 P_2 + \&c.) \quad\text{.............................(1)},$$

where all the coefficients A_1, A_2, &c. are small. Similarly we may express the surface density in the form

$$\rho = D(1 + B_1 P_1 + B_2 P_2 + \&c.) \quad\text{.............................(2)}.$$

If the conductor were accurately spherical, the expression for ρ would be of the form $D(1 + B_1 \cos\theta)$ (Art. 406, Ex. 4). It follows that when the surface is nearly spherical the coefficients B_2, B_3 &c. are small, but B_1 is not necessarily small.

Proceeding as in Art. 420, we make the potential at an internal point R whose coordinates are (r', θ') equal to a constant K.

$$\therefore \int \frac{\rho r^2 d\omega}{\sqrt{(r^2 - 2rr'q + r'^2)}} + Mr'\cos\theta' = K \quad\text{.......................(3)},$$

where q is the cosine of the angle between the radii vectores r, r'. Expanding and equating the several powers of r' to zero, we find

$$\int \frac{\rho\, d\omega}{r^{n-1}} Q_n = 0 \quad\text{or}\quad -M\cos\theta' \quad\text{..........................(4)}$$

according as $n > 1$ or $= 1$. Here Q_n is a Legendre's function of q.

To find the constants B_1, B_2, &c. it will be sufficient to put the point R in some convenient positions. Let us place R on the axis, then $q = p$, the Legendre's function Q_n becomes P_n, and $\cos\theta' = 1$. We then have when $n=1$

$$D\int(1 + B_1 P_1 + B_2 P_2 + \dots) P_1 d\omega = -M \quad\text{.......................(5)}.$$

Since $\int P_m P_1 dp = 0$, this gives $\qquad B_1 = -3M/4\pi D$ ....(6).

When $n > 1$ we have, since B_2 &c., A_1 &c. are small

$$\int\{1 + B_1 P_1 + \&c.\}\{1 - (n-1) A_1 P_1 - \&c.\} P_n dp = 0,$$

$$\therefore \int\{1 + B_2 P_2 + \&c. - (n-1) A_1 P_1 - \&c.\} P_n dp$$
$$- B_1 (n-1) \int\{A_1 P_1 + A_2 P_2 + \&c.\} P_1 P_n dp = 0 \quad\text{......(7)}.$$

The first line presents no peculiarity and reduces to $\{B_n - (n-1) A_n\} 2/(2n+1)$. Since $P_1 = p$ the integral in the second line may be written $\Sigma A_\kappa \int P_\kappa P_n p\, dp$. Now by Art. 273 $\qquad (n+1) P_{n+1} - (2n+1) p P_n + n P_{n-1} = 0$,

$$\therefore (2n+1) \int P_\kappa P_n p\, dp = (n+1) \int P_\kappa P_{n+1} dp + n \int P_\kappa P_{n-1} dp.$$

This is zero except when $\kappa = n \pm 1$. We then have

$$\int P_{n+1} P_n p\, dp = \frac{2(n+1)}{(2n+1)(2n+3)}, \quad \int P_{n-1} P_n p\, dp = \frac{2n}{(2n+1)(2n-1)}.$$

The latter of these results follows also from the first by writing $n-1$ for n. The second line of the equation (7) becomes

$$-B_1 \frac{n-1}{2n+1} \cdot \left\{ A_{n+1} \frac{2(n+1)}{2n+3} + A_{n-1} \frac{2n}{2n-1} \right\}.$$

Finally we have, when $n > 1$,

$$B_n = (n-1) A_n + (n-1) B_1 \left\{ \frac{n+1}{2n+3} A_{n+1} + \frac{n}{2n-1} A_{n-1} \right\} \text{..........(8).}$$

If E' be the quantity of electricity on the surface we have

$$E' = \int \rho r^2 d\omega = 4\pi D a^2 (1 + \tfrac{2}{3} A_1 B_1),$$

$$\therefore\ 4\pi D a^2 = E' (1 - \tfrac{2}{3} A_1 B_1) \text{..............................(9).}$$

Substituting in (2) the values B_1, B_n and D given in (6), (8) and (9) we find the value of the surface density ρ when the surface of the conductor is given.

The potential at the origin is $K = 4\pi D a (1 + \frac{1}{3} A_1 B_1)$.

422. Sphere with a ring. Ex. 1. A uniform circular wire (radius b), charged with electricity of line density $-e$, surrounds an uninsulated concentric spherical conductor (radius a). Prove that the electrical density at any point of the surface of the conductor is

$$\frac{e}{2a} \left\{ 1 - 5 \cdot \frac{1}{2} P_2 \left(\frac{a}{b}\right)^2 + 9 \frac{1 \cdot 3}{2 \cdot 4} P_4 \left(\frac{a}{b}\right)^4 - 13 \frac{1 \cdot 3 \cdot 5}{2 \cdot 4 \cdot 6} P_6 \left(\frac{a}{b}\right)^6 + \&c. \right\}.$$

Ex. 2. A uniform circular wire (centre C), charged with electricity of line density $-e$, influences an uninsulated spherical conductor (centre O), the plane of the wire being perpendicular to OC. Prove that the electrical density at any point R of the surface of the conductor is

$$\frac{e \sin \alpha}{2a} \Sigma (2n+1) P_n (\cos \alpha) P_n (\cos \theta) \left(\frac{a}{b}\right)^n,$$

where Σ implies summation from $n=0$ to $n=\infty$. Also a is the radius of the sphere, b the distance of any point M on the rim of the ring from O, α the angle subtended at O by any radius of the ring and θ the angle OR makes with the axis OC of the ring.

The potential of the ring at any point Q on the axis referred to O as origin is

$$V_1 = \frac{M}{\sqrt{(b^2 - 2br \cos \alpha + r^2)}} = \frac{M}{b} \Sigma P_n (\cos \alpha) \left(\frac{r}{b}\right)^n,$$

and $M = -2\pi b e \sin \alpha$. The potential at any point S not on the axis is found by introducing the factor $P_n(\cos \theta)$ into the general term, where θ is the angle COS. The potential V_2 of the spherical layer is given in Art. 294. The sum of the two potentials being zero, the value of Y_n follows at once.

423. Orthogonal spheres. *The boundary of an insulated conductor is formed by two orthogonal spheres. Find the law of distribution of a charge of electricity*.*

* The problem of finding the law of distribution of electricity on two orthogonal spheres when acted on by an electrical point is solved in Maxwell's *Treatise on Electricity &c.* He also gives the solution for spheres intersecting at an angle π/n, for three and also four orthogonal spheres.

Let A, B be the centres, a, b the radii and let AB cut the plane DD' of intersection of the spheres in C. Then, as before, the distances of A, B, C from D are a, b, c. Let mass particles E, E', E'' be placed at A, B, C such that

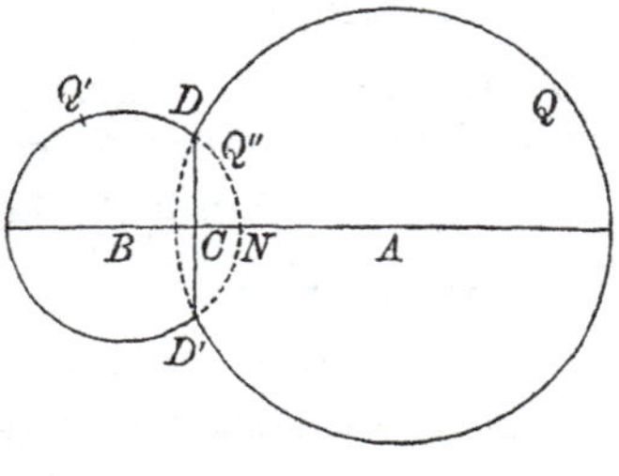

$$\frac{E}{a} = \frac{E'}{b} = \frac{-E''}{c} = e.$$

Since the sphere A is a level surface of zero potential of the particles at the inverse points B, C (Art. 397), it is a level surface of potential e of all the three particles. In the same way the other sphere is also a level surface of the same three particles and is at the same potential.

Using Green's theorem, we see that the quantity of electricity $(a + b - c)\,e$, if distributed properly over the whole surface, will be in equilibrium at potential e, (Art. 395).

The normal force at any point Q on the sphere A due to both the points B, C, has been proved to be proportional to $1/CQ^3$ and also to $1/BQ^3$ (Art. 401). The normal force due to the particle at A is E/a^2. We have therefore

$$4\pi\rho = \frac{e}{a} + \frac{H}{CQ^3} = \frac{e}{a} + \frac{K}{BQ^3}$$

where H, K are some constants. Since two sheets of a level surface intersect in the circle DD', the normal force and therefore ρ vanishes when Q is at D (Art. 122), that is when $CQ = c$ or $BQ = b$. The density may therefore be written in either of the forms

$$4\pi\rho = \frac{e}{a}\left\{1 - \left(\frac{c}{CQ}\right)^3\right\} = \frac{e}{a}\left\{1 - \left(\frac{b}{BQ}\right)^3\right\}.$$

The density at any point Q' on the other sphere is given by

$$4\pi\rho' = \frac{e}{b}\left\{1 - \left(\frac{c}{CQ'}\right)^3\right\} = \frac{e}{b}\left\{1 - \left(\frac{a}{AQ'}\right)^3\right\}.$$

424. We may also consider the solid bounded by the *convex portion* of the sphere A and the *concave portion* DND' of the other sphere. The quantity on the solid is then ea, the potential is e, and the electricity is acted on by the external electrified points $E' = eb$, $E'' = -ec$. The densities are given by the same formulæ as before, except that the sign of that on the concave portion must be changed, because the normal force outwards into the non-conductor (Art. 369) tends towards the centre of the sphere B instead of from the centre, as on the convex portion of that sphere.

425. Ex. When both portions are convex the quantities Q, Q' of electricity on the spheres A, B respectively are

$$Q=\tfrac{1}{2}e\{a-c+b+c(a^2-b^2)/ab\},$$
$$Q'=\tfrac{1}{2}e\{a-c+b-c(a^2-b^2)/ab\}.$$

When one portion, as DND', is concave, the electricity on that portion is

$$Q''=\tfrac{1}{2}e\{a+c-b-c(a^2-b^2)/ab\}.$$

426. *To find the law of distribution of electricity on a conductor bounded by the convex portions of two orthogonal spheres when acted on by an external electrical point.*

The two orthogonal planes xOy, yOz in the left-hand figure are the planes of zero potential of four equal particles A_1, A_2,

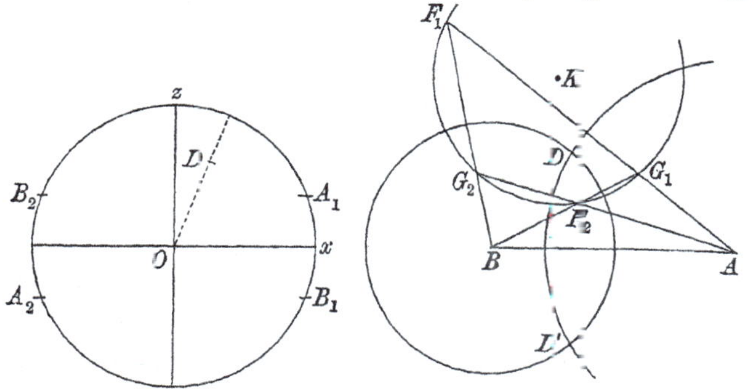

B_1, B_2; A_1, A_2 being of positive and B_1, B_2 of negative mass, see Art. 414. Let us invert this with regard to any point L. Consider first the section by the plane xOz. The straight lines Ox, Oz invert into orthogonal circles which intersect in D and in another point D' lying in DO produced. The radii a, b of these circles are arbitrary because D is any point. Let their centres be A and B as represented in the right-hand figure. The circle $A_1B_1A_2B_2$ inverts into another circle cutting the two former orthogonally and (being symmetrical about DOD') has its centre K in DD'. The radius of this circle is such that the perimeter passes through the inverse point of the arbitrary point A_1.

Let the points A_1, A_2 invert into F_1, F_2 and the points B_1, B_2 into G_1, G_2; all these four points lie on the circle whose centre is K. Since the plane xy is a level surface of zero potential of A_1,

B_1, the inverse sphere (say the sphere whose centre is A) is a level surface of zero potential of F_1, G_1, Art. 179. It thence follows that F_1, G_1 are inverse points with regard to that sphere. In the same way F_2, G_2 are inverse with regard to the same sphere, while F_1, G_2 and G_1, F_2 are inverse with regard to the sphere B. Thus F_1G_1A, F_1G_2B, G_2F_2A, G_1F_2B are straight lines. It also follows that if F_1 is external, the other three points F_2, G_1, G_2 are inside one or other of the two spheres A and B.

The ratio of the masses m, m' at any two inverse points Q, Q' is known by the rule $m'/m = DQ'/k$, Art. 168. The quantities of electricity at F_1, F_2, G_1, G_2 are therefore proportional to their distances from D. Let these be

$$E_1 = e\,.\,DF_1, \quad E_2 = e\,.\,DF_2, \quad E_1' = -e\,.\,DG_1, \quad E_2' = -e\,.\,DG_2.$$

The potential at D of each electrical point is therefore numerically the same. We may also use the rule (proved in the footnote to Art. 397) that *the squares of the quantities of electricity* which occupy points inverse to a sphere, and make the sphere to be of zero potential, *are proportional to the distances of those points from the centre.* Thus $E_1^2/E_1'^2 = AF_1/AG_1$; $E_2^2/E_1'^2 = BF_2/BG_1$ and so on.

If we take the convex portions of the spheres A, B to be the boundary of a solid conductor, that boundary will be a level surface of zero potential of the four particles at F_1, G_1, F_2, G_2. Hence the quantity of electricity $Q = e\,(DF_2 - DG_1 - DG_2)$ will be in equilibrium under the influence of a quantity $E_1 = e\,.\,DF_1$ placed at F_1 if distributed according to Green's law.

427. *The surface density at any point P* on the sphere whose centre is A is found by considering the two doublets F_1, G_1 and F_2, G_2. We have by Art. 401

$$4\pi\rho a = -\frac{E_1\alpha^2}{(F_1P)^3} + \frac{E_2'\beta^2}{(G_2P)^3}$$

where α^2 and β^2 are the products of the segments of chords of that sphere drawn from F_1 and G_2. Since ρ must vanish when P is any point D of the intersection of two sheets of a level surface, we see that

$$4\pi\rho a = -E_1\alpha^2\left\{\frac{1}{(F_1P)^3} - \left(\frac{G_2D}{F_1D}\right)^3\frac{1}{(G_2P)^3}\right\}.$$

Since D lies on the sphere B with regard to which F_1, G_2 are inverse points, we may write $G_2D/F_1D = b/F_1B$, Art. 397. Also

$a^2 = (F_1A)^2 - a^2$ where a and b are the radii of the spheres whose centres are A, B.

Let the position of the influencing point F_1 be at an infinite distance from the sphere. The electricity at F_1 is then infinite but its potential, viz. E_1/DF_1, becomes the constant e. The conductor being at zero potential, the sum of the potentials of the electricities at the three remaining points G_1, G_2, F_2 is therefore $-e$. The positions of these points are evidently A, B and C, where C is the intersection of AB and LD'. We thus fall back on the case of a solid conductor charged with a quantity of electricity $e(-DA - DB + DC)$ and at potential $-e$; (Art. 423).

428. If we *insulate the conductor* and give it such a charge that the potential becomes e, we have, by superimposing the density found in Art. 423,

$$4\pi\rho a = e\left\{1 - \left(\frac{b}{BP}\right)^3\right\} - Ea^2\left\{\frac{1}{(F_1P)^3} - \left(\frac{b}{F_1B}\right)^3 \frac{1}{(G_2P)^3}\right\}.$$

429. The rule to find the distribution of electricity on two orthogonal spheres at zero potential may be summed up in the following manner. The point F_1 being given, we seek (1) the inverse points of F_1 with regard to the two spheres A and B, let these be G_1, G_2; (2) the inverse point of G_1 with regard to the sphere B or the inverse point of G_2 with regard to the sphere A, let this be F_2. These four points, any F being taken with any G, form two doublets. The sphere is a level surface of zero potential of each doublet. The ratios of the quantities of electricity at the points of each doublet, and the resulting surface density due to each, follow from the elementary rules given in Arts. 397, 401. The electricity at any G has an opposite sign to that at any F.

430. Ex. An uninsulated conductor consists of a sphere and an infinitely large and infinitely thin plane passing through the centre B of the sphere. If it be exposed to the influence of a given charge of electricity at the point F_1 where F_1B is perpendicular to the plane, prove that G_1 being a point on F_1B produced such that BG_1 is equal to BF_1, the superficial density at any point P on the hemispherical surface nearest to F_1 is proportional to $\frac{1}{F_1P^3} - \frac{1}{G_1P^3}$.

[Math. Tripos, 1877.]

The infinite plane may be regarded as the limiting case of an orthogonal sphere. We then follow the rule in Art. 429. The inverse point of F_1 with regard to the plane is G_1, the inverse points of F_1, G_1 with regard to the sphere are G_2, F_2. The given system of sphere and plane is a level surface of zero potential of these four points. We use Green's method as explained in Art. 401.

431. Geometrical properties. Ex. 1. Prove (1) that the centre of each of the three orthogonal circles lies in the radical axis of the other two, and that the orthocentre of the triangle ABK formed by joining the centres is the radical centre of the circles. Prove (2) that the diagonals of the quadrilateral $F_1G_1F_2G_2$ intersect in the orthocentre of ABK.

The first results have been proved in Art. 426 for the centre K and the two circles whose centres are at A and B, and are therefore true for all the circles. The diagonals intersect on the polar lines of A and B, and since the circles are orthogonal this is also the intersection of the radical axes.

Ex. 2. Prove (1) that $\frac{1}{E_1^2}+\frac{1}{E_2^2}-\frac{1}{E'^2_1}-\frac{1}{E'^2_2}=0$. Prove (2) that, if particles whose masses are proportional to $1/E_1^2$, $1/E_2^2$, $-1/E'^2_1$, $-1/E'^2_2$ are placed at the points F_1, F_2, G_1, G_2, the sum of their moments about every straight line is zero. Prove (3) that the centre of gravity of $1/E_1^2$ and $1/E_2^2$ coincides with that of $1/E'^2_1$ and $1/E'^2_2$ and also with the orthocentre of ABK.

We notice that the centre of gravity of each of the doublets $1/E_1^2$, $-1/E'^2_1$ and $1/E_2^2$, $1/E'^2_2$ is at A, Art. 397. Thus the centre of gravity of all four particles is at A. Similarly it is at B and this is impossible unless the results (1) and (2) are true. To prove the third result we take moments about the diagonals of the quadrilateral $F_1G_1F_2G_2$.

432. Ex. A conductor is formed by the outer surfaces of two equal spheres, the angle between the radii at a point of intersection being $2\pi/3$. Prove that the capacity of the conductor is $\frac{5\sqrt{3}-4}{2\sqrt{3}}a$, where a is the radius. [Coll. Ex. 1899.]

This result follows by inverting with regard to A the second figure of Art. 414. The inverse of the electrical point A contributes only the constant potential E/k to the inverse figure (Art. 180). Omitting this point, the inverse of the rest of the system is in equilibrium at potential $-E/k$. By Art. 170 the mass of any portion of either system is equal to k times the potential at A of the corresponding portion of the other system. In this way without drawing the inverse figure we find both the quantity of electricity on the spheres, and its potential. The ratio is the capacity required.

The capacity of the inverse system is therefore k^2V/Q where Q is the quantity of electricity on the original system and V its potential at the centre of inversion. In our case the point A in Art. 414 bisects the arc xz and $k=a$. Also $Q=-E$ and V is twice the potential at A of B plus twice that of A' plus that of B'.

433. *The boundary of a conductor is formed by the external boundary of three spheres which have a common circular intersection, each sphere making an angle $\pi/3$ with the next in order.* To find the law of distribution on this conductor we invert the right-hand figure in Art. 414 just as we inverted the left-hand figure of that article when we required the distribution on two orthogonal spheres (Art. 426).

Let the plane of the paper contain the centre of inversion D and be perpendicular to the common intersection Oy of the three planes. These planes invert into spheres whose centres C_1, C_3, C_2 lie on a straight line perpendicular to DO. Let the planes Ox, Oz which bound the conductor invert into the spheres whose centres are C_1, C_2, the third plane, which is entirely in the conductor, inverting into the sphere whose centre is C_3. In the inverse figure therefore the centres of the outer spheres are C_1, C_2. Since these centres lie on the perpendiculars drawn from D to the planes, the angles C_1DC_3, C_3DC_2 are each $\pi/3$. These spheres have a common circle of intersection and D is any point on that circle.

434. If the position of the centre of inversion D is arbitrary the six electrical points in the figure invert into six F_1, G_1, F_2, G_2, F_3, G_3 which lie on the circle inverse to that containing A, B, &c. and the general results are very similar to those obtained in Art. 426. If we place D on the circle $BA'B'A''B''$ (say between A and z) our results will correspond to those found in Art. 423 by the use of Green's method. Let us consider concisely this last case as presenting some novelty.

Since D lies on the circle $ABA'B'$ &c. the arcs AA', $A'A''$ &c. subtend at D angles each equal to $\pi/3$; hence in the inverse figure also *the angles subtended by* F_1F_2, F_2F_3, G_1G_2, G_2G_3, C_1C_3 *and* C_3C_2 *at* D *are each equal to* $\pi/3$. So again F_1G_1, F_2G_2, F_3G_3 subtend equal angles at D. The six electrical points F_1, G_1, &c. *now lie on the diameter* $C_1C_2C_3$. Let a radius vector starting from DA turn round D, it evidently passes in order through the points F_1, G_1, C_1; F_2, G_2, C_3; F_3, G_3, C_2. The electrical points and the centres in the inverse figure *are therefore arranged from right to left in this order*. By considering the triangles C_1DC_3, C_3DC_2 we see that the three radii are connected by the equation $1/r_3 = 1/r_1 + 1/r_2$. In the same way if (ξ_1, ξ_2, ξ_3), (η_1, η_2, η_3) are the distances of (F_1, F_2, F_3), (G_1, G_2, G_3) from D we have $1/\xi_2 = 1/\xi_1 + 1/\xi_3$, $1/\eta_2 = 1/\eta_1 + 1/\eta_3$. The perpendicular p from D on the straight line $C_1C_2C_3$ is given by $p\sqrt{(r_1^2+r_2^2+r_3^2)} = \frac{1}{2}\sqrt{3}r_1r_2$.

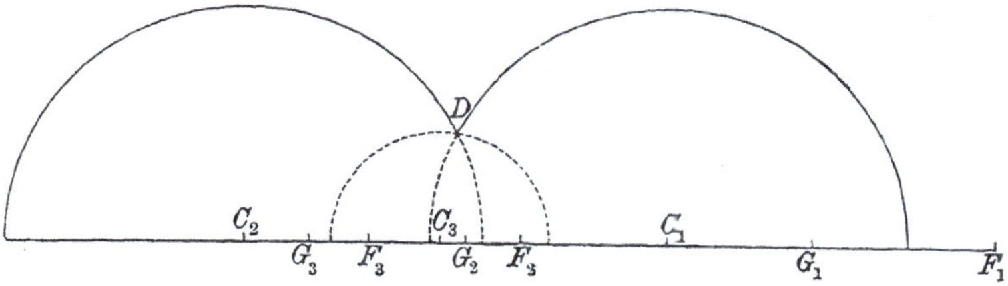

The points (F_1, G_1), (F_2, G_3), (F_3, G_2) are inverse points with regard to the point C_1; (F_3, G_3), (F_1, G_2), (F_2, G_1) are inverse with regard to C_3, and (F_2, G_2), (F_1, G_3), (F_3, G_1) are inverse with regard to C_2. The arrangement of the suffixes suggests an obvious rule to find the inverse of any point with regard to any sphere. The point F_1 being arbitrarily taken outside the spheres C_1, C_2, all the other five are within the boundary.

The quantities of electricity at the points F_1, F_2, G_1 &c. are respectively $E_1 = e \,.\, DF_1$, $E_2 = e \,.\, DF_2$, $E'_1 = -e \,.\, DG_1$ &c. by Art. 169; the potentials at D of the six electrified points are therefore numerically equal.

Since each sphere is a surface of zero potential of the six points F_1, G_1, &c. we may apply Green's theorem. In this way we can find *the law of distribution on the surface formed (say) by the two spheres whose centres are* C_1, C_2 *when acted on by an electrical point situated at any external point* F_1 *on the diameter* $C_2C_3C_1$.

435. Let us place the point F_1 at an infinite distance from the spheres. Since the attraction of F_1 is then zero (though the potential is finite) we may remove this point from the system. We now have the case of *an insulated conductor bounded as before by the spheres* C_1, C_2 *and charged with a given quantity of electricity.*

The points G_1, G_2, G_3 now coincide with C_1, C_3, C_2 respectively. Also since F_1F_2, F_2F_3 each subtend an angle $\pi/3$ at D, the triangle F_2F_3D is equilateral. The potential at D of F_1 is $e \,.\, DF_1/DF_1$ and is therefore equal to e. When the point F_1 is removed from the system (which was at zero potential) the potential V_0 of the remaining five is $-e$. The quantity of electricity on the two external spheres is the sum of the electricities at the five points and is therefore the sum of $-er_1$, $-er_2$, $-er_3$, $2e \,.\, DF_2$. The capacity is therefore

$$\frac{C}{V_0} = r_1 + r_2 + \frac{r_1r_2}{r_1+r_2} - \frac{2r_1r_2}{\sqrt{(r_1^2+r_2^2+r_1r_2)}}.$$

436. Three orthogonal spheres. Let ABC be any triangle; a, b, c the lengths of its sides. Let AF, BG, CH be the perpendiculars drawn from A, B, C on the opposite sides. Let O be the orthocentre. Let us describe three spheres with centres A, B, C and radii α, β, γ such that the spheres taken two and two are orthogonal. Then since the square of the distance between the centres of two orthogonal spheres is the sum of the squares of the radii, we have

$$\alpha^2+\beta^2=c^2,\quad \beta^2+\gamma^2=a^2,\quad \gamma^2+\alpha^2=b^2.$$

Let the chord of intersection of the circles whose centres are A, B made by the plane of the paper intersect AB in S. Then

$$AS^2-BS^2=\alpha^2-\beta^2=b^2-a^2=AH^2-BH^2.$$

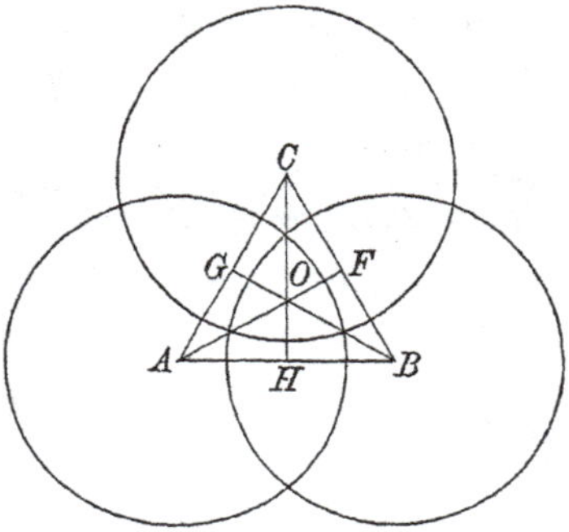

The point S therefore coincides with H. The three chords of intersection of the circles, taken two and two, are therefore the three perpendiculars AF, BG, CH. If the lengths of these chords are respectively $2f$, $2g$, $2h$, we have $af=\beta\gamma$, each of these being twice the area of the triangle whose base is BC and altitude f. Similarly $bg=\gamma\alpha$, $ch=\alpha\beta$.

A circle can be drawn about $CFOG$, hence

$$AO\,.\,AF=AG\,.\,AC=bc\cos A$$
$$=\tfrac{1}{2}(b^2+c^2-a^2)=\alpha^2.$$

Thus the systems of points (O, F), (G, C), (H, B) are each inverse with regard to the sphere A. Similar results hold for the points in each of the lines through B and C.

Let us place at the points A, B, C; F, G, H, quantities of electricity respectively equal to $e\alpha$, $e\beta$, $e\gamma$; $-ef$, $-eg$, $-eh$, as explained in Art. 423. Also, since F and O are inverse points with regard to the sphere A, we place at O a quantity of electricity $\Omega=ef\,.\,\alpha/AF$, (Art. 397). Since $AF\,.\,a=2\Delta$ where Δ is the area of the triangle ABC, we find $\Omega=e\alpha\beta\gamma/2\Delta$. It appears that Ω is a symmetrical function of the radii of the spheres.

It follows that any sphere, as A, is a level surface of zero potential of the particles placed respectively at (B, H), (F, O), (C, G) while its potential due to the particle placed at its centre is e. *Each of three spheres is a level surface of potential e of the seven particles placed at A, B, C, F, G, H and O.*

Let the external surfaces of the three orthogonal spheres be the boundary of an insulated conductor charged with a quantity E of electricity, then the law of distribution may be found by Green's method. If ρ be the surface density at any point Q on the external surface of the sphere A, we have by Art. 401

$$4\pi\rho\alpha=e\left\{1-\left(\frac{\beta}{BQ}\right)^3-\left(\frac{\gamma}{CQ}\right)^3+\frac{ft^2}{FQ^3}\right\}$$

where t is the tangent drawn from F to the sphere A. If we wish to express our results in terms of the radii α, β, γ, we may prove that

$$t^2=\frac{\beta^2\gamma^2}{\beta^2+\gamma^2},\qquad 4\Delta^2=\alpha^2\beta^2+\beta^2\gamma^2+\gamma^2\alpha^2.$$

The potential is e and the quantity E is

$$E=e\{\alpha+\beta+\gamma-f-g-h+\alpha\beta\gamma/2\Delta\}.$$

437. The law of distribution on three orthogonal spheres may also be determined very simply by inversion. The three coordinate planes xOy, yOz, zOx are level surfaces at zero potential of eight points, four of which are represented by

$A_1B_1A_2B_2$ in the figure of Art. 426 and the other four are on the opposite side of the plane of xz. The coordinates of these are $\pm x$, $\pm y$, $\pm z$ and the charges numerically equal. After inversion with regard to any point D the planes become orthogonal spheres. We may thus find *the law of distribution on the external surface of three orthogonal spheres at potential zero when acted on by an external electrical point F_1.*

The three coordinate planes and a sphere whose centre is the origin are level surfaces at zero potential of sixteen points, viz., the eight described above and their inverse points with regard to the sphere. By inverting this system with regard to any point D we find *the distribution on four orthogonal surfaces at potential zero when acted on by an external electrical point F_1.*

By proceeding as in Art. 427, we deduce the law of distribution *when the conductor is insulated* and not acted on by an external electrical point. Finally, by superimposing the two distributions thus arrived at, we obtain the law of distribution *when the conductor is insulated and acted on by the external electrical point F_1.*

438. Theory of a system of conductors. Let A_1, $A_2, \ldots A_n$ be a system of insulated conductors, each being external to all the others. Let $p_{11}, p_{12}, \ldots$ be the potentials due to a charge unity given to A_1, the others being uncharged. In the same way let $p_{21}, p_{22}, \ldots$ be the potentials when a charge unity is given to A_2 alone, and so on. If we give to A_1 alone a charge E_1 or to A_2 alone a charge E_2, &c. these potentials will be respectively multiplied by E_1, E_2, &c. Superimposing these states of equilibrium, we see that the potentials inside A_1, A_2, &c. when charged with E_1, E_2, &c. are respectively

$$\left.\begin{aligned} V_1 &= p_{11}E_1 + p_{21}E_2 + \ldots \\ V_2 &= p_{12}E_1 + p_{22}E_2 + \ldots \\ \&c. &= \&c. \end{aligned}\right\} \quad \ldots\ldots\ldots\ldots\ldots\ldots(1).$$

If we now solve these equations we have a second set of linear equations which we represent by

$$\left.\begin{aligned} E_1 &= q_{11}V_1 + q_{21}V_2 + \ldots \\ E_2 &= q_{12}V_1 + q_{22}V_2 + \ldots \\ \&c. &= \&c. \end{aligned}\right\} \quad \ldots\ldots\ldots\ldots\ldots\ldots(2).$$

The coefficients p_{11}, p_{12}, &c. and q_{11}, q_{12}, &c. depend only on the forms and relative positions of the conductors in the field and are independent of the charges given to them.

The coefficients q_{11}, q_{22}, &c. (in which the two numbers forming the suffix are the same) are called *the electric capacities* of the bodies A_1, A_2, &c. *The capacity of a conductor may be defined to be its charge when its own potential is unity and that of every other conductor in the field is zero.*

The coefficients q_{12}, q_{23}, &c. (in which the numbers in the suffix are different) are called *the coefficients of induction.* Any one of them, as q_{rs}, denotes the charge on A_s when A_r is raised to potential unity, the potentials of all the conductors except A_r being zero.

The coefficients p_{11}, p_{12}, p_{23}, &c. are called *the coefficients of the potential.* Any one of them as p_{rs} denotes the potential of A_s when a charge unity is given to A_r, the charges on all the other conductors being zero.

Since the dimensions of potential are quantity/distance, it follows that every coefficient of potential is the reciprocal of a length. For the same reason every coefficient of induction has the dimensions of a length.

439. *To prove that* $p_{rs}=p_{sr}$ *and* $q_{rs}=q_{sr}$. Let the conductors $A_1 \ldots A_n$ when the charges are $E_1 \ldots E_n$ and the potentials $V_1 \ldots V_n$ be called system I. Let the same conductors when the charges are $E_1' \ldots E_n'$ and the potentials $V_1' \ldots V_n'$ be called system II. Let us treat these as independent coexistent systems.

The mutual work between two systems has been proved in Art. 59 to be equal to the sum of the products of each element of mass of either system by the potential of the other system at that element. In the body A_r each element of electricity in one system is to be multiplied by the potential of the other system at that body, and the product is either E_rV_r' or $E_r'V_r$. We may therefore form the equation

$$E_1V_1'+E_2V_2'+\ldots=E_1'V_1+E_2'V_2+\ldots \quad \ldots\ldots\ldots(3)$$

which may be shortly written $\Sigma EV'=\Sigma E'V$.

Let us now put each of the electricities E_1, E_2, &c., E_1', E_2', &c., *except* E_r *and* E_s', equal to zero. Then by equations (1), $V_s=p_{rs}E_r$, $V_r'=p_{sr}E_s'$. The equation (3) then gives $p_{rs}=p_{sr}$.

In the same way if we put each of the potentials V_1, V_2, &c., V_1', V_2' &c. *except* V_r *and* V_s' equal to zero we deduce from (2) and (3) $q_{rs}=q_{sr}$.

Ex. 1. Three small conducting spheres, whose radii are r_1, r_2, r_3, are placed with their centres at the corners of a triangle whose sides a, b, c are very much greater than the radii. Prove the following approximate relations

$$\frac{a^2-r_2r_3}{a^2r_2r_3q_{11}}=\frac{-(ab-cr_3)}{abcr_3q_{12}}=\frac{-(ac-br_2)}{abcr_2q_{13}}=\frac{1}{r_1r_2r_3}-\frac{1}{a^2r_1}-\frac{1}{b^2r_2}-\frac{1}{c^2r_3}+\frac{2}{abc}.$$

Proceeding as in Art. 374 we find that the potentials V_1, V_2, V_3, at the centres of the spheres are given by three linear equations of the form $V_1=E_1/r_1+E_2/c+E_3/b$.

These correspond to equations (1) of Art. 438. Solving these we find E_1 expressed as a linear function of V_1, V_2, V_3, the three coefficients are respectively q_{11}, q_{12}, q_{22}.

Ex. 2. Two insulated electrified spheres (radii r_1, r_2) are at a considerable distance c from each other; prove that the coefficients of potential and induction are approximately given by

$$r_1 p_{11} = c p_{12} = r_2 p_{22} = 1,$$
$$r_2 q_{11} = -c q_{12} = r_1 q_{22} = r_1 r_2 (1 + r_1 r_2/c^2).$$

440. *The lines of force.* Consider the lines of force which intersect the surface of a conductor. Since at any point of the surface $4\pi\rho = -dV/dn$, it is clear that the potential decreases or increases *outwards* along these lines according as they intersect the conductor at a point of positive or negative electricity, (Art. 114).

Let a point P travel along a line of force in such a direction that the potential at P continually decreases. *The line of force is said to issue from or terminate at a conductor* according as the point P crosses its surface in an outward or inward direction.

It follows that a line of force can issue from a conductor only at a point of positive electricity and will then either proceed to an infinite distance or terminate at a point of negative electricity on some conductor of lower potential.

If a line of force proceed from one conductor to another, it joins points A, B on the two conductors which are oppositely electrified.

441. *If a tube of force intersect two conductors, the quantities of electricity at the two ends are equal and of opposite signs.*

Divide the given tube into elementary tubes; let the areas at the extremities A, B of any one of these be $d\sigma$, $d\sigma'$. Let the forces at A, B measured outwards from the conductors be F, F'; then $Fd\sigma = -F'd\sigma'$, (Art. 127). Since $4\pi\rho = F$, $4\pi\rho' = F'$, we have $\rho d\sigma = -\rho' d\sigma'$.

442. *The conductor of greatest positive potential can have only positive electricity on its surface.* For, if any element of its surface were negatively electrified, a line of force could terminate at that element. Such a line must have issued from a conductor of greater positive potential. Similarly the conductor of greatest negative potential can have only negative electricity on its surface. See Art. 380.

443. *To prove that all the coefficients of the potential (p_{11}, p_{12}, &c.) are positive and that the coefficient p_{rs} is less than either p_{rr} or p_{ss}.*

Let the body A_r be charged with *a positive unit of electricity* and let all the others be uncharged. Then $V_r = p_{rr}$ and $V_s = p_{rs}$, by Art. 438. The body A_r cannot be entirely covered with negative electricity and is therefore not the body of greatest negative potential, Art. 442. Any other conductor A_s has both positive and negative electricity on its surface and cannot be the body of greatest positive or greatest negative potential. The charged body A_r must therefore be the conductor of greatest positive potential, and there is no conductor of greatest negative potential. Hence all the conductors are at positive potential and $p_{rr} > p_{rs}$.

Let the body A_s be placed in a hollow excavated in A_r and completely surrounded by it, then, since A_s is uncharged, there is no development of electricity either on its surface or on the inside of the shell A_r, Art. 389. The potential throughout the interior of A_r is p_{rr} and hence in our present notation $p_{rs}=p_{rr}$. In the same way, if A_s is enclosed by a shell A_t, then $p_{rs}=p_{rt}$.

The case in which A_s is enclosed by one of the other bodies is thus only a limiting case of the theorem and is not an exception.

444. *To prove that q_{rr} is positive and q_{rs} negative, and that the sum of the series $S = q_{1r} + q_{2r} + \dots + q_{rr} + \dots + q_{nr}$ is positive.*

Let the body A_r be charged to potential unity, all the others being at zero potential. The charges given to the conductors A_1, A_2, &c. are therefore q_{1r}, q_{2r}, &c. (Art. 438). The body A_r is the conductor of greatest positive potential, its charge q_{rr} is therefore positive, (Art. 442).

The body A_s is at zero potential. If there were a point of positive electricity on its surface a line of force could issue from it and must terminate at some point of lower potential, but there are no such points. The body A_s is therefore covered with negative electricity, that is q_{rs} is negative.

The unoccupied space outside the system is bounded by the surfaces of the conductors and by a sphere of infinite radius. Hence the potential at every point of this space lies between the greatest and least potential on the boundary, (Art. 116). These potentials are respectively unity and zero. The potential of the system at a very distant point is the same as if the whole quantity of electricity were collected into its centre of gravity (Art. 109) and its sign is therefore the same as that of the series S. The sum of this series must therefore be positive.

If A_s is enclosed by any body A_t and both are at potential zero, no line of force can pass between A_s and the shell A_t. There is therefore no electricity on the body A_s, (Art. 440), and in this case the charge $q_{rs}=0$.

If A_s is enclosed by A_r and A_r be at potential unity, A_s at potential zero, all the lines of force between A_r and A_s must issue from A_r and arrive at A_s. The body A_s is therefore charged only with negative electricity (Art. 440) and q_{rs} is negative.

445. Ex. Prove that when r and s are unequal

$$p_{1r}q_{1s}+p_{2r}q_{2s}+\ldots+p_{nr}q_{ns}=0,$$

and when $r=s$, the sum is unity. Thence show that the series represented by S in Art. 444 lies between 0 and $1/p_{rr}$.

The first two results follow from Art. 438, by putting $E_r=1$, and every other $E=0$. The third follows from the first two, since $p_{rr}>p_{rs}$.

446. *To find the mutual potential energy W of a system of conductors.* It has been proved in Art. 61 that W is equal to half the sum of the products of each element of mass by the potential at that element. As in Art. 439 this product for the body A_r is E_rV_r. We therefore have

$$W=\tfrac{1}{2}(E_1V_1+E_2V_2+\ldots)=\tfrac{1}{2}\Sigma EV \text{ ....(4).}$$

By substituting from equations (1) and (2) of Art. 438 we see that this may be written in either of the forms

$$\left.\begin{aligned}W&=\tfrac{1}{2}p_{11}E_1^2+p_{12}E_1E_2+\ldots\\&=\tfrac{1}{2}q_{11}V_1^2+q_{12}V_1V_2+\ldots\end{aligned}\right\} \text{ ....(5).}$$

447. Ex. 1. *Prove analytically that the expression for W is always positive.*

Since q_{rs} is negative, let $q_{rs}=-\beta_{rs}$. Hence by Art. 444 $q_{rr}>\beta_{1r}+\beta_{2r}+\&c.$ It follows from the expression (5) in Art. 446 that

$$\begin{aligned}2W&>(\beta_{12}+\beta_{13}+\ldots)V_1^2+(\beta_{21}+\beta_{23}+\ldots)V_2^2-2\beta_{12}V_1V_2\pm\&c.\\&>\beta_{12}(V_1-V_2)^2+\beta_{13}(V_1-V_3)^2+\ldots\end{aligned}$$

Ex. 2. A given charge is distributed over a number of conductors so that the potential energy of the system when in electrical equilibrium is least. Prove that the conductors are at the same potential. [Math. T. 1897.]

Make the expression (5) for W in Art. 446 a minimum with the condition that ΣE is given.

Ex. 3. **Energy of condensers.** *Two conducting surfaces are separated from each other by a plate of some non-conducting substance so as to form a condenser; as described in Art. 417. Find the potential energy.*

Let β, β' be the potentials of the conductors; ρ, ρ' the surface densities. Let dS be an element of area of either surface, θ the thickness of the conductor at this element. The potential energy due to this element is (by Art. 446)

$$dW=\tfrac{1}{2}\beta\rho dS+\tfrac{1}{2}\beta'\rho' dS \text{ ....(1).}$$

Since $4\pi\rho$ is equal to the fall of the potential divided by the thickness, we have

$$4\pi\rho=(\beta-\beta')/\theta, \quad 4\pi\rho'=(\beta'-\beta)/\theta \text{ ....(2).}$$

The capacity per unit of area, if measured by the ratio of the quantity of electricity on either conductor to the difference of the potentials, is

$$C_1=\rho/(\beta-\beta') \text{ ....(3).}$$

Using the equations (1) and (2) we can express dW in terms of either $\beta-\beta'$ or ρ.

We find $$dW=(\beta-\beta')^2\frac{dS}{8\pi\theta}=2\pi\rho^2\theta dS.$$

The value of W may then be found by integration. If θ is constant at all points of an area S, and Q the quantity of electricity on that area, we have

$$W=2\pi\rho^2\theta S=2\pi Q^2\theta/S.$$

In the case of *a spherical conductor* separated from a concentric conducting shell by a *thin non-conductor* (Art. 392) we have $S=4\pi a^2$. The potential V at the centre is $\frac{Q}{a}-\frac{Q}{a+\theta}=\frac{Q\theta}{a^2}$, the capacity C is $Q/V=a^2/\theta$. The potential energy is therefore $W=\frac{Q^2\theta}{2a^2}=\frac{Q^2}{2C}=\frac{a^2V^2}{2\theta}$.

As a second example, let the *condenser be formed by a cylindrical conductor* separated from a concentric cylindrical shell by a *thin non-conductor*, (Art. 419). The area of a unit of length is $S=2\pi a$. The capacity C' per unit of length is $2\pi a\rho/(\beta-\beta')$ which by (2) reduces to $a/2\theta$. The energy *per unit of length* is

$$W=\frac{Q^2\theta}{a}=\frac{Q^2}{2C'}.$$

448. Junction of conductors. Ex. 1. Two conductors A_1, A_2, of a system are joined together by a fine wire. Prove that the capacity of the united bodies is $q_{11}+2q_{12}+q_{22}$. Prove also that this is less than the sum of the capacities before the junction. [Coll. Ex.]

Let the conductors be charged with such quantities of electricity E_1, E_2, &c. that the potentials of A_1, A_2 are equal. By joining these no change is made in the distribution of the electricity. The total quantity on the united bodies is E_1+E_2, and the n equations of Art. 438 become the following $n-1$ equations

$$\begin{aligned} E_1+E_2&=(q_{11}+2q_{12}+q_{22})V_1+(q_{13}+q_{23})V_3+\dots \\ E_3&=(q_{13}+q_{23})V_1+q_{33}V_3+\dots \\ \&c.&=\&c. \end{aligned}$$

The results follow at once, since q_{12} is negative.

Ex. 2. Five equal uncharged and insulated conducting spheres are placed with their centres at the angular points of a regular pentagon. Another charged sphere is moved so as to touch each in succession at the point nearest the centre of the pentagon. Prove, that if $e_1 \dots e_5$ are the charges on the spheres when they have been each touched once

$$\begin{vmatrix} e_2-e_1, & e_1, & 0 \\ e_3-e_1, & e_2, & e_1 \\ e_4-e_1, & e_3, & e_1+e_2 \end{vmatrix}=0, \qquad \begin{vmatrix} e_2-e_1, & e_1, & 0 \\ e_3-e_1, & e_2, & e_1 \\ e_5-e_1, & e_1+e_4, & e_2+e_3 \end{vmatrix}=0.$$

[Coll. Ex. 1901.]

Let $A_1 \dots A_5$ be the fixed spheres, A_6 the moveable one. When A_6 is close to A_1, but not touching it, we have six equations expressing $V_1 \dots V_6$ in terms of any charges $E_1 \dots E_6$ which may be given to them, (Art. 438). When A_1 and A_6 touch, E_1 and E_6 are so modified that $V_1=V_6$, but the sum E_1+E_6 remains unaltered. Equating the potentials V_1 and V_6 we see that E_1 is a linear function of $E_2 \dots E_6$. Let this linear relation be

$$E_1=\alpha E_6+\beta(E_2+E_5)+\gamma(E_3+E_4).$$

Since the five spheres are equal and arranged in a regular figure, this relation will hold at each successive contact, provided E_1 always represents the electricity on the sphere which is being touched. We therefore have just after the contacts in order have occurred,

$$e_1=\alpha E_6, \quad e_2=\alpha(E_6-e_1)+\beta e_1, \quad e_3=\alpha(E_6-e_1-e_2)+\beta e_2+\gamma e_1,$$
$$e_4=\alpha(E_6-e_1-e_2-e_3)+\beta e_3+\gamma(e_2+e_1),$$
$$e_5=\alpha(E_6-e_1-e_2-e_3-e_4)+\beta(e_1+e_4)+\gamma(e_2+e_3).$$

Eliminating α, β, γ from these five equations we obtain the two results to be proved.

449. Introduction of a conductor. An insulated uncharged conductor B is introduced into the system of conductors A_1, A_2, &c. Prove that the coefficient of potential p_{rr} of any one of the others on itself is diminished.

Let the body B be brought into its place as an uncharged non-conductor and let it suddenly become a conductor. At this instant the potential energy of the system, viz. $\frac{1}{2}\Sigma EV$, is not altered, because the E of the new body is zero. The electricity is not now in equilibrium and must tend to assume a new arrangement. It is a dynamical principle that when a system is in *stable equilibrium* the potential energy is a minimum. It follows that in the new position of equilibrium the energy is less than before.

To separate the effect on p_{rr} from that on the other coefficients, let the conductor A_r alone have a charge, all the others, as well as the new body B, being uncharged. The energy before the introduction of B was $\frac{1}{2}E_r p_{rr}$, and after that event became $\frac{1}{2}E_r p'_{rr}$. The new value of the coefficient of the potential, viz. p'_{rr}, is therefore less than p_{rr}.

450. Potential Energy. Ex. 1. *A conductor having a charge Q and being at potential V_0 is acted on by a quantity E of electricity situated at an external point B; in this state the potential at an external point B' is $V_{B'}$. The same conductor with a charge Q' and at a potential V_0' when acted on by E' placed at B' has a potential V_B' at B. Prove that $Q'V_0+E'V_{B'}=QV_0'+EV_B'$.*

This is the mutual work of the two states described above when regarded as different systems, see Art. 439.

Ex. 2. An uncharged insulated conductor is acted on by a quantity E of electricity situated at an external point B. Prove that the potential at any external point B' is a symmetrical function of the coordinates of B and B'.

This theorem is also true if the conductor is uninsulated, for we may join it to earth by a fine wire and include the earth as part of the system.

The first result follows from Ex. 1 by putting $Q=0$, $Q'=0$, $E=E'$.

Ex. 3. The locus of a point B at which a given quantity E of electricity must be placed to develop a given quantity Q of electricity in an uninsulated conductor, is that level surface of the same conductor (when insulated, charged to potential V_0' and not acted on by any external point) at which the potential is $-QV_0'/E$.

451. A circular disc. *To find the distribution of electricity on a circular disc when acted on by an external electrical point B situated in its plane*.*

The electric density at any point Q on either side of an insulated circular disc is $\rho=\dfrac{V_0}{2\pi^2}\dfrac{1}{(QR\,.\,QR')^{\frac{1}{2}}}$ where R, R' are

* The problem of finding the law of distribution of electricity on a circular disc and spherical bowl when influenced by an electrical point was first solved by Sir W. Thomson, see section xv. of the *reprint* of his papers. In the *Quarterly Journal* for 1882 Ferrers found the potential due to the bowl at any point of space. He uses the method of spherical harmonics. In the same *Journal* 1886, Gallop applied Bessel's functions to find the distribution on a circular disc. He also investigates the distribution on a spherical bowl and finds the capacity of the bowl; for this purpose he uses the method of inversion.

the intersections of a chord BQ with the circle. The internal potential is V_0 and the quantity M of electricity is $M = 2aV_0/\pi$, (Art. 382).

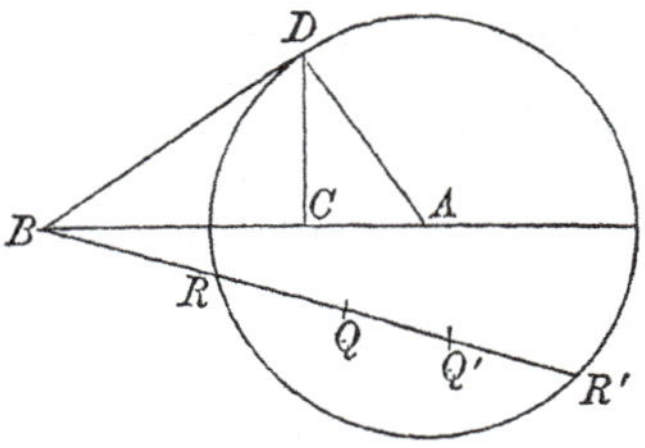

If we invert this with regard to an external point B, with a radius of inversion k equal to the tangent BD, we shall obtain the law of distribution of electricity on the same disc when acted on by an electrical point at B.

Let Q' be the point inverse to Q, then since R, R' also are inverse points,

$$\frac{QR}{Q'R'} \cdot \frac{QR'}{Q'R} = \frac{k^2}{BQ'.BR'} \frac{k^2}{BQ'.BR} = \frac{k^2}{BQ'^2}$$

by Art. 172. The surface density ρ' at Q' is given by

$$\rho' = \rho \left(\frac{k}{BQ'}\right)^3 = \frac{V_0 k^2}{2\pi^2} \frac{1}{BQ'^2} \cdot \frac{1}{(Q'R'.Q'R)^{\frac{1}{2}}}.$$

The potential at any point P' within the disc is $V_0 k/BP'$. Put $V_0 k = E$, then the potential of a quantity $-E$ situated at B together with that of the distribution

$$\rho' = \frac{E}{2\pi^2} \frac{1}{BQ'^2} \left(\frac{AB^2 - a^2}{a^2 - AQ'^2}\right)^{\frac{1}{2}} \text{..................(1)}$$

is zero at all points within the disc. Here we have written $k^2 = AB^2 - a^2$, $Q'R'.Q'R = a^2 - AQ'^2$ where A is the centre of the disc and a the radius.

The expression (1) gives the required surface density at any point Q' on one side of the disc when the internal potential is zero, and the electricity at B is $-E$.

452. *To find the quantity M' of electricity on the inverse disc* we use the rule $M' = kV_1$ where V_1 is the potential of the original disc at the centre of inversion, Art. 170. This gives by Art. 384 $M' = kM\phi/a$, where ϕ is the angle subtended by any radius of the disc at the apex of the confocal spheroid through B. Since $M = 2aV_0/\pi$ and $E = V_0 k$, we have $M' = 2E\phi/\pi$. Let a', c' be the semi-axes of the confocal which passes through B, then $\tan\phi = a/c'$, $c'^2 = a'^2 - a^2$ and $a' = AB$. Hence *ϕ is also half the angle subtended by the disc at the electrified point B,* i.e. $\phi = DBA$.

453. *To find the potential at any external point P' in the plane of the disc.* The potential at P of the original disc is $V = \frac{M\phi}{a} = \frac{2V_0}{\pi}\sin^{-1}\frac{a}{a'}$, where $a' = AP$ is the semi-major axis of the confocal through P, (Art. 384). The potential V' at P' of the inverted disc is therefore

$$V' = \frac{2V_0}{\pi}\sin^{-1}\left(\frac{a}{CP'}\cdot\frac{BC\,.\,BP'}{k^2}\right)\cdot\frac{k}{BP'}$$

where k is the length of the tangent BD, and C, the inverse point of A, is the foot of the ordinate of D, see Art. 172.

454. *To find the distribution of electricity on a plane circular disc, centre A, when acted on by a quantity $-E$ of electricity situated at a point O on the axis.*

Let us cover the area of the plane outside the disc (regarded as a non-conductor) with a layer of electricity whose surface density at any point B is $\rho = -\frac{Eh}{2\pi}\frac{1}{OB^3}$ and let this layer be fixed in the plane (Art. 412). Then if Q be any point on the conducting disc, the induced density at Q is (by Art. 451)

$$\rho' = \frac{Eh}{2\pi}\iint\frac{x\,d\theta\,dx}{OB^3}\cdot\frac{1}{2\pi^2}\frac{1}{BQ^2}\left(\frac{x^2-a^2}{a^2-r^2}\right)^{\frac{1}{2}}$$

where $x = AB$, $r = AQ$, θ is the angle QAB. We now substitute

$$OB^2 = x^2 + h^2, \qquad BQ^2 = x^2 + r^2 - 2xr\cos\theta,$$

where $h = OA$. We first integrate with regard to θ between the limits 0 and 2π, using the integral $\int\frac{d\theta}{1 - e\cos\theta} = \frac{2\pi}{\sqrt{(1-e^2)}}$. To effect the integration with regard to x, write $x = h\tan\psi$ and express the result in terms of $\cos\psi$. The ordinary rules of the integral calculus then show that we should put $(h^2+a^2)\cos^2\psi = h^2 - y^2$. The limits for x being a to ∞, those for y are 0 to h. We thus find

$$\rho' = \frac{Eh}{2\pi^2}\cdot\frac{t - \tan^{-1}t}{(h^2+r^2)^{\frac{3}{2}}}, \qquad t^2 = \frac{h^2+r^2}{a^2-r^2}.$$

The result is that the potential due to the forced distribution ρ outside the disc together with that due to the distribution ρ' on each side of the disc is zero at all internal points.

Now by Art. 412 an electrical point $-E$ situated at O and an infinite plane whose density is that represented above by ρ (but with its sign changed) exert no attraction at all points on the side of the plane opposite to O, and the sum of their potentials at all such points is zero.

Superimpose this second electrical system on the first; then the forced distributions outside the disc cancel each other. The sum of the potential due to $-E$ situated at O and that due to the electricity on the two sides of the disc is zero at all points within the conducting substance.

The densities on the sides most remote from and nearest to O are respectively

$$\rho' = \frac{Eh}{2\pi^2}\frac{t - \tan^{-1}t}{(h^2+r^2)^{\frac{3}{2}}}, \qquad \rho'' = \rho' + \frac{Eh}{2\pi}\frac{1}{(h^2+r^2)^{\frac{3}{2}}}.$$

These formulæ represent the density at any point Q when the internal potential is zero and the disc is acted on by an electrified point $-E$ situated at O.

Here $t^2 = \dfrac{h^2+r^2}{a^2-r^2}$. It is easily seen that $t = \cot\psi$, where ψ is half the angle subtended at O by the chord drawn at Q perpendicular to the diameter AQ.

455. *To find the potential of the electrified disc at any external point P, and also the quantity of electricity on the disc.*

Consider two discs whose surface densities are respectively

$$\rho_1 = \frac{1}{(a^2-r^2)^{\frac{1}{2}}}, \quad \rho_2 = \frac{2m}{(h^2+r^2)^{\frac{3}{2}}}\left(t - \tan^{-1} t + \frac{\pi}{2}\right).$$

By differentiation we find that $\dfrac{d\rho_2}{da} = \dfrac{d\rho_1}{da}\dfrac{2m}{a^2+h^2}$ and that at the rims where $r=a$ and $\tan^{-1} t = \frac{1}{2}\pi$, the densities also have the ratio $2m$ to a^2+h^2. Let V_1 and V_2 be the potentials of the discs at external points similarly situated. Now dV_1/da is the sum of the potentials of a disc whose density is $d\rho_1/da$ and of an annulus round its rim. It immediately follows that $\dfrac{dV_2}{da} = \dfrac{dV_1}{da}\dfrac{2m}{a^2+h^2}$.

Now by Art. 384, V_1 is the potential of the electricity on *one side* of a circular disc charged with a quantity $M = 4\pi a$, hence $V_1 = 2\pi\phi$. If we put $m = Eh/2\pi^2$, V_2 becomes the potential of a circular area whose density is the sum of the densities on the two sides of the disc. We therefore have

$$V_2 = \frac{2Eh}{\pi}\int_0^a \frac{1}{a^2+h^2}\frac{d\phi}{da}\,da,$$

where ϕ is the angle subtended by any radius of the disc at the apex of the confocal spheroid drawn through P.

When the point P lies in the plane of the disc, the integration is easy. Let x be the abscissa of P, then x is also the semi-axis major of the confocal through P and $\sin\phi = a/x$. We therefore have

$$V_2 = \frac{2Eh}{\pi}\int \frac{d\phi}{h^2 + x^2\sin^2\phi} = \frac{2E}{\pi}\frac{1}{(h^2+x^2)^{\frac{1}{2}}}\tan^{-1}\left\{\frac{a}{h}\left(\frac{x^2+h^2}{x^2-a^2}\right)^{\frac{1}{2}}\right\}.$$

When x is infinite, this takes the simple form $V_2 = \dfrac{1}{x}\cdot\dfrac{2E}{\pi}\tan^{-1}\dfrac{a}{h}$. Now at a great distance, potential is mass divided by distance; the quantity of electricity on the disc is therefore $\dfrac{2E}{\pi}\tan^{-1}\dfrac{a}{h}$. This is the same as $2E\phi_1/\pi$ where ϕ_1 is half the angle subtended at the electrified point O by any diameter of the disc.

When the point P is on the axis, we have $\tan\phi = a/z$ where z is the ordinate of P. The potential is then

$$V_2 = \frac{2E}{\pi}\frac{1}{z^2-h^2}\left\{z\tan^{-1}\frac{a}{h} - h\tan^{-1}\frac{a}{z}\right\}.$$

When $z=h$, this expression takes the form 0/0. We easily find however that the potential at the electrified point O is

$$V_2 = \frac{E}{\pi}\left(\frac{a}{a^2+h^2} + \frac{1}{h}\tan^{-1}\frac{a}{h}\right).$$

When the point P has a position defined by any values of x, z, both the process of integration and the final result are somewhat complicated. The whole of the work is given by Gallop in the *Quarterly Journal*, vol. XXI.

456. Spherical bowl. *To find the distribution of electricity on an insulated spherical segment with a plane rim.*

The electrical distribution on the bowl may be deduced by inversion from that on a circular disc at zero potential with a quantity of electricity $-E$ at a point O on the axis.

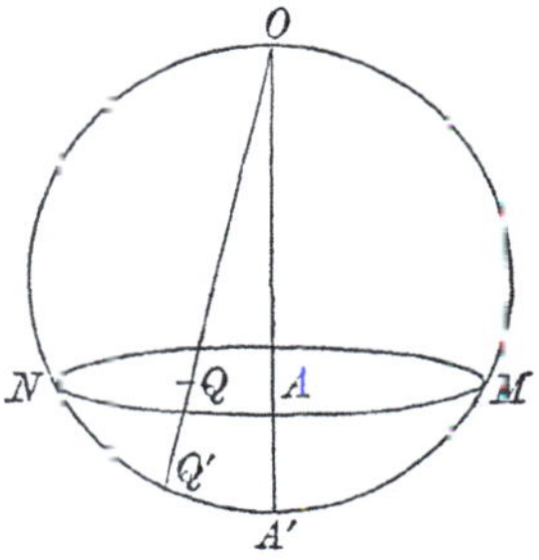

Let MN be the rim of the disc, O the centre and $OM=k$ the radius of inversion. We regard O as the centre of a sphere of small radius ϵ. This sphere inverts into a large sphere of radius k^2/ϵ. The quantity M' on the inverted sphere is given by $M'=kV_1$ (Art. 180) and is evidently equal to $-Ek/\epsilon$. The attraction at any internal point is therefore zero but the potential is $-E/k$.

The disc inverts into the segment $MA'N$, the sides nearest to or farthest from O corresponding to the convex and concave sides of the bowl. To deduce the density at Q' from that at Q we use the formula of Art. 169 as applied to surfaces. Since $k/OQ'=OQ/k$ and $OQ^2=h^2+r^2$ we deduce that the surface densities at any point Q' on the concave and convex sides are respectively

$$\rho_1=\frac{E}{2\pi^2}\frac{h}{k^3}(t-\tan^{-1}t),\qquad \rho_{11}=\rho_1+\frac{Eh}{2\pi k^3}.$$

The sum of the potentials of the electricity on the bowl and of that on the sphere of infinite radius being zero, the internal potential V_0 of the electricity on the bowl alone is E/k.

Let $A'Q'=r'$, $A'M=a'$, $OM=k$, and let the diameter OA' of the sphere be f. We then have since $hf=k^2$

$$r^2=AQ^2=\frac{h^2r'^2}{f^2-r'^2},\qquad t^2=\frac{h^2+r^2}{a^2-r^2}=\frac{f^2-a'^2}{a'^2-r'^2}.$$

The densities at any point on the concave and convex sides of the insulated bowl then take the forms

$$\rho_1=\frac{V_0}{2\pi^2 f}(t-\tan^{-1}t),\qquad \rho_{11}=\rho_1+\frac{V_0}{2\pi f},$$

where V_0 is the internal potential.

457. *To find the quantity M' of the electricity on the bowl*, we use the rule $M'=kV_1$, Art. 170. We have therefore merely to write $E=kV_0$ in the expression for the potential of the disc at O (Art. 455) and to multiply the result by k. *Let 2α be the angle subtended at the centre of the sphere by any radius of the rim*, then $a=h\tan\alpha$ and $h=f\cos^2\alpha$. The quantity M' is therefore given by

$$\frac{M'}{V_0}=\frac{f}{2\pi}(\sin 2\alpha+2\alpha).$$

The potential V' of the bowl at any point P' may be deduced from that of the disc at the inverse point P. The result takes a simple form when P' lies on the unoccupied part of the sphere. We then have

$$V'=\frac{2V_0}{\pi}\tan^{-1}\frac{a'}{(P'M\,.\,P'N)^{\frac{1}{2}}},$$

where $a'=A'M=f\sin\alpha$.

458. Ex. Prove that the density at any point Q' of a spherical bowl at zero potential when acted on by a quantity $-E'$ of electricity at any point B' on the unoccupied part of the sphere is

$$\rho'=\frac{E'}{2\pi^2}\frac{1}{(B'Q')^2}\left(\frac{OM^2-OB'^2}{OQ'^2-OM^2}\right)^{\frac{1}{2}}.$$

See the figure of Art. 456. This follows from the result in Art. 451 by inversion with regard to a point O on the axis.

459. Electricity on two spheres*. *Two electrified conducting spheres are in presence of each other; it is required to find the resultant force due to their mutual action.* The spheres may be either insulated or maintained at a constant potential, say, by being joined to a distant large reservoir of electricity by a fine wire. The investigation depends on the following theorem.

The electricity on a sphere (radius a) is *maintained at constant potential* and is in equilibrium under the action of any number of electrical points. Another electrical point A, charged with a quantity E of electricity, is placed at a distance x from the centre of the sphere. The new distribution of electricity may be represented by the addition of a layer on the sphere such that its potential plus that of the electrical point A is zero throughout the interior. The potential of such a layer at all external points is the same as that of an electric particle $E' = -Ea/x$ placed at the image or inverse point of A. The increase in the quantity of electricity on the sphere is then E', Art. 402.

If the sphere is *insulated*, the additional layer representing the change produced by A must be such that its mass is zero. The potential of this layer at any external point is equal to the sum of the potentials of two electric particles. One of these has a mass $E' = -Ea/x$ and is to be placed at the image of A, the other has an equal and opposite mass and is to be placed at the centre. The potential inside the sphere is then increased by $-E'/a$ or E/x.

It is evident that the former case is less complicated than the latter. We shall therefore in the first instance suppose that both the spheres are maintained at constant potential, and finally deduce the case of insulation from the former.

460. Let the radii of the spheres be a, b, and let the distance between the centres A_0, B_0 be c. Since c is necessarily greater than either a or b, we can express the force in a convergent series by regarding a/c and b/c as small quantities of the first order. Let the given potentials inside the spheres A, B be u, v. If the distance c were very great the quantities of electricity on the spheres would be $ua = E$ and $vb = F$, and the mutual force of repulsion would be EF/c^2.

The electrical point E placed at A_0 will disturb the electricity on the sphere B. The external effect of this disturbance is represented by a mass particle placed at the image A_0' of A_0 with regard to the sphere B. Since this mass is proportional to E we represent it by Ep_0'. In the same way the effect of the electrical point F is represented by a mass particle Fq_0' placed at B_0' the image of B_0 with regard to the sphere A. For another approximation we seek the images of A_0', B_0' and so on continually.

To fix our ideas, let 1, p_0', p_1, p_1', &c. denote the masses of the series of which the first term is a mass unity placed at the centre A_0. Then p_1, p_2, &c. are within

* The problem of determining the distribution of electricity over two spheres in presence of each other was attacked by Poisson in 1811, who expressed the results by definite integrals, see *Mém. de l'Institut*. There is a solution founded on the method of successive images by Kelvin, *Phil. Mag.* 1853, reproduced in his *Papers on Electrostatics and Magnetism*, page 86. In Maxwell's *Electricity*, edition of 1892, page 281, there is a short discussion of Kirchhoff's results by J. J. Thomson. He also gives references to other papers on this subject. The principle of successive influences was first enunciated by Murphy in his treatise on Electricity, 1833. In the case of two equal spheres whose distance apart is 100 times either radius he finds the difference of densities at the ends of the symmetrical diameter.

the sphere A; p_0', p_1', &c. are within the sphere B. Let f_0 f_1, f_2, &c. denote the distances of 1, p_1, p_2, &c. from A_0; f_0', f_1', &c. denote the distances of p_0', p_1', &c. from B_0. Then $f_0=0$, $f_0'=b^2/c$ and so on. In the same way, if a unit of mass is placed at B_0, let 1, q_0', q_1, &c. denote the masses, and g_0, g_0', g_1, g_1', &c. the distances of the successive points of the corresponding series from B_0 and A_0 alternately. Then $g_0=0$, $g_0'=a^2/c$, &c. We obviously have the following equations

$$\left.\begin{aligned} f_n' &= \frac{b^2}{c-f_n}, & p_n' &= \frac{-bp_n}{c-f_n} \\ f_{n+1} &= \frac{a^2}{c-f_n'}, & p_{n+1} &= \frac{-ap_n'}{c-f_n'} \end{aligned}\right\} \quad\text{.......................(1).}$$

The corresponding relations for the points of the other series are obtained by interchanging a, f, p with b, g, q.

We notice that all the masses p_n, q_n are independent of the electrical conditions of the spheres and are functions of a, b, c only. If we regard a/c, b/c as small quantities of the first order, p_n and q_n are small quantities of the order $2n$, while p_n', q_n' are of the order $2n+1$. The distances f_n, f_n', g_n, g_n are all of the second order. We also notice that the distance between the masses p_s, p_t is f_s-f_t, the distance between p_t, q_t is $c-f_s-g_t$, and so on.

The whole repulsion between each sphere and the other is equal to the repulsive force exerted by the fictitious masses inside one sphere on those within the other sphere. It is therefore represented by

$$X=\Sigma\Sigma\left\{\frac{EFp_sq_t}{(c-f_s-g_t)^2}+\frac{EFp_s'q_t'}{(c-f_s'-g_t')^2}+\frac{E^2p_sp_t'}{(c-f_s-f_t')^2}+\frac{F^2q_sq_t'}{(c-g_s-g_t')^2}\right\}\text{.....(2),}$$

where the summations extend from $s=0$ to ∞ and $t=0$ to ∞, and $p_0=1$, $q_0=1$.

The total quantities of electricity on the spheres are

$$E'=E\Sigma p_n+F\Sigma q_n', \quad F'=E\Sigma p_n'+F\Sigma q_n \quad\text{....................(3),}$$

where the summations extend from $n=0$ to ∞.

It follows that (2) also represents *the mutual force X between the spheres when insulated and charged with quantities E', F' of electricity.*

461. When the spheres are not very close to each other it is sufficient to take a few terms only of this doubly infinite series. *Let us reject quantities of the order EF/c^2 when multiplied by $(a/c)^4$ or $(b/c)^4$.* In this case we require only the repulsive forces between the points E, Fq_0', Ep_1, Fq_1' inside the sphere A and the points F, Ep_0', Fq_1, Ep_1' within the sphere B. Taking any two of these we see (since all the f's and g's are of the second order) that their distance apart may be regarded as equal to c, except in the case of the particles E and Ep_0' and the particles F and Fq_0'. The force between E and Ep_0' is

$$\frac{E^2p_0'}{(c-f_0')^2}=\frac{E^2p_0'}{c^2}+\frac{2E^2p_0'}{c^2}\frac{f_0'}{c}\text{,.................................(4),}$$

with a similar expression for that between F and Fq_0'. The whole repulsion is therefore

$$X=\frac{E'F'}{c^2}+\frac{2E^2}{c^2}\frac{p_0'f_0'}{c}+\frac{2F^2}{c^2}\frac{q_0'g_0'}{c}\text{..............................(5),}$$

where

$$\left.\begin{aligned} E'&=E+Fq_0'+Ep_1+Fq_1' \\ F'&=F+Ep_0'+Fq_1+Ep_1' \end{aligned}\right\} \quad\text{.............................(6).}$$

It is evident that E' and F' are the quantities of electricity on the spheres.

Since the masses 1, p_0', p_1, p_1' occupy successive inverse points we have when terms of the fourth order are neglected

$$\left.\begin{aligned} f_0' &= \frac{b^2}{c}, \quad f_1 = \frac{a^2}{c-f_0'} = \frac{a^2}{c}, \quad f_1' = \frac{b^2}{c-f_1} = \frac{b^2}{c} \\ p_0' &= -\frac{b}{c}, \quad p_1 = \frac{-ap_0'}{c-f_0'} = \frac{ab}{c^2}, \quad p_1' = \frac{-bp_1}{c-f_1} = -\frac{ab^2}{c^3} \end{aligned}\right\} \quad \ldots\ldots\ldots\ldots(7);$$

these results follow directly from (1). Substituting these expressions in (5) and again rejecting all terms of the fourth order, we find

$$X = \frac{EF}{c^2}\left(1 + \frac{3ab}{c^2}\right) - \frac{E^2}{c^2}\left\{\frac{b}{c} + \frac{2b^2(a+b)}{c^3}\right\} - \frac{F^2}{c^2}\left\{\frac{a}{c} + \frac{2a^2(a+b)}{c^3}\right\} \ldots\ldots(8),$$

where $u = E/a$, $v = F/b$ are the given potentials of the two spheres.

462. To find the force of repulsion when both the spheres are insulated we notice that the expression (5) gives the force between the spheres when charged with the quantities E', F' of electricity and that their potentials are respectively $u = E/a$, $v = F/b$. It follows immediately from (5) that

$$X = \frac{E'F'}{c^2} - \frac{E'^2}{c^2}\frac{2b^3}{c^3} - \frac{F'^2}{c^2}\frac{2a^3}{c^3} \quad \ldots\ldots\ldots\ldots\ldots\ldots\ldots\ldots(9).$$

463. When the spheres are close to each other the method of finding the functions p_n, q_n, &c. by continued approximation becomes laborious. If we put $p_n = 1/P_n$ and eliminate f_n' and p_n' from the equations (1) we arrive at the equation of differences

$$P_{n+1} + P_{n-1} = \frac{c^2 - a^2 - b^2}{ab} P_n.$$

The solution is obviously

$$P_n = Ah^n + Bh^{-n}, \text{ where } h + \frac{1}{h} = \frac{c^2 - a^2 - b^2}{ab}.$$

We shall suppose that h is the root which is less than unity. To find the constants A, B we have by (2) the conditions $P_0 = 1$, $P_1 = \dfrac{c^2 - b^2}{ab}$. In the same way we find that P_n' satisfies the same equation of differences, with the conditions

$$P_0' = -\frac{c}{b}, \quad P_1' = -\frac{c}{ab^2}(c^2 - a^2 - b^2).$$

The reader will find methods of reducing the doubly infinite series for the force X to a single series, and also a discussion of the case in which the two spheres are in contact in Kelvin's *Papers on Electrostatics, &c.*, page 89.

464. Ex. 1. Two conducting spheres touch each other externally and are charged with electricity. Prove that the density at the point of contact is zero. [Use Art. 142.] [Murphy.]

Ex. 2. A conducting sphere, of radius a, having an electric charge E, is in front of a large plane conducting surface connected to earth, its centre being at a distance c from this surface, which is large compared with a. Prove that the sphere experiences an attraction towards the plane which is approximately equal to

$$\frac{E^2}{4c^2}\left(1 + \frac{a^3}{2c^3}\right).$$

[St John's Coll. 1897.]

Place on the other side of the plane at the same distance a second sphere of equal radius and let its charge be $-E$. The required attraction is the force X, given by (9), which one sphere exerts on the other (Art. 461).

Ex. 3. Two equal conducting insulated spheres of radius a are placed with their centres at a distance c apart in a uniform field of force, of intensity F, and whose direction is at right angles to the line joining the centres of the spheres. Show that, if the spheres are initially uncharged, their mutual repulsion will be

$$\frac{3F^2a^6}{c^4}\left[1-2\left(\frac{a}{c}\right)^3-8\left(\frac{a}{c}\right)^5+\ldots\ldots\right].\quad \text{[Math. Tripos, 1900.]}$$

The force F acting alone would cause a distribution of electricity on each sphere whose surface density is $\sigma=ky$, where $4\pi ak=3F$ (Art. 406, Ex 4), and the straight line joining the centres of the two spheres is the axis of x. The potential at any external point of a thin layer of surface density ky placed on a sphere is the y component of the repulsion of a solid sphere, of density ka, (Art. 92). This potential, again, is the same as that of a small doublet, or quasi-magnet, whose moment is a^3F placed at the centre with its axis parallel to the axis of y, (Art. 316).

Such a doublet (strength m, length l), if placed at the centre A of one sphere, would change the distribution of electricity on the other sphere. By Art. 459, the changes produced by each mass m acts at external points like a second mass particle $m'=-ma/c$, placed at the inverse point of the particle m. These form an inverse doublet of strength m' and length $l'=la^2/c^2$. This inverse doublet has therefore a moment $-a^3F(a/c)^3$ and is placed at the inverse point of the centre A, with its axis parallel to that of the first doublet.

To find approximately the mutual action of the two spheres, we consider each to be occupied by two doublets. The force exerted by one broadside doublet on another is proved in Art. 320 to be $X=3MM'/r^4$. The force exerted by one of the larger doublets on the other is therefore $3a^6F^2/c^4$. The force exerted by each large doublet on the opposite small one is $3a^3F\{-a^3F(a/c)^3\}/r^4$, where $r=c-a^2/c$. This when doubled reduces to $-6a^9F^2(c^2+4a^2)/c^9$. The force exerted by one small doublet on the other is of an order higher than the terms given in the enunciation. Adding these together we arrive at the result to be proved.

Ex. 4. Two spheres (centres A, B; $AB=c$; radii a, b) are charged with electricity and mutually influence each other. Let $f\left(\frac{r}{a}\right)$ and $\frac{a}{r}f\left(\frac{a}{r}\right)$ be the potentials of the sphere A at any internal and external point respectively, the point being situated on the line AB (Art. 294). Prove that f must satisfy the equation

$$af\left(\frac{r}{a}\right)-\frac{a^2b}{c^2-b^2-cr}f\left(\frac{ac-ar}{c^2-b^2-cr}\right)=h-\frac{bk}{c-r},$$

where h and k are the potentials of the two spheres.

If the spheres are in contact, deduce

$$zf(1-z)-\frac{mz}{m+z}f\left(1-\frac{mz}{m+z}\right)=\frac{hz}{a}-\frac{hmz}{a\{m+(1-m)z\}},$$

where $m=b/(a+b)$ and $r/a=1-z$. Prove also that a solution of this functional equation is

$$zf(1-z)=\frac{mh}{a}\int_0^1\frac{1-t^{1-m}}{1-t}\cdot t^{m/z}\frac{dt}{t}.$$

Deduce the potential of the sphere A at any point P not on the axis. See Art. 178.

To prove these results, let F be the function corresponding to f for the other sphere. Equate the potentials inside the spheres to h and k. Then eliminate F. See Poisson's two memoirs, *Mém. de l'Institut, &c.*, 1811, pages 1 and 163. Also Plana, *Mem. de &c. Torino*, ser. II. vol. VII., 1845, and vol. XVI., 1854.

Magnetic Induction.

465. When magnetism is induced in a neutral body A by the influence of a magnetised body B, it is supposed that each element dv of the volume of A becomes magnetic*. Let R be the resultant *magnetic force* (Art. 342) on a positive unit pole situated in the element, due to the influencing body B and the induced magnetism in A. In an isotropic body the axis of magnetisation of the element dv is in the line of action of the force R. The intensity I is nearly proportional to the force R provided that force is not very large. We therefore put $I = kR$. When k is positive the body is said to be *paramagnetic* and the direction of magnetisation coincides with that of the force F, when k is negative the body is *diamagnetic* and these directions are opposite. The value of k for soft iron is positive and great, but for bismuth it is negative and very small. Thus for soft iron k may vary under different circumstances from 10 to nearly 200, but for bismuth (which is one of the most highly diamagnetic substances known) k is about 1/400000. The coefficient k is called *the magnetic susceptibility*; it is also called *Neumann's coefficient.*

466. Let U be the magnetic potential of the magnetism of the influencing body B and Ω that of the induced magnetism in A. Let $V = U + \Omega$ be the potential due to all causes. Let (l, m, n) be the direction cosines of the direction of magnetisation of any element dv of the body A. It immediately follows that

$$Il = -k\frac{d(U+\Omega)}{dx},\ Im = -k\frac{d(U+\Omega)}{dy},\ In = -k\frac{d(U+\Omega)}{dz}\ \ldots(1).$$

We may in Poisson's manner *represent* the potential due to the induced magnetism by that of a distribution of *fictitious* matter throughout the volume and over the surface of the body A. The density ρ of the former is given by

$$\rho = -\left\{\frac{d(Il)}{dx} + \frac{d(Im)}{dy} + \frac{d(In)}{dz}\right\} = k(\nabla^2 U + \nabla^2\Omega).$$

Here we have introduced *the condition that k is constant for*

* The mathematical theory of induced magnetism was first given by Poisson, *Mémoires de l'Institut*, 1824. The difference between his theory and that of Weber cannot be discussed here. The reader will find the fundamental principles of induced magnetism explained in the reprint of Kelvin's papers. The theory of Faraday and Maxwell, that the dielectric is the seat of a peculiar kind of stress, does not come within the limits of a treatise on attractions.

the body A. Since the element dv is outside the body B and inside A we have

$$\nabla^2 U = 0, \quad \nabla^2 \Omega = -4\pi\rho; \quad \therefore (1 + 4\pi k)\,\rho = 0.$$

It follows that *when the magnetic susceptibility is constant, the volume density ρ is zero. The potential of the magnetism induced in the homogeneous isotropic body A at any internal or external point may therefore be represented by that of an imaginary layer of matter on the surface of that body. The surface density σ of this layer is known by Poisson's proposition to be* $\sigma = I\cos\theta$, see Art. 339.

If F is the normal component of force at any point P close to the surface but in the substance of the body, the surface density at P is $\sigma = kR\cos\theta = \pm kF$, the upper or lower sign being used according as F is measured positively from P in direction pointing outwards or inwards from the boundary.

467. *The actual distribution of induced magnetism is both solenoidal and lamellar.* Since $\rho = 0$ the condition that the magnetism is solenoidal is satisfied, (Art. 349).

The level surfaces due to the acting forces are defined by $U + \Omega = c$. Each element of the body is magnetised at right angles to the level surface which passes through it, and, since $I = kR$, the intensity is inversely proportional to the normal distance between two consecutive level surfaces c and $c + dc$ (Art. 46). The distribution is therefore lamellar, Art. 351.

468. The boundary condition. Let F_1, F_2 be the normal components of the magnetic force due to all causes at points P_1, P_2 respectively just inside and just outside the stratum but situated on the same normal. Let these forces be measured positively in each medium from the stratum on its boundary. Then by Arts. 142, 466,

$$F_1 + F_2 = 4\pi\sigma, \qquad \sigma = -kF_1 \quad \text{..............(2)}$$

From these we deduce the equation

$$(1 + 4\pi k)\,F_1 + F_2 = 0 \quad \text{................(3)}$$

469. When two substances, both of which are susceptible of induced magnetism, are separated by a surface S the conditions at the boundary are slightly altered. Let k_1, k_2 be their respective susceptibilities, F_1, F_2 the normal components of the magnetic force in the two substances at any point P of the boundary S, each being measured positively from S.

The surface of each substance is bounded by a fictitious layer whose surface densities are respectively

$$\sigma_1 = - k_1 F_1, \qquad \sigma_2 = - k_2 F_2 \quad \text{................(4).}$$

The minus sign is used because each F is measured inwards as explained above.

We then have by Green's theorem (Art. 142)

$$4\pi(\sigma_1 + \sigma_2) = F_1 + F_2 \quad \text{................(5).}$$

Eliminating σ_1, σ_2 we deduce the condition

$$(1 + 4\pi k_1) F_1 + (1 + 4\pi k_2) F_2 = 0.$$

The coefficient $1 + 4\pi k$ is called *the magnetic permeability* and is often represented by the letter μ. The equation then takes the form

$$\mu_1 F_1 + \mu_2 F_2 = 0 \quad \text{......................(6).}$$

It is often convenient to measure the normal forces F_1, F_2 in the same direction. *Let either direction of the common normal to the separating surface be chosen as the positive direction*, we deduce from (6) the following theorem. *The normal forces just within the two substances at any point of the boundary* (when there is no charge on the boundary) *are inversely as the permeabilities of the substances.*

When the body is not susceptible of magnetisation $k = 0$ and therefore $\mu = 1$. In a paramagnetic body k is positive and μ is greater than unity. In a diamagnetic body k is negative and μ is less than unity.

470. In some applications of this theory to electricity the *separating surface S is also occupied by a thin layer of matter capable by its repulsion of inducing polarisation in the two media.* This layer is to be regarded as part of the influencing body. Let ρ be its surface density.

We then have by Green's theorem (Art. 142)

$$4\pi(\rho + \sigma_1 + \sigma_2) = F_1 + F_2 \quad \text{..............(7),}$$

as in Art. 469. Eliminating σ_1, σ_2 by using the fundamental equations $\sigma_1 = - k_1 F_1$, $\sigma_2 = - k_2 F_2$ we arrive at the generalised equation

$$\mu_1 F_1 + \mu_2 F_2 = 4\pi\rho \quad \text{....................(8).}$$

All the conditions are included in the two statements briefly expressed by $\sigma = - kF$ and the equation (8).

471. Let H represent the resultant magnetic force and B the magnetic induction at any point P respectively. When the magnetism of the body is due solely to induction, the direction of magnetisation coincides in direction with that of the magnetic force H, (Art. 455). It follows that the force of induction B (being the resultant of the magnetic force H and a magnetic force $4\pi I$, Art. 342) must also coincide in direction with that of the magnetic force. We therefore have $B = H + 4\pi I$, and since $I = kH$, this gives $B = \mu H$.

The equation (6) of Art. 469 then asserts that *the normal component of the magnetic induction at P is unaltered in magnitude when P passes from one medium into another*, the components being measured in the same direction along the normal.

472. We know by Art. 144, that *the tangential component of the magnetic force is unaltered in magnitude when P passes from one medium into another*, the components being measured in the same direction along the tangent. The magnetic potential at P is also unaltered, (Art. 145).

Let H_1, H_2 be the resultant magnetic forces in the two media at any point P of the boundary; θ_1, θ_2 the angles their directions make with the normal at P, then

$$H_1 \sin\theta_1 = H_2 \sin\theta_2, \quad \mu_1 H_1 \cos\theta_1 = \mu_2 H_2 \cos\theta_2,$$

$$\therefore \tan\theta_1/\mu_1 = \tan\theta_2/\mu_2.$$

When therefore a line of magnetic force passes from one medium into another in which the permeability is greater than in the first its direction is bent away from the normal.

473. Specific inductive capacity. In the problems on electricity which have been hitherto solved in this treatise the non-conducting medium or *dielectric* which surrounds the conductors has been supposed to be air or some other gas. But the capacities thus determined do not agree with experiment when some solid non-conductor is substituted for the air. In this case the elements of the solid become excited in such a manner that each assumes a polarity analogous to the magnetic polarity induced in the substance of a piece of soft iron under the influence of a magnet. To take account of this state, called *polarisation*, we apply the same analysis as that used for induced magnetism.

We suppose each element dv of the dielectric to become an

elementary doublet whose poles are occupied by equal and opposite quantities of electricity. The direction of polarisation is that of the electric force F due to all causes and the intensity is $I = kF$. This polarity is then replaced by a fictitious stratum of electricity on the surface of the dielectric whose repulsive force at any point is equal to that of the polarised dielectric. One effect of the repulsion of this stratum is to alter the potential of each conductor and therefore to change its capacity.

The coefficient $1 + 4\pi k$ is called *the specific inductive capacity* of the dielectric and is generally represented by the letter K. It is evidently analogous to the permeability μ in the theory of magnetism. The two however differ in this particular; the specific inductive capacity of a dielectric is very approximately independent of the intensity of the electric force, while the permeability is not an absolute constant but varies with the magnetic force when that force is not small. The reader will find in J. J. Thomson's *Electricity and Magnetism* (Art. 154) a diagram which clearly exhibits the variations of μ produced by changes in the magnitude of the magnetic force.

A short table is given in Deschanel's treatise (edited by Everett) Art. 158 of the corresponding values of the magnetic force H, intensity I, and permeability μ, for a specimen of soft iron.

$H=$	0·3,	1·4,	3·5,	4·9,	10·2,	78,	585.
$I=$	3,	32,	574,	917,	1173,	1337,	1530.
$\mu=$	128,	299,	2070,	2350,	1450,	215,	34.

The values differ in different specimens. We notice that as the magnetic force increases, μ is at first nearly constant, then rapidly increases and arrives at a maximum and again decreases. The value of μ depends also on the temperature. At first it increases slowly with the temperature but at such high temperatures as 600° to 800° the rate of increase is very rapid. It then begins to decrease as rapidly as it rose.

The specific inductive capacities of the following substances are taken from J. J. Thomson's treatise, (Art. 67). Solid paraffin 2·29, sulphur 3·97, flint glass 6·7 to 7·4, distilled water 76, alcohol 26.

474. Effect of the substitution of a solid dielectric for air. Let there be any number of closed conductors A_1, A_2, &c. separated from each other by air as the non-conductor. Let E_1, E_2, &c. be the charges on the conductors, U_1, U_2, &c. the constant internal potentials, ρ_1, ρ_2, &c. the surface densities at any points Q_1, Q_2, &c. on the several conductors.

When the conductors are separated by a dielectric of specific inductive capacity K we represent the repulsions and attractions of the dielectric by equivalent strata placed on the boundaries of the conductors. Let their surface densities be respectively σ_1, σ_2, &c. We suppose that the dielectric is uniform so that there are no equivalent strata in the field except those on the surfaces of the conductors. If the dielectric have a boundary at an infinite distance both the force and the density vanish at that boundary.

Let us assume as a trial solution that $\sigma_1 = \lambda\rho_1$, $\sigma_2 = \lambda\rho_2$, &c. where λ is an unknown constant multiplier which is the same at every point of every conductor. The sum of the potentials of all the conductors at any point P in the field will then be changed by the introduction of the dielectric in the constant ratio 1 to $1+\lambda$. The potentials U_1, U_2, &c. will also be changed in the same ratio and will remain constant. The conditions of equilibrium will therefore not be disturbed (Art. 372).

The test that the trial assumption leads to a correct solution is that all the boundary conditions can be satisfied by the same constant value of λ. The conditions at the boundary of any conductor A are given in Art. 470. These are

$$\sigma = -kF, \quad KF + K'F' = 4\pi\rho,$$

where K has been written for μ. In our case, F and F' are the normal forces respectively just inside the dielectric and just inside the conductor. The latter being zero, we have

$$\sigma = -kF, \quad KF = 4\pi\rho, \quad \sigma = \lambda\rho.$$

Eliminating F and σ, and remembering that $1 + 4\pi k = K$, we have at once $1 + \lambda = 1/K$. Similar equations apply at the boundary of each conductor and give the same value of λ.

The result is that *the distribution of real electricity on the surfaces remains unaltered, but the potential inside each conductor is changed by the attractions and repulsions of the dielectric and reduced to $1/K$th part of what it was when the separating medium was air.*

475. *To find the change of force at any point.* Since the surface density of each equivalent stratum is λ times that of the real electricity at the same point, the force X' at any point P in the field, due to both the equivalent strata and the real electricity, must coincide in direction with the force X at the same point P due to the real electricity alone, and the magnitudes are such that $X' = (1+\lambda)X$. We therefore have $X' = X/K$.

If, when the dielectric is introduced to replace the air, *the potentials of the conductors are kept unaltered*, the charges of real electricity are increased in the ratio 1 to K and the force *at* any point will then be unaltered. The force which one conductor exerts on another will be increased in the ratio 1 to K.

The potential energy is $W=\frac{1}{2}\Sigma Vm$ (Art. 61) where m is the quantity of electricity on the conductor whose potential is V. It follows that the energy will be divided or multiplied by K according as the charges or the potentials are kept unaltered.

476. The case in which *one conductor A is entirely surrounded by a shell formed by another conductor B needs some special attention.* We suppose at first that there are no other conductors in the field. The separating medium being in the first instance air there is a distribution of electricity on the external surface S of A and the internal surface S' of B. Let the surface densities at any points Q, Q' be respectively ρ and ρ'. If the conductor B has no external boundary, but extends to infinite distances, the distributions on S and S' are such that the sum of their potentials is constant throughout all space external to S' and is the same as at an infinite distance. The potential at every point external to S' is therefore zero and the charges on S, S' are equal and opposite. We may now remove any portion we please of the neutral matter outside the surface S' and reduce the conductor B to a finite size.

In this state of the system, there is no electricity on the external boundary of the shell B. The potential of the system is zero within the substance of the conducting shell B and equal to some constant α within the conductor A. See Art. 386.

When *the whole space between A and B is filled with a dielectric*, we represent its repulsions by those of equivalent strata placed on the surfaces S, S'. Assuming, as before, that their densities are $\sigma=\lambda\rho$, $\sigma'=\lambda\rho'$, where λ is some constant, we find that the conditions at the boundary of A (viz. $\sigma=-kF$, $KF=4\pi\rho$) give immediately $1+\lambda=1/K$. The conditions at the other boundary give the same value of λ.

The result is that the distributions of real electricity on S and S' remain unaltered, but the potentials inside A and B are reduced to $1/K$th part of what they were when the medium was air. The potential inside B was zero and remains zero. The potential inside A becomes α/K.

The capacity of the conductor A (being measured by the ratio of the charge to the potential, when the conductor B is at potential zero, Art. 371) *is therefore K times as great as when the two conductors were separated by air.*

477. *Effect of external conductors.* Let us next suppose that the external surface S'' of the shell B is charged with electricity and that other conductors are placed in the field outside S''. These additions to the system will not disturb the equilibrium of the charges on the surfaces S, S', but will increase the potential throughout the interior of S'' by some constant β. Supposing the conductors A and B to be separated by air, the potentials inside B and A become β and $\alpha+\beta=\alpha'$.

The system thus formed (as explained in Arts. 389, 390) consists of two parts which are *independent of each other*. Let us therefore fill the space between the shell B and the conductor A with a dielectric of inductive capacity K, leaving the conductors outside the shell still separated by air. The distributions of electricity on S and S' are not disturbed by this change, the potential inside B remains equal to β, but that within A becomes $\alpha''=\beta+a/K$ where $a=\alpha'-\beta$. *The difference between the potentials of A and B is therefore decreased in the ratio 1 to K.*

The capacity of the conductor A (if measured by the ratio of the charge on A to *the difference* of the potentials of A and B) *is therefore K times as great as when A and B were separated by air.*

478. A plane dielectric. Two conducting plates of infinite extent are placed with their nearest plane faces A and B parallel to each other and at a distance θ. A plate C, of specific inductive capacity K and thickness t is introduced into the intervening space with its two faces parallel to the planes A and B, the space on each side of C being occupied by air. Find the effect of the introduction of the dielectric C on the capacity and potential energy of the system.

Let a, b be the distances of the faces L, L' of the dielectric C from the planes A, B, L being the nearest to A and L' to B, then $\theta=a+t+b$.

Let ρ and $\rho'=-\rho$ be the surface densities of the charges on the planes A and B. Let σ and $\sigma'=-\sigma$ be the surface densities of the strata on L and L' which are equivalent to the polarity of the dielectric.

At a point P between the planes A and L the force F, measured from A towards B, is constant and equal to $4\pi\rho$ (Art. 22). The constant force F', measured in the same direction, at a point R between L and L' is found from the condition that the induction is unchanged when P crosses the boundary of the dielectric (Art. 469), hence $KF'=F$. At a point Q between L' and B the force is again $F=4\pi\rho$.

Let α, β be the potentials at the planes A, B, and λ, λ' those at L, L'. The force at a point P distant x from A is $-dV/dx=F$, $\therefore V=\alpha-Fx$. Similar reasoning applies to the points Q and R. We have therefore

$$\lambda=\alpha-Fa, \qquad \lambda'=\lambda-F't, \qquad \beta=\lambda'-Fb.$$

Adding these three equations together and substituting for F, F' their values, we find

$$\begin{aligned}\beta-\alpha&=-4\pi\rho\,(a+b+t/K)\\ &=-4\pi\rho\,(\theta-t+t/K)\qquad\qquad(1).\end{aligned}$$

The capacity C (when measured by the ratio of the charge on either of the conductors A, B to *the difference* of their potentials) is given by

$$\frac{1}{C}=4\pi\left\{\theta-\left(1-\frac{1}{K}\right)t\right\}.$$

We notice that this is independent of the position of the dielectric C.

If the whole space between the plates A, B is filled with air, we have $t=0$ and the capacity is $1/4\pi\theta$. The capacity is therefore increased by the introduction of the dielectric C. When the dielectric C fills the whole space between the plates A, B, we have $t=\theta$ and the capacity is K times as great as when the separating medium was air.

The potential energy per unit of area due to the charges $\pm\rho$ on the plates is by Art. 61,
$$W=\tfrac{1}{2}\Sigma EV=\tfrac{1}{2}\rho\alpha-\tfrac{1}{2}\rho\beta=\tfrac{1}{2}\rho\,(\alpha-\beta).$$

We may express this result either in terms of ρ or $\alpha-\beta$. We have by (1),

$$W=2\pi\rho^2\,(\theta-t+t/K)=\frac{(\alpha-\beta)^2}{8\pi}\,\frac{1}{\theta-t+t/K}.$$

It follows that the introduction of the dielectric decreases or increases the potential energy according as the charge ρ or the difference of potentials is kept unaltered.

The force per unit of area which one conductor A exerts on the other B is $\tfrac{1}{2}F\rho$ (Art. 143). Since $F=4\pi\rho$ this becomes $2\pi\rho^2$. The force is therefore not changed by the introduction of the dielectric C provided the charges are kept unaltered.

If the difference of the potentials is kept unaltered, we substitute for ρ from equation (1). The force per unit of area is then $(\beta-\alpha)^2/8\pi(\theta-t+t/K)^2$.

479. A cylindrical dielectric. The outer and inner boundaries of two conductors A, B are infinite co-axial circular cylinders whose radii are a, b. A co-axial circular cylindrical dielectric C of specific inductive capacity K is introduced into the space between A and B, the rest of the space being filled with air. To find the effect of the shell C on the capacity and potential energy.

Let ρ, ρ' be the densities of the charges on the surfaces A, B of the conductors; σ, σ' those of the strata on C whose repulsions represent the forces due to the dielectric. Let the radii of the two surfaces L, L' of the shell C be a', b'; L being nearer A than B. Let α, β be the potentials at the conductors, λ, λ' those at the surfaces L, L'.

The repulsion of any one of these cylinders at an internal point is zero. At an external point the force varies inversely as the distance r from the axis and is equal to $2m/r$ where m is the charge per unit of length (Arts. 55, 56). For the cylinder A, $m=2\pi\rho a$.

The force at any point P between A and L is $4\pi\rho a/r$. Putting $r=a'$, and using the rule that the product of the force and K is unaltered when P passes into the dielectric (Art. 469), we see that the force just outside L is $4\pi\rho a/a'K$. The force at any point R between L and L' is therefore $4\pi\rho a/r'K$ where r' is the distance of R from the axis. Similarly the force at a point Q between L' and B is $4\pi\rho a/r''$ where r'' is the distance of Q from the axis.

We now find by easy integrations

$$\lambda-\alpha=-4\pi\rho a\log\frac{a'}{a},\qquad \lambda'-\lambda=-\frac{4\pi\rho a}{K}\log\frac{b'}{a'},\qquad \beta-\lambda'=-4\pi\rho a\log\frac{b}{b'}.$$

Adding these together we have

$$\beta-\alpha=-4\pi\rho a\left(\log\frac{a'}{a}+\frac{1}{K}\log\frac{b'}{a'}+\log\frac{b}{b'}\right)\ldots\ldots\ldots\ldots(1).$$

The capacity C per unit of length (measured by the ratio of the charge on A to the difference of potentials) is given by

$$\frac{1}{2C}=\log\frac{b}{a}-\left(1-\frac{1}{K}\right)\log\frac{b'}{a'}.$$

Since the whole quantity of matter given by Poisson's equivalent strata is zero (Art. 340), we have $\sigma a'+\sigma'b'=0$. Also since the potential of the whole system at any point external to B is constant, the quantity $(\rho a+\rho'b+\sigma a'+\sigma'b')\log r$ is independent of r, and this is impossible unless $\rho a+\rho'b=0$. *The charges on the conductors A and B are therefore equal and opposite.*

The potential energy per unit of length (Art. 61) is given by

$$W=\tfrac{1}{2}(2\pi\rho a\,\alpha+2\pi\rho' b\,\beta)=\pi\rho a(\alpha-\beta),$$

which can be expressed in terms of either the charge or the difference of potentials, by substituting from (1).

480. *A repelling point of mass E is placed at a point A in a medium of inductive capacity K; prove that the potential at any point P distant r from A is E/Kr.*

The point may be regarded as the limit of a small sphere of equal mass, radius a, whose specific inductive capacity is unity.

This sphere is then the inner boundary of the dielectric and the equivalent distribution on its surface must be taken into the account. The force due to the charge E at all points just inside the surface is E/a^2 and, a being small, all other forces in the field may be neglected. Since there is no real electricity on the sphere, the normal forces on each side are inversely as the specific inductive capacities (Art. 469). The force at all points just outside the sphere but within the substance of the dielectric is therefore $F = E/Ka^2$. The surface density σ of the stratum on the sphere is $\sigma = -kF$, and is therefore uniform. *The resultant repulsion of the charge E together with that of the uniform stratum is therefore* E/Kr^2 *at all points external to the sphere* (Art. 64).

If another charge of mass E' be at a point B distant r from A, we replace it by a small sphere of mass E' and radius b. The force on the sphere E' due to a uniform distribution of attracting matter on this sphere is zero, (Art. 65). *When therefore two point-charges, separated by a uniform dielectric, repel each other, the force is* EE'/Kr^2.

481. Problems on dielectrics. To find the effect of induction on a dielectric we have generally to begin with a trial solution. Sometimes we assume the density of the equivalent stratum on the boundary S of the dielectric to be an unknown constant multiple (say λ) of some quantity suggested by a corresponding problem when the dielectric is air (Art. 474). We can then deduce the potentials on each side of the equivalent stratum and determine the constant λ by using some one of the forms of the boundary condition.

In other cases it is more convenient to assume some expressions for the potentials Ω, Ω' due to the repulsions of the dielectrics; these must be suggested by the circumstances of the case. They must obviously satisfy the following conditions, (1) the functions Ω, Ω' must satisfy Laplace's equation at all points not occupied by attracting matter, and be finite and continuous each on its own side. If the medium on one side of S extend to infinity, the potential corresponding to that side must be zero at an infinite distance. (2) The two functions Ω, Ω' must be such that at every point of S,

$$\Omega = \Omega', \quad \mu_1 \frac{d}{d\nu}(\Omega + U) + \mu_2 \frac{d}{d\nu'}(\Omega' + U) = 0,$$

where $d\nu$, $d\nu'$ are elements of the normal to S, measured positively from S, and U is the potential of the influencing body. These are called *Poisson's conditions.* When Ω has been found, the magnitude and direction of the induced polarity follow from equations (1) of Art. 466. The surface density of the equivalent strata can be found by (4) of Art. 469.

482. Ex. *Find the polarisation induced in two media of capacities K_1, K_2 separated by a plane and acted on by an electric charge E situated at a point B which is in the first medium at a distance $BM=h$ from the separating plane.*

First solution. Produce BM to C and make $MC=h$, see the figure of Art. 412. Let (r_1, r_1'), (r_2, r_2') be the distances of any two points P_1, P_2 in the two media from B, C respectively. Assume as a trial solution that the potentials due to all causes at P_1, P_2 are

$$V_1=\frac{E}{K_1}\frac{1}{r_1}+\frac{M}{r_1'},\quad V_2=\frac{N}{r_2} \quad\text{.........................(1)},$$

where M, N are two unknown constants. These potentials are finite at all points unoccupied by matter, zero at infinity, and satisfy Laplace's equation. They must also satisfy the boundary conditions $V_1=V_2$ and $K_1F_1+K_2F_2=0$, at all points which make $r_1=r_1'=r_2$. We find by resolution

$$F_1=\left(-\frac{E}{K_1}+M\right)\frac{h}{r^3},\quad F_2=\frac{Nh}{r^3} \quad\text{.........................(2)},$$

$$\therefore \frac{E}{K_1}+M=N,\quad K_1\left(-\frac{E}{K_1}+M\right)+K_2N=0.$$

These equations give M and N, we therefore have

$$V_1=\frac{E}{K_1}\left(\frac{1}{r_1}+\frac{K_1-K_2}{K_1+K_2}\frac{1}{r_1'}\right),\quad V_2=\frac{2E}{K_1+K_2}\frac{1}{r_2}.$$

From these the values of the components of polarisation Il, Im, In follow at once, Art. 466.

The density σ of the equivalent layer on the boundary plane at a point P distant r from either B or C is given by

$$\sigma=-k_1F_1-k_2F_2=-\frac{1}{4\pi}\cdot\frac{K_2-K_1}{(K_2+K_1)K_1}\cdot\frac{2hE}{r^3}.$$

In forming the trial solution (1) we may assume that the potential at a point P_1 in the first medium is the sum of the potentials due to the electric point B and any imaginary electric points properly placed in *the other medium.* No electric point (other than the real point B) in the first medium can be used, because the potential would then be infinite at that point. Similarly in forming a trial potential at P_2 in the second medium, any suitable imaginary points situated in the first medium, but none in the second, may be used.

Second solution. Instead of assuming some values for the potentials V_1, V_2, we may take as our trial assumption some form for the density σ of the equivalent layer on the plane. By referring to Art. 412 we are led to the assumption $\sigma=\lambda\frac{2Eh}{4\pi r^3}$. The repulsion due to the stratum at any point P on either side is the same as that of a charge λE situated at a point (B or C) on the side *opposite to P.* The normal forces on each side of the separating plane (measured from that plane) due to the electric point at B and the stratum are therefore

$$F_1=\left(-\frac{1}{K_1}+\lambda\right)\frac{Eh}{r^3},\quad F_2=\left(\frac{1}{K_1}+\lambda\right)\frac{Eh}{r^3}.$$

The boundary condition $K_1F_1+K_2F_2=0$ gives at once $\lambda=-\frac{K_2-K_1}{(K_2+K_1)K_1}$. Since this value of λ is constant the trial solution is verified.

The value of σ thus found obviously agrees with that found in the first solution. The potentials are evidently $V_1=\frac{E}{K_1r_1}+\frac{\lambda E}{r_1}$, $V_2=\frac{E}{K_1r_2}+\frac{\lambda E}{r_2}$.

483. Effect of the substitution of a dielectric shell for some of the air. A conductor A is surrounded by another B at zero potential, the space between being occupied by air. Charges E and $-E$ being given to these bodies respectively, let V_0 be the potential inside A. Let a shell C in the space between A and B be bounded by two equipotential surfaces L, L' of the charges on A and B, L being the nearest to A. Let U, U' be the potentials at these surfaces. *If a dielectric of capacity K be substituted for the air in the shell C* (the rest of the space between A and B being still occupied by air) *the whole effect of the dielectric is to diminish the potential in the interior of A by* $\left(1-\frac{1}{K}\right)(U-U')$, see Art. 476. This theorem is due to Kelvin [reprint, &c. Art. 45].

When the separating medium is air the potential of the system at the interior surface S' of the conductor B and at every point without its surface is zero (as explained in Art. 476) while the potential at the surface and within the interior of A has some constant value α.

Let us place on the surfaces L, L' indefinitely thin layers whose surface densities σ, σ' are respectively given by $4\pi\sigma=\lambda F$ and $4\pi\sigma'=-\lambda F'$ where F, F' are the normal components of force due to the charges on A and B, both forces being measured from A towards B. The total masses of these layers are respectively λE and $-\lambda E$ (Art. 156).

Since L is a level surface, the potential due to the charge on it at any external point P is λV_1, where V_1 is the potential at P due to the charge E on A; and at an internal point Q its potential is $\lambda(U-V_2)$ where V_2 is the potential at Q due to the charge $-E$ on B (Art. 156). Similar remarks apply to the layer on the surface L' except that the sign of λ is altered.

Hence at any point P external to both L and L', the effect of the introduction of one stratum is to increase the potential by λV_1 and the effect of the other is to decrease the potential by the same amount. The potential at any external point is therefore unaltered.

At any point Q internal to both L and L' the potential is increased by the sum of $\lambda(U - V_2)$ and $-\lambda(U' - V_2)$. The potential is therefore increased by the constant quantity $\lambda(U - U')$.

At any point R between L and L' the potential is increased by the sum of λV_1 and $-\lambda(U' - V_2)$. The potential is therefore increased by $\lambda(V - U')$, where V is the potential at R due to the given charges on A and B, i.e. $V = V_1 + V_2$.

The introduction of these strata therefore increases the potential inside A by a constant quantity and does not alter the potential within the substance of B. *The electric equilibrium of the two conductors is therefore not disturbed.*

The layers placed on L and L' will be the equivalent strata of the dielectric C if the densities σ, σ' are respectively equal to $-k$ times the normal components of force due to all causes at points just within the two boundaries of the dielectric each measured from its own stratum. The potential at a point R just outside L being $V + \lambda(V - U')$ the outward normal force (obtained by differentiation) is $(1 + \lambda)F$. We therefore have the two equations

$$\sigma = -k(1 + \lambda)F, \quad 4\pi\sigma = \lambda F.$$

Hence $1 + \lambda = 1/K$. The conditions at the other boundary give the same value of λ.

The effect of the introduction of the dielectric is not to alter the level surfaces, but to decrease the potential α in the interior of A by a known quantity.

Since no restriction has been placed on the size of the external conductor B, we may replace it by a sphere of infinite radius. The charge on its surface being finite, we may then eliminate that conductor altogether from the field. Kelvin's theorem may therefore be applied when the shell C surrounds a single conductor A, provided the boundaries of C are equipotential surfaces.

484. Let us now suppose that the shell B and the conductor A are placed in a field of constant potential (see Art. 477), so that the potential at every point is increased by the same quantity β. The electrical equilibrium is not disturbed, but the potentials inside the conducting matter of B and A (when separated by air) become respectively β and $\beta + \alpha = \alpha'$. Each of the potentials U, U' is also increased by β, but their difference is not altered.

After the introduction of the dielectric shell C, the potential inside B remains β, while that inside A becomes $\alpha' - \left(1 - \frac{1}{K}\right)(U - U')$. *The capacity C* (if measured by the ratio of the charge Q on A to the *difference* of the potentials of A, B) *is then given by*

$$\frac{Q}{C}=\alpha'-\beta-\left(1-\frac{1}{K}\right)(U-U').$$

When *the dielectric fills the whole space between A and B*, we have $U=\alpha'$, $U'=\beta$; so that the effect of introducing the dielectric is *to multiply the capacity by K.*

The potential energy of the system is by Art. 61 equal to $\frac{1}{2}\Sigma EV$. The effect of introducing the dielectric shell C is to diminish the internal potential of the shell A, leaving that of B unaltered. The potential energy is therefore diminished by

$$\tfrac{1}{2}E\left(1-\frac{1}{K}\right)(U-U').$$

485. Ex. 1. A spherical shell (whose inner radius is c) and a solid concentric conducting sphere (radius a) are charged with quantities $\pm E$ of electricity. The space between is filled with two dielectrics separated by a third sphere of radius b. Prove that the capacity γ is given by $\frac{1}{\gamma}=\left(\frac{1}{a}-\frac{1}{b}\right)\frac{1}{K_1}+\left(\frac{1}{b}-\frac{1}{c}\right)\frac{1}{K_2}$. [Coll. Ex.]

This result follows at once from Kelvin's theorem (Art. 483). Let Q be the charge on the sphere of radius a. When the separating medium is air, the potentials V and U at the surfaces of the spheres a and b are

$$V=Q/a-Q/c,\quad U=Q/b-Q/c.$$

The effect of the dielectric is to reduce the potential within the sphere a to the value

$$V'=V-\left(1-\frac{1}{K_1}\right)(V-U)-\left(1-\frac{1}{K_2}\right)U.$$

The capacity required is Q/V'.

Ex. 2. A spherical conductor of radius a is surrounded by a concentric spherical conducting shell of radius b and the space between is filled with a dielectric of specific inductive capacity $\mu e^{-p^2}/p^3$ (where $p=r/a$) at a distance r from the centre. Prove that the capacity of the condenser so formed is $2\mu a/(e^c-e)$ where $c=b^2/a^2$. [Coll. Ex. 1896.]

Ex. 3. Prove that the capacity of two parallel plates, separated by air and placed at a distance apart equal to θ, will be increased n-fold by introducing between them a slab of substance whose specific inductive capacity is K and thickness $\frac{n-1}{n}\frac{K\theta}{K-1}$ where $n<K$. [Coll. Ex. 1900.]

Ex. 4. A condenser is formed of two parallel plates, whose distance apart is h, one of which is at zero potential. The space between the plates is filled with a dielectric whose specific inductive capacity K increases uniformly from one plate to the other. Prove that the capacity per unit area of the condenser is

$$\frac{K_2-K_1}{4\pi h}\Big/\log\frac{K_2}{K_1},$$

where K_1 and K_2 are the values of K at the surfaces of the plates, the inequality of the distribution at the edges of the plates being neglected. [Math. Tripos, 1899.]

Ex. 5. Three closed surfaces 1, 2, 3 are equipotentials of an electric field; if an air condenser is constructed with faces 1, 2, its capacity is A; with faces 2, 3 the capacity is B; if with 1, 3, the capacity is C. Prove $\frac{1}{C}=\frac{1}{A}+\frac{1}{B}$.

If a dielectric K fill the space 1, 2 and one K' fill 2, 3, prove that the capacity of the condenser having 1, 3 for faces is $\frac{1}{C}=\frac{1}{AK}+\frac{1}{BK'}$. [St John's Coll. 1898.]

Ex. 6. A condenser consists of two confocal ellipsoids, the squares of whose semi-axes are respectively a^2, b^2, c^2 and a^2+u, &c. If the dielectric be air and β the capacity for electricity, prove that

$$\frac{2}{\beta}=\int_0^u \frac{du}{\{(a^2+u)(b^2+u)(c^2+u)\}^{\frac{1}{2}}}.$$

If the dielectric be a solid arranged in ellipsoidal shells confocal with the conductors, and such that the specific inductive capacity of each shell is inversely proportional to the volume of the enclosed ellipsoid, prove that the capacity is $2Kabc/u$, where K is the specific inductive capacity of the innermost layer.

[St John's Coll. 1879.]

486. Ex. 1. A charge E is placed at a distance f from the centre of a sphere of S.I.C. K and outside the sphere. Prove that the potential at any point inside the sphere at a distance r from the centre is $\frac{E}{f}\Sigma\frac{2n+1}{Kn+n+1}\left(\frac{r}{f}\right)^n P_n$, where the summation extends from $n=0$ to ∞. [Coll. Ex. 1897.]

Let $\sigma=\Sigma A_n P_n$ represent the surface density of the charge equivalent to the polarity of the dielectric. We write P_n instead of Y_n because the system is symmetrical about the straight line joining the charged point to the centre of the sphere. The potentials due to this stratum are given in Art. 294 at points inside and outside the sphere. Adding to these the potentials of the external charge and using the equation (6) of Art. 469 we obtain the result to be proved.

Ex. 2. A sphere of S.I.C. K is placed in air, in a field of force due to a potential X_n (before the introduction of the sphere) referred to rectangular axes through the centre of the sphere, where X_n is a solid harmonic of degree n. Prove that the potential inside the sphere is $\frac{2n+1}{n+1+Kn}X_n$. [Coll. Ex. 1898.]

Ex. 3. Find the potential at any point when a sphere of specific inductive capacity K is placed in air in a field of uniform force.

A circle has its centre on the line of force which passes through the centre of the sphere and its plane perpendicular to this line of force. Prove that if the plane of the circle does not cut the sphere, the presence of the sphere increases the induction through the circle in the ratio $1+2\frac{K-1}{K+2}\sin^3\alpha$ to 1, where 2α is the angle of the enveloping cone drawn from any point on the circumference of the circle to the sphere. [Coll. Ex. 1896.]

Proceeding as in Ex. 1 we find that the potential due to all causes at any point outside the sphere is $V'=F\left(x-\frac{K-1}{K+2}\frac{a^3\cos\theta}{r^2}\right)$ where F is the given force of the field. The flux of force through the circle is then $-\int\frac{dV'}{dx}2\pi y dy$, Art. 107.

Ex. 4. A circular wire is situated in a uniform magnetic field, with its plane at right angles to the lines of force; prove that the effect of introducing into the middle of it a sphere of soft iron of permeability μ, which exactly fits its section, is to increase the induction through it in the ratio of 3 to $1+2/\mu$.

[By Art. 471 the induction is μ times the flux.] [St John's Coll. 1896.]

Ex. 5. A spherical shell of radii a, b ($b>a$) and specific inductive capacity K is placed in a field of uniform force F; prove that, if F_1 is the force in the space within both spheres, $\frac{F}{F_1}=1+\frac{2(K-1)^2}{9K}\left(1-\frac{a^3}{b^3}\right)$. [Coll. Ex. 1899.]

Ex. 6. An infinite solid with a plane face is acted on by a small magnet, of unit moment, situated at a point E outside the solid, the axis of the magnet being perpendicular to the plane face. Prove that the magnetic potential at any point P within the solid is $\dfrac{2\cos\theta}{(\mu+1)r^2}$ where $r=EP$, θ is the angle EP makes with the axis of the magnet and μ is the permeability of the solid. [Coll. Ex. 1897.]

We represent the repulsion of the solid by that of a thin stratum of variable density σ on its surface. The normal force at a point Q close to that surface is due ultimately to the repelling matter in the neighbourhood of Q and is therefore $2\pi\sigma$. If Z be the normal force due to the magnet, the condition at the boundary is

$$(2\pi\sigma+Z)\mu+(2\pi\sigma-Z)=0.$$

This gives by Art. 516 $\sigma=\dfrac{1}{2\pi}\dfrac{\mu-1}{\mu+1}\dfrac{d}{dz}\left(\dfrac{z}{r'^3}\right)$ where $r'=EQ$ and z is the distance of E from the plane. The potential due to a stratum z/r'^3 is given in Art. 412, that due to σ is then deduced by differentiation as explained in Art. 93, Ex. 3. Finally the given result is obtained by adding the potential of the magnet itself.

Ex. 7. A sphere of specific inductive capacity K and of radius a is held in air with its centre O at a distance c from a point A where there is a positive charge E. Prove that the resultant attraction on the sphere is

$$\tfrac{1}{2}\beta E^2\left(\frac{a}{c}\right)^3\left[\frac{1+\beta}{c^2-a^2}+\frac{2c^2}{(c^2-a^2)^2}-\frac{c}{a^3}(1-\beta^2)\left(\frac{a}{c}\right)^\beta\int_0^{a/c}\frac{x^{2-\beta}\,dx}{1-x^2}\right],$$

where $\beta=(K-1)/(K+1)$. [Math. Tripos, Part II. 1897.]

The potential at an internal point is given in Art. 486, Ex. 1, thence the surface density σ of the stratum equivalent to the polarity of the dielectric may be found by an obvious differentiation, Art. 466. If R be the distance of any elementary area dS of the sphere from A, the resultant force on the sphere is $X=\displaystyle\int\frac{E}{R^2}\frac{c-ap}{R}\sigma dS$. The expansion of $\dfrac{1-ph}{R^3}=\dfrac{1}{c^3}\Sigma(n+1)P_nh^n$, where $h=a/c$, is found by differentiating that for c/R (Art. 264) with regard to h. The integrations can be effected at sight by using Arts. 288 and 289. The series thus found for X agrees with that obtained by expanding in powers of a/c the result given in the enunciation.

Ex. 8. The space between two concentric conducting spheres is filled on one side of a diametral plane with dielectric of specific capacity K and on the other side with dielectric of specific capacity K'. The inner sphere is of radius a and has a charge E. Prove that the force on it perpendicular to this diametral plane is

$$\frac{K-K'}{(K+K')^2}\frac{E^2}{2a^2}. \qquad \text{[Coll. Ex. 1901.]}$$

The potential V in either dielectric is $\Sigma\left(A_nr^n+\dfrac{B_n}{r^{n+1}}\right)P_n$, but since V must be independent of θ both when $r=a$ and $r=b$ we find $V=A+B/r$. Since V has the same value on both sides of the diametral plane (Art. 481) for all values of r between $r=a$ and $r=b$, this formula, with the same values of A and B, gives the potential in both dielectrics. By Art. 470, we find that the real densities ρ, ρ' on the two halves of the sphere are given by $4\pi\rho=KB/a^2$, $4\pi\rho'=K'B/a^2$. Since $2\pi(\rho+\rho')a^2=E$, we find $\dfrac{\rho}{K}=\dfrac{\rho'}{K'}=\dfrac{E}{2\pi a^2}\dfrac{1}{K+K'}$. The pulling force on an element ρdS is $\frac{1}{2}\rho dS\,.\,(-dV/dr)$, which reduces to $2\pi\rho^2dS/K$. We now write $dS=2\pi a^2\sin\theta\,.\,d\theta$ and multiply by $\cos\theta$ to resolve the force parallel to x. The integral from $\theta=0$ to

$\frac{1}{2}\pi$ gives the resolved force on half the sphere. Interchanging K and K' we have the resolved force on the other half. The difference is the force required.

Ex. 9. A dielectric hemisphere of radius a and inductive capacity K is placed with its base in contact with the plane boundary of an otherwise unlimited conductor. Prove that the potential at any point of the field outside both the conductor and the dielectric is $V' = -4\pi\sigma\cos\theta\left(r - \frac{K-1}{K+2}\frac{a^3}{r^2}\right)$, where the origin is at the centre of the hemisphere, and σ is the surface density of the charge on the plane conductor at a great distance from the hemisphere. [Coll. Ex.]

487. Magnetic shells. Ex. 1. An iron shell (radii a, b, $a > b$) is placed in a field of uniform magnetic force f. Find the induced magnetism and the force X inside the hollow.

Put $\rho = \Sigma Y_n$, $\rho' = \Sigma Z_n$ for the surface densities on the spheres. Their potential within the material is

$$\Omega = 4\pi a\,\Sigma\frac{Y_n}{2n+1}\left(\frac{r}{a}\right)^n + \frac{4\pi b^2}{r}\Sigma\frac{Z_n}{2n+1}\left(\frac{b}{r}\right)^n.$$

The boundary conditions to be satisfied are

$$\rho = -kF_1 = -k\left(-f\cos\theta + \frac{d\Omega}{dr}\right), \qquad \rho' = -k\left(f\cos\theta - \frac{d\Omega}{dr}\right),$$

where a and b are to be written for r respectively after the differentiations have been effected. These show that $Y_n = 0$, $Z_n = 0$ except when $n = 1$. We find

$$Y_1 = \frac{3kf\cos\theta}{N}\left\{3 + 8\pi k\left(1 - \left(\frac{b}{a}\right)^3\right)\right\}, \qquad Z_1 = \frac{-9kf\cos\theta}{N},$$

where $N = 9(1+4\pi k) + 2(4\pi k)^2\left\{1 - \left(\frac{b}{a}\right)^3\right\}$. The potential V and force X inside the hollow, due to all causes, are

$$V = \frac{4\pi}{3}(Y_1 + Z_1)r - fx, \qquad X = \frac{9\mu f}{9\mu + (\mu-1)^2 \cdot 2(1-(b/a)^3)}.$$

Ex. 2. A solid uniform sphere (radius a) is placed in a uniform field of force whose potential is $-fx$, say the magnetic force of the earth. Prove that the potential of the induced magnetism at all external points is the same as that of a concentric simple magnet whose moment is $a^3 f\frac{\mu-1}{\mu+2}$.

Ex. 3. A small magnet of moment M is placed at the centre of an iron shell, radii a, b. Prove that the potential at any point external to the shell, due to all causes, is $\frac{M\cos\theta}{r^2}\,\frac{9\mu}{9\mu + 2(\mu-1)^2(1-p^3)}$ where $p = b/a$ and μ is the permeability of the shell. Thence show that if μ is great and p not nearly equal to unity, the potential is zero. In this case the induced magnetism on the shell neutralises that of the magnet at all external points.

488. *To find the surface integral of the magnetic induction through any closed surface S.*

To find the component of the magnetic induction at any point P in a direction PN, we construct a disc-like cavity at P which has its plane normal to PN. The normal component of the induction is then the same as the actual normal component of force at P due to all causes, (Art. 343).

To find the surface integral of the magnetic induction we remove a thin layer of matter all over the surface S or, at least, over that part of S which lies within the magnetic body. We shall now apply Gauss' theorem to the repelling matter situated within the internal boundary of this empty shell, i.e. within the surface S.

Since each magnetic molecule has two equal and opposite poles, and no magnet lies partly within and partly without the empty shell, the algebraic sum of the magnetic matter within the surface S is zero. For the sake of generality, let us suppose that there may be other repelling particles (besides the magnetism) situated within S. Let M be their total mass.

Let H be the magnetic force, B the magnetic induction and I the intensity of magnetisation at any point P of S. Let θ, θ' and i be the angles their directions respectively make with the outward normal at P. Then

$$B\cos\theta' = H\cos\theta + 4\pi I\cos i \quad\ldots\ldots\ldots\ldots(1).$$

Applying Gauss' theorem to the surface S, we have

$$4\pi M = \int B\cos\theta' dS \quad\ldots\ldots\ldots\ldots(2),$$

$$= \int (H\cos\theta + 4\pi I\cos i)\, dS \quad\ldots\ldots\ldots\ldots(3).$$

489. *Another proof.* We may also arrive at these results very easily, if we first replace the magnetism by Poisson's solid and superficial distributions. Let ρ' be the density of the solid distribution, $I\cos i$ the surface density. If the surface S lie wholly within the magnetic body, the superficial distribution on the body will be outside S. We then have by Gauss' theorem

$$4\pi\left(M + \int\rho' dv\right) = \int H\cos\theta\, dS \quad\ldots\ldots\ldots\ldots(\alpha).$$

Since Poisson's rule applies also to any portion of a magnetic body (Art. 340) we have also

$$\int\rho' dv + \int I\cos i\, dS = 0 \quad\ldots\ldots\ldots\ldots(\beta),$$

where the surface integral extends over the surface S. Eliminating ρ' we have

$$4\pi M = \int (H\cos\theta + 4\pi I\cos i)\, dS \quad\ldots\ldots\ldots\ldots(3).$$

If the surface S intersect the boundary of the magnetic body, we suppose $I=0$ at all points of S which are outside the body.

We must also include on the left-hand side of (α) that portion of the superficial density on the body which lies within S; let this portion be called J. At the same time we must add J to the left-hand side of (β), since $\int I\cos i\, dS$ only extends over that portion of the surface S which lies within the body. When therefore we eliminate $\int\rho' dv$, the quantity J also disappears and we again arrive at (3).

490. If there is no repelling matter besides the magnetism, $M = 0$. We then find that *the surface integral of the magnetic induction across any closed surface S is zero.*

491. If the magnetism is wholly induced we have $B=\mu H$, and $\theta'=i$, (Art. 471). We then have

$$4\pi M=\int \mu H \cos i\, dS \quad \text{.....................(4).}$$

We infer that *in a dielectric of specific capacity K the outward flux across a closed surface S of K times the normal force is equal to 4π times the repelling mass inside.* This is also called the outward induction across the surface S.

492. Let us apply the modified form (4) of Gauss' theorem to a Cartesian element of volume of a dielectric. The value of the right-hand side of (4) for the two faces perpendicular to x is (as explained in article 108) $d(KX)\,dy\,dz$. Treating the other faces in the same way and writing $M=\rho\, dx\,dy\,dz$, we find

$$4\pi\rho=\frac{d(KX)}{dx}+\frac{d(KY)}{dy}+\frac{d(KZ)}{dz}.$$

If we use the potential V, this becomes

$$-4\pi\rho=\frac{d}{dx}\left(K\frac{dV}{dx}\right)+\frac{d}{dy}\left(K\frac{dV}{dy}\right)+\frac{d}{dz}\left(K\frac{dV}{dz}\right) \quad \text{...(5),}$$

where ρ is the density of any real repelling matter which may occupy the space S independently of the Poisson volume density ρ' due to the presence of the heterogeneous dielectric.

If we write $I=kH$, (Art. 471), the equation (β) takes the form

$$\int \rho' dv+\int kH\cos\theta\, dS=0.$$

Applying this also to a Cartesian element we have

$$\rho'=\frac{d}{dx}\left(k\frac{dV}{dx}\right)+\frac{d}{dy}\left(k\frac{dV}{dy}\right)+\frac{d}{dz}\left(k\frac{dV}{dz}\right) \quad \text{.................(6),}$$

where ρ' is the density of Poisson's solid distribution.

The equation (α) becomes in the same way

$$-4\pi(\rho+\rho')=\frac{d^2V}{dx^2}+\frac{d^2V}{dy^2}+\frac{d^2V}{dz^2} \quad \text{.........................(7).}$$

Since $K=1+4\pi k$, any one of these three equations follows from the other two.

493. *To deduce the condition at the common boundary of two dielectrics from the modified Gauss equation* (4).

Let a thin stratum of repelling matter of surface density m separate two dielectric media of capacities K, K'; see Art. 470. We follow the same reasoning as in Art. 147, but writing KX, KY, KZ for X, Y, Z. If we take x normal to the separating surface we then have

$$(K'X'-KX)\,dy\,dz+\left(\frac{d(KY)}{dy}+\frac{d(KZ)}{dz}\right)t\,dy\,dz=4\pi m\,dy\,dz.$$

In the limit this becomes $\qquad K'X'-KX=4\pi m,$

where X, X' are the normal forces on each side of the separating stratum, both measured in the same direction, viz. from the medium K towards the medium K'. This of course is the same as the result arrived at in Art. 470.

We may put this argument in another way. *Let us enquire what form the equation* (5), viz.
$$\frac{d}{dx}(KX)+\frac{d}{dy}(KY)+\frac{d}{dz}(KZ)=4\pi\rho \quad\text{.......................}(5),$$
assumes when the specific inductive capacity changes from K to K' at any surface. Taking x normal to the surface we notice that dV/dx increases rapidly on crossing the surface, while dV/dy, dV/dz do not. The left-hand side of (5) is therefore ultimately reduced to its first term. Integrating from $x=0$ to $x=t$, we have
$$K'X'-KX=4\pi\rho' t=4\pi m.$$

494. *As an example*, let us consider the problem solved in Art. 482. At all points in the medium which contains the point charge E, the density $\rho=0$, except at that charge, while in the other medium $\rho=0$ at all points. We may therefore take as the trial values of the potential
$$V_1=\frac{L}{r}+\frac{M}{r_1'},\qquad V_2=\frac{N}{r_2},$$
since these satisfy equation (5) of Art. 492 at all points at a finite distance from E. To find L we apply (5) (or equation (4) of Art. 491 from which (5) was derived) to the points of space near the charge E. To avoid the difficulties of infinite terms, we shall choose the equation (4). Taking as the surface S a sphere whose centre is at E and whose radius is a, we have
$$4\pi E=K\int H\cos i\,dS=K\int\left(\frac{L}{a^2}-\frac{M}{r_1'^2}\cos i'\right)a^2d\omega,$$
where i' is the angle r_1' makes with the normal to the sphere. In the limit, when a is very small, we reject the term containing M. We immediately have $L=E/K$, and the solution may then be continued as in Art. 482.

We notice that, when a is not very small, the term containing M is zero by Gauss' theorem (Art. 106) because the point C from which r_1' is measured lies outside the surface S.

495. *To deduce from the extended form of Gauss' theorem an expression for the potential of an electric system.*

By Art. 61 the potential energy of a system of repelling particles is
$$W=\tfrac{1}{2}\Sigma Vm=\tfrac{1}{2}\int V\rho\,dv,$$
where V is the potential and ρ the density at the element of volume dv. If there be no repelling particles within the element, then for that element $\rho=0$. The integration extends throughout the volume of some closed surface S within which all the repelling particles lie. Substituting for ρ its value given in Art. 492, we have
$$W=-\frac{1}{8\pi}\iiint V\left\{\frac{d}{dx}\left(K\frac{dV}{dx}\right)+\frac{d}{dy}\left(K\frac{dV}{dy}\right)+\frac{d}{dz}\left(K\frac{dV}{dz}\right)\right\}dx\,dy\,dz,$$
where K is the specific inductive capacity of the medium which occupies the element $dxdydz$.

We now integrate each term by parts, following Green's method, (Art. 149). We have

$$\int V\frac{d}{dx}\left(K\frac{dV}{dx}\right)dx=\left[KV\frac{dV}{dx}\right]-\int K\left(\frac{dV}{dx}\right)^2 dx,$$

where the square brackets imply that the term is to be taken between the limits of integration. These are represented by A to B in the figure of Art. 149. Treating all the terms in the same way, we have

$$8\pi W=-\int KV\frac{dV}{dn}d\sigma+\int K\left\{\left(\frac{dV}{dx}\right)^2+\left(\frac{dV}{dy}\right)^2+\left(\frac{dV}{dz}\right)^2\right\}dv.$$

If the integration extend throughout a sphere of large radius R, the product VdV/dn is of the order $1/R^3$ while $d\sigma$ is of the order R^2. The surface integration therefore vanishes when the integration extends throughout all space. We thus find

$$W=\frac{1}{8\pi}\int K\left\{\left(\frac{dV}{dx}\right)^2+\left(\frac{dV}{dy}\right)^2+\left(\frac{dV}{dz}\right)^2\right\}dv.$$

Ex. Find the potential energy of the system described in Art. 478.

We have $8\pi W=\int KF^2dx$, where $F=-dV/dx$ and W is the energy per unit of area.

Between A and L, $F=4\pi\rho$, $K=1$ and the limits of integration are $x=0$ to a. Between L and L', $F=4\pi\rho/K$, and the limits are $x=a$ to $a+t$. Between L' and B, $F=4\pi\rho$, $K=1$ and the limits are $x=a+t$ to θ. Outside A and B, $F=0$. Effecting these integrations and adding the results, we arrive at the result given in Art. 478.

In the same way the energy of the cylindrical condenser described in Art. 479, is given by

$$8\pi W=\int_a^{a'}F^2dv+\int_{a'}^{b'}(F^2/K)dv+\int_{b'}^{b}F^2dv,$$

where $F=4\pi\rho a/r$ and $dv=2\pi r\,dr$. This evidently reduces to the result given in the article just referred to.

THE BENDING OF RODS.

Introductory Remarks.

1. OUR object in this chapter is to discuss the stretching, bending, and torsion of a thin rod or wire. We may define a rod as a body whose boundary is a tubular surface, of small section. The surface is therefore generated by the motion of a small plane area whose centre of gravity describes a certain curve and whose plane is always normal to the curve. The curve is generally called *the central axis* or *central line* of the rod.

The rod or wire is to be so thin, that, *so far as the geometry of the figure is concerned*, it may be regarded as a curved line having a tangent and an osculating plane. Although this limitation will be generally assumed it will be seen in the sequel that some of the theorems apply to rods of considerable thickness. *It is not proposed to enter into the general theory of the elasticity of solid bodies*, except where it is necessary for the elucidation of the point under discussion, and even then the reference will be restricted as far as possible to the most elementary considerations.

2. In general the deformation of the body will be regarded as very small, so that each element of the body is only slightly strained from its natural shape. It will therefore be assumed that the whole effect, when properly measured, of any number of disturbing causes may be obtained by superimposing their separate effects.

3. By reference to Art. 142 of the first volume of this treatise, it will be seen that the action across any section C of a thin rod AB consists of a force and a couple. On this is founded the mathematical distinction between a string and a rod. The action across any section of the former is a force, called its tension, which acts along the tangent to the string, Vol. I., Art. 442. In the case of a rod the force may act at any angle to the tangent and there is in addition a couple.

4. Let P be any point of a body, let a closed plane curve be described round P of indefinitely small area, and let this area be ω. If the body is a fluid it is the fundamental principle of hydrostatics that the action between the fluid on one side and the fluid on the other side of the area ω consists of a force whose direction is perpendicular to the plane of the area. It is thence deduced that the magnitude of this force or pressure is the same for all inclinations to the horizon of the elementary curve provided its area remains unaltered. If the body is an elastic solid, the action across the plane is also a force, but its direction is not necessarily perpendicular to the plane of the area and its magnitude is not necessarily the same for all inclinations of the plane.

In discussing the mechanics of a rod, its cross section, though very small, is not to be regarded as infinitely small. If we divide any section into elementary areas, the action across each element will be an elementary force, and the resultant of all these will be, in general, a force and a couple, Vol. I., Art. 142.

The Stretching of Rods.

5. *To determine the simple stretching of a straight rod by a force applied at one extremity, the other being held fast.*

The relation which exists between the force and the extension of the rod has already been discussed in the first volume of this treatise under the name of Hooke's law. If l_1, l be the unstretched and stretched lengths of the rod, ω the area of the section of the unstretched rod, $T\omega$ the tension, then $\dfrac{l-l_1}{l_1}=\dfrac{T}{E}$, where E is a constant depending on the material of the rod and is usually called Young's modulus.

When a rod is stretched we know by common experience that its breadth and thickness are also altered. These lateral changes follow a law similar to Hooke's law except that the modulus E is not necessarily the same as that for extension. The study of these lateral contractions belongs properly to the theory of elasticity and only a simple case will be considered here.

6. The substance of a homogeneous body is called *isotropic* when the properties of a solid of *any* given form and dimensions cut from it are the same whatever directions its sides may have in the body. The substance is called *æolotropic* when the properties of the solid depend on the directions which its sides have in the body. We shall suppose that the material of which the rod is composed is isotropic.

7. Theory of a stretched rod. Let the unstretched rod form a cylinder with a cross section of any form and size. When stretched the rod becomes thinner, so that the several particles undergo lateral as well as longitudinal displacements. There is one fibre or line of particles which is undisturbed by the lateral contraction. Let this straight line, which we may regard as the central line, be taken as the axis of x, and let the origin be at the fixed extremity of the rod. We suppose that the stretching forces at the two ends are distributed over the extreme cross sections in such a manner that after the rod is stretched these sections continue to be plane and perpendicular to the central axis. It will appear from the result that the force at each end should be equally distributed over the area.

Let x, y, z be the coordinates of any particle P in the unstrained solid, $x+u$, $y+v$, $z+w$ the coordinates of the same particle P' of matter in the deformed body. Then u, v, w are such functions of x, y, z that the equations of equilibrium of all the elements of the solid are satisfied. We shall now prove that if we take $u=Ax$, $v=-By$, $w=-Bz$ all the equations of equilibrium may be satisfied by properly choosing the constants A and B. According to this supposition the external boundary of the stretched rod will be a cylinder and the particles of matter which occupy any normal cross section of the unstrained rod will continue to lie in a plane perpendicular to the axis when the rod is stretched.

Let $PQRS$ be any rectangular element of the unstrained solid having the faces PQ and RS perpendicular to the central axis. By the given conditions of the question this element assumes in the strained solid a form $P'Q'R'S'$ in which all the angles are still right angles and the sides parallel to their original directions. The direction of the stress across each face of the strained element is therefore perpendicular to that face. To measure these forces we refer each to a unit of area. Let N_x, N_y, N_z be the forces, so referred; let these act on the three faces which meet

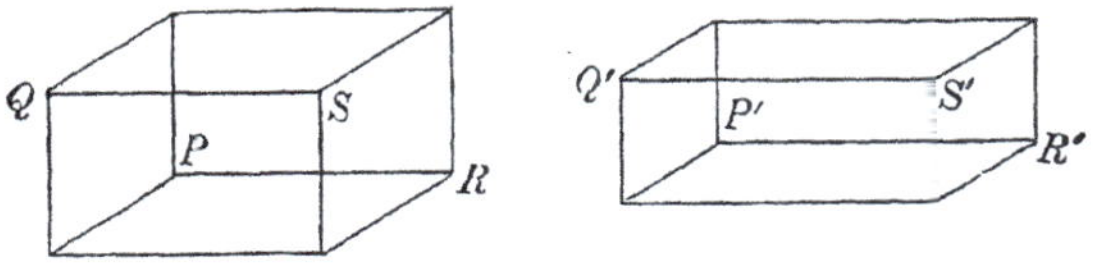

at the corner P' and are respectively perpendicular to the axes of x, y, z; we shall regard these forces as positive when (like the tension of a string) they pull the matter on which they act, and as negative when (like a fluid pressure) they push.

Let a, b, c and $a(1+\alpha)$, $b(1+\beta)$, $c(1+\gamma)$ be the sides of the element before and after the deformation. Then N_x, N_y, N_z are functions of α, β, γ, see Art. 489, Vol. I. We shall expand these functions in ascending powers of α, β, γ and since we here confine our attention to a first approximation, we shall neglect all the higher powers of α, β, γ. Assuming the lowest powers in the expansion to be the first, we have
$$N_x=\kappa\alpha+\lambda(\beta+\gamma),$$
the coefficients of β and γ being the same because the medium is isotropic. For the same reason the stress N_y must be the same function of β and γ, α, that N_x is of α and β, γ. Thus
$$N_y=\kappa\beta+\lambda(\alpha+\gamma).$$
In the same way N_z may be derived from N_x by interchanging α and γ. To make these more symmetrical, it is usual to write them in the form
$$N_x=2\mu\alpha+\lambda(\alpha+\beta+\gamma), \qquad N_y=2\mu\beta+\lambda(\alpha+\beta+\gamma), \qquad N_z=2\mu\gamma+\lambda(\alpha+\beta+\gamma).$$

The constants λ and μ are the same as those chosen by Lamé to measure the elastic properties of a solid; see his *Leçons sur la théorie mathématique de l'élasticité des corps solides.*

8. In the problem under consideration the sides dx, dy, dz of the unstrained element become $dx+du$, $dy+dv$, $dz+dw$. It follows that $\alpha=\frac{du}{dx}$, $\beta=\frac{dv}{dy}$, $\gamma=\frac{dw}{dz}$. Substituting the assumed values of u, v, w, we have

$$N_x=2\mu A+\lambda(A-2B),\qquad N_y=-2\mu B+\lambda(A-2B),\qquad N_z=-2\mu B+\lambda(A-2B).$$

These values are independent of x, y, z, so that the opposite faces of any element *wholly internal* are acted on by equal and opposite forces. *It follows that every internal element is in equilibrium.*

Consider next the elements which have one or more of their faces on the boundary of the rod. Such faces must be parallel to the central axis and in a vacuum are not acted on by any pressure. It is therefore necessary for their equilibrium that the constant forces represented by N_y and N_z should be zero. We therefore have
$$\frac{B}{A}=\frac{\lambda}{2(\lambda+\mu)},\qquad N_x=\frac{(3\lambda+2\mu)\mu}{\lambda+\mu}A.$$

Since Ax is the extension, By the contraction of a rod of length x and breadth y and N_x is the stretching force per unit of area of the section, it follows that

$$\frac{\text{increase of length}}{\text{original length}}=\frac{\lambda+\mu}{\mu(3\lambda+2\mu)}N_x,\qquad \frac{\text{decrease of breadth}}{\text{original breadth}}=\frac{\lambda}{2\mu(3\lambda+2\mu)}N_x.$$

Comparing the first of these with the statement of Hooke's law given in Vol. I. Art. 489, we see that the constant E, usually called Young's modulus, is the reciprocal of the coefficient of N_x. If E' be the corresponding coefficient for the decrease of breadth we have
$$E=\frac{\mu(3\lambda+2\mu)}{\lambda+\mu},\qquad E'=\frac{2(\lambda+\mu)}{\lambda}E.$$

It follows from this solution that when a rod has been stretched, each fibre (or column of particles parallel to the central axis) is stretched and contracted independently of the others and exerts no action on the neighbouring fibres. The total force required to produce a given extension is therefore independent of the form of the cross section provided its area remains unaltered.

In this investigation the action across each of the six faces of the element is normal to that face. In many problems in elastic solids this simplicity does not exist and there are tangential actions also across the faces. For the discussion of these questions the reader is referred to *A Treatise on the Mathematical Theory of Elasticity*, by A. E. H. Love, 1892.

9. Ex. 1. Show that E and $\frac{1}{2}E'$ are the forces which would stretch a rod of unit section to twice its original length and half its original breadth respectively. Show also that E' is greater than $2E$.

If the stretching tension be N_x, v the volume, δv the increase of volume, prove that
$$\frac{\delta v}{v}=\frac{E'-2E}{EE'}N_x.$$

Ex. 2. If the side faces of the rod are exposed to a uniform normal pressure equal to p per unit of area, prove that the force required to produce a given extension is less than that in a vacuum by $\lambda p/(\lambda+\mu)$ per unit of area of cross section.

The Bending of Rods.

10. *To form the equations of equilibrium of a thin inextensible rod bent in one plane.*

First Method. In this method we consider the conditions of equilibrium of a finite portion of the rod or wire. The method has been used in Vol. I. Arts. 142—147 to determine the stress at any point of a rod naturally straight and slightly bent by the action of given forces, and the same reasoning may be applied to rods whose natural forms are curved.

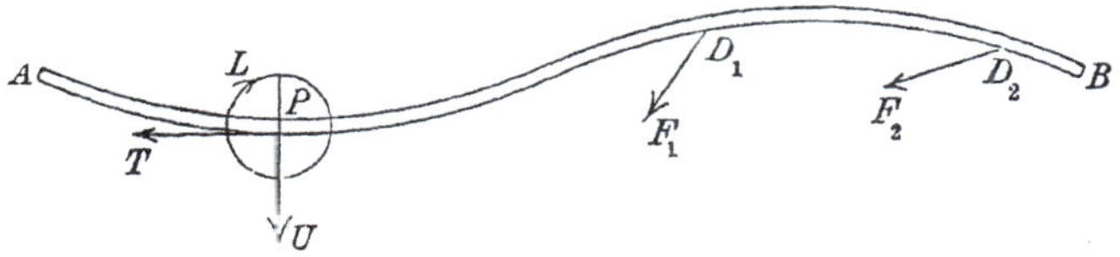

Let P be any section of a thin rod APB regarded as a curved line. Let T and U be the resolved parts of the stress force along the tangent and normal at P, and let L be the stress couple. These represent the mutual action of the two parts AP, PB of the rod on each other. These stresses are then obtained by considering the conditions of equilibrium of the portions AP, PB separately. Let F_1, F_2 &c. be forces acting at the points D_1, D_2 &c. of the portion PB in directions making angles δ_1, δ_2 &c. with the tangent at P. Taking any directions along the tangent and normal at P as positive, let T and U act on the portion PB in these directions; we then have by resolution

$$T + \Sigma F \cos \delta = 0, \quad U + \Sigma F \sin \delta = 0.$$

In the same way if p_1, p_2 &c. be the perpendiculars from P on the lines of action of the forces, we have by moments $L + \Sigma Fp = 0$. These three equations determine T, U and L when the form of the curve is known.

11. *Second Method.* In this method we form the equations of equilibrium of an elementary portion of the rod or wire.

Let PQ be any element of the rod and let the arc s be measured from some fixed point D on the rod up to P in the direction AB, so that $s = DP$. Let the stress forces of AP on PB be represented by a tension T acting, when positive, in the direction PA and a shear U acting in the direction opposite to the radius of curvature PC. Then the stress forces of QB on QA are represented by $T+dT$ in the direction QB and $U+dU$ in the direction QC, these directions being represented in the figure by the double arrow heads. Let the stress couple at P on PB be represented by L, the

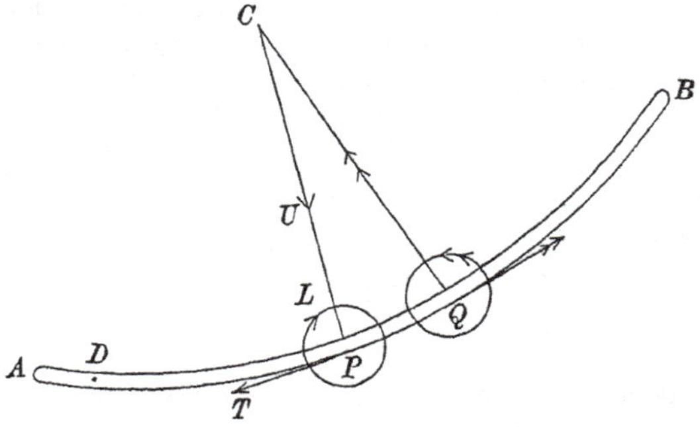

positive direction being indicated by the arrow head on the circle at P; then the stress couple at Q on AQ is represented by $L+dL$ acting in the opposite direction, i.e. in that indicated by the double arrow heads on the circle at Q. Let Fds, Gds be the impressed forces on the element PQ resolved in the direction of the tangent PQ and normal PC, taken positively when acting respectively in the directions in which the arc s and the radius of curvature ρ are measured. Let $d\psi$ be the angle between the tangents at P and Q, and let ψ be so measured that ψ and s increase together.

Resolving the forces in the direction of the tangent and normal at P, we have

$$-T+(T+dT)\cos d\psi-(U+dU)\sin d\psi+Fds=0,$$

$$-U+(U+dU)\cos d\psi+(T+dT)\sin d\psi+Gds=0.$$

In the limits these become

$$dT-Ud\psi+Fds=0\ldots\ldots\ldots\ldots(1),$$

$$dU+Td\psi+Gds=0\ldots\ldots\ldots\ldots(2).$$

Also taking moments about P

$$-L+(L+dL)+(U+dU)\,ds+\tfrac{1}{2}\,Gds\,(\tfrac{1}{2}\,ds)=0,$$

$$\therefore\ dL+Uds=0\ldots\ldots(3).$$

Writing $d\psi = ds/\rho$, these equations take the form

$$\left.\begin{aligned} \frac{dT}{ds} - \frac{U}{\rho} + F &= 0 \\ \frac{T}{\rho} + \frac{dU}{ds} + G &= 0 \\ \frac{dL}{ds} + U \qquad &= 0 \end{aligned}\right\} \quad \ldots\ldots\ldots\ldots\ldots\ldots(4).$$

If each element of the rod is acted on by an impressed couple, as well as by the impressed forces Fds, Gds, it must be taken account of in the equation of moments. Let Ids be its moment taken positively when the couple acts on the element PQ in the opposite direction to the couple L. We then add Ids to Uds in equation (3) and therefore add I to the left-hand side of the last of equations (4).

The positive directions. The positive direction of the couple L at P on that part of the rod towards which the arc s is measured is opposite to that in which the angle $d\psi = ds/\rho$ is measured. The positive direction of the shear U on the same part of the rod is opposite to that in which the radius of curvature is measured. The positive direction of the tension at P on the same part of the rod is opposite to that in which the arc s is measured.

12. When we compare the advantages of the two methods of solution we notice that the second gives differential equations which must be integrated, and the constants must be determined by the conditions at the extremities. On the other hand the first method, though it gives expressions for T, U and L, introduces into the equations the action of all the forces on the finite arc PE. When, therefore, the form of the strained curve is so well known that we can calculate the resolved parts and the moments of the impressed forces the first method gives the required stresses at once. When however the form of the strained curve is very different from that when unstrained, and is itself unknown, the second method presents several advantages over the first.

13. Experimental Results. When a thin rod or wire is bent under the action of forces we have to determine not merely the components of stress, i.e. T, U and L, but also the form of the strained rod. The equations of equilibrium found above supply three equations, so that a fourth is required to make up the necessary number. For this purpose we have recourse to experiment, Vol. I., Art. 148. If ρ_1, ρ are the radii of curvature at any point P before and after the deformation, the stress couple L is given by

$$L = K\left(\frac{1}{\rho} - \frac{1}{\rho_1}\right) \ldots\ldots\ldots\ldots\ldots\ldots(5),$$

where K is some constant depending on the material of which the rod is made and on the section at P. It is usually called the *flexural rigidity* of the rod. This expression for L agrees very well with the results of experiment when the change of curvature is not very great.

Since the moment L represents the product of a force and a length, it is evident that the dimensions of K are represented by a force multiplied by the square of a length. If E be Young's modulus for the material of the rod and ω the area of the section, $E\omega$ will represent a force, so that the constant K is often written in the form $K = E\omega k^2$, where k is a length.

It will be shown further on that *in certain cases* ωk^2 is the moment of inertia of the area of the normal section about a straight line drawn through its centre of gravity perpendicular to the plane of bending. This result does not agree so well with experiment as that represented by (5).

14. It is hardly necessary to remind the reader of the remarks made in Vol. I. Art. 490, on *the limits to the laws of elasticity*. When the stretching or bending of the rod exceeds a certain limit, the rod does not tend to return to its original form, but assumes a new natural state different from that which it had at first. In all the reasoning in which the equation (5) is used, it is assumed that the bending is not so great that the limit of elasticity has been passed.

15. The theoretical considerations which tend to prove the truth of the equation (5) *depend on the theory of elasticity and therefore lie somewhat outside the scope of the present chapter.* As however this theory clears up some of the difficulties which belong to the bending of rods, it does not seem proper wholly to pass it over. One case can be presented in a simple form, and that case will be discussed a little farther on after the use of the equation (5) has been explained.

16. The work of bending an element. *To find the work done by the stress couple L when the curvature of an element of the rod is increased from its natural value $1/\rho_1$ to the value $1/\rho_2$.*

Let PQ be an element of the central line and let ds be its length. As PQ is being bent, let ψ be the angle between the tangents at its extremities; let ρ be its radius of curvature. If ψ_1 be the value of ψ when the rod has its natural form, the stress couple L is

$$L = K\left(\frac{1}{\rho} - \frac{1}{\rho_1}\right) = K\frac{\psi - \psi_1}{ds}.$$

The work done by the couple L when ψ is increased by $d\psi$ is $-Ld\psi$, (see Vol. I. Art. 292). The negative sign is given to the expression because, as explained in Art. 11, L is measured in the direction opposite to that in which ψ is measured. The whole work done by the couple when ψ is increased from ψ_1 to ψ_2 is therefore equal to $$-\tfrac{1}{2}K \cdot \frac{(\psi_2-\psi_1)^2}{ds}.$$

Replacing ψ_2, ψ_1 by their values in terms of ρ_2, ρ_1, we see that the work Wds done by the couple L may be written in either of the forms $$Wds = -\tfrac{1}{2}K\left(\frac{1}{\rho_2}-\frac{1}{\rho_1}\right)^2 ds = -\frac{L^2 ds}{2K}.$$

If the change of curvature at every point of the rod is known, the whole work done by the stress couples in the rod may be found by integrating the first of these expressions along the length of the rod. *If however the change of curvature is unknown, and the couple is given,* the work is found by integrating the latter expression.

Resilience. Resilience denotes the quantity of work that a spring, or elastic body, gives back when strained to some stated limit and then allowed to return to the condition in which it rests when free from strain. The word "resilience" used without special qualifications may be understood to mean *extreme resilience* or the work given back by the spring after being strained to the extreme limit within which it can be strained again and again without breaking or taking a permanent set. See Kelvin's article on "Elasticity" in the *Encyc. Brit.* 1878.

17. Deflection of a straight rod. *A heavy rod, originally straight, rests on several points of support A, B, C &c. arranged very nearly in a horizontal straight line, and is slightly deflected both by its own weight, and by a weight W attached to a point H between B and C. It is required to explain the method of finding the deflection at any point of the rod and to determine the relations which exist between the stresses at successive points of support.*

Let A, B, C be three successive points of support. These are so nearly on the same level that the distances $AB=a$, $BC=b$, may be regarded as equal to their projections on a horizontal straight line. To simplify the process of taking moments the order of the letters used is exhibited in the upper figure, as if they were all strictly in a horizontal line, instead of being only very nearly so.

Let x be measured horizontally from B in the direction BC. The rod, when bent by its weight, will assume the form of some curve which differs very slightly from the nearly straight line ABC. Let y be the ordinate at any point Q, between B and C, *measured positively upwards*, from a horizontal straight line drawn through B and let the radius of curvature be positive when the concavity is upwards. The stress couple at the point Q is K/ρ; when ρ is positive the fibres of the under part of the rod are stretched while those above are compressed, hence the stress couple at Q acts on QC in the clock direction and on BQ in the opposite direction. Let the shear at Q be U and let its positive direction when acting on QC be downwards.

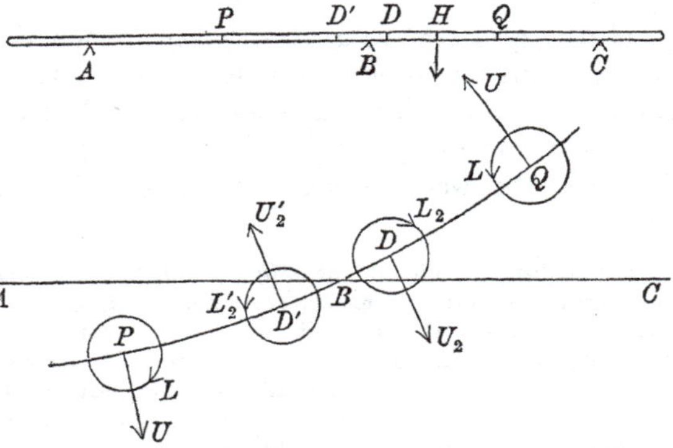

Let L_2 and U_2 be the couple and shear at a point D indefinitely near to B on its right-hand side. Let w be the weight of the rod per unit of length, then the weight of DQ is wx, and this weight acts at the centre of gravity of DQ. Let $BH = \xi$. Taking moments about Q for the finite portion DQ of the rod, we have

$$\frac{K}{\rho} = L_2 - U_2 x - \tfrac{1}{2} w x^2 - W(x - \xi) \quad \text{...........(1).}$$

The term containing W is to be omitted when Q is on the left-hand side of H, i.e. when $x < \xi$.

In forming the right-hand side of this equation the rod has been supposed to be straight and horizontal, because the deflections are so small that only a very small error is made by neglecting the curvature. If this were not so, the shear would not be vertical, and the arm of its moment would be different from that used in the equation. In the same way the thickness of the rod has been neglected, and in all its geometrical relations the rod is regarded as if it were a line coincident with its central axis, Art. 1.

The rod is supposed to be of such material that *a considerable effort is required to produce a slight curvature; the coefficient K is therefore large.* On the left-hand side of the equation all the small terms cannot be rejected because these are multiplied by K. It is however sufficient, in a first approximation, to retain only the largest of these small terms. We therefore put

$$\frac{1}{\rho} = \pm \frac{d^2y}{dx^2}\left\{1 + \left(\frac{dy}{dx}\right)^2\right\}^{-\frac{3}{2}} = \pm \frac{d^2y}{dx^2}.$$

The upper sign must be taken because ρ is measured positively when the concavity is upwards, and in this case dy/dx is increasing and therefore d^2y/dx^2 is positive.

The general rule followed in these problems is, (1) *that all terms not containing K are formed on the supposition that the rod has its natural shape,* (2) *that in all terms containing K as a factor only the first power of the deflection y is retained.* The natural shape in our case is a horizontal straight line.

18. The equation (1) now takes the form

$$K\frac{d^2y}{dx^2} = L_2 - U_2x - \tfrac{1}{2}wx^2 - W(x-\xi) \ldots\ldots\ldots\ldots(2)$$

where x is restricted to lie between $x=0$ and $x=b$ and the term containing W is to be omitted when $x<\xi$. Let L_2' and U_2' be the stress couple and shear at a point D' indefinitely near B on its left-hand side, and let R_2 be the pressure of the point of support B on the rod upwards. Consider the equilibrium of the small portion $D'D$ of the rod. The stress couples and the stress forces at the extremities act on this element in the directions opposite to those represented in the figure, the weight wds acts downwards and the pressure R_2 upwards. We have, by taking moments about D', and resolving,

$$\left.\begin{aligned} L_2' - L_2 = U_2ds - \tfrac{1}{2}w(ds)^2 + R_2 . BD' = 0 \\ U_2' - U_2 - R_2 = -wds = 0 \end{aligned}\right\}.$$

Hence in the limit

$$L_2' = L_2, \qquad U_2' - U_2 = R_2 \ldots\ldots\ldots\ldots\ldots\ldots(3).$$

If we take a point P between A and B so that BP represents a negative value of x, we have $K\dfrac{d^2y}{dx^2} = L_2' - U_2'x - \tfrac{1}{2}wx^2$ ...(4),

where x is restricted to lie between $x=0$ and $x=-a$. Since $L_2' = L_2$ this equation differs from (2) only in having U_2' written for U_2, the term in W disappearing naturally.

Lastly, if U be the shear at any point of the rod we have by equation (3) of Art. 11 $\quad U = -\dfrac{dL}{dx} = -K\dfrac{d^3y}{dx^3}$ ....(5).

It is evident that the two arcs AB, BC of the rod must have the same tangent at B and therefore the same value of dy/dx. It follows from the first of equations (3) that the stress couples on each side of B are equal; *the two arcs have therefore the same curvature.* But the shears on each side of B differ by the pressure R_2, and therefore *there is an abrupt change in the value of* d^3y/dx^3 *at a point of support.*

These equations are sufficient to determine the stresses when the terminal conditions are known. But the integrations must be effected for each span separately and the conditions at the points of junction allowed for. To shorten the mathematical labour we require *some method of passing over a point of discontinuity.* This is effected by the theorem or equation of the three moments, by which a relation is found between the stress couples at any three successive points of support.

19. Equation of the three moments*. Let us integrate (2) over the length BQ of the rod. The limits for every term, except the last, are $x = 0$ to x, and for the last term $x = \xi$ to x. We thus have

$$K\left(\frac{dy}{dx} - \beta\right) = L_2x - \tfrac{1}{2}U_2x^2 - \tfrac{1}{6}wx^3 - \tfrac{1}{2}W(x-\xi)^2 \text{ ......(6)}$$

where β is the inclination of the rod at B to the horizon. Integrate again,

$$K(y - \beta x) = \tfrac{1}{2}L_2x^2 - \tfrac{1}{6}U_2x^3 - \tfrac{1}{24}wx^4 - \tfrac{1}{6}W(x-\xi)^3 \text{......(7).}$$

Eliminating U_2 between (2) and (7) and writing $L = Kd^2y/dx^2$, we have

$$6K(y - \beta x) = (L + 2L_2)x^2 + \tfrac{1}{4}wx^4 + W\xi(x-\xi)(2x-\xi) \text{ ...(8).}$$

This equation holds at any point Q between H and C. When Q lies between B and H, the term with W is to be omitted.

Since U_2 does not appear in the equation, it also holds when Q

* The theorem of the three moments in its first form is due to Clapeyron, *Comptes Rendus*, 1857, Tome XLV.; but it has been greatly extended since then. A sketch of these changes is given by Heppel in the *Proceedings of the Royal Society*, 1870, vol. XIX. The extension to include the case of variable flexural rigidity is due to Webb, *Proceedings of the Camb. Phil. Soc.* 1886, vol. VI. The allowance for the yielding of the supports is given by K. Pearson, *Messenger of Mathematics*, 1890, vol. XIX.

lies between A and B, provided x is then regarded as negative. Let y_1 and y_3 be the altitudes of the points of support A and C above B. The equation (8) becomes when $x = b$ and $x = -a$

$$6K(y_3 - \beta b) = (L_3 + 2L_2)b^2 + \tfrac{1}{4}wb^4 + W\xi(b-\xi)(2b-\xi)$$
$$6K(y_1 + \beta a) = (L_1 + 2L_2)a^2 + \tfrac{1}{4}wa^4.$$

Eliminate β and we find

$$6K\left(\frac{y_1}{a} + \frac{y_3}{b}\right) = L_1 a + 2L_2(a+b) + L_3 b$$
$$+ \tfrac{1}{4}w(a^3+b^3) + W\frac{\xi(b-\xi)(2b-\xi)}{b} \quad\text{............(9).}$$

Here w is the weight of a unit of length of the rod. If the spans AB, BC have unequal values of w, say w_1 and w_3, we write $\frac{1}{4}(w_1 a^3 + w_3 b^3)$ for the fourth term on the right-hand side.

This important relation between the stress couples at any three successive points of support is usually called *the equation of the three moments.* By help of this equation, when the stress couples at two of the points of support are known, the stress couples at all the points may be found.

To find the shear at the point B of support, we take moments about either C or A. We then have

$$L_3 = L_2 - U_2 b - \tfrac{1}{2}wb^2 - W(b-\xi) \quad\text{............(10)}$$
$$L_1 = L_2 + U_2' a - \tfrac{1}{2}wa^2 \quad\text{..........................(11)}$$

which may also be derived from (2) by putting $x = b$ and $x = -a$. The pressure R_2 on the point of support may then be found by (3).

If the point H at which the weight W is attached lie between A and B instead of B and C, we reverse the positive direction of x. Let the distance $BH = \xi'$ measured positively from B towards A. The last term of (9) must then be replaced by

$$W\xi'(a-\xi')(2a-\xi')/a.$$

This may also be derived from the last term of (9) by writing $-\xi'$ for ξ and $-a$ for b.

20. The equation of the three moments when written in the form (9) may be regarded as *the relation between the ordinates y_1, y_3 of any two points and the stress couples L_1, L_3 at those points.* It may be used, for example, to find the deflection y_3 at the free end of a rod where $L_3 = 0$.

21. If the rod rest on n points of support, the equation of the three moments supplies $n-2$ equations connecting the n stress couples $L_1, L_2, \ldots L_n$ at the points of support. Two more equations are therefore necessary to find the n couples, and these may be deduced from the conditions at the extremities.

If one end of the rod is free, and at a distance c from the nearest point of support, the stress couple L_n at that point of support is found, by taking moments about it, to be $L_n = -\frac{1}{2}wc^2$.

If an extremity rest on a point of support the stress couple at that point is zero.

If an extremity be built into a wall so that the tangent to the rod at that point is fixed in a horizontal position we may imagine that the fixture is effected by attaching that end of the rod to two points of support indefinitely close together. The required condition at that end then follows at once from the equation of the three moments. Let L_{n+1} be the couple at the wall, L_n that at the nearest point of support and let c be the distance, then writing $a = c$, $b = 0$ in the equation of the three moments we have

$$L_n + 2L_{n+1} + \tfrac{1}{4}wc^2 = 0.$$

The pressures on the points of support may be obtained by combining equations (10), (11) and (3). If R_2 be the pressure on the rod measured *upwards* at B, we find by eliminating U_2, U_2'

$$\frac{L_1}{a} - L_2\left(\frac{1}{a} + \frac{1}{b}\right) + \frac{L_3}{b} = R_2 - \tfrac{1}{2}w(a+b) - W\frac{b-\xi}{b} \ldots (12).$$

The case in which $W = 0$ has also been attained in Vol. I. Art. 145.

The weight W has been included in the equation of the three moments to facilitate the calculations. It is however evident that we may regard the point of the rod to which the weight W is attached as a point of support at which the pressure is known. Such a point may be included in the equation of the three moments as one of the three consecutive points A, B, C. The deflection at each of these points being unknown, the extended equation of the three moments fails to determine the stress couple. But the pressure being known, the equation (12) gives an additional equation connecting the stress couples, and the extended equation of the three moments then gives the deflection.

Yielding of the supports. In some cases the points of support are the tops of vertical columns which are themselves elastic. Let the bases of the columns be on the same level, h_1, h_2 &c., z_1, z_2 &c. their original altitudes and their altitudes under pressure; σ_1,

σ_2 &c. their sectional areas, then for any column $h - z = hR/E\sigma$. We have therefore the additional equations

$$y_1 = z_1 - z_2 = \frac{R_2}{p_2} - \frac{R_1}{p_1}, \quad y_3 = \frac{R_2}{p_2} - \frac{R_3}{p_3}, \text{ \&c.}$$

where $p_1 = E_1\sigma_1/h_1$ &c. and E is Young's modulus.

22. Ex. 1. A uniform rod, weight W, is supported at its extremities; the deflection at its middle point is observed and found to be h. Show that the value of the constant K for the rod is given by $48h \,.\, K = 5a^3W$, where $2a$ is the length of the rod. If the inclination to the horizon of the tangent at either end of the rod be observed by a level and found to be θ, show that the value of K is also given by $K = a^2W/6\theta$.

This example shows how the value of K may be found by experiment for any given rod.

Ex. 2. A uniform heavy rod is supported at its extremities A, C and at its middle point B; A and C are at the same level and B such that the pressures on the three supports are equal. Prove that the depth of B below AC is 7/15ths of the whole central deflection of the beam AC when supported only at its ends.

This example shows that when a long heavy bridge is supported on three columns of equal strength, their summits ought not to be on the same level.

Ex. 3. A heavy rod rests on a series of points of support which are very nearly in a horizontal line. Let A, B be any consecutive two of these points, a their distance apart, y_1, y_2 their altitudes above a horizontal plane. Let L_1, L_2 be the stress couples, θ_1, θ_2 the inclinations of the rod to the horizon at A, B. Prove that

$$K(\tan\theta_2 - \tan\theta_1) = \tfrac{1}{2}(L_1 + L_2)\,a + \tfrac{1}{12}wa^3,$$
$$K(y_2 - y_1 - a\tan\theta_1) = \tfrac{1}{6}(2L_1 + L_2)\,a^2 + \tfrac{1}{24}wa^4.$$

The stress couples having been found, the first of these equations enables us to find the inclination of the rod at any point of discontinuity when the inclination at some point is known. The second determines the inclination at any one point.

Ex. 4. A uniform slightly elastic rod rests on five supports in the same horizontal line, two at the ends and one at each of the points found by dividing the rod into four equal parts. The second and fourth supports from either end are now removed. Prove that the ratio of the new to the old pressure on an end support is as 21 : 11. [Coll. Ex. 1893.]

Ex. 5. A uniform bridge of weight W' formed of a single uniform plank is supported at its ends: a man of weight W stands on the bridge at a point whose distances from the ends are a and b. Prove that the deflection just under the man is

$$ab\,\{W'(a^2 + 3ab + b^2) + 8Wab\}/24\,(a+b)\,E,$$

where E is the bending modulus. [Coll. Ex. 1893.]

Ex. 6. A naturally straight weightless wire of flexural rigidity C has its ends A and B built in horizontally at the same level, and is slightly bent by a weight W attached to it at a point Q. Prove that the deflection y at a point P in AQ is given by the equation

$$Cy = \frac{1}{6}\frac{W}{AB^3}\,BQ^2 \,.\, AP^2\,(3AQ \,.\, BP - BQ \,.\, AP).$$

Find the points of inflection of the axis of the wire and show that the point at which the axis is horizontal is in the longer segment, and that its distance from the corresponding support is bisected by one of the inflections. [St John's Coll. 1893.]

Ex. 7. A uniform heavy beam rests on three points of support, A and C at its ends and B at the middle. The middle support is at first so placed at a depth y_1

below AC that the beam is entirely supported by A and C. The support B is then gradually raised to a height y_2 above AC such that the beam is wholly supported by B. Prove that as B is being raised, the pressure at B is proportional to the height raised. Prove also that the ratio $y_1 : y_2$ is equal to $5 : 3$. [Fidler's *Treatise on Bridge Construction*, 1893.]

23. Britannia Bridge. Ex. 1. A uniform heavy beam ABC is supported at its extremities A, C and at its middle point B, and the three points are in one horizontal line. Prove that 3/16ths of the weight is supported at either end and 5/8ths at the middle point. We notice that the pressure at the middle support is more than three times that at either end.

Prove also that the stress couple is a maximum at a point which divides either span in the ratio of 3 : 5, but the stress couple at either of these points is 9/16ths of the stress couple at the central point of support. Prove that the latter is equal to the stress couple at the middle point of a beam supported at each end whose length is equal to that of either span.

Prove that there is a point of contrary flexure in each span dividing it in the ratio 1 : 3.

Ex. 2. A uniform beam is supported at its extremities and at two other points dividing the beam into three equal spans, all the four points being on the same level. Prove that the pressures on the supports are in the ratios 4 : 11 : 11 : 4.

Ex. 3. A uniform beam $ABCDE$ is supported at its extremities A, E and at three points B, C, D, all five being on the same horizontal line. To assimilate this problem in some measure to the case of the Britannia Bridge the two middle spans are supposed to be twice the lengths of the outside ones, i.e. $BC=CD=2AB=2DE$. Prove that the pressures on A, B, C are in the ratios 4 : 27 : 34.

The examples in this article are taken from a treatise on *The Britannia and Conway Tubular Bridges* by Edwin Clark, resident engineer, 1850.

The tubes AB, BC, CD, DE, which form the four spans of the Britannia Bridge, were raised separately into their proper places and then rigidly connected into one long tube. The connections at B and D were such that the tubes adjacent to each had a common tangent. The junction at C was however so arranged that the tangents to BC and CD should make a small angle with each other. The object of this was to diminish the inequality between the pressure on C and that on either B or D. It was found convenient to make the angle between the tangents equal to $2\tan^{-1}\cdot 002$. In Example 3, given above, the analytical condition to be satisfied at C is that the tangents to BC and CD should be continuous, but in the bridge the condition is that these tangents should make a known small angle with each other.

24. Ex. 1. A rod without weight is supported at its extremities A, C and at some other point B, all three being in the same horizontal line. Given weights P, Q are suspended at the points D, E, bisecting AB and BC. Show that the inclination to the horizon of the tangent at A and the deflection y at the weight P are given by

$$32(a+b)K\tan\alpha = -Pa^2(a+2b)+Qab^2,$$

$$768(a+b)Ky = -P(7a+16b)a^3+9Qa^2b^2,$$

where $AB=a$, $BC=b$.

It appears from this result that when the point of support B bisects AC and $Q=3P$ the tangent at A should be horizontal. Moseley describes three experiments with different rods supported on knife edges by which this curious result has been verified. See his *Mechanical Principles of Engineering and Architecture*, 1855.

Ex. 2. A uniform thin rod of length $2(a+b)$ rests on two points of support in a horizontal line whose distance apart is $2a$. Show that if the middle point and the two free ends are on the same horizontal line, b/a must be the positive root of the cubic $3r^3+9r^2-8r-5=0$.

Ex. 3. A uniform heavy rod rests on any number of points of support in the same horizontal line. Let A, B, C, D, E be any consecutive five of these, and let their distances apart be a, b, c, d. Prove that the pressures R_2, R_3, R_4, at B, C, D are connected by the linear relation $\alpha R_2+\beta R_3+\gamma R_4=\frac{1}{4}w\delta$, where

$$\alpha=a^2(b+c)(c+d)(b+c+d),$$
$$\beta=(a+b)(c+d)\{b^2(d+2c)+2bc(a+d)+c^2(c+2b)+ad(b+c)\},$$
$$\gamma=d^2(b+c)(a+b)(a+b+c),$$
$$\delta=(a+b)(b+c)(c+d)(b+c+d)(a+b+c)(a-b-c+d).$$

Ex. 4. Prove that the deflection y at any point Q between B and C, in Ex. 3, is given by

$$-6Kby=BQ\,.\,CQ\,\{L_2(b+CQ)+L_3(b+BQ)+\tfrac{1}{4}wb\,(b^2+BQ\,.\,CQ)\}.$$

Ex. 5. A wire is bent into the form of a circle of radius c, and the tendency at every point to become straight varies as the curvature. Show that, if it be made to rotate about any diameter with a small angular velocity ω, it will assume the form of an ellipse whose axes are $2c\left(1\pm\frac{m\omega^2c^4}{12\mu}\right)$, m being the mass of a unit of length, and μ/c the couple necessary to bend the straight line into the circle. [Math. T. 1868.]

Ex. 6. A heavy elastic flexible wire originally straight is soldered perpendicularly into a vertical wall. If the deflection is not small prove that the difference between the tension at any point P and the weight of a portion of the wire whose length is the height of P above the free end is proportional to the square of the curvature at P. [May Exam.]

Ex. 7. A flexible wire is pushed into a smooth tube forming an arc of a circle, and lies in a principal plane of the tube; prove that it will only touch it in a series of isolated points, and that if it only touch the inner circumference at one point, the pressure there will be $4E\cos\alpha(\sin\alpha-\sin\gamma)/a^2\sin^2\alpha$, where a is the inner radius of the tube, 2α the angle subtended at the centre by the wire, γ the angle at which either end of it meets the wire, and E the coefficient of flexibility. [Math. T. 1871.]

Ex. 8. Three very slightly flexible rods are hinged at the extremities so as to form a triangle, and are attracted by a centre of force attracting according to the law of nature situated in the centre of the inscribed circle. Show that the curvature of any side, as AB, at the point of contact of the inscribed circle varies as

$$\frac{\cos\frac{1}{2}A+\cos\frac{1}{2}B-\cos\frac{1}{2}C}{\cos\frac{1}{2}C}.$$

Ex. 9. Equal distances AB, BC, CD are measured along a light rod which is supported horizontally by pegs at B, D below the rod and C above. A weight is now hung on at A, producing at that point a deflection. Find how much B must be moved horizontally towards A that the deflection may be unaltered when the peg D is removed. [Coll. Exam. 1888.]

25. Ex. 1. A uniform heavy rod rests symmetrically on $2m+1$ supports placed at equal distances apart, and the altitudes are such that the weight of the rod is equally distributed over the supports. Show that the altitude y_p of the support distant pa from the middle point, is given by

$$\frac{24(2m+1)K}{wa^4}(y_p-y_0)=(2\beta-1)p^4-\{(2\beta-1)(6m^2-1)+(6\beta^2-1)(2m+1)\}p^2,$$

where a is the distance between two consecutive points of support and βa is the length of the rod beyond either of the terminal supports.

We first see by taking moments about the pth support that the stress couple L_p at that point is a quadratic function of p. The extended equation of the three moments is $L_p+4L_{p+1}+L_{p+2}+\frac{1}{2}wa^2=(y_{p+2}-2y_{p+1}+y_p)\,6K/a^2$.

By an easy finite integration, or by the rules of algebra, it follows that y_p is a biquadratic function of p. Since there can be no odd powers of p, we have

$$y_p-y_0=Ap^4+Bp^2.$$

The values of A and B are then found by applying the equation of the three moments to any two convenient spans.

Ex. 2. A uniform heavy rod rests on m supports placed at the same level at equal distances a from each other, one being at each end. Prove that the stress couple at the nth point of support is

$$L_n=\tfrac{1}{12}wa^2\left\{-1+\frac{(1-k^{m-1})h^{n-1}-(1-h^{m-1})k^{n-1}}{h^{m-1}-k^{m-1}}\right\},$$

where h and k are the roots of $h^2+4h+1=0$. Prove also that the pressure on the nth support is $aR_n=3wa^2-6L_n$ except when $n=1$ or m.

The equation of the three moments is an equation of differences and may be solved in the usual manner by assuming $L_n=A+Bh^n+Ck^n$. The constants B, C are determined by the conditions that $L_n=0$ when $n=1$ and $n=m$. It is also evident that $h=-\tan\frac{1}{12}\pi$, $k=-\cot\frac{1}{12}\pi$.

26. Flexural rigidity not constant. If the rod is not uniform the equation of the three moments takes a more complicated form. We shall first suppose the flexural rigidity K to vary from point to point of the rod, but the weight per unit of length to remain constant. We start as before with the equation (2), Art. 18. Let us multiply this equation by $(b-x)/K$ and integrate over the length BQ. Since

$$\int_0^b (b-x)\frac{d^2y}{dx^2}\,dx=y_3-b\beta,$$

where y_3 and β have the same meaning as before, we find

$$y_3-b\beta=\int_0^x (L_2-U_2x-\tfrac{1}{2}wx^2)\frac{b-x}{K}\,dx-W\int_\xi^x\frac{(x-\xi)(b-x)}{K}\,dx\;\ldots\ldots(\text{I.}).$$

Substituting for U_2 from (10), this becomes

$$y_3-b\beta=L_3B+L_2B'+wB''+W\,.\,G,$$

where $$B=\int_0^b\frac{x(b-x)}{bK}\,dx,\quad B'=\int_0^b\frac{(b-x)^2}{bK}\,dx,\quad B''=\int_0^b\frac{x(b-x)^2}{2K}\,dx,$$

$$G=(b-\xi)B-\int_\xi^b\frac{(x-\xi)(b-x)}{K}\,dx.$$

The left-hand side of (I.) is the elevation of the point C of support above the tangent at B. The equation obtained by integrating over the length AB is similarly

$$y_1+a\beta=L_1A+L_2A'+wA'',$$

where A, A', A'' are obtained from B, B', B'' by writing a for b.

Eliminate β and we have

$$\frac{y_1}{a}+\frac{y_3}{b}=L_1\frac{A}{a}+L_2\left(\frac{A'}{a}+\frac{B'}{b}\right)+L_3\frac{B}{b}+w\left(\frac{A''}{a}+\frac{B''}{b}\right)+W\frac{G}{b},$$

which is the equation of the three moments when the flexural rigidity is not uniform.

When the weight w per unit of length is not constant, we may include the weight in the term W. We put $W=w\,dx$ and integrate that term throughout each of the spans.

27. A bent bow. *A uniform inextensible rod, used as a bow, is slightly bent by a string tied to its extremities. It is required to find its form.*

Taking the string as the axis of x, the statical equation is evidently

$$\frac{K}{\rho}=\pm K\frac{d^2y}{dx^2}=Ty \qquad\qquad (1),$$

where T is the tension of the string. Let A, B be the extremities of the rod, C a point on the rod at which the tangent is parallel to the string. Let OC be the axis of y. Then since dy/dx vanishes when $x=0$ and decreases algebraically as x increases, d^2y/dx^2 is negative. In forming (1), ρ has been taken as positive, we must therefore give the second term the negative sign. Putting $T=Kn^2$ for brevity, the equation gives

$$y=h\cos nx \qquad\qquad (2),$$

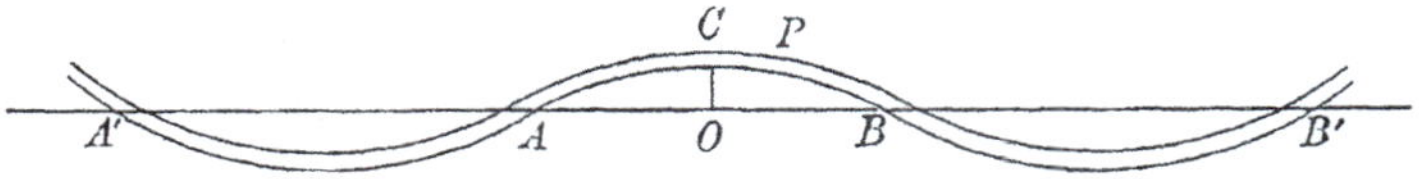

where h is the versine of the arc formed by the bow. It is obvious that unless the conditions of the problem make h small, we cannot reject the terms containing $(dy/dx)^2$ in the expression for ρ in equation (1).

The form of the curve given by the equation (2) is sketched in the diagram. It appears therefore that the bow may take the form ACB, the string being attached at A and B. It may also take the form ACB' with the string attached at A and B' and so on.

28. We may easily find a second approximation to the solution of the differential equation. This is perhaps necessary, for, owing to the smallness of the inclination of the rod to the string, if the ordinates near B were slightly decreased, a considerable change might be made in the distance OB.

If we substitute for ρ its full value, the differential equation becomes

$$-\frac{d^2y}{dx^2}=n^2y\left\{1+\left(\frac{dy}{dx}\right)^2\right\}^{\frac{3}{2}} \qquad\qquad (3).$$

Expanding the right-hand side we have

$$\frac{d^2y}{dx^2} + n^2y = -\tfrac{3}{2}n^2y\left(\frac{dy}{dx}\right)^2.$$

We see that the terms on the right-hand side are of the third order of small quantities. We therefore assume as a trial solution

$$y = k\cos cnx + Bk^3\cos 3cnx \quad \text{..............(4)}$$

where k is a small quantity analogous to h, and c, B are as yet undetermined constants. Substitute in the differential equation and neglect all powers of k above the third, we then have

$$(1-c^2)\,n^2k\cos cnx + (1-9c^2)\,Bk^3n^2c^2\cos 3cnx$$
$$\doteq -\tfrac{3}{2}n^2(k\cos cnx)(k^2c^2n^2\sin^2 cnx)$$
$$= -\tfrac{3}{8}n^4k^3c^2\{\cos cnx - \cos 3cnx\}.$$

The equation is therefore satisfied if we put

$$1-c^2 = -\tfrac{3}{8}n^2k^2c^2; \quad \therefore c = 1 + \tfrac{3}{16}n^2k^2, \quad B = -\tfrac{3}{64}n^2.$$

The solution to the third order of small quantities is therefore

$$y = k\cos cnx - \tfrac{3}{64}n^2k^3\cos 3cnx \quad \text{..............(5)}$$

where c exceeds unity by the small quantity $\frac{3}{16}n^2k^2$. Let, as before, h represent the distance OC; we have $y=h$ when $x=0$, hence

$$h = k - \tfrac{3}{64}n^2k^3 \quad \text{.....................(6).}$$

Let the lengths of the string and the rod be $2a$ and $2l$, then when $x=a$, $y=0$, and the *least value of* a is given by $cna = \frac{1}{2}\pi$. We also have

$$l = \int_0^a \left\{1 + \left(\frac{dy}{dx}\right)^2\right\}^{\frac{1}{2}} dx = a\,(1 + \tfrac{1}{4}n^2k^2c^2) \text{...........(7)}$$

when terms above the order k^3 are neglected. Eliminating a, we have

$$l = \frac{\pi}{2n}(1 - \tfrac{3}{16}n^2k^2)(1 + \tfrac{1}{4}c^2n^2k^2) = \frac{\pi}{2n}(1 + \tfrac{1}{16}n^2k^2) \quad \text{...(8)}$$

when the fourth powers of k are neglected.

Since $n^2 = T/K$ we have $\left(\dfrac{T}{K}\right)^{\frac{1}{2}} = \dfrac{\pi}{2l}\left(1 + \dfrac{1}{16}\dfrac{\pi^2k^2}{4a^2}\right)$.

Hence
$$T = \frac{\pi^2K}{4l^2}\left(1 + \frac{\pi^2h^2}{32l^2}\right) \quad \text{...................(9)}$$

when the fourth powers of h/l are rejected. *This equation determines the tension necessary to produce a given deflection* $OC = h$.

29. Let us regard the half CB of the bow as *a uniform rod having one end* C *and the tangent at* C *fixed while the other end* B *is acted on by a force* T *whose direction is parallel to the tangent at* C.

Let the length l be given, then the equation (8) shows that k is imaginary unless l exceeds $\pi/2n$. Let $n_0 = \pi/2l$ and let $T_0 = Kn_0^2$ be the corresponding value of the force T. It follows that *the rod cannot begin to bend unless the force exceeds* T_0, *where* $T_0 = K\pi^2/4l^2$.

Let $T = T_0(1+\xi)$ where ξ is a small quantity, then

$$n = n_0(1 + \tfrac{1}{2}\xi).$$

The equation (8) gives

$$n = n_0(1 + \tfrac{1}{16}n_0^2k^2); \qquad \therefore k^2 = 32l\,\xi/\pi^2.$$

Since $dk/d\xi$ is infinite when $k=0$, we see that k (and therefore also h) increases much more rapidly than the force does. *A slight increase in the force makes a considerable change in the value of k.*

30. When the terms containing dy/dx are included in equation (1), we have

$$-K\frac{y''}{(1+y'^2)^{\frac{3}{2}}} = Ty \quad \text{.................(10)},$$

where accents denote differential coefficients with regard to x. Multiplying by y' and integrating, we find

$$1 - \cos\psi = \tfrac{1}{2}n^2(h^2 - y^2) \quad \text{................(11)},$$

where ψ is the acute angle made by the tangent at any point P with the string of the bow.

Let $y = h\cos\phi$, then $\sin\tfrac{1}{2}\psi = \tfrac{1}{2}nh\sin\phi$. The equation may be written in the form $d\psi/ds = n^2y$. Put $e = \tfrac{1}{2}nh$ substitute for y and ψ and integrate between the limits $s=0$ to $s=l$, we then have

$$nl = \int_0^{\frac{1}{2}\pi} \frac{d\phi}{(1 - e^2\sin^2\phi)^{\frac{1}{2}}} \quad \text{..................(12)}.$$

If the length l and the force T are given, $n^2 = T/K$ is also known. This equation then determines e and therefore h.

The integral (12) is lessened by writing unity for the denominator. We then have $nl > \tfrac{1}{2}\pi$. Since $n^2 = T/K$ it immediately follows that *the tension or force must exceed the value of* $\pi^2K/4l^2$. This is the result already arrived at in Art. 29, and it has now been proved without the use of series. The equations (8) and (9) of Art. 28 may be obtained by expanding the integral (12) in powers of e^2 and neglecting all powers of e above the second.

31. The importance of the case considered in Art. 29 lies in *its application to the theory of thin vertical columns.* The rod may be regarded as a vertical column having the tangent at its lower end C fixed in a vertical position, while a weight, much greater than that of the column, is supported on the upper extremity. It appears from what precedes that if the weight on the summit is gradually increased, the column will remain erect, without bending, until the weight becomes nearly equal to a certain quantity depending on the flexibility and dimensions of the column.

Since the constant K is equal to $E\omega k^2$ (Art. 13) it follows that the bending weight, for columns of the same kind, varies as the fourth power of the diameter directly, and as the square of the length inversely. This result is usually called Euler's* law.

Columns yield under pressure in two ways, first the materials may be crushed, and secondly the column may bend and then break across. In some cases both effects may occur at once. If the column is short it follows from Euler's law that the bending weight is large, so that short columns yield by crushing. Long columns on the other hand break by bending and are not crushed.

Many experiments have been made to test the truth of Euler's law. The results have not been altogether confirmatory, possibly because Euler's law applies only to uniform thin columns, in which the central line in the unstrained state is a vertical straight line. For the details of these experiments we must refer the reader to works on engineering. See also Mr Hodgkinson's *Experimental researches on the strength of pillars*, Phil. Trans. 1840.

In this investigation we have supposed that the weight has been placed centrically over the axis of the column. The weight of the column itself has also been neglected and no allowance has been made for the shortening of the column due to the weight it has to support.

32. Heavy columns. Ex. 1. A vertical column in the form of a paraboloid of latus rectum $4m$ with its vertex upwards is fixed in the ground. Show that it will bend under its own weight when slightly displaced if the length be greater than $\pi(2Em/w)^{\frac{1}{2}}$, where w is the weight of a unit of volume, E the weight which would stretch a bar of the same material and unit area to twice its natural length.

Ex. 2. A vertical cylindrical column of radius r is fixed in the ground. Show that it will bend under its own weight if its length be greater than $c^{\frac{2}{3}}\left(\frac{9Er^2}{16w}\right)^{\frac{1}{3}}$, where c is the least root of $J_{-\frac{1}{3}}(c)=0$.

Let A be the area, r the radius of a section of the column (supposed to be thin and straight) at a distance x from the base C, then (Art. 13), $K=EAk^2$. When the

---

* Euler, *Berlin Memoirs*, 1757. *Petersburg Commentaries*, 1778. Lagrange, *Acad. de Berlin*, 1769. Poisson, *Traité de Mécanique*, 1833. See also Thomson and Tait, vol. I. Art. 611, where some figures are given. Also the *Proceedings of the Roy. Irish Acad.* 1873, where Sir R. Ball notes an error in Poisson's analysis. In the *Proc. London Math. Soc.* 1893, vol. XXIV., Prof. Love discusses the stability of columns. A discussion of Euler's theory is contributed to the Canadian Society of Civil Engineers, 1890, by C. F. Findlay, C.E.

column is a paraboloid $Ak^2 = \frac{1}{4}\pi r^4$ and $r^2 = 4m(l-x)$, when the column is a cylinder Ak^2 is constant. In the figure of Art. 27, let x', y' be the coordinates of any point P' between P and B. Taking moments about P the differential equation is seen to be

$$-K\frac{d^2y}{dx^2} = \int_x^l wA'dx' \,.\, (y-y'),$$

where A' is the area of the section at P'. Differentiating this equation with regard to x, we find after some reduction

$$\frac{d^2}{d\xi^2}\left(\xi\frac{dy}{d\xi}\right) + \beta^2\left(\xi\frac{dy}{d\xi}\right) = 0,$$

where $\xi = l - x$, $\beta^2 = w/2mE$, and the column is supposed to be a paraboloid.

At the free end where $\xi = 0$, we have $\xi dy/d\xi = 0$ and, since the stress couple is there zero, $d^2y/d\xi^2 = 0$. At the base where $\xi = l$ we have $dy/d\xi = 0$ and this leads to the condition that the column cannot begin to bend unless $\beta > \pi$.

When the column is a cylinder, the differential equation becomes

$$\frac{d^3y}{d\xi^3} + \frac{4w}{Er^2}\xi\frac{dy}{d\xi} = 0,$$

which may be reduced to Bessel's form. To effect this put $dy/d\xi = \xi^\lambda z$, $\rho = \xi^\mu$, we then see that $\lambda = \frac{1}{2}$, $\mu = \frac{3}{2}$.

Both these examples are due to Prof. Greenhill, *Proceedings of the Camb. Phil. Soc.* 1881, vol. IV.

33. Theory of a bent circular rod. *A uniform thin straight rod without weight is bent without tension into the form of a circular arc of great radius; it is required to find the stress couple at any point P.* See Art. 15.

We shall obtain a particular solution of this problem by making an hypothesis which simplifies the process, and which we afterwards verify by showing that all the equations of equilibrium are satisfied.

We assume (1) that all filaments of matter parallel to the length of the rod are bent into circles with their centres on a straight line perpendicular to the plane of bending. This straight line will be referred to as the *axis of bending*. We assume (2) that the particles of matter which in the unstrained rod lie in a normal section continue to lie in a plane when bent, (3) that this plane is normal to the system of circles above described.

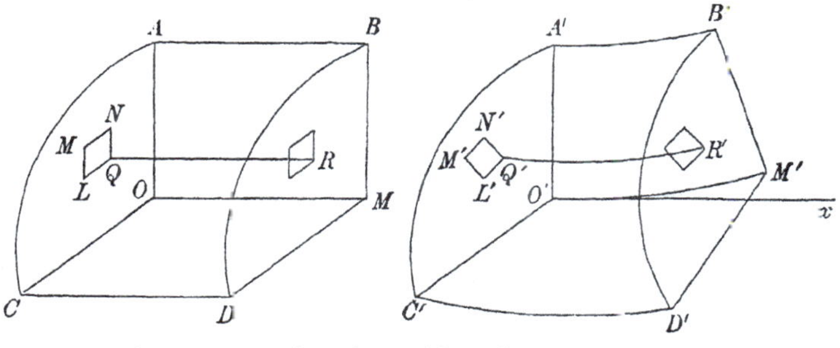

Let $ABCD$ be a short length of the straight rod bounded by two normal planes AOC, BMD. To examine the small changes which this length undergoes we take the plane AOC as that of yz and let some perpendicular straight line OM be the axis of x. To avoid confusing the figure only the lines on the positive octant have been drawn. Let the plane of xz be the plane of bending, so that the axis of y is parallel to the axis of bending. Thus OA is the axis of z, OC that of y. Let QR be any elementary filament parallel to the axis of x, let $(0, y, z)$, (x, y, z) be the coordinates of Q and R. Let the positions of these points and lines in the bent rod be denoted by corresponding letters with accents. According to the hypothesis $A'O'C'$, $B'M'D'$ are normal to all the filaments of the bent rod, and (when produced) these planes intersect in the axis of bending. Any filament, such as $Q'R'$, is a circular arc whose unstretched length is OM.

The rod being bent without tension, the filaments near $A'B'$ are compressed while those on the opposite side of the rod are extended. There is therefore some surface such that the filaments which lie on it have their natural length. This surface is usually called *the neutral surface*, and the lines on it parallel to the length of the rod are called *neutral lines*. Since the filaments on this surface are circular arcs of the same length with their centres on the axis of bending, the neutral surface is a cylinder which cuts the plane of yz in a straight line parallel to the axis of bending. Let the origin O' be taken on the neutral surface, the axis of x is therefore a tangent to a neutral line, and the unstretched length of every filament, such as $Q'R'$, is equal to OM or $O'M'$. Let ρ be the radius of curvature of this neutral line. Since the rod is thin, all the linear dimensions of the mass $ABCD$ are small compared with ρ.

When the unstretched length QR has been compressed or stretched into the length $Q'R'$, it remains sensibly parallel to the axis of x, but its distances from the planes xz, xy may have been altered. Let these distances be $y'=y+v$, $z'=z+w$, and let the stretched length $Q'R'$ be $x'=x+u$. Since R' lies in a plane normal to the neutral line at M', we have $$x'=(\rho-z-w)\sin\frac{x}{\rho}=x-\frac{(z+w)\,x}{\rho}.$$

The difference $x'-x$ represents the stretch of the fibre QR whose unstretched length is x. The tension per unit of sectional area is therefore equal to $-E\dfrac{z+w}{\rho}$. When the rod is only slightly deformed by the bending (as in Art. 17) the displacement w must be small compared with z. We may then, as a first approximation, equate the tension to $-Ez/\rho$.

Since the rod has been bent without altering its length, the resultant tension across the section AOC is zero, and we have

$$\iint (Ez/\rho)\,dy\,dz=0.$$

It immediately follows that the centre of gravity of the section lies in the plane of xy. The neutral surface therefore passes through the centre of gravity of every normal section. *In a cylindrical rod therefore, bent without tension, the central line is also a neutral line.*

Since the elementary tensions have no components parallel to the axes of y or z, it follows that the shear is zero.

If L be the moment about the axis of y of the tensions which act across the section AOC, measured positively from z to x, we have

$$L=\iint z^2\,dy\,dz\,.\,\frac{E}{\rho}=E\,\frac{\omega k^2}{\rho},$$

where ωk^2 is the moment of inertia of the sectional area about the axis of y, i.e. about a straight line drawn through the centre of gravity of the section perpendicular to the plane of bending, see Art. 13. Since the rod is a uniform cylinder bent into a circular arc, the corresponding couples about $O'C'$, $M'D'$ balance each other.

In the same way the moment about the axis of z of the tensions which act across the section AOC is $\iint yz\,dy\,dz\,.\,E/\rho$. This couple cannot be balanced by the equal couple about $M'B'$ because their axes are not parallel. It is therefore necessary that this moment should vanish. It follows that the rod will not remain in the plane of bending unless the product of inertia of the area of the normal section about the axis of y and any perpendicular straight line in its plane is zero. In other words, the plane of bending must be perpendicular to a principal axis of the section at its centre of gravity.

34. If we suppose, as already explained in Art. 8, that each fibre or filament of the rod is contracted or extended in the same manner as if it were separated from the rest of the rod, the mutual pressures of these filaments transverse to the length of the rod and also the tangential actions are zero. Each element of the rod is therefore in equilibrium, and the surface conditions are also satisfied. Each filament is slightly displaced, like those discussed in Art. 3, and slightly turned round. These displacements are those represented by v, w, and are such that, when the fibres are stretched independently of each other, the body remains continuous.

The expressions for the coordinates $y'=y+v$, $z'=z+w$, of Q' in terms of the coordinates y, z of Q may be deduced from the theorems given in Art. 8. It follows from that article that when the filament QR is stretched into the filament $Q'R'$ by a tension N_x, the rectangular base $QLMN$ remains rectangular and similar to its original form, and is of such size that corresponding sides are connected by the relation $(Q'L'-QL)/QL=-N_x/E'$.

Let ϕ be the angle which the side $Q'L'$ makes with the axis of y, measured positively from z to y; then

$$Q'L'\cos\phi=\frac{dy'}{dy}dy=\left(1+\frac{dv}{dy}\right)dy, \qquad -Q'L'\sin\phi=\frac{dz'}{dy}dy=\frac{dw}{dy}dy.$$

Rejecting the squares of the small quantities v, w and remembering that $QL=dy$, we have

$$\frac{dv}{dy}=-\frac{N_x}{E'}, \qquad -\tan\phi=\frac{dw}{dy}.$$

Treating the side $Q'N'$ in the same way, we have $\dfrac{dw}{dz}=-\dfrac{N_x}{E'}$, $\quad \tan\phi=\dfrac{dv}{dz}$.

Substituting for N_x its value $-E(z+w)/\rho$, and neglecting w/ρ as before, we find by integration

$$v=\frac{E}{E'}\frac{yz+f(z)}{\rho}, \qquad w=\frac{E}{2E'}\frac{z^2+F(y)}{\rho}.$$

Equating the two values of $\tan\phi$ and substituting for v and w we find that

$$-\tfrac{1}{2}F'(y)=y+f'(z).$$

It follows that $f(z)=az+b$, and therefore

$$v=\frac{E}{E'}\frac{(y+a)z+b}{\rho}, \qquad w=\frac{E}{2E'}\frac{z^2-(y+a)^2-c}{\rho}.$$

The terms containing b and a^2-c represent a translation of the section as a whole, those containing the first powers of y, z represent a rotation through an angle $Ea/E'\rho$. If neither of these displacements exist, we may omit these terms.

The expressions thus found for u, v, w, give the displacements of Q referred to the axes $O'M'$, $O'A'$, $O'C'$. They also give those of R referred to corresponding axes with M' for origin. The displacements of R referred to the axes with O' as origin are therefore given by

$$u=-\frac{xz}{\rho}, \qquad v=\frac{E}{E'}\frac{yz}{\rho}, \qquad w=\frac{x^2}{2\rho}+\frac{E}{E'}\frac{z^2-y^2}{2\rho},$$

where x, y, z are the coordinates of R.

35. If the section of the beam is a rectangle having the sides *EF*, *GH* perpendicular to the plane of bending, we see by examining the expression for v and w that these sides become curved when the rod is bent, and that they have their convexities

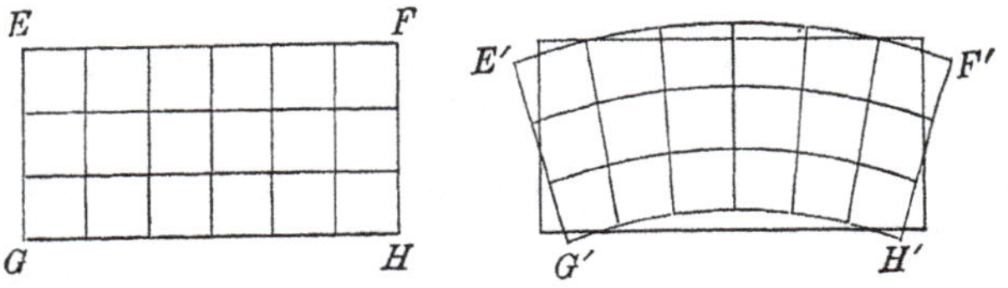

turned towards the centre of curvature of the rod. The sides *EG*, *FH* which before bending were parallel to the plane of bending remain straight lines but are inclined to the plane of bending and tend outwards on the concave side of the rod.

The expressions found in Art. 34 for the displacements u, v, w agree with those given by Saint-Venant for one case of bending. *But what has been said in that article is not to be taken for a complete discussion of his problem ;* for that the reader should consult a treatise on the theory of Elasticity.

The second and third assumptions of Art. 33 are included in the first, either if the circle is complete, or if proper forces are applied at the extremities of the arc. The first assumption may be regarded as following from the statement in the enunciation that the rod is uniform, without weight and bent into the form of a circular arc.

In the theory of Bernoulli and Euler these assumptions are applied to the case of any *thin* rod *. The theory thus extended leads to the result that the bending moment is proportional to the curvature and this result agrees with experiment. But other results of the theory are not so nearly in agreement with facts. To obtain a correct theory it is necessary to have recourse to the general equations of equilibrium of an elastic solid. In this treatise the expression for the bending moment is intended to rest on experiment (Art. 13), and the bending of a circular arc has been considered merely as the simplest example of the theory of elasticity.

36. Airy's Problem. In using standards of length two considerations have attracted attention, (1) the application of supports in such a manner as to produce no irregularities of flexure and (2) the application of such supports as will permit the expansive or contractive effects of temperature. The importance of the former was made known by Kater, that of the latter by Baily. Freedom of expansion is usually secured by supporting the body on rollers. Excessive flexure is avoided by making the rollers rest on levers which are so arranged that the weight of the body is either equally distributed over the points of support or distributed in such ratios as may be thought proper.

The flexure is so small that the mere curvature of the central line does not produce a sensible alteration of its length. *If however the measured length is marked on the upper surface of the measuring rod*, this length may be either stretched or shortened by the curvature of the central line. There may therefore be a small error in each length measured by the rod, which would be multiplied indefinitely when the whole distance measured is great. The problem is to determine how this

* Prof. Pearson shows in *The Quarterly Journal* for 1889 that the results of the Bernoulli-Eulerian theory give fairly approximate formulæ for the stress and strain of beams whose diameter is one-tenth, or less, of their length.

error may be avoided. Airy's principle is that the extension of each element of the upper surface of the measuring rod is proportional to the bending moment L. He therefore infers *that the supports of the rod should be so arranged that* $\int L\,dx = 0$, the limits of integration being from one end of the measured length to the other.

We may deduce the correctness of this principle from the theory given in Art. 33. The extension of the filament QR has been shown to be approximately $QR\,(z/\rho)$ where ρ is the radius of curvature of the central line and z the distance from the central line of the projection of QR on the plane of bending. If then z be the half thickness of the rod, the extension of an element dx on the surface is $z\,dx/\rho$. Since $L = K/\rho$, it immediately follows that the extension of any element on the surface of a uniform rod is proportional to the bending moment.

Ex. 1. A bar, of length a, is supported at two points symmetrically placed, and the marks defining the extremities of the measured length are close to its ends; prove that the distance between the points of support should be $a/\sqrt{3}$.

Ex. 2. A standard of length a is supported on m rollers placed at equal distances, and the weight is equally distributed over the rollers. The measuring marks are placed at distances e from the ends. If D be the distance between two consecutive rollers, prove that $D\sqrt{(m^2 - 1)} = a\sqrt{(1 - 8e^3/a^3)}$.

Memoirs of the Royal Astronomical Society, Vol. XV., 1846 and *Monthly Notices*, Vol. VI., 1845.

37. Bending of Circular rods. *The natural form of a thin inextensible rod is a circular arc; supposing it to be slightly flexible, it is required to find the deviation from the circular form produced by any forces*.*

Let AB be the arc of the circle when undeformed, O its centre, a its radius. Let P be any point on the circle, P' the corresponding point on the rod when bent. Let a, θ be the polar coordinates of P; $a(1+u)$, $\theta+\phi$ those of P', referred to O as origin.

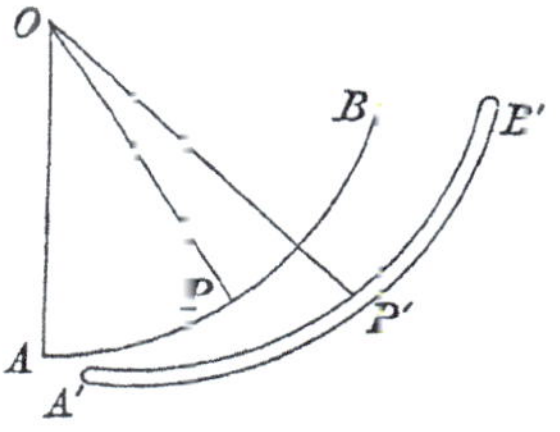

If ρ be the radius of curvature at P', we have by a theorem in the differential calculus
$$\frac{1}{\rho} - \frac{1}{a} = -\frac{1}{a}\left(u + \frac{d^2u}{d\theta^2}\right) \ldots (1),$$
where the squares of u are neglected. *Let us represent either side of this equation by* q_1/a.

If the central line be extensible, let ds_1 and ds be the unstretched and stretched lengths of an element of arc, then
$$ds_1 = a\,d\theta, \qquad (ds)^2 = (a\,du)^2 + a^2(1+u)^2(d\theta + d\phi)^2.$$

* The case of a circular arc is important because the periods of its vibrations both when inextensible and extensible, can be found. See the second volume of the Author's *Rigid Dynamics*, where also the expression for the work of the stresses is found in a different manner.

Neglecting the squares of small quantities, this gives

$$ds = a(1+u)\,d\theta + a d\phi.$$

If p be the proportional elongation of the elementary arc

$$p = \frac{ds - ds_1}{ds_1} = \frac{d\phi}{d\theta} + u \quad \text{.....................(2).}$$

If the rod is inextensible, we have $p = 0$.

The equations of equilibrium of an inextensible rod may be formed by either of the methods described in Arts. 10, 11. Taking, for example, the three equations marked (4) in Art. 11, and joining them to

$$L = K\frac{q}{a}, \qquad p = 0 \quad \text{.....................(3),}$$

we have five equations to find T, U, L, u, ϕ in terms of θ.

38. If the rod is slightly extensible as well as flexible, the equations become somewhat changed. The arc ds in the equations of equilibrium in Art. 11 means now the stretched length of the element, while F and G represent the impressed forces referred to a unit of length of the stretched rod. The equation $p=0$ must also be replaced by another connecting p with the tension.

The relations which connect L and T with p and q are perhaps most easily deduced from the expression for the work done by the stresses when the rod is deformed. If Wds_1 be the work done by the stresses when the element is stretched and bent, we have

$$Wds_1 = -\tfrac{1}{2}ds_1\left(Hp^2 + \frac{Kq^2}{a^2}\right) \quad \text{..........................(4),}$$

where H and K are the constants of tension and flexural rigidity. This result follows at once from those given in Art. 16 of this volume and in Art. 493 of Vol. I., *when we assume that the work due to a deformation of bending is independent of that of stretching.*

From this expression for W we may deduce the values of T and L. Keeping one end P' of an element $P'Q'$ fixed, let the element be further stretched, without altering the curvature, so that its length ds becomes ds', then $dp = \frac{ds' - ds}{ds_1}$. The work done by the tension T at the end Q' is $-T\,(ds' - ds)$, and that done by the couple at Q' is $-L\frac{ds' - ds}{\rho}$. The sum of these is $dW . ds_1$. We therefore have

$$-\left(T + \frac{L}{\rho}\right) = \frac{dW}{dp} \quad \text{...................................(5).}$$

Next let the element, without altering its length, receive an increase of curvature so that the radius of curvature is changed from ρ to ρ'; then $\frac{dq}{a} = \frac{1}{\rho'} - \frac{1}{\rho}$. The tension at Q' does no work, while the work of the couple L at Q' is $-L\left(\frac{1}{\rho'} - \frac{1}{\rho}\right)ds$.

Also $ds = (1+p)\,ds_1$,

$$\therefore -\frac{L}{a} = \frac{dW}{dq}\frac{1}{1+p} \quad \text{..............................(6).}$$

These expressions give for a slightly extensible and flexible rod

$$L = \frac{Kq}{a}, \qquad T = Hp - \frac{K}{a^2}q \quad \text{.........................(7).}$$

The equations of equilibrium found in Arts. 10 and 11 when joined to (7) supply five equations from which L, T, U, u, ϕ may be found

39. Ex. One end of a heavy, slightly flexible inextensible wire, in the form of a circular quadrant, is fixed into a vertical wall, so that the plane of the wire is vertical and the tangent at the fixed end horizontal. Assuming that the change of curvature at any point is proportional to the moment of the bending couple there, prove that the horizontal deflection at the free end is $\pi wa^4/8E$, where E is the flexural rigidity, w the weight of a unit of length, and a the radius of the circle.

[Trin. Coll. 1892.]

Let A be the free end of the rod, B the end fixed into the wall, O the centre. Taking moments about any point P for the side PA, Art. 10, we arrive at

$$\frac{E}{a^3w}\left(u+\frac{d^2u}{d\theta^2}\right)=-\sin\theta+\theta\cos\theta,$$

where $AOP=\theta$, and $OP=a(1+u)$. The constants of integration are determined from the conditions that u and $du/d\theta$ vanish at B, and the deflection required is the value of au when $\theta=0$.

40. *To find the work when a thin rod, whose central line in the natural state is a circle of radius a, is stretched and bent so that the central line becomes a circle of radius ρ, by a method analogous to that used in Art. 33 for a straight rod.*

The figure of Art. 33 may be used in what follows, except that the lines OM, AB, CD must be supposed to be small arcs of circles.

Let OM be an element of the central line of the unstrained solid, $O'M'$ the same element when the rod is deformed. Let the tangents to OM, $O'M'$ be the axes of x and x', and let the planes of xz, $x'z'$ be the planes of the circles. Let QR be any filament parallel to OM, $Q'R'$ its position in the strained rod. Let y, z; y', z' be the coordinates of Q, R; Q', R', each referred to its own set of axes.

If ds_1, ds be the lengths of OM, $O'M'$ and $1+p$ stand for ds/ds_1 as before, the tension of $O'M'$ per unit of area is Ep. If $d\sigma_1$, $d\sigma$ be the lengths of QR, $Q'R'$, we have

$$d\sigma_1=ds_1\left(1-\frac{z}{a}\right),\qquad d\sigma=ds\left(1-\frac{z'}{\rho}\right)\quad\text{.................(1)},$$

and the resultant tension of all the fibres which cross the area $dydz$ is therefore

$$Edydz\left(\frac{d\sigma}{d\sigma_1}-1\right)\quad\text{.........................(2)}.$$

The work done by this tension when the filament is pulled from its unstretched length $d\sigma_1$ to the length $d\sigma$, is

$$-\tfrac{1}{2}Edydz\left(\frac{d\sigma}{d\sigma_1}-1\right)^2 d\sigma_1\quad\text{.........................(3)}.$$

The difference $z'-z$ is a small fraction of z; *for a straight rod it has been shown to be of the order* z^2/ρ, Art. 34. As a first approximation we take $z'=z$. Substituting for ds/ds_1 and for $1/\rho$ their values $1+p$ and $(1+q)/a$, and neglecting all powers of z/a above the second, we find that the work is

$$-\tfrac{1}{2}Edydz\left[p^2-\{p^2+2pq(1+p)\}\frac{z}{a}+q^2(1+p)^2\left(\frac{z}{a}\right)^2\right]ds_1\quad\text{...........(4)}.$$

Integrating this over the area ω of the section, and remembering that O is the centre of gravity of the area, we have for the whole work

$$Wds_1=-\tfrac{1}{2}E\omega\left[p^2+\frac{k^2}{a^2}q^2(1+p)^2\right]ds_1\quad\text{.......................(5)};$$

when the higher powers of p, q are rejected this becomes

$$Wds_1=-\tfrac{1}{2}E\omega\left[p^2+\frac{k^2}{a^2}q^2\right]ds_1\quad\text{.........................(6)}.$$

41. In the same way we find that the tension of the fibres which cross the area $dydz$ is

$$Edydz\left[p-q\,(1+p)\left\{\frac{z}{a}+\left(\frac{z}{a}\right)^2\right\}\right] \quad \text{........................(7)}.$$

Remembering that O is the centre of gravity of the section, we find by an obvious integration that the resultant tension T and the resultant couple L are given by

$$T=E\omega\left\{p-\frac{k^2}{a^2}q\,(1+p)\right\}, \qquad L=E\omega a\frac{k^2}{a^2}q\,(1+p) \quad \text{.........(8)}.$$

These reduce to the forms given in Art. 38 when the product pq is neglected.

42. If we examine the expressions for the work, tension and couple given by equations (6) and (8) of Arts. 40, 41 we see that they contain two constants of elasticity, viz. $E\omega$ and $E\omega k^2$. These were represented by the letters H, K in the corresponding expressions in Art. 38.

When the rod is such that the constant of elasticity $E\omega$ is infinite or very great, a small change in the proportional extension p alters the product $E\omega p$ very considerably. If, therefore, the tension is finite or not very great, p must be very nearly equal to zero. It follows that in all the *geometrical relations* of the figure we may regard p as equal to zero. At the same time the product $E\omega p$ which occurs in the tension is not to be regarded as zero, but as a quantity analogous to the singular form $\infty\,.\,0$. If the tension is finite, the term Ep^2 which occurs in the work is zero.

Since the other constant of elasticity, viz. $E\omega k^2/a^2$, is not necessarily large in thin rods, it does not follow that q must be small, because $E\omega$ is large.

Rods in which $E\omega$ is very great are said to be inextensible. Such rods may be bent, and the bending couple is proportional to the change of curvature.

43. Very flexible rod. When the flexibility of the rod is such that it may be made to pass through several small rings not nearly in one straight line the integrations of the differential equation become more intricate. To simplify the problem we suppose that though weights may be attached to any points, the rod itself is without weight.

Let A, B, C &c. be a series of small smooth rings through which the rod is passed. Let the stress couple at A be L_1, and let T_1, U_1 be the tension and shear at the same point. Let L_2, T_2, U_2 be the corresponding stresses at B and so on.

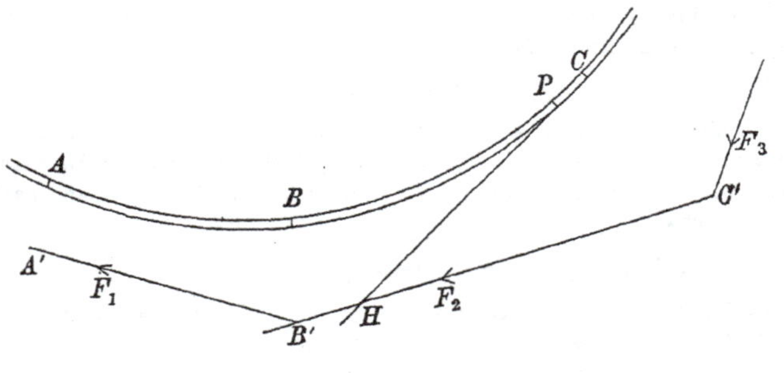

The stress L_1, T_1, U_1 acting at A may be reduced to a single resultant F_1 acting along some straight line $A'B'$, whose position is found in Vol. I. Art. 118. If P' be any point between the rings A and B, the stress at P' must be equivalent to the same force, for otherwise the portion AP' of the *light* rod would not be in equilibrium. In the same way the stress at every point of the rod between the rings B and C is equivalent to a single resultant F_2 acting along some other straight line $B'C'$, and so on for every portion of the rod lying between contiguous rings. These straight lines may be called *the lines of pressure*. We shall suppose the forces F_1, F_2, &c. when positive to be pulling forces, so that for instance the action of AP on PB is equivalent to the force F_1 acting in the direction $B'A'$.

The stress forces at points one on each side of any ring, as B, being F_1 and F_2, it follows that the pressure on the ring B is the resultant of F_1 acting along $B'A'$ and F_2 reversed and thus made to act along $B'C'$. *The pressure at B therefore acts along $B'B$, and this line is a normal to the rod at B.*

Let us consider the equilibrium of any portion BP of the rod, where P is a point between B and C. Let ψ be the angle the tangent PH makes with $B'C'$, and let $B'C'$ be the axis of ξ. Let η be the perpendicular distance of P from that axis. Let L, T, U be the stress couple, tension and shear at P. Then

$$T = F_2 \cos\psi, \qquad U = -F_2 \sin\psi, \qquad L = F_2\eta \quad\text{.....................(1).}$$

Taking moments about P for the portion BP we have $\dfrac{K}{\rho} = K\dfrac{d\psi}{ds} = F_2\eta$ ......(2).

Multiplying both sides by $\sin\psi = d\eta/ds$ and integrating, we find $-2K\cos\psi = F_2\eta^2 - C$. This result may be written in the form

$$2KF_2\cos\psi + F_2^2\eta^2 = I \quad\text{..................................(3),}$$

where I is a constant for the portion BP of the rod. We notice that in this equation $F_2\cos\psi$ is the tension and $F_2\eta$ the stress couple at the point P.

A similar equation holds for each portion of the rod which lies between contiguous rings. If P move along the rod and pass through the ring C, the tension and stress couple undergo no sudden changes of value, though the shear is altered discontinuously. It follows that $F_2\cos\psi$ and $F_2\eta$ are the same on both sides of C and that therefore I is the same for both portions of the rod. *The constant I has therefore the same value throughout the whole length of the rod.*

If one extremity of the rod is free, let A be the ring nearest the free end. The tension and the stress couple at A are therefore zero; hence, by equation (3) *the value of I is zero.* In this case, since the stress at A is reduced to the shear only, the line of pressure between the rings A and B is the normal at A.

Since $\rho\cos\psi = (ds/d\psi)(d\xi/ds) = d\xi/d\psi$, we have by (2)

$$\frac{d\xi}{d\psi} = \frac{K\cos\psi}{F_2\eta} = \frac{K\cos\psi}{(I - 2KF_2\cos\psi)^{\frac{1}{2}}} \quad\text{..............................(4),}$$

where ξ is measured positively opposite to the direction of F_2. Putting $\psi = \pi - 2\theta$, we reduce this to the difference of two elliptic integrals,

$$F_2\xi = i\int (1 - c^2\sin^2\theta)^{\frac{1}{2}}\, d\theta - \frac{I}{i}\int \frac{d\theta}{(1 - c^2\sin^2\theta)^{\frac{1}{2}}},$$

where $i^2 = I + 2KF_2$ and $c^2i^2 = 4KF_2$.

44. To show that these results supply a sufficient number of equations, let us suppose, as an example, that both ends of the rod are free and that it has been made to pass through five small rings at A, B, C, D, E.

Beginning at the ring A, the line of pressure $A'B'$ is the normal at A; let θ be the angle it makes with any fixed straight line in the plane of the rings. Taking AB' as the axis of ξ and A as origin, the coordinates of B, viz. ξ, η, are known functions of θ. The equations (3) and (4) give ξ, η, in terms of ψ and F_1, the constant in (4) being determined from the condition that when $\xi = 0$ the value of ψ is known, viz. in this case ψ is a right angle. Equating these two values of ξ and η

we have two equations to determine F_1 and the value of ψ at B. The tangent at B having been found, the normal BB' can be drawn and the position of B determined.

In the figure of Art. 43 we have $F_1 \sin A'B'B = F_2 \sin BB'C'$. When therefore we repeat the process just described and take $B'C'$ as a second axis of ξ and the foot of the perpendicular from B on $B'C'$ as the origin, with the object of finding F_2 and the value of ψ at C, we really have sufficient equations to find the angle $BB'C'$ also.

In the same way we next take $C'D'$ as the axis of ξ and finally $D'E'$. But since this last line of pressure must be the normal at E, the value of ψ at E must be a right angle. This supplies a final equation from which θ may be found.

Ex. A light rod DE is made to pass through two small rings A, C in the same horizontal line at a distance apart equal to $2b$, and has a weight W applied at a point B so that the vertical through B bisects AC at right angles. If 2ϕ be the angle between the normals at A and C prove that

$$2\cos\phi\,(\cot\phi)^{\frac{1}{2}} + (\cos\phi)^{\frac{1}{2}}\int_0^{\phi}(\sin\phi)^{\frac{1}{2}}\,d\phi = b\operatorname{cosec}\phi\left(\frac{W}{K}\right)^{\frac{1}{2}}.$$

On rods in three dimensions.

45. Measures of Twist. Let PK be a normal to the central line of an elastic rod at any point P, and let K lie on the outer boundary of the rod, the portion PK is called a *transverse* of the rod. This name is due to Thomson and Tait.

Let P, P', P'' &c. be a series of adjacent points on the central line of the unstrained rod, and let each of the arcs PP', $P'P''$ &c. be infinitesimal. Any transverse PK having been drawn at the first of these points, let the plane KPP' intersect the normal plane at P' in a second transverse $P'K'$. Let the plane $K'P'P''$ intersect the normal plane at P'' in a third transverse, and so on. We thus obtain a series of transverses, *any consecutive two of which lie in a tangent plane to the central line.*

If the rod when unstrained is straight and cylindrical it is obvious that all the transverses thus drawn lie in a plane passing through the central line. It is also clear that the extremities K, K' &c. of the transverses then trace out a straight line on the surface of the rod parallel to the central line.

Let these transverses be fixed in the material of the rod and move with it when the rod is strained. The normal section at P of the rod being fixed, let the elements lying between the normal planes at P, P', P'' &c. be twisted round the tangents PP', $P'P''$ &c. respectively, so that the points K, K', K'' &c. trace out a spiral line on the outer boundary of the rod. The twist of the elementary portion of the rod which lies between the normal planes at P, P' is measured by the infinitesimal angle which the transverse $P'K'$ makes with the plane KPP', or, what is ultimately the same, by the angle which the planes KPP', $PP'K'$ make with each other. If the arc PP' of the central line be ds, and if the angle which the planes KPP', $PP'K'$ make with each other be $d\chi$, the ratio $d\chi/ds$ represents the twist of the portion ds of the rod referred to a unit of length, and is usually called *the twist at P*.

It is sometimes useful to so choose the transverses PK, $P'K'$ &c. in the unstrained rod that the angle which the planes KPP', $PP'K'$ make with each other has any convenient value. Let $d\chi_1$ be this angle and let $d\chi_1 = \tau_1 ds$, then τ_1 is an arbitrary function of the arc s. If $d\chi$ or τds be the corresponding angle in the strained rod, the twist is measured by $\tau - \tau_1$.

46. Resolved Curvature. Let a straight rod be strained by bending, so that the central line takes the form of a curve of double curvature. If $d\epsilon$ be the angle between the normal planes to the central axis at P, P', the curvature at P is measured by the ratio $d\epsilon/ds$, and the central line is said to be curved in the osculating plane.

It is sometimes more convenient to resolve the curvature in two directions at right angles. Let the normal planes at P, P' intersect each other in a straight line CO, then CO intersects the osculating plane at right angles in some point C. Since PC, $P'C$ are two consecutive normals lying in the osculating plane, the point C is the centre of the circle of curvature; let $CP = \rho$. Let us now draw a plane through the tangent PP' to the central line making an arbitrary angle ϕ with the osculating plane, and let this plane cut CO in Q. Then since PQ, $P'Q$ are two consecutive normals to the central line, the point Q is the centre of a circle of curvature drawn in the plane QPP'. If the radius PQ of this circle be R,

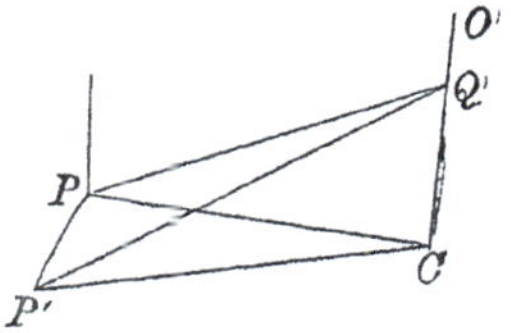

we have from the right-angled triangle QCP, $\frac{1}{R} = \frac{1}{\rho}\cos\phi$.

It follows that *the curvature in a plane drawn through the tangent may be deduced from the curvature in the osculating plane by the same rule that we use in statics to resolve a force.*

47. Let us draw two planes through the tangent at P to the central line, and let these be at right angles to each other. Let the resolved curvatures of the central line in these planes be called κ and λ. Then the curvature in the osculating plane is $\sqrt{(\kappa^2+\lambda^2)}$, and the tangent of the angle the osculating plane makes with the first of these planes is λ/κ.

These two planes intersect the normal plane at P to the central line in two straight lines at right angles. Let these be PK, PL, the straight line PK being perpendicular to the plane in which the resolved curvature is κ.

The three straight lines PK, PL, PP' thus form a convenient system of orthogonal axes to which we may refer that part of the rod which lies in the immediate neighbourhood of P. The resolved curvatures of the central line in the planes perpendicular to PK, PL, being κ, λ and the twist about PP' being τ, it follows that in passing from the point P to P' the three axes are screwed into positions $P'K'$, $P'L'$, $P'P''$ by a combination (1) of the rotations κds, λds, τds about the axes PK, PL, PP', and (2) of a translation of the origin P along the tangent to P'. It should be noticed that each of the three quantities κ, λ, τ is of -1 dimension as regards space.

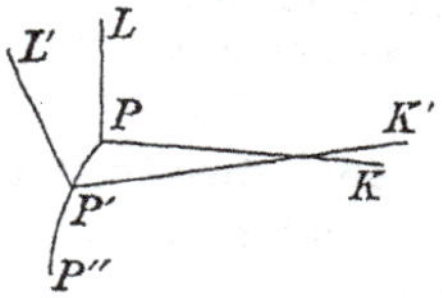

The quantities κ, λ are the resolved curvatures of the strained rod and are the same as the resolved bendings produced by the forces, only when the unstrained rod is straight. To find the bending produced by the external forces when the unstrained rod is itself curved we must subtract from κ, λ the resolved curvatures of the unstrained rod.

48. Since κds, λds, τds are rotations about the axes of reference, we know by the parallelogram of angular velocities that they may be resolved about other axes by the parallelogram law. If then we wish to refer the strains to a different set of axes, say PK_1, PL_1, PT_1, we change κ, λ, τ into κ_1, λ_1, τ_1 by the usual formulæ for the transformation of coordinates or for the resolution of forces. In this way we may refer the bending and twist in the neighbourhood of P to any arbitrary system of axes having the origin at P. These generalized axes may be screwed from their positions at the origin P to those at P' by the three rotations $\kappa_1 ds$, $\lambda_1 ds$, $\tau_1 ds$ and the translation ds along the tangent.

49. In many of the applications of analytical geometry to physical problems it is found advantageous to make the coordinate axes moveable, so that they may always be in the most convenient position. Thus if a point travel along the strained rod and successively occupy the positions P, P', P'', &c., the axes change their directions in space. To specify the motion of these axes we may either use a second system of axes fixed in space or we may refer the motion to the moving axes themselves in the manner above described. The first method requires the use of the formulæ of transformation of axes which are often complicated, in the second we avoid the introduction of a second system of axes. Moving axes are of great importance in Dynamics and are also of much use in discussing the geometrical properties of curves and surfaces. For these applications the reader is referred to the second volume of the Author's *Treatise on Rigid Dynamics*.

50. Ex. 1. A straight line is marked on the surface of a thin unstrained cylindrical rod, parallel to the central line. If the rod is bent along any curve on a spherical surface so that the marked line is laid in contact with the spherical surface, show that the twist is zero.

If the rod is laid on a cylindrical surface so that the marked line is in contact with the cylinder, show that the twist is $\sin\alpha\cos\alpha/a$, where a is the radius of the cylinder and α is the angle the rod makes with the axis of the cylinder. Both these results are given by Thomson and Tait, Art. 126.

If P, P' be two consecutive points on the central line, the transverses PK, $P'K'$ are normals to the surface. The first result follows, because the transverses pass through the centre of the sphere, so that the angle between the planes KPP', $PP'K'$ is zero. Since the radius of curvature at any point of a helix lies on the normal to the cylinder on which the helix is drawn, the second result follows from the ordinary expression for the radius of geometrical torsion.

Ex. 2. A straight thin rod has a straight line marked along one side. If the rod is bent and laid on a surface so that this line lies in contact with a geodesic, show that the twist at any point P is $\Delta\sin\theta\cos\theta$, where Δ is the difference of the curvatures of the principal sections of the surface at P and θ is the angle the rod makes with either line of curvature.

51. Relations of stress to strain. Let P be any point on the central line; the mutual action of the parts of the rod on each side of the normal section at P can be reduced to a force and a couple with any convenient point of that section as base.

Let three rectangular axes be taken at the point P to which we may refer the strains and stresses in the neighbouring portion of the rod. Let K, L, T be the components of the stress couple about these axes. If the unstrained rod is straight, let κ, λ, τ be the resolved parts of the curvature and twist about the axes; if the unstrained rod is itself curved, then κ, λ, τ represent the changes in the curvature and twist produced by the external forces.

We shall now assume the two following principles*:—

(1) that the changes in the twist and curvature of the rod in

* See Thomson and Tait, 1883, Art. 591.

the neighbourhood of P are independent of the force and are functions only of the couple;

(2) that the couples K, L, T are linear functions of the strains κ, λ, τ.

These assumptions are necessary because we do not in this place enter into the theory of elasticity.

If we suppose, as usual, that the strains are so small that we may neglect all powers but the lowest which enter into the equations, the second principle is equivalent to the assumption that when K, L, T are expanded in powers of κ, λ, τ the lowest powers in the series are the first.

52. Since the three couples K, L, T are each expressed in terms of κ, λ, τ by a different linear equation, it might be supposed that we shall have to deal with nine constants. But if the elastic forces form a conservative system we may reduce these to six by using the work function.

Let Wds be the work function of an element of the rod bounded by the normal sections at P, P'. Supposing the end P fixed, let one strain, say λ, become $\lambda + d\lambda$, the other two remaining unaltered. Since the element of the rod has been rotated about the axis of the couple L through an angle equal to $d\lambda \,.\, ds$, the work done by the couple L is $Ld\lambda ds$, while that done by each of the couples K and T is zero. We therefore have $dsdW = Ld\lambda ds$. Similar expressions hold when K and T are increased by $d\kappa$ and $d\tau$, so that in general

$$K = dW/d\kappa, \qquad L = dW/d\lambda, \qquad T = dW/d\tau.$$

Since K, L, T are linear functions of κ, λ, τ it follows that W is a quadratic function of κ, λ, τ, i.e.

$$W = \tfrac{1}{2}(A\kappa^2 + B\lambda^2 + C\tau^2 + 2a\lambda\tau + 2b\tau\kappa + 2c\kappa\lambda).$$

$$\therefore\ K = A\kappa + c\lambda + b\tau, \quad L = c\kappa + B\lambda + a\tau, \quad T = b\kappa + a\lambda + C\tau.$$

53. We have already seen that if we refer the strains to another set of axes the quantities κ, λ, τ are changed by the ordinary formulæ for transformation of coordinates, Art. 48. Since a homogeneous quadratic expression can always be cleared of the terms containing the products of the variables, it follows that by a proper choice of the axes of reference the expressions for W, and therefore those for K, L, T may be reduced to the simplified forms

$$W = \tfrac{1}{2}(A_1\kappa_1^2 + B_1\lambda_1^2 + C_1\tau_1^2),$$

$$K_1 = A_1\kappa_1, \qquad L_1 = B_1\lambda_1, \qquad T_1 = C_1\tau_1.$$

These axes are called *the principal axes of stress* and the constants A_1, B_1, C_1, are *the principal flexure and torsion rigidities.*

In what follows it will generally be assumed that the tangent to the central line at P is one principal axis of stress at P; this is of course the axis of torsion. If also the constants of rigidity for the other two principal axes are equal, we have

$$W = \tfrac{1}{2}A(\kappa^2 + \lambda^2) + \tfrac{1}{2}C\tau^2,$$

where the suffixes have been dropped as being no longer required.

The expression for the work is not complete if the rod is extensible, for we have not yet taken account of the extension or stretching, of the element PP' of the rod. This additional term is given in Vol. I. on the supposition that the tension obeys Hooke's law. It will not be required in the problems considered in this chapter.

54. Helical twisted rods. *A uniform thin rod, naturally straight, whose principal stress axes at any point are the central line and any two perpendicular axes, is bent into the form of a helix of given angle and receives at the same time a given uniform twist. It is required to find the force and couple which must be applied at one extremity, the other being fixed, that the rod may retain the given strains.*

Let APQ be an arc of the helix, A the fixed extremity, Q the terminal at which the forces are applied. Let AMB be a circular section of the cylinder on which the helix lies, OZ the axis of the cylinder.

The mutual action of the portion AP of the helix and the portion PQ consists of a force and a couple. From the uniformity of the figure it is clear that the force and couple must be the same *in magnitude* wherever the point P is taken on the helix, and that *their direction and axis* respectively must make the same angles with the principal axes of the curve at P.

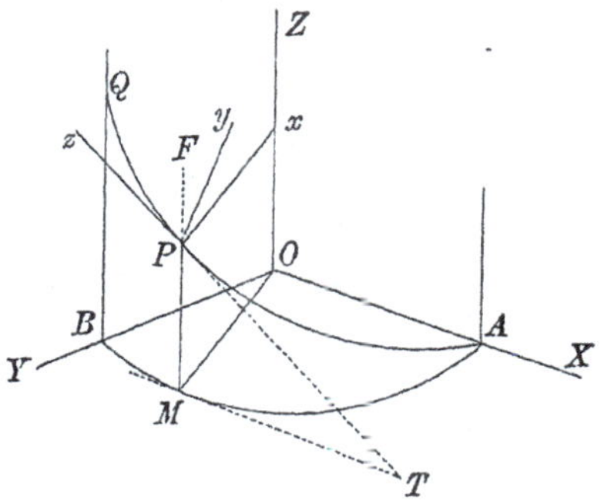

The stress force at P may be resolved into two

components, one acting along the generating line MPF of the cylinder and the other acting parallel to the plane XY. The latter, if it is not zero, must be in equilibrium with the component at Q parallel to the same plane. This however is impossible, because as P is moved along the helix the direction of the component at P makes always the same angles with the principal axes at P and is therefore changed, while that of the component at Q remains unaltered. Both these components must therefore be zero. It follows that *the resultant stress force at any point P must act along the generator through that point.*

Let R be the stress force at any point of the rod. The force R may be transferred to the axis OZ of the cylinder by introducing the couple Ra acting in the plane $OZFM$. *The force R thus becomes independent of the position of P.*

Let us now turn our attention to the stress couples at P. Let Px be drawn perpendicular to the axis of the cylinder and let TPz be a tangent at P to the helix. Then by the known properties of the curve, the plane TPx is the osculating plane at P. Let Py be the binormal. If $\rho = 1/\kappa$ be the radius of curvature of the helix, the strains round Py and Pz are respectively κ and τ, each being measured in the positive direction round the axes, i.e. from z to x and x to y. There is no strain round Px because the rod is naturally straight. If A and C are the constants of flexure and torsion, the corresponding stress couples are $A\kappa$ and $C\tau$. These couples may be resolved into two components, one having the generator PF for axis and the other having its axis parallel to the plane of XY.

Let the resultant of the latter couple and of the couple Ra be called H. The couple H at P together with the force R acting along OZ must be in equilibrium with the corresponding reversed couple H' and the reversed force R at Q. The forces are equal and opposite, hence the couples H, H' must be in equilibrium. Since the axis of the couple H always makes the same angle with OM, its direction is altered when the point P is moved along the helix while that of the couple Q is fixed. Equilibrium cannot exist for all positions of P unless both H and H' are zero. *The stress at P is therefore equivalent to a wrench whose force acts along the axis OZ of the cylinder and whose couple acts in a plane perpendicular to that axis.*

Consider the equilibrium of the portion AP of the helix. The fibres of the rod which are nearest to OZ are compressed and those more remote are stretched. Hence the arc QP tends to turn AP round the binormal Py in the direction z to x. Also, as P travels along the wire in the direction APQ the positive direction of the torsion is from x to y, hence the twist couple exerted by PQ on AP is in the same direction, viz. x to y.

The stress couples which act at P on the portion AP of the rod are therefore $L=0$, $M=A\kappa$, $N=C\tau$ round Px, Py, Pz respectively. These together with the force at P are equivalent to a wrench, let G be the couple measured clockwise round OZ and let R be the force acting along OZ. By equating the moments of these about MP and also about a parallel to MT drawn through P, we find that

$$G = A\kappa \cos\alpha + C\tau \sin\alpha,$$

$$Ra = -A\kappa \sin\alpha + C\tau \cos\alpha.$$

Here R tends to pull out the spiral AP, and G to twist it round OZ from A to B.

These equations determine R and G when the angle α of the spiral, the curvature κ and the twist τ of the material are known.

By giving the proper twist, we can make $G=0$ and then the spring can be maintained in the spiral form by a force R only.

55. Spiral Springs. *A thin rod or wire, whose natural form is a given helix and whose principal axes of stress at any point are the tangent to the central line and any two perpendicular axes, is bent into the form of another given helix. It is required to find the forces and couples which must be applied at one end, the other being fixed, that the rod may retain the given form.*

Let a_1, a be the radii of the cylinders on which the unstrained and strained helices lie; α_1, α the angles of the helices. Let the axes of the two cylinders be coincident and let it be taken as the axis of Z, the plane of XY being perpendicular to it.

Let P, P' &c. be a series of consecutive points on the central line of the unstrained rod and let $P\xi$, $P'\xi'$ &c. be the principal normals at these points. The angle between the consecutive planes $\xi PP'$, $PP'\xi'$ is $ds \sin\alpha_1 \cos\alpha_1/a_1$ where ds is the arc PP'. Let $P\eta$, $P'\eta'$ be the binormals at the same points, then the curvature of the unstrained rod, measured, as in Art. 47, round the binormal, is $ds \cos^2\alpha_1/a_1$. Let $P\zeta$, $P'\zeta'$ be the tangents to the helix taken positively in the direction in which s is measured.

Let the axes of ξ, η be fixed in the material of the rod and be the transverses of reference. When the rod is strained, let Px, Py, Pz be the principal normal, binormal and tangent at P, then $P\zeta$ coincides with Pz, and $P\xi$ makes some angle ϕ with Px. The figure of Art. 54 may be used to represent the strained position of the rod, the axes $P\xi$, $P\eta$, $P\zeta$ are not drawn but may easily be supplied by the description just given.

The stress at the point P of the strained rod consists of (1) a force which we may suppose to be resolved into two components, one along the generating line of the cylinder and the other parallel to the plane of XY. (2) A couple $C(\tau - \tau_1)$, whose axis is the tangent Pz, and two couples $A\kappa$ and $-A\kappa_1$, whose axes are Py and $P\eta$ respectively; where

$$\tau_1 = \frac{\sin \alpha_1 \cos \alpha_1}{a_1}, \quad \kappa_1 = \frac{\cos^2 \alpha_1}{a_1}, \quad \kappa = \frac{\cos^2 \alpha}{a},$$

and τds is the elementary angle between the planes $\xi PP'$, $PP'\xi'$ in the strained rod.

Examining first the stress force, we find, as in Art. 54, that the component parallel to the plane of XY is zero. *The stress force at every point P therefore acts along the generating line of the cylinder*; let this force be R, and let it be transferred to the axis of the cylinder by introducing a couple Ra.

Taking next the stress couples, we find by the same reasoning that the component about any axis parallel to the plane of XY is zero. Let us first equate to zero the moment about Px; since Px is perpendicular to Py, Pz and to the axis of the couple Ra, and makes with $P\eta$ an angle $\frac{1}{2}\pi + \phi$, we have $\kappa_1 \sin \phi = 0$. *Since κ_1 is not zero (as it was in Art. 54), it follows that $\phi = 0$.* The axes $P\xi$ and Px therefore coincide and *the couples $A\kappa$ and $-A\kappa_1$ have a common axis Py, viz. the binormal of the strained helix.* The angle τds is also equal to the angle between the consecutive osculating planes to the strained helix, i.e. $\tau = \sin \alpha \cos \alpha / a$.

Equating to zero the moment about a perpendicular to the plane passing through Px and the generator of the cylinder, we have

$$Ra = -A \sin \alpha (\kappa - \kappa_1) + C \cos \alpha (\tau - \tau_1) \quad \text{.........(1)}.$$

Equating the moment about a generator to the corresponding moment at the terminal we have

$$G = A \cos \alpha (\kappa - \kappa_1) + C \sin \alpha (\tau - \tau_1) \quad \text{...........(2)}.$$

The curvatures and torsions are

$$\kappa_1 = \frac{\cos^2 \alpha_1}{a_1}, \quad \kappa = \frac{\cos^2 \alpha}{a}, \quad \tau_1 = \frac{\sin \alpha_1 \cos \alpha_1}{a_1}, \quad \tau = \frac{\sin \alpha \cos \alpha}{a}.$$

56. If the spiral spring have a great many turns so that α_1 and α are both small, we have when the squares of α_1, α are neglected

$$Ra = -A\alpha\left(\frac{1}{a} - \frac{1}{a_1}\right) + C\left(\frac{\alpha}{a} - \frac{\alpha_1}{a_1}\right)$$

$$G = A\left(\frac{1}{a} - \frac{1}{a_1}\right).$$

If there be no couple G but only a force at each end pulling the spiral out, we deduce from these equations that $a = a_1$, so that the spring occupies a cylinder of the same radius as before the strain. We also have

$$Ra = C\frac{\alpha - \alpha_1}{a},$$

which is independent of the constant of flexure. It appears therefore that *the spring resists the pulling out chiefly by its torsion.* It is stated by both Saint-Venant and Thomson and Tait, that this result was first obtained by Binet in 1815.

Let l be the length of the spiral spring, h the elongation of its axis produced by the force R; then

$$l \sin \alpha - l \sin \alpha_1 = h.$$

Rejecting as before the squares of α_1 and α we find that $R = C \cdot \frac{h}{la^2}$. This expression determines the force required to produce a given elongation in a given spring of small angle.

57. Equations of Equilibrium. *To form the general equations of equilibrium* in three dimensions of a strained rod.*

Let P, P' &c. be a series of consecutive points on the central line of the unstrained rod. Let a series of transverses PK, $P'K'$ &c. be drawn such that the angle of twist τ_1 is either zero or some arbitrary function of the arc s. Taking the transverse PL perpendicular to PK, let the resolved curvatures about these lines be λ_1 and κ_1. If these transverses are the principal flexure and torsion axes at each point of the rod they form a convenient

* The general equations of a rod in Cartesian coordinates may be found in the *Treatise on Natural Philosophy* by Thomson and Tait, 1879. The intrinsic equations, or those referred to moving axes, are given in the *Treatise on Statics* by Minchin, 1889.

system of coordinate axes. If not let some other system of axes, $P\xi$, $P\eta$, $P\zeta$, be chosen which are connected with the transverses PK, $P'K'$ &c. in a known manner. Let θ_1, θ_2, θ_3 be the resolved parts of κ_1, λ_1, τ_1 about the axes $P\xi$, $P\eta$, $P\zeta$. Then, as explained in Art. 48, the axes at P may be brought into positions parallel to those at P' by rotations $\theta_1 ds$, $\theta_2 ds$, $\theta_3 ds$ about themselves.

When the rod is strained the axes $P\xi$, $P\eta$, $P\zeta$ will move with the material of the rod and assume new positions in space. Let these be Px, Py, Pz. Let $\omega_1 ds$, $\omega_2 ds$, $\omega_3 ds$ be the rotations by which the axes at P in the strained rod are brought into positions parallel to those at P'. The differences $(\omega_1 - \theta_1)\,ds$, $(\omega_2 - \theta_2)\,ds$, $(\omega_3 - \theta_3)\,ds$ may be used to measure the strains produced by the external forces.

Let R_1, R_2, R_3; L_1, L_2, L_3 be the stress forces and couples which act at P on the element PP' in front of P and let them be estimated as positive when they act in the negative directions of the axes at P. Then $R_1 + dR_1$ &c., $L_1 + dL_1$ &c., are the corresponding forces and couples at P' and act on the element PP', behind P', in the positive directions of the axes at P'. Besides these the element is acted on by the impressed forces $F_1 ds$, $F_2 ds$, $F_3 ds$ and the impressed couples (if any) $G_1 ds$, $G_2 ds$, $G_3 ds$.

Since R_1, R_2, R_3 are the components of a vector, viz. the stress force at P, the differences of the resolved parts at P and P' along the same set of axes are given by the rule for resolving vectors*; we therefore have

$$\frac{dR_1}{ds} - \omega_3 R_2 + \omega_2 R_3 + F_1 = 0 \quad \text{..............(1)},$$

$$\frac{dR_2}{ds} - \omega_1 R_3 + \omega_3 R_1 + F_2 = 0 \quad \text{..............(2)},$$

$$\frac{dR_3}{ds} - \omega_2 R_1 + \omega_1 R_2 + F_3 = 0 \quad \text{..............(3)}.$$

* The following proof of the rule is the same as that given in the second volume of the Author's *Rigid Dynamics*.

Describe a sphere of unit radius whose centre is at P and let the axes Px, Py, Pz cut its surface in x, y, z. Let parallels to the corresponding axes at P' drawn through P cut the surface in x', y', z'. Thus we have two spherical triangles xyz, $x'y'z'$, all whose sides are quadrants. Also x, y, z are brought into coincidence with x', y', z' by the combined effect of the rotations $\omega_1 ds$, $\omega_2 ds$, $\omega_3 ds$ about Px, Py, Pz respectively.

Let U, V, W be the components of the vector at P in the directions of the axes x, y, z; $U + dU$, &c. the components of the vector at P' along the axes x', y', z'. The difference of the resolved parts along the axis of x is then

$$(U + dU)\cos xx' + (V + dV)\cos xy' + (W + dW)\cos xz' - U.$$

In the same way since L_1, L_2, L_3 are the components of a vector, we have

$$\frac{dL_1}{ds} - \omega_3 L_2 + \omega_2 L_3 + G_1 + \mu R_3 - \nu R_2 = 0 \quad \text{.........(4)}$$

$$\frac{dL_2}{ds} - \omega_1 L_3 + \omega_3 L_1 + G_2 + \nu R_1 - \lambda R_3 = 0 \quad \text{.........(5)}$$

$$\frac{dL_3}{ds} - \omega_2 L_1 + \omega_1 L_2 + G_3 + \lambda R_2 - \mu R_1 = 0 \quad \text{.........(6)},$$

where λ, μ, ν are the direction cosines of the arc PP' referred to the axes at P.

The relations between the couples L_1, &c., and the strains $\omega_1 - \theta_1$, &c., may be deduced from the expression for the work W given in Art. 52, by writing $\omega_1 - \theta_1$, &c. for κ, τ, λ. Supposing for the sake of brevity that the axes are the principal flexure and torsion axes, we have

$$L_1 = A(\omega_1 - \theta_1), \quad L_2 = B(\omega_2 - \theta_2), \quad L_3 = C(\omega_3 - \theta_3) \text{...(7)}.$$

If the axes are the tangent at P to the central line and two perpendicular axes, we have $\lambda = 0$, $\mu = 0$ and $\nu = 1$; but in all cases λ, μ, ν are known from the given conditions of the rod.

We thus have nine equations to determine the quantities R_1, R_2, R_3; L_1, L_2, L_3; ω_1, ω_2, ω_3. If the rod is extensible there will be another equation supplied by Hooke's law.

58. The meaning of these equations will be made clear if we apply them to the simpler case in which the rod is uniform and when unstrained is straight and without twist. In this case $\theta_1 = 0$, $\theta_2 = 0$, $\theta_3 = 0$, and ω_1, ω_2, ω_3 are the components of the curvature and twist. Let us also take the tangent PT as the axis of x and the principal flexure axes PK, PL as axes of y and z

---

The rotations about Px, Py cannot alter the arc xy, but the rotation about Pz will move y' away from x by the arc $\omega_3 ds$. In the same way the rotations about Px and Pz cannot alter the arc xz, but the rotation about Py will move z' towards x by the arc $\omega_2 ds$. Therefore

$$xy' = xy + \omega_3 ds, \qquad xz' = xz - \omega_2 ds.$$

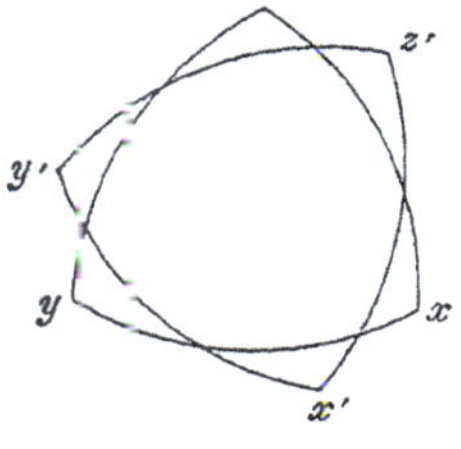

Also the cosine of the arc xx' differs from unity by the square of a small quantity. Substituting we find that the difference of the resolved parts along the axis of x is $\quad dU - V\omega_3 ds + W\omega_2 ds.$

If U, V, W stand for R_1, R_2, R_3 we join to this the force $F_1 ds$; equating the result to zero and dividing by ds, we obtain the first of the six equations. If U, V, W stand for L_1, L_2, L_3 we add the couple $G_1 ds$ and the moments of the forces $R_1 + dR_1$ &c. acting at P. We thus obtain in the same way the fourth of the six equations.

so that $\lambda=1$, $\mu=0$, $\nu=0$. The equations (1) to (6) then take the forms

$$\frac{dR_1}{ds}-\omega_3R_2+\omega_2R_3+F_1=0, \qquad \frac{dL_1}{ds}-\omega_3L_2+\omega_2L_3+G_1 \qquad =0,$$

$$\frac{dR_2}{ds}-\omega_1R_3+\omega_3R_1+F_2=0, \qquad \frac{dL_2}{ds}-\omega_1L_3+\omega_3L_1+G_2-R_3=0,$$

$$\frac{dR_3}{ds}-\omega_2R_1+\omega_1R_2+F_3=0, \qquad \frac{dL_3}{ds}-\omega_2L_1+\omega_1L_2+G_3+R_2=0,$$

and the equations (7) become $L_1=A\omega_1$, $L_2=B\omega_2$, $L_3=C\omega_3$.

From these modified equations we can immediately deduce the equations of the bending of a rod in two dimensions as given in Art. 11. Let the plane of the rod be the plane of xy and as before let the axis of x be the tangent, then $\omega_1=0$, $\omega_2=0$, $\omega_3=1/\rho$. The stress forces R_1, R_2 are respectively T, U, while $R_3=0$; the stress couples $L_1=0$, $L_2=0$, $L_3=K\omega_3$. Also $G_1=0$, $G_2=0$, $G_3=0$ and in the notation of Art. 11, $F_1=F$, $F_2=G$, $F_3=0$. The two first and the sixth equations immediately reduce to those given in Art. 11, while the third, fourth and fifth are identities.

59. Supposing the rod to be uniform and when unstrained to be straight and without twist, we find, by eliminating L_1, L_2, L_3, three equations of the form

$$A\frac{d\omega_1}{ds}-(B-C)\,\omega_2\omega_3=R_2-G_1.$$

These are the same as the three equations of the motion of a rigid body about a fixed point (with s instead of t). This analogy is due to Kirchhoff, but we cannot properly discuss it here.

ASTATICS.

On Astatic Couples.

1. THE conditions of astatic equilibrium in two dimensions have already been investigated in the first volume of this treatise. We have now to consider what other conditions are necessary when the body is displaced in any manner in three dimensions*.

* The subject of Astatics appears to have been first studied by Moebius, who published his results in his *Lehrbuch der Statik*, 1837. Moigno also, in his *Statique* has discussed the subject at great length. Minding in the fifteenth volume of Crelle's *Journal* gave the theorem that, whenever the body is so placed that the forces admit of a single resultant, that resultant intersects two conics fixed in the body. Many proofs have been given of this curious theorem we may mention that by Darboux, Tait's proof by quaternions modified by Minchin; Larmor's proof with the use of the six coordinates of a line.

Darboux published in the *Mémoires de la Société des Sciences physiques et naturelles de Bourdeaux*, t. II. [2e Série], 2e Cahier, a very long paper on this subject. In contradiction to Moebius, he showed that when one point of a body is fixed there are in general four positions, and only four, in which the body can be placed so that the forces are in equilibrium. These he called the initial positions of the body. His investigation is rather long and a different proof is given here. He also introduced the idea of a central ellipsoid analogous to the momental ellipsoid used in discussing moments of inertia. This result is given in Art. 14 of the text, and the general lines of his argument have been followed in that article. By the use of this ellipsoid he gave a geometrical turn to the proof of Minding's theorem, but it remained rather complicated. Extending the theory by considering all positions of the body, he showed that Poinsot's central axis formed a complex of the second order, such that each straight line is the intersection of two perpendicular tangent planes to the conics used by Minding. The first part of this result was subsequently arrived at by Somoff in 1879.

The theorems on Astatics given by Moigno may be found in his *Leçons de Mécanique Analytique*, 1868, which he tells us are chiefly founded on the methods of Cauchy. As his demonstrations are different from those given in this treatise, it may be useful to indicate the plan of his work. First, by a transformation of axes, he obtains the twelve equations of equilibrium given in Art. 11. Thence he deduces the conditions that a system of forces can be astatically reduced to a single force by considering what single force can be in equilibrium with the system. Supposing these conditions not to be satisfied, he shows that the system can be reduced to two forces provided two conditions are satisfied. These conditions agree with the two last determinantal equations given in Art. 73. He next shows that the system can always be reduced to a force and two couples and that the point of application of the force may be arbitrarily chosen on a plane fixed in the body. This plane is defined to be the central plane. He then shows that if the arbitrary point is properly chosen the directions of the forces and of the arms may be simplified in the manner described in Art. 27. This point is defined to be the central point. Proceeding next to consider the case in which the body is so placed that the forces admit of a single resultant, he shows that that single resultant must intersect two conics fixed in the body. He next discusses the case in which the equilibrium is astatic only for displacements of the body round a given axis following the same plan as before, he enquires into the conditions that the system can be reduced to one, two or three forces. He concludes with an application to magnetic forces and investigates the positions of the central plane and central point.

2. We shall suppose, as before, that each force acting on the body retains the same direction in space, the same magnitude and continues to act at the same point of the body, for all displacements.

The forces of a couple remain parallel, equal, and unaltered in magnitude as the body is moved, but the length of the arm is not necessarily the same. Let A, B be the points of application of the forces, then the distance AB is unaltered, and is called the *astatic arm of the couple.* If in any position of the body the inclination of the astatic arm to the forces is θ, the arm of the couple is $AB \sin \theta$.

The product of either force into the astatic arm is called the *astatic moment of the couple.* The astatic moment is of course unaltered by any change in position of the body. Representing the astatic moment by K, the actual moment in any position of the body is $K . \sin \theta$.

The angle θ which the astatic arm makes with the force is called the *astatic angle of the couple.*

Two couples are said to have the *same astatic effect* when they are equivalent in all positions of the body.

For the sake of brevity the couple whose force is P and astatic arm is AB is represented by the symbol (P, AB).

3. *The astatic effect of a couple is not altered if we replace it by another having the same astatic moment, the astatic arms being parallel, and the forces acting in the same direction in space as before.*

Let the astatic arm AB be moved to a new position $A'B'$ in the body. The extremities of the astatic arm of a couple are fixed in the body and move with it; thus as the body is displaced, AB and $A'B'$ continue to be parallel to each other. The astatic angles of the two couples continue therefore to be equal to each other. Since the astatic moments are equal, it follows that the actual moments of the couples are equal. The two couples are therefore equivalent.

It may be noticed that we cannot in general turn the astatic arm of a couple through any angle in the manner explained in Vol. I. Art. 92; for the planes of the couples may not remain parallel to each other, unless the displacements of the body are restricted to be parallel to the original plane of the couples.

4. *To find the astatic resultant of two couples whose forces are parallel but whose astatic arms are inclined at any angle.*

Let AB, $A'B'$ be the astatic arms of the couples, the forces at A, A' being supposed to act in the same direction in space. Through any point O draw OL, OM to represent the directions of AB, $A'B'$ and let the lengths of OL, OM be proportional to the astatic moments of the couples. We shall now prove that the diagonal ON of the parallelogram described on OL, OM will represent in direction the astatic arm of the resultant couple and in length the magnitude of the astatic moment of that couple.

Let the straight lines OL, LN be fixed in the body. By Art. 3 the two couples may be replaced by two others having OL, LN for their astatic arms and having the four forces all equal. The two forces acting at L being equal and opposite may be removed, so that the two given couples are equivalent to two equal and opposite forces acting respectively at O and N. These two forces constitute a single couple having ON for its astatic arm and having its astatic moment proportional to the length of ON. The proposition is therefore proved.

From this proposition we infer that the theorems used to compound forces apply also to compound the astatic arms of couples having their forces parallel. It is hardly necessary to add that the forces of the resultant couple are parallel to those of the two constituents.

5. *To find the astatic resultant of two couples whose astatic arms are parallel but whose forces are inclined at any angle.*

Let AB, $A'B'$ be the parallel astatic arms of the couples, both AB, $A'B'$ pointing in the same direction in the body. Through any point O draw OC parallel to AB and also two straight lines OL, OM parallel to the forces at A and A' and proportional to the astatic moments of the couples. We shall prove that the diagonal ON of the parallelogram OLM represents the moment of the resultant couple, the plane of the couple is parallel to the plane NOC, and the astatic arm is in the direction of OC.

Let the couples be referred to a common astatic arm along OC, the forces at O are then represented by OL and OM. Proceeding as in Art. 4 the results stated are easily seen to be true.

6. Working rule. Uniting these two propositions we may construct a rule to resolve or compound couples

When the forces are parallel we resolve or compound lengths, measured along the astatic arms and proportional to the astatic moments, by the parallelogram law, the new forces being supposed to act parallel to their former directions.

When the arms are parallel we resolve or compound lengths, measured along the directions of the forces and proportional to the astatic moments, by the parallelogram law, the new arms being parallel to their former directions.

7. There is one resolution of a couple which will be found useful afterwards.

Let Ox, Oy, Oz be any set of Cartesian axes, not necessarily rectangular. Let (x, y, z) be the coordinates of any point D, and let $OD = r$. Then a couple whose astatic arm is r and forces $\pm P$ may be resolved into three other couples whose astatic arms are situated in the axes of coordinates and whose lengths are equal to x, y, z. The forces of these couples are parallel to that of the original couple and their astatic moments are Px, Py, Pz.

Let us now take any three points A, B, C on the axes and let $OA = a$, $OB = b$, $OC = c$. These three couples may be replaced by three others having OA, OB, OC for their astatic arms. It follows that any force P acting at any point D may be replaced by four parallel forces acting at any four points A, B, C and O whose magnitudes are respectively equal to Px/a, Py/b, Pz/c and $P(1 - x/a - y/b - z/c)$.

Conversely, since these four parallel forces may be compounded into a single force equal to their sum and acting at the centre of gravity of A, B, C, O, it is evident that they are equivalent to the force P acting at the point (x, y, z). See Vol. I., Art. 80.

8. *Two couples cannot be astatically compounded together into a single resultant couple unless either the four forces are parallel or the two astatic arms are parallel.*

If possible let three couples be in astatic equilibrium. Transfer these parallel to themselves so that one force of each couple acts at the point O. Let OA, OB, OC be the astatic arms, let OP, OQ, OR be the directions of the forces. Then as the body is displaced, OA, OB, OC are fixed in the body, OP, OQ, OR are fixed in space.

If the four forces of any two of the three couples are parallel, the forces of their resultant couple are also parallel to them, by Art. 4. Thus equilibrium could not exist unless all the six forces were parallel to each other. In what follows, we may therefore suppose that no two of the three lines OP, OQ, OR are coincident. In the same way no two of the three arms OA, OB, OC are coincident.

Place the body so that OC, OR are in one straight line. Since in this position the couples (P, OA), (Q, OB) are in equilibrium, the planes POA, QOB coincide. Thus OA, OB lie in the plane POQ and continue to lie in that plane as the body is turned round OC. It follows that the axis OC must be perpendicular to this plane and therefore to both OA and OB. Similarly OA is perpendicular to both OB and OC.

Supposing as before that OC, OR are in one straight line, it is clear that the body may be turned round OC until OA coincides with OP. The axis OB must then coincide with OQ, for otherwise equilibrium could not exist. Summing up, the axes OA, OB, OC are at right angles and the body can be so placed that the forces of the respective couples act along their astatic axes.

Referring to the figure of Art. 76, Vol. I., we see that if the couple (P, OA) is a stable couple, the couple (Q, OB) must be unstable, for otherwise they would not act in opposite directions when the body is rotated about OC. Similarly by rotating the body about OB we see that (R, OC) is an unstable couple. Therefore (R, OC) cannot balance (Q, OB) when the body is rotated about OA. The three couples cannot therefore be in equilibrium in all positions of the body.

The Central Ellipsoid.

9. *To reduce any number of forces astatically to a single force and three couples.*

Let the forces be P_1, P_2, &c. and let their points of application be M_1, M_2, &c. respectively. Let Ox, Oy, Oz be any axes, not necessarily rectangular, which are fixed in the body and move with it. Let (x, y, z) be the coordinates of the point of application M of any one force P, and let $OM = r$.

Take three arbitrary points A, B, C on the axes of coordinates; let $OA = a$, $OB = b$, $OC = c$. By Art. 7 the force P acting at x, y, z, is equivalent to an equal and parallel force acting at O, together with three astatic couples whose arms are OA, OB, OC respectively, whose astatic moments are Px, Py, Pz and whose forces are parallel to P.

In this way all the forces may be brought to act at the origin parallel to their original directions. These may be compounded together into a single force, whose magnitude and direction in space are the same for all positions of the body. Let us represent this force by R.

Each force P will also give a couple having OA for its astatic arm. Compounding the forces at the extremities of this common arm, all these couples reduce to a single couple. The arm OA of this couple is fixed in the body while the magnitude and direction in space of the forces are the same for all positions of the body. Let us represent the magnitude of either of its forces by F.

The couples having OB, OC, for their astatic arms may be treated in the same way. Their astatic arms also are fixed in the body, while the magnitude and direction in space of the forces are always the same. Let these forces be G and H.

Summing up, we see that a system of forces can be reduced to a principal force R acting at any assumed base point O, together with three couples (F, OA), (G, OB) and (H, OC), having their astatic arms arranged along any three assumed straight lines OA, OB, OC fixed in the body and not all in one plane.

It may be seen that this reasoning, as far as we have gone, is the same as that used in the corresponding proposition when the body is fixed in space (Vol. I., Art. 257). The difference is, that when the body has only one position in space these three couples may be compounded into a single couple. But no single couple can be found which is equivalent to these, when the body may assume any position in space (Art. 8).

10. Consider any one position of the forces and of the body. In this position let X, Y, Z, be the components along the axes of any force P. To find the resultant force R, we bring all these P's to act at the base O. The force R is therefore the resultant of $\Sigma X, \Sigma Y, \Sigma Z$ acting at O along the axes. To avoid the continual recurrence of the symbol Σ it will be convenient to represent these components by X_0, Y_0, Z_0.

To find the force F we seek the resultant of all the forces similar to Px/a acting at A. The force F is therefore the resultant of the three forces $\Sigma Xx/a, \Sigma Yx/a, \Sigma Zx/a$ acting at A parallel to the axes. In the same way the forces G and H are the resultants of $\Sigma Xy/b, \Sigma Yy/b, \Sigma Zy/b$ and of $\Sigma Xz/c, \Sigma Yz/c, \Sigma Zz/c$. It will be found convenient to represent the summations ΣXx, ΣXy &c. by the symbols X_x, X_y, &c.

In this way the three couples (F, a), (G, b), (H, c) are resolved into nine elementary couples whose astatic moments are represented by the constituents of either of the following determinantal figures

$$\begin{aligned} \text{couple } (F, a) &= \Sigma Xx,\ \Sigma Yx,\ \Sigma Zx = X_x,\ Y_x,\ Z_x \\ \text{couple } (G, b) &= \Sigma Xy,\ \Sigma Yy,\ \Sigma Zy = X_y,\ Y_y,\ Z_y \\ \text{couple } (H, c) &= \Sigma Xz,\ \Sigma Yz,\ \Sigma Zz = X_z,\ Y_z,\ Z_z \end{aligned}$$

where the common arms of the three couples in the first, second and third rows are OA, OB, OC respectively. Thus the small letter or suffix indicates the axis on which the astatic arm is situated, while the large letter indicates the direction of the force. This convenient notation is the same as that used by Darboux.

These will be referred to afterwards as the nine elementary

couples. Together with X_0, Y_0, Z_0, the three force components, we thus have twelve elementary quantities for each base point.

For the sake of brevity we shall represent the couples (F, a). (G, b), (H, c) by the symbols K_x, K_y, K_z.

As we are chiefly concerned with the astatic moments of the couples, the forces and arms are separately of only slight importance. It is often convenient to *choose the arms of all the couples to be unity and positive.* The signs of the forces alone then determine the signs of the moments. In other cases it is found advantageous to *make the forces of all the couples equal to the force R.* The forces then divide out of the equations, leaving relations between lengths only.

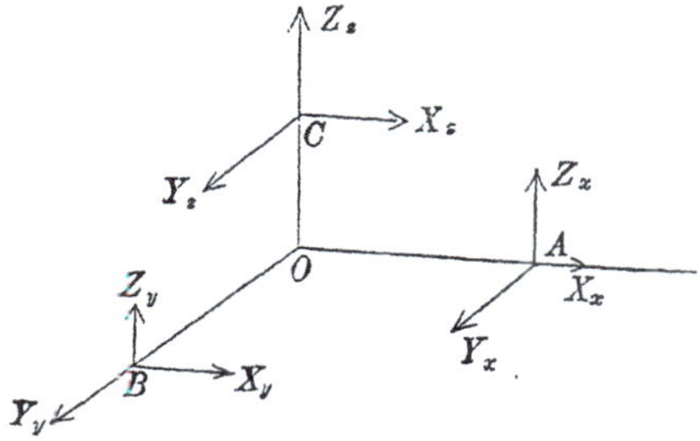

It will be found useful to remember that the direction ratios of any one of the forces F, G, H are proportional to the constituents of the corresponding row of the determinantal figure. An interpretation of the symbols when taken in columns will be found later on.

The figure represents the relation of the elementary couples to the axes. To avoid complication the forces at O are omitted. The directions of the forces at the extremities A, B, C of the astatic arms are shown by the arrow-head, while each arrow-head is marked by the astatic moment of the corresponding couple.

11. Conditions of equilibrium. *If a system of forces be in astatic equilibrium each of the twelve elements is zero.*

Resolving parallel to the axes we have $X_0 = 0$, $Y_0 = 0$, $Z_0 = 0$.

Taking moments about the axes of coordinates we have

$$Z_y - Y_z = 0, \quad X_z - Z_x = 0, \quad Y_x - X_y = 0.$$

But the body must be in equilibrium in all positions. Instead of turning the body round any axis, let us turn every force in the opposite direction round a parallel axis through its point of application. First let the rotation be about an axis parallel to x through a right angle. The X forces are all unchanged, but the Y forces now act parallel to z in the positive direction while the Z forces act parallel to y in the negative direction. Hence, writing Y for Z and $-Z$ for Y in the equations of moments already found, we have

$$Y_y + Z_z = 0, \quad X_z - Y_x = 0, \quad -Z_x - X_y = 0.$$

Joining these to the preceding equations we find $Z_x = 0$, $X_y = 0$, $X_z = 0$, $Y_x = 0$, i.e. every constituent with an x in it (except X_x) is zero.

In the same way by turning the system round y we find that all the constituents are zero except X_x, Y_y, Z_z. But we also find that $Y_y + Z_z = 0$, $Z_z + X_x = 0$, $X_x + Y_y = 0$. Hence each of the three X_x, Y_y, Z_z is also zero. *Thus all the twelve elements are zero.*

That these conditions of equilibrium are sufficient as well as necessary follows at once from the previous article. Thus, since the force F is the resultant of X_x/a, Y_x/a, Z_x/a, it is clear that F is zero. Similarly G and H are zero. Since X_0, Y_0, Z_0, are zero the principal force R is zero, so that the body is in equilibrium in all positions.

We may however also arrive at the same result independently. The body and forces in any one position being referred to axes x, y, z, let the twelve elements be zero. The axes x, y, z remaining fixed in space, let the body be moved about the origin into any other position, and let the coordinates of the point (x, y, z) become (x', y', z'). Since x, y, z are linear functions of x', y', z' whose coefficients are independent of the coordinates, it is evident that the twelve elements $\Sigma Xx'$ &c. are also zero. The six statical equations of equilibrium referred to in Art. **11** are therefore satisfied in this new position of the body.

12. *If two systems of forces be referred to the same origin and axes they cannot be astatically equivalent unless the twelve elements are equal each to each.*

Let the twelve elements of the two systems be X_x &c., X_x' &c. If we reverse the forces of the second system, the two systems together would be in equilibrium. Hence $X_x - X_x' = 0$, &c. $= 0$.

Thus all the elements are equal each to each.

13. Ex. 1. If the same system of forces can be astatically represented in either of two ways, viz. (1) by three forces (F, G, H) acting at (A, B, C) or (2) by three other forces (F', G', H') acting at (A', B', C'), prove that (unless the system can be reduced to two astatic forces instead of three) the planes ABC, $A'B'C'$ must coincide.

Let us first suppose that the three forces F, G, H, are not all parallel to one plane. Take the plane $A'B'C'$ as the plane of xy. We have X_z, Y_z, Z_z, the same for both systems. But since the ordinates of the points of application of F', G', H', are zero, each of the three X_z, Y_z, Z_z must be zero. Consider the equation $Z_z=0$. Place the body in such a position that the forces F, G act parallel to the plane $A'B'C'$. This is possible since a plane can be drawn parallel to any two straight lines. Then by hypothesis the direction of H will not be parallel to the plane $A'B'C'$. The components of the forces F, G, parallel to the axis of z are now zero. Hence Z_z must be zero for the single force H. Thus either $H=0$ or the ordinate of its point of application is zero. Supposing F, G, H to be all finite, it follows that C lies in the plane $A'B'C'$. By similar arguments we prove that the other points A, B also lie in the same plane $A'B'C'$.

Next, let us suppose all the three forces F, G, H are parallel to one plane. In this case one of the forces as H can be resolved into two components f and g parallel to F and G respectively. Each of the two sets of parallel forces (f, F) and (g, G) can be replaced by a single force at its centre of parallel forces. The system F, G, H can therefore be reduced to two astatic forces.

Ex. 2. If a system of forces F, G, H, acting at the corners of a triangle ABC, can be reduced to two astatic forces F', G' acting at two points A', B', then either the forces F, G, H are all parallel to one plane or the triangle ABC is evanescent.

We need only to examine the case in which F, G, H are all finite, for, if one be zero, the other two are necessarily parallel to one plane.

The system F', G' can be regarded as the limiting case of a triangle of forces F', G', H' acting at the corners of a triangle $A'B'C'$ where H' is zero and the position of C' is arbitrary. If then the forces F, G, H are not all parallel to the same plane it would follow from Ex. 1 that all the corners A, B, C lie in the plane $A'B'C'$. But this is impossible since C' is an arbitrary point, unless the triangle ABC is evanescent and lies in the straight line $A'B'$.

14. The central ellipsoid. A base point O having been chosen, the rectangular axes Ox, Oy, Oz are arbitrary. We shall now show that there is one system of axes which will enable us to analyse the system of forces more simply than any other.

Let Ox', Oy', Oz' be a second system of axes also fixed in the body. Let A', B', C' be points taken arbitrarily on these axes, let their distances from O be a', b', c'. Let F', G', H' be the forces which act at A', B', C'. We shall suppose both systems of axes to be rectangular.

As the body is moved about, the forces F', G', H' keep their directions in space unaltered, so that as regards the body the points of application and the magnitude of each force are the only elements fixed. Let us then find the magnitude of the force F' which acts at A', the forces at O, A, B, C being regarded as given. To effect this we shall resolve the arms of each of the nine elementary couples along OA', OB', OC', keeping the forces unaltered. We shall reserve for examination only those components whose arms are along OA'.

Let (l, m, n) be the direction cosines of the axis Ox'. Then the groups of couples (X_x, X_y, X_z); (Y_x, Y_y, Y_z); (Z_x, Z_y, Z_z) yield three component couples having their forces parallel to X, Y, Z respectively. Their astatic moments are (Art. 6),

$$X_xl+X_ym+X_zn=L_1,$$
$$Y_xl+Y_ym+Y_zn=L_2,$$
$$Z_xl+Z_ym+Z_zn=L_3.$$

These couples have a common arm OA' and their forces are at right angles. Compounding them we have

$$(F'a')^2=(X_xl+X_ym+X_zn)^2+(Y_xl+Y_ym+Y_zn)^2+(Z_xl+Z_ym+Z_zn)^2.$$

The direction cosines of the force F' are proportional to the three moments L_1, L_2, L_3.

We notice that this expression for $F'a'$ contains only the direction cosines of OA', and does not depend on the position of OB' or OC', except only that these must be at right angles to OA'. We are thus able to consider the couple whose arm is OA' apart from those whose arms are OB' and OC'.

Let us measure along OA' a length OP', such that OP' is inversely proportional to the astatic moment of the couple whose arm is OA'. For convenience we shall suppose the product of OP' and this astatic moment to be unity. Thus $OP'.F'a'=1$. Let $OP'=\rho$, and let ξ, η, ζ, be the coordinates of P' referred to the original axes Ox, Oy, Oz. Then $\xi=l\rho$, $\eta=m\rho$, $\zeta=n\rho$. We therefore find for the locus of P' the quadric

$$1=(X_x\xi+X_y\eta+X_z\zeta)^2+(Y_x\xi+Y_y\eta+Y_z\zeta)^2+(Z_x\xi+Z_y\eta+Z_z\zeta)^2.$$

15. This quadric may be regarded as defined by a statical property, viz. if any radius vector be taken as the axis Ox', the astatic moment of the corresponding couple (F', a') is measured by the reciprocal of that radius vector. It follows that whatever coordinate axes Ox, Oy, Oz are chosen we must have the same quadric. The equations of the quadric when referred to different sets of axes may be different, but the quadric itself is always the same. The quadric is therefore to be regarded as fixed in the body. Any point of the body may be chosen as the base O, and every such base has a corresponding quadric whose centre is at the base. This quadric is called *the central ellipsoid of that point.* It is also called *Darboux's ellipsoid.*

16. Let us represent the astatic moment of the couple whose astatic arm is directed from a given base along the radius vector ρ by the symbol K_ρ. In the same way the astatic moments, Fa,

Gb and Hc, of the couples whose astatic arms are directed along the axes will be represented by K_x, K_y, K_z. With this notation we have

$$X_x^2 + Y_x^2 + Z_x^2 = F^2a^2 = K_x^2,$$
$$X_y^2 + Y_y^2 + Z_y^2 = G^2b^2 = K_y^2,$$
$$X_z^2 + Y_z^2 + Z_z^2 = H^2c^2 = K_z^2;$$
$$X_yX_z + Y_yY_z + Z_yZ_z = K_yK_z \cos \alpha,$$
$$X_zX_x + Y_zY_x + Z_zZ_x = K_zK_x \cos \beta,$$
$$X_xX_y + Y_xY_y + Z_xZ_y = K_xK_y \cos \gamma;$$

where α, β, γ are the angles between the directions of the forces (G, H), (H, F), (F, G) of the couples K_x, K_y, K_z.

Expanding the squares in the equation of the central ellipsoid at the origin, it may be written in the form

$$K_x^2\xi^2 + K_y^2\eta^2 + K_z^2\zeta^2 + 2K_yK_z \cos \alpha\eta\zeta + 2K_zK_x \cos \beta\zeta\xi + 2K_xK_y \cos \gamma\xi\eta = 1.$$

Also if K' be the moment of the couple corresponding to the arm OA', whose direction cosines are l, m, n, we have

$$K'^2 = K_x^2l^2 + K_y^2m^2 + K_z^2n^2 + 2K_yK_zmn \cos \alpha + 2K_zK_xnl \cos \beta + 2K_xK_ylm \cos \gamma.$$

It may be useful to state the rule by which the signs of any of the astatic moments K_x, K_y, K_z are determined. The directions of the forces being fixed in space, there is for each line of action a positive and a negative direction determined by reference to some axes fixed in space. The astatic arms are measured in the body, and for each of these also there is a positive and a negative direction. Now imagine the couple moved parallel to itself until either extremity of its astatic arm is placed at the origin, so that one force acts at the origin. The moment is then the product of the astatic arm into the other force, when each is taken with its proper sign.

17. Ex. 1. Show that the discriminant of the central ellipsoid at the origin is equal to $(6VFGH)^2$, where V is the volume of the tetrahedron $OABC$.

Prove also that the minors of the coefficients of ξ^2, η^2, ζ^2 in the discriminant are $(K_yK_z \sin \alpha)^2$, $(K_zK_x \sin \beta)^2$ and $(K_xK_y \sin \gamma)^2$, respectively.

If parallels to the directions of the forces F, G, H are drawn from the centre of a sphere to cut the surface, the arcs joining the points of intersection form a spherical triangle whose sides are α, β, γ. If θ, ϕ, ψ be the opposite angles, the minors of the coefficients of $\eta\zeta$, $\zeta\xi$, $\xi\eta$ in the discriminant are respectively

$$-K_yK_zK_x^2 \sin \beta \sin \gamma \cos \theta, -K_zK_xK_y^2 \sin \gamma \sin \alpha \cos \phi \text{ and } -K_xK_yK_z^2 \sin \alpha \sin \beta \cos \psi.$$

Ex. 2. An astatic arm OP moves about any given base point O so that its corresponding astatic moment is constant. Show that OP traces out a cone in the body coaxial with the central ellipsoid at O.

Ex. 3. If Ox, Oy, Oz be any rectangular axes meeting at a fixed origin O, K_x, K_y, K_z the corresponding astatic moments, prove that $K_x^2 + K_y^2 + K_z^2$ is invariable for all such axes.

Since this expression is the first invariant of the central ellipsoid at O the property follows at once. It also follows from the geometrical property of an ellipsoid, that the sum of the squares of the reciprocals of three diameters at right angles is constant.

18. If we refer the central ellipsoid to its principal diameters as axes of reference, the equation loses the terms containing the products of the coordinates. If F, G, H represent the forces of the three couples for this position of the axes, the equation is

$$F^2a^2\xi^2 + G^2b^2\eta^2 + H^2c^2\zeta^2 = 1.$$

The quadric is therefore in general an ellipsoid. If one of the three forces is zero, i.e. if one of the couples is absent, the quadric reduces to a cylinder.

Since the terms containing the products $\xi\eta$, $\eta\zeta$, $\zeta\xi$ are absent, it follows that if the three forces F, G, H are all finite, their directions are at right angles to each other. If one force is zero, the other two must be at right angles.

Summing up, we see that *whatever point of the body we choose as base, there are always three straight lines at right angles, fixed in the body, such that, when these are taken as the astatic arms of the couples, the forces of the couples act in directions at right angles to each other and are fixed in space.*

In this way we have for each base point two convenient systems of rectangular axes, one fixed in the body, viz. the astatic arms of the couples, the other fixed in space, viz. the directions of the forces.

The axes fixed in the body are called the *principal axes of the base.* The couples are then called the *principal couples.*

19. The initial position. The base point O being regarded as fixed, and the body referred to principal axes, it is evident that we may turn the body about O until the system of axes fixed in the body coincides in position with the system fixed in space.

The peculiarity of this position of the body is that the forces of each of the three couples act along the astatic arm of that couple. The moments of the couples are therefore zero. The forces P_1, P_2, &c. of the given system reduce to the single resultant R whose line of action passes through the given base.

This is called an *initial position* of the body and the couples are then said to be in their *zero positions.*

The body being placed in an initial position, it is clear that if we turn it round any one of the astatic arms through two right angles, the same property will recur again, i.e. the force of each couple will act along its astatic arm. Thus any base being given there are at least four corresponding initial positions.

Though in all these four positions of the body the two systems of axes coincide in position, yet the positive direction of an axis of one system may be the same as either the positive or the negative direction of an axis of the other system. It is usual to choose the positive directions of one system so that in one of these four positions of the body the two systems of axes may have the same positive directions as well as coincide in position. This initial position is called the *positive initial position.*

20. When the body is placed in a positive initial position the nine elementary couples described in Art. 10 are reduced to

$$\begin{array}{lll} X_x & Y_x = 0 & Z_x = 0, \\ X_y = 0 & Y_y & Y_z = 0, \\ X_z = 0 & Y_y = 0 & Z_z. \end{array}$$

The equation to the central ellipsoid then takes the simple form
$$X_x^2 \xi^2 + Y_y^2 \eta^2 + Z_z^2 \zeta^2 = 1.$$

If (l, m, n) be the direction cosines of any other arm OA' the direction cosines of the force F' acting at its extremity are proportional to
$$X_x l, \quad Y_y m, \quad Z_z n,$$
and
$$(F'a')^2 = X_x^2 l^2 + Y_y^2 m^2 + Z_z^2 n^2.$$

Thus the direction and magnitude of F' have been found. If the body is now moved into any other position, F' continues to act in the same direction in space and therefore continues to make the same angles with F, G, H that it made in the initial position.

21. *There are no other positions besides the four initial positions in which a body can be placed so that the system of forces may reduce to a single resultant which passes through the given base, except when the central ellipsoid at the given base point is a surface of revolution.*

Let OA, OB, OC be the principal axes at the given base O. Let OF, OG, OH be three straight lines at right angles drawn parallel to the forces of the corresponding couples. In order to use conveniently the formulæ of spherical trigonometry we suppose these axes to cut the surface of a sphere whose centre is at O in the six points A, B, C, F, G, H. The planes of the couples are the planes which contain the astatic arms and the forces, and are therefore the planes of the spherical arcs AF, BG, CH. If their astatic moments are $K_x = Fa$, $K_y = Gb$, $K_z = Hc$ their moments in any position of the body are $K_x \sin AF$, $K_y \sin BG$ and $K_z \sin CH$.

When the body is in an initial position the spherical triangles coincide. Starting from this position, the body may be brought into any other by turning it round some axis OI. If this axis intersect the sphere in I, the spherical arcs IA, IB, IC are respectively equal to IF, IG, IH, and if 2ω is the angle of rotation, the angles AIF, BIG, CIH are each equal to 2ω. Join AF, BG, CH by arcs of great circles and draw the perpendicular arcs IL, IM, IN.

If this position of the body can be one of equilibrium when the base is fixed, the three couples must balance each other. Resolving the axis of each of these along and perpendicular to OI, the moments of the three latter components are respectively $K_x \sin AF \cos IL$, $K_y \sin BG \cos IM$, and $K_z \sin CH \cos IN$. Since the three components are in equilibrium, these moments must be proportional to $\sin MIN$, $\sin NIL$, $\sin LIM$, that is to $\sin BIC$, $\sin CIA$ and $\sin AIB$.

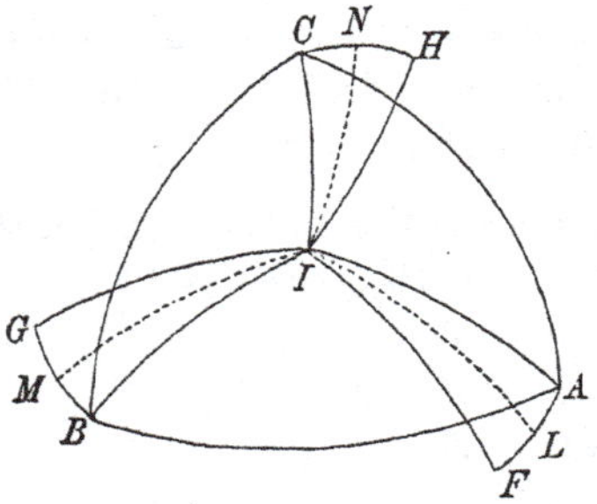

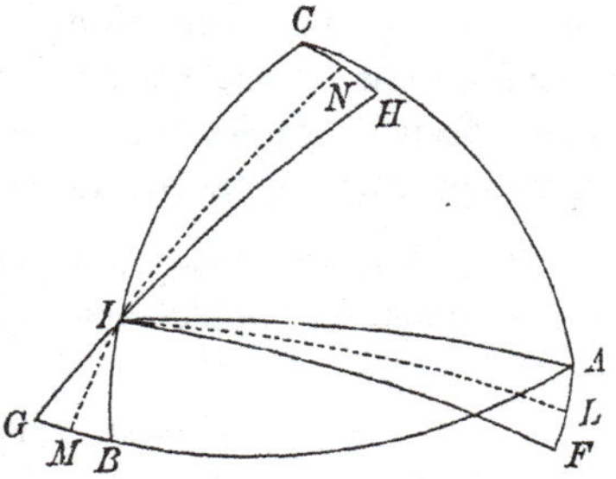

For brevity let α, β, γ represent the arcs IA, IB, IC. Since BC is a right angle we have
$$\cos\beta\cos\gamma + \sin\beta\sin\gamma\cos BIC = \cos BC = 0,$$
$$\begin{aligned}\therefore \sin^2\beta\sin^2\gamma\sin^2 BIC &= \sin^2\beta\sin^2\gamma - \cos^2\beta\cos^2\gamma\\ &= 1 - \cos^2\beta - \cos^2\gamma\\ &= \cos^2\alpha.\end{aligned}$$
Again, $$\sin AF\cos IL = 2\sin\tfrac{1}{2}AF\cos\alpha = 2\sin\alpha\cos\alpha\sin\omega.$$
Similar expressions hold for the other angles.

Substituting these values in the condition of equilibrium, and dividing out the common factors, we have $K_x^2 = K_y^2 = K_z^2$. Thus the proposed position of the body cannot be one of equilibrium when the base is fixed unless the ellipsoid is a sphere.

This argument assumes that none of the factors divided out are zero. We must therefore examine separately the case in which I lies on one of the principal planes. If I lies on BC, the first component is zero, and the other two are $K_y \sin BG \cos IM$ and $K_z \sin CH \cos IN$. The condition of equilibrium is that these moments should be equal; hence $$K_y^2\sin^2\beta\cos^2\beta = K_z^2\sin^2\gamma\cos^2\gamma.$$
Since β and γ are complementary, this requires that $K_y^2 = K_z^2$, i.e. the ellipsoid is one of revolution.

Lastly, if I is at the point C, each of the three component couples is zero. The component having OI for its axis is then the sum or difference of the couples $K_x \sin 2\omega$, $K_y \sin 2\omega$. Since this component also must vanish we again have $K_x^2 = K_y^2$, i.e. the ellipsoid is one of revolution.

22. Ex. 1. The body being placed in a positive initial position, prove that the direction of F' is parallel to the normal to the ellipsoid $X_x\xi^2 + Y_y\eta^2 + Z_z\zeta^2 = 1$ drawn at the point where OA' cuts the ellipsoid. This ellipsoid is called the second central ellipsoid of Darboux.

Ex. 2. The body being placed in a positive initial position, a straight line OQ is drawn from the base parallel and proportional to the force F' for all positions of OA' in the body. Prove that the locus of Q is the ellipsoid
$$\left(\frac{\xi}{X_x}\right)^2 + \left(\frac{\eta}{Y_y}\right)^2 + \left(\frac{\zeta}{Z_z}\right)^2 = 1.$$
This is called the third central ellipsoid of Darboux.

Prove also that, if the arms OA', OB', OC' be at right angles, the corresponding forces F', G', H' are parallel to a system of conjugate diameters in the third ellipsoid. This and the last example are due to Darboux.

Ex. 3. When the body is in a positive initial position for any base, prove that the direction of the force corresponding to any astatic arm OA' is parallel to the eccentric line of OA' in the central ellipsoid of the given base.

The Central Plane and the Central Point.

23. *To compare the central ellipsoids at different points of the body.*

Suppose the forces to be referred to any base O and any axes Ox, Oy, Oz, and that the nine elementary couples and the three force-components are known for these axes. We shall now find the corresponding quantities when some point O', whose coordinates are (p, q, r), is taken as the base.

Through O' we draw axes $O'x'$, $O'y'$, $O'z'$ parallel to (x, y, z). The nine elementary couples may be transferred to these new axes without any change (Art. 3). But the three force-components will introduce new couples. By Art. 9 the component X_0 acting at O may be transferred to the origin O' if we introduce the new couples $(X_0, -p)$, $(X_0, -q)$, $(X_0, -r)$, the coordinates of O referred to O' being $(-p, -q, -r)$. Similar reasoning applies to the components Y_0, Z_0. Hence we have for the nine elementary couples at O'

$$\begin{array}{lll} X_x' = X_x - X_0p, & Y_x' = Y_x - Y_0p, & Z_x' = Z_x - Z_0p, \\ X_y' = X_y - X_0q, & Y_y' = Y_y - Y_0q, & Z_y' = Z_y - Z_0q, \\ X_z' = X_z - X_0r, & Y_z' = Y_z - Y_0r, & Z_z' = Z_z - Z_0r. \end{array}$$

The equation of the central ellipsoid at O' is therefore, by Art. 14,

$$\begin{aligned} &\{(X_x - X_0p)\,\xi' + (X_y - X_0q)\,\eta' + (X_z - X_0r)\,\zeta'\}^2 \\ &+ \{(Y_x - Y_0p)\,\xi' + (Y_y - Y_0q)\,\eta' + (Y_z - Y_0r)\,\zeta'\}^2 \\ &+ \{(Z_x - Z_0p)\,\xi' + (Z_y - Z_0q)\,\eta' + (Z_z - Z_0r)\,\zeta'\}^2 = 1\,; \end{aligned}$$

the origin of the running coordinates ξ', η', ζ' being O'.

24. If the principal force R is zero, we have $X_0 = 0$, $Y_0 = 0$, $Z_0 = 0$. Int his case the central ellipsoid at O' is the same as that at O. Thus the central ellipsoids at all base points are similar and similarly situated.

25. **The Central Plane.** If the principal force R be not zero

the form of the central ellipsoid will depend on the position of the base point. We notice that the three planes

$$(X_x - X_0 p)\,\xi' + (X_y - X_0 q)\,\eta' + (X_z - X_0 r)\,\zeta' = 0,$$
$$(Y_x - Y_0 p)\,\xi' + (Y_y - Y_0 q)\,\eta' + (Y_z - Y_0 r)\,\zeta' = 0,$$
$$(Z_x - Z_0 p)\,\xi' + (Z_y - Z_0 q)\,\eta' + (Z_z - Z_0 r)\,\zeta' = 0$$

are conjugate planes.

If the central ellipsoid is a cylinder the conjugate planes pass through the axis of the cylinder, and the equations to the three conjugate planes are then not independent. We thus have the determinantal equation

$$\begin{vmatrix} X_x - X_0 p, & X_y - X_0 q, & X_z - X_0 r, \\ Y_x - Y_0 p, & Y_y - Y_0 q, & Y_z - Y_0 r, \\ Z_x - Z_0 p, & Z_y - Z_0 q, & Z_z - Z_0 r, \end{vmatrix} = 0 \ldots (1).$$

This equation may be written in the form

$$\begin{vmatrix} X_0 & X_x & X_y & X_z \\ Y_0 & Y_x & Y_y & Y_z \\ Z_0 & Z_x & Z_y & Z_z \\ 1 & p & q & r \end{vmatrix} = 0 \quad \ldots\ldots\ldots\ldots\ldots (2).$$

When p, q, r are regarded as the running coordinates, this is evidently the equation to a plane. *The peculiarity of this plane is that, if any point on it is chosen as base, the central ellipsoid is a cylinder.* This plane is called *the central plane.*

26. Since the central ellipsoid at every point is fixed in the body the locus of base points at which the ellipsoid is a cylinder is also fixed. *The central plane is therefore fixed in the body.* In discussing its properties we may put the body into any position we please.

Take any point O on the central plane as base, and let the body be placed in an initial position. By Art. 20 all the nine elementary couples, except X_x, Y_y, Z_z, are zero. Since the ellipsoid is a cylinder one of the three X_x, Y_y, Z_z is also zero, say $X_x = 0$. Substituting in the second form of the equation to the central plane given in Art. 25, we see that it becomes $pX_0 Y_y Z_z = 0$. If any one of the three X_0, Y_y, Z_z is zero, the equation to the plane is indeterminate, but if all these are finite, the equation to the central plane is $p = 0$. It follows therefore that *the infinite axis of the central ellipsoid at any point of the central plane is perpendicular to that plane.*

27. This leads to a simplified reduction of the forces P_1, P_2, &c. Let us take the base of reference O at any point of the central plane, and the principal diameters of the central cylinder as axes of coordinates. The moment of that principal couple whose astatic axis is along the infinite axis of the cylinder is measured by the reciprocal of that axis, and is therefore zero. Thus all the forces have been reduced to two couples (instead of three) and a force R. The astatic arms of the couples lie in the central plane and the forces of one couple are perpendicular to those of the other.

28. **The Central Point.** It has been proved in Art. 9 that a system of forces may be reduced to a principal force R at the base of reference and three couples having their arms directed along any three straight lines at right angles. Let us now enquire if a base O' can be found such that each of the forces of the couples is perpendicular to the principal force.

If one system of axes $O'A$, $O'B$, $O'C$ at any base O' possess this property, then every system of axes at that base will also possess the same property. To prove this, let $O'A'$, $O'B'$, $O'C'$ be any other such system of axes. To deduce the forces at A', B', C' from those at A, B, C, we resolve the arms OA, OB, OC in the directions OA', OB', OC' and transfer the forces parallel to themselves, see Art. 6. Since each of the forces at A, B, C is perpendicular to the force R, it follows that the forces at A', B', C', which are compounded of these, are also perpendicular to R.

Let Ox, Oy, Oz be any given rectangular axes, and let p, q, r be the coordinates of O'. Through O' draw a system of axes $O'x'$, $O'y'$, $O'z'$ parallel to Ox, Oy, Oz. Then, by what has just been proved, the couples corresponding to these axes must have their forces perpendicular to R. If the nine corresponding elementary couples are X_x' &c., the conditions of perpendicularity are

$$X_0X_x' + Y_0Y_x' + Z_0Z_x' = 0,$$

and two similar equations obtained by writing y and z for x in the suffixes. Substituting for X_x', &c. their values given in Art. 23,

$$R^2p = X_0X_x + Y_0Y_x + Z_0Z_x,$$
$$R^2q = X_0X_y + Y_0Y_y + Z_0Z_y,$$
$$R^2r = X_0X_z + Y_0Y_z + Z_0Z_z.$$

Since these give only one set of values for p, q, r there is but one point which possesses the given property. This point is called *the central point.*

29. *The central point lies on the central plane.* To prove this let us consider the principal axes at the central point. Since the forces of the three couples are at right angles to each other, they cannot all, if finite, be perpendicular to the principal force. One of these must therefore vanish. The central ellipsoid is therefore a cylinder, i.e. the central point lies on the central plane.

That the central point lies in the central plane may also be proved by substituting its coordinates in the equation (2) of the central plane found in Art. 25. These coordinates p, q, r are given in Art. 28, and a simple inspection shows that the equation is satisfied.

Thus it appears that *there is a certain point, lying on the central plane, such that the forces of the two principal couples at that point are at right angles to each other and to the principal force. This point is called the central point.*

The central point in the three-dimensional theory has not the same signification as the central point defined in Vol. I., Art. 160, with reference to two dimensions. In the latter the displacements of the body are confined to one plane, and for such displacements the single resultant always passes through a central point fixed in the body. In the former the displacements are unrestricted so that the lines of action of the forces do not necessarily remain in one plane.

The preceding theorems on the central plane and central point are generally given in treatises on Astatics, though the demonstrations in each may be different.

30. *We may express the formulæ for the coordinates of the central point in the form of a working rule.*

As already explained in Art. 9 the forces are represented by P_1, P_2, &c. Their points of application are M_1, M_2, &c. and their coordinates are (x_1, y_1, z_1), (x_2, y_2, z_2), &c. Also let the direction cosines of P_1, P_2, &c. be respectively (a_1, b_1, c_1), (a_2, b_2, c_2), &c. Then

$$X_x = P_1a_1x_1 + P_2a_2x_2 + \dots \qquad X_0 = P_1a_1 + P_2a_2 + \dots$$

$$Y_x = P_1b_1x_1 + P_2b_2x_2 + \dots \qquad Y_0 = P_1b_1 + P_2b_2 + \dots$$

$$Z_x = P_1c_1x_1 + P_2c_2x_2 + \dots \qquad Z_0 = P_1c_1 + P_2c_2 + \dots$$

Let θ_{12}, θ_{13}, &c., be the inclinations of the forces (P_1, P_2), (P_1, P_3) &c. Then

$$\cos\theta_{12} = a_1a_2 + b_1b_2 + c_1c_2, \text{ \&c.}$$

Substituting in the expression for p, Art. 28, we have

$$p = \frac{P_1Q_1x_1 + P_2Q_2x_2 + \dots}{P_1Q_1 + P_2Q_2 + \dots}$$

where $$Q_1 = P_1 + P_2 \cos\theta_{12} + P_3 \cos\theta_{13} + \dots$$

$$Q_2 = P_1 \cos\theta_{12} + P_2 + P_3 \cos\theta_{23} + \dots$$

&c. &c.

It is evident that Q_1 is the sum of the resolved parts of all the forces in the direction P_1, Q_2 is the sum of the resolved parts in the direction P_2, and so on.

The equation just arrived at is the common formula for the centre of gravity of weights P_1Q_1, P_2Q_2 &c. Similar equations hold for q and r. Hence we have this rule. *To find the central point of any number of forces, we first multiply each force by the sum of the resolved parts of all the forces along the direction of that force. We then place weights proportional to these products at the points of application of the forces. The centre of gravity of these weights is the central point required.*

31. Ex. Show that the equation to the central plane, referred to any axes, when expressed in terms of the forces and their mutual inclinations takes the form

$$L_x\xi + L_y\eta + L_z\zeta = M$$

where $$M = \Sigma P_1P_2P_3V_{123} \begin{vmatrix} x_1, & x_2, & x_3 \\ y_1, & y_2, & y_3 \\ z_1, & z_2, & z_3 \end{vmatrix} \text{ and } V_{123} = \begin{vmatrix} a_1, & a_2, & a_3 \\ b_1, & b_2, & b_3 \\ c_1, & c_2, & c_3 \end{vmatrix}.$$

The coefficient L_x is derived from M by writing unity for each of the x's in the determinant, L_y is derived from M by writing unity for each y, and so on.

To prove this, we start with the equation (2) of the central plane given in Art. 25 and make the same substitutions as in Art. 30. On writing down the determinant it will be seen that the determinants L_x, L_y, L_z may be obtained from the determinant M by the rule just stated. The determinantal sum M when expanded takes the form of a series of products of triplets of the forces. To find the coefficient of $P_1P_2P_3$ we put all the other forces equal to zero; the determinant then assumes the known form of the product of the two determinants just written down.

32. Summary. It will be convenient if we now sum up shortly the gradual steps made in reducing a system of forces to its simplest equivalents.

1. In Art. 9 the forces were reduced to a force R at an arbitrary base point O together with three couples whose arms Ox, Oy, Oz are arbitrary.

2. In Art. 18 it was shown that at the arbitrary base the arms Ox, Oy, Oz could be chosen at right angles to each other so that the forces of each couple are at right angles to the forces of the other two couples. These arms are called the principal axes at O and are fixed in the body.

3. In Art. 25 it was shown that, if the base point O is placed anywhere on a certain plane fixed in the body, the forces can be reduced to the single force R together with *two* couples. The arms of these couples are at right angles and lie in

the plane. The forces also of each couple are perpendicular to those of the other. This plane is called the central plane.

4. In Art. 28 it was shown that if the base point is placed at a certain point on the central plane the forces of the couples are perpendicular to the force R. Thus the forces of the original system can finally be reduced to a force R together with two couples whose arms are at right angles and such that the forces of each couple are not only perpendicular to those of the other but are also perpendicular to the force R. This base point is called the central point.

The principal axes at the central point are two straight lines lying in the central plane and a third, perpendicular to that plane. The two former are called *the central lines* of the central plane. The latter is sometimes called the central axis. But it must not be confused with Poinsot's central axis with which it coincides only when the body is properly placed. It bears indeed a certain resemblance to Poinsot's central axis, for the system is reduced to a force and two couples (instead of one) such that the forces of the couples are perpendicular to the force.

33. Analogy to Moments of Inertia. Ex. 1. If K be the astatic moment of the couple corresponding to any astatic arm OP drawn from the central point O, prove that the astatic moment K' of the couple corresponding to any parallel arm $O'P'$ drawn from any point O' is given by $K'^2 = K^2 + R^2p^2$ where p is the projection of OO' on either astatic arm.

Thus, a motion of the base O in a direction perpendicular to the astatic arm does not alter the magnitude of the astatic moment, but a motion along the arm from the central point increases the moment.

Ex. 2. If K_1, K_2, K_3 be the astatic moments corresponding to the principal astatic axes Ox, Oy, Oz drawn from any point O, prove that the astatic moment K corresponding to any arm OP making angles α, β, γ with the axes is given by

$$K^2 = K_1^2 \cos^2\alpha + K_2^2 \cos^2\beta + K_3^2 \cos^2\gamma.$$

It appears from these two propositions that the theory of astatic moments of couples has an analogy to the theory of moments of inertia. The square of the astatic moment about an arm drawn from O in any direction OP corresponds to the moment of inertia of a rigid body with regard to a *plane* drawn through O perpendicular to OP. By noticing this correspondence we may deduce the analogous propositions in the two theories one from the other.

It is clear from these two propositions that the mass of the rigid body is analogous to the square of the principal force R, and that the centre of gravity must be at the central point. For any base in the central plane the moment of the couple whose astatic arm is perpendicular to that plane is zero, hence the rigid body must be a lamina whose plane is the central plane of the forces.

The analogy may be made more distinct by adding another proposition. Let O be the central point, Oy, Oz the principal astatic axes in the central plane, Ox that perpendicular. The astatic moment K about any axis OP, whose direction cosines are l, m, n, is given by

$$K^2 = K_2^2 m^2 + K_3^2 n^2 \qquad \text{(1)}.$$

Let a lamina be placed in the plane of yz with its centre of gravity at O, having the axes of x, y, z for its principal axes of inertia; and let K_2^2, K_3^2 be its moments of inertia at the origin with regard to the planes respectively perpendicular to the axes of y and z. The equation (1) then shows that K^2 is the moment of inertia of the lamina with regard to a plane drawn through O perpendicular to OP.

Let O' be any other point whose coordinates are ξ, η, ζ, and let $O'P'$ be parallel to OP. The astatic moment K' at O' corresponding to the arm $O'P'$ is given by

$$K'^2 = K^2 + R^2p^2 \quad \text{.............................(2)},$$

where p is the projection of OO' on OP. This is also the formula which gives the moment of inertia of the lamina with regard to a plane drawn through O' perpendicular to $O'P'$, provided R^2 is the mass of the body.

It follows that the moment of inertia of the lamina with regard to a plane drawn through any point O' perpendicular to any straight line $O'P'$ represents the square of the astatic moment at the base O' for the arm $O'P'$.

Since the moments of inertia for all arms through O' represent the squares of the astatic moments for the same arms, it follows that they have the same maxima and minima and are connected together by the same rules. The principal axes of inertia at O' are therefore the same in direction as the principal astatic axes at O'.

That the principal astatic moments at O' are the normals to the confocals (4) of Art. 34, and that the astatic moments are the three values of M given by the cubic, follow at once from the properties of the principal axes of inertia, see *Rigid Dynamics*, Vol. I. Art. 56.

Since the moments of inertia of the lamina about the axes of y and z are respectively K_3^2 and K_2^2, it follows that the lamina might take the form of a homogeneous elliptic disc, whose semi-axes of y and z are respectively $2K_2/R$ and $2K_3/R$, and whose mass is R^2. The boundary is therefore similar to the imaginary focal conic.

The Confocals.

34. *To investigate the mode in which the central ellipsoids at different bases are arranged about the central point.*

Let the central point be chosen as the origin and the principal diameters of the central ellipsoid as axes of coordinates. Let the infinite axis be the axis of x, then the plane of yz is the central plane.

As we are enquiring into the positions of the neighbouring central ellipsoids, and as these are fixtures in the body, we may put the body itself into any position we may find convenient. Let it be placed in its positive initial position with the central point as the base.

In this position all the nine elementary couples are zero, except Y_y and Z_z. Also $X_0 = R$, $Y_0 = 0$, $Z_0 = 0$. The central ellipsoid at the origin is $\quad Y_y^2\eta^2 + Z_z^2\zeta^2 = 1$. ....(1).

The central ellipsoid at any point O' whose coordinates are p, q, r, is

$$Y_y^2\eta'^2 + Z_z^2\zeta'^2 + R^2(p\xi' + q\eta' + r\zeta')^2 = 1 \quad \text{........(2)},$$

where (ξ', η', ζ') are referred to axes meeting at O' parallel to the axes x, y, z, Art. 23.

Let an astatic arm $O'A'$ move about O' so that the corresponding couple (F', OA') has a constant astatic moment equal to M, and in any position let (l, m, n) be its direction cosines. Then, since the moment M (Art. 14) is the reciprocal of the corresponding radius vector of the central ellipsoid, we see that l, m, n are connected together by the relation

$$Y_y^2m^2 + Z_z^2n^2 + R^2(pl + qm + rn)^2 = M^2;$$

$$\therefore M^2l^2 + (M^2 - Y_y^2)\,m^2 + (M^2 - Z_z^2)\,n^2 = R^2(pl + qm + rn)^2 \ldots(3).$$

Now, after division by R^2, the left-hand side of equation (3) expresses the square of the perpendicular drawn from the central point on a tangent plane to the ellipsoid

$$\frac{\xi^2}{M^2} + \frac{\eta^2}{M^2 - Y_y^2} + \frac{\zeta^2}{M^2 - Z_z^2} = \frac{1}{R^2} \ldots\ldots\ldots\ldots(4);$$

and the right-hand side of (3) expresses the square of the perpendicular from the central point on a plane through O' parallel to that tangent plane. The equation (3) therefore shows that this tangent plane passes through O'. Hence we infer *that if $O'A'$ move about O', so that the corresponding astatic moment is constant and equal to M, then $O'A'$ is always perpendicular to a tangent plane drawn from O' to touch the confocal* (4).

These tangent planes all touch the enveloping cone of the confocal (4), and the axis $O'A'$ traces out the reciprocal cone of this enveloping cone. These two cones are known to be co-axial and their axes (Art. 17, Ex. 2) are in the same directions as those of the central ellipsoid at O'.

If M is so chosen that the confocal (4) passes through the point O', the enveloping cone becomes the tangent plane and therefore the cone traced out by $O'A'$ reduces to the normal at O'.

Hence *the principal diameters of the central ellipsoid at any point O' are the three normals to the three quadrics which pass through O' confocal to the quadric* (4). *Also the astatic moments of the three corresponding couples are the values of M given by the cubic* (4) *when we write for ξ, η, ζ the coordinates of O'.*

35. *Instead of using the three confocals we may use any one of them, say the ellipsoid.* By known properties of solid geometry the three normals at any point O' are (1) the normal to the ellipsoid, (2) parallels to the principal diameters of the section of the ellipsoid diametral to OO'.

Let M_1, M_2, M_3 be the three values of M given by the cubic (4), M_1 being the greatest. Let D_2, D_3 be the lengths of the principal semidiameters of the section of the ellipsoid, D_2 being parallel to the normal at O' to the confocal M_2, and D_3 parallel to the normal to M_3. Then it is known by solid geometry that

$$D_2^2 = M_1^2 - M_2^2, \qquad D_3^2 = M_1^2 - M_3^2.$$

Thus M_2, M_3 are known in terms of M_1 and quantities connected with the ellipsoid.

36. As these confocals play an important part in the theory of astatic forces, it is necessary to state distinctly their position.

Let the body be referred to the central point as origin, and the principal diameters of the central cylinder as axes, the plane of yz being the central plane. Let K_2, K_3 be the astatic moments of the couples whose astatic arms are along y and z. These astatic moments are the same for all positions of the body and are represented by Y_y and Z_z when the body is in its initial position. The equation to the confocals is therefore

$$\frac{\xi^2}{M^2} + \frac{\eta^2}{M^2 - K_2^2} + \frac{\zeta^2}{M^2 - K_3^2} = \frac{1}{R^2}.$$

The focal conics of these are obtained in the usual manner by putting $M = K_2$, $\eta = 0$; $M = K_3$, $\zeta = 0$; and $M = 0$, $\xi = 0$. We thus have

$$\frac{\xi^2}{K_2^2} - \frac{\zeta^2}{K_3^2 - K_2^2} = \frac{1}{R^2}, \qquad \eta = 0;$$

$$\frac{\xi^2}{K_3^2} + \frac{\eta^2}{K_3^2 - K_2^2} = \frac{1}{R^2}, \qquad \zeta = 0;$$

$$-\frac{\eta^2}{K_2^2} - \frac{\zeta^2}{K_3^2} = \frac{1}{R^2}, \qquad \xi = 0.$$

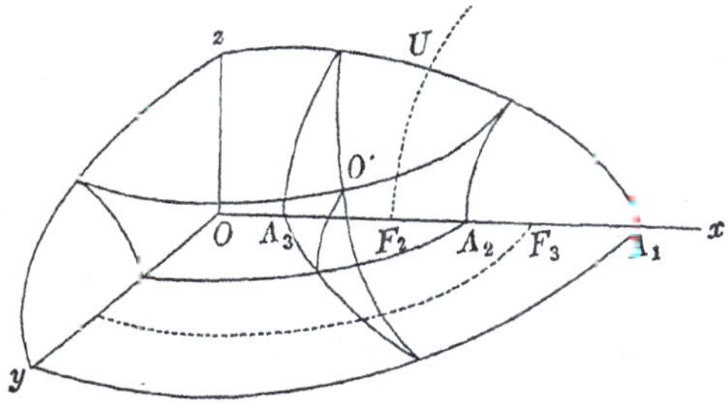

If we take as the standard case $K_3 > K_2$, the first is a hyperbola, the second an ellipse, and the third is imaginary. The two first are represented in the diagram by the dotted lines. *These*

conics will be referred to as the focal conics, and a straight line intersecting both conics may be called a focal line.

The figure represents the positive octant of a set of confocal quadrics intersecting in O'. The semi x-axes are represented by OA_1, OA_2, OA_3 and are respectively equal to M_1/R, M_2/R, M_3/R. As is well known the vertices F_3, F_2 of the two focal conics lie between A_1, A_2 and A_3. We have $OF_2 = K_2/R$, $OF_3 = K_3/R$.

If $K_2 = 0$, the ellipsoid and the hyperboloid of one sheet are surfaces of revolution. The hyperboloid of two sheets reduces to any two planes through Oz, and the hyperbolic conic becomes the axis of z. The central plane is now indeterminate and is any plane through the astatic arm of K_3.

If both $K_2 = 0$ and $K_3 = 0$, the ellipsoid becomes a sphere, one hyperboloid is a right cone, and the other any two planes through the axis of the cone.

37. Theorem on focal lines. *A straight line is drawn from any point P on one focal conic to any point Q on the other, it is required to prove that*

$$R^2\rho^2 = K_2^2 a_2^2 + K_3^2 a_3^2,$$

where a_1, a_2, a_3 are the direction cosines of PQ, and ρ is the perpendicular distance from the origin.

We know that the tangent planes drawn through any right line to the two confocals which that line touches are at right angles to each other, see Salmon's *Solid Geometry*, Art. 172. Since the focal conics are evanescent confocals, the planes through PQ and the tangents at P and Q to the conics are at right angles. If p, p' are the perpendiculars on these planes, l, m, n; l', m', n' their direction cosines, we have

$$R^2p^2 = K_2^2l^2 - (K_3^2 - K_2^2)\,n^2, \qquad R^2p'^2 = K_3^2l'^2 + (K_3^2 - K_2^2)\,m'^2.$$

$$\therefore\ R^2\rho^2 = R^2(p^2 + p'^2) = K_2^2(l^2 + n^2 - m'^2) + K_3^2(l'^2 + m'^2 - n^2).$$

Since the straight lines p, p' and PQ are mutually at right angles, this becomes

$$K_2^2(1 - m^2 - m'^2) + K_3^2(1 - n^2 - n'^2) = K_2^2a_2^2 + K_3^2a_3^2.$$

The theorem may be more easily proved by taking as the coordinates of P and Q (x, y, z) and (x', y', z') where

$$Rx = K_2\sec\theta, \qquad Ry = 0, \qquad Rz = (K_3^2 - K_2^2)^{\frac{1}{2}}\tan\theta,$$

$$Rx' = K_3\cos\phi, \qquad Ry' = (K_3^2 - K_2^2)^{\frac{1}{2}}\sin\phi, \qquad Rz' = 0.$$

The direction cosines a_2, a_3 and the length ρ may then be found by elementary formulæ, and it will be seen that the relation to be proved is satisfied.

It follows from this theorem that every focal line is a generator of the right circular cylinder whose radius is ρ and whose axis passes through the common centre of the conics and is parallel to the focal line.

Ex. 1. Show that four real focal lines can be drawn parallel to a given straight line.

Let a generator parallel to the given straight line travel round the hyperbolic conic and trace out a cylinder. This will cut the plane of the other conic in a hyperbola. Each branch of this hyperbola passes inside the elliptic conic, because it goes through the focus; it therefore cuts the ellipse in two points.

Ex. 2. If a straight line PQ intersect one focal conic and if its distance from the central point be ρ, where ρ is given in the theorem above, show that that straight line will intersect the other conic also.

If possible let PQ intersect one focal conic in P and not intersect the other. Describe two cylinders whose bases are the focal conics and whose generators are parallel to PQ. By Ex. 1 these intersect in four lines, and each of these four is also a generator of the right circular cylinder whose radius is ∞. Now by supposition PQ lies on one of the elliptic cylinders and also on the circular cylinder, hence these two quadric cylinders intersect each other in five lines, which is impossible.

Ex. 3. The locus of all the straight lines drawn from any given point P on the hyperbolic conic to intersect the elliptic conic is a right cone, the tangent of whose semi-angle is $(K_3^2 - K_2^2)/K_3Rz$ where z is the ordinate of P.

Ex. 4. Show that four real focal lines can be drawn through a given point P, and that they are the intersections of the two quadric cones

$$\frac{(p\zeta - r\xi)^2}{K_3^2} + \frac{(q\zeta - r\eta)^2}{K_3^2 - K_2^2} = \frac{\zeta^2}{R^2} \qquad \frac{(p\eta - q\xi)^2}{K_2^2} - \frac{(q\zeta - r\eta)^2}{K_3^2 - K_2^2} = \frac{\eta^2}{R^2}$$

where (p, q, r) are the coordinates of P and ξ, η, ζ are referred to parallel axes meeting at P.

Ex. 5. Prove that the circular sections of the central ellipsoid whose centre is at O' are perpendicular to the generating lines at O' of the hyperboloid of one sheet. [Darboux.]

Ex. 6. If the base is situated on one of the principal planes at the central point, show that one principal axis at that base is perpendicular to that plane and the astatic moment of the corresponding couple is the same for all base points in that plane.

Ex. 7. If the base is situated on one of the principal axes at the central point, prove that the three principal axes at the base are parallel to those at the central point.

Ex. 8. If a straight line is a principal axis at every point of its length, prove that it is one of the principal axes at the central point.

Ex. 9. Find the locus of the base point O' at which the central ellipsoid is a surface of revolution.

In order that two of the three quantities M_1, M_2, M_3, in Art. 35 may be equal we must have either $D_2=0$ or $D_2=D_3$. In the first case O' lies on the elliptic focal conic. In the second case O' is at an umbilicus U and the locus is therefore the hyperbolic focal conic. In both cases the unequal axis is a tangent to the focal conic.

The same results follow from the equation to the central ellipsoid in the form

$$Y_y^2\eta^2 + Z_z^2\zeta^2 + R^2(p\xi + q\eta + r\zeta)^2 = 1,$$

see Art. 34. By applying the usual analytical conditions that this is a surface of revolution we obtain the required relation between p, q, r.

Arrangement of Poinsot's central axes.

38. In whatever position the body is placed relatively to the forces it has been shown in Vol. I. that the forces acting on the body can be simplified into a single force, acting along a straight line called by Poinsot the central axis, and a couple round that axis. As the body takes different positions relative to the forces

Poinsot's axis also moves relatively to both. In order to determine the arrangement of Poinsot's axes for all possible positions of the body and forces it will be convenient to have two systems of axes, one fixed in the body and the other fixed relatively to the forces.

Let the axes fixed in the body be the principal axes at the central point. These we shall represent by Ox, Oy, Oz. Following the same notation as before, the forces are represented by the astatic couples (G, b), (H, c), whose astatic arms are placed along y and z, together with a force R acting at O. The astatic moments of these couples are represented by K_2, K_3 respectively. Let the axes fixed in space be parallel to the forces R, G, H These are represented by Ox', Oy', Oz'. We shall sometimes speak of them as the *axes of the forces*.

Let the direction cosines of either set of axes relatively to the other be given by the diagram. The positive directions of these axes are so chosen that by turning one set round the common origin the positive directions of x, y, z may be made to coincide with those of x', y', z'. The advantage of this choice is, that in the determinant of direction cosines every constituent is equal to its minor with the proper sign as given by

	x	y	z
R	a_1	a_2	a_3
G	b_1	b_2	b_3
H	c_1	c_2	c_3

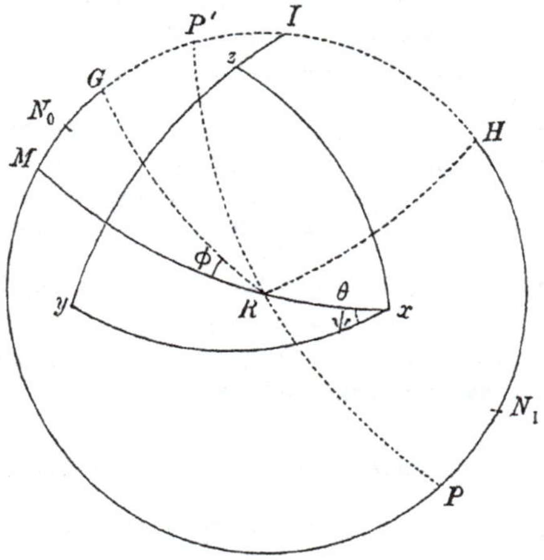

the ordinary rules of determinants. Without losing the simplicity of the other relations of these constituents, we thus avoid any ambiguity of sign in the minors.

In the figure the axes are represented in the manner usually adopted in spherical trigonometry. The axes Ox, Oy, Oz and Ox', Oy', Oz' cut the sphere in x, y, z and R, G, H respectively; the angles being represented by arcs of great circles. The Eulerian angular coordinates of R referred to x are $\theta = xR$, $\psi = yxR$, $\phi = MRG$. Since the angle between any two planes is equal to the arc joining their poles, it is easy to see that $zIG = \theta$, $Iz = \psi$, $IH = \phi$.

39. *To find the position of Poinsot's axis referred to the axes of the forces, and also the moment of the forces about it.*

Let Px'' be the required Poinsot's axis, Γ the moment of the couple round it. The axis Px'' is parallel to Ox', let its coordinates referred to x', y', z', be η', ζ'.

The couples K_2, K_3 have their astatic arms on the axes y, z, and their forces parallel to y', z'. To refer these couples to the axes x', y', z' we resolve the arms and move the forces parallel to themselves (Art. 6). Thus we replace the two couples by six others whose arms are arranged along the axes of x', y', z'. In the figure the forces at O are omitted to avoid complication, the arrows indicate the directions of the other forces of each of the six couples; and each arrow-head (as in Art. 10) is marked by the astatic moment of the corresponding couple.

By hypothesis all these couples together with a force R acting at O are equivalent to the couple Γ round Px'' and a force equal to R acting along Px''. Taking moments about the axes Ox', Oy', Oz' we have

$$\Gamma = K_3 b_3 - K_2 c_2 \quad\text{.....................(1)},$$

$$R\zeta' = -K_3 a_3 \quad\text{.........................(2)},$$

$$R\eta' = -K_2 a_2 \quad\text{.........................(3)}.$$

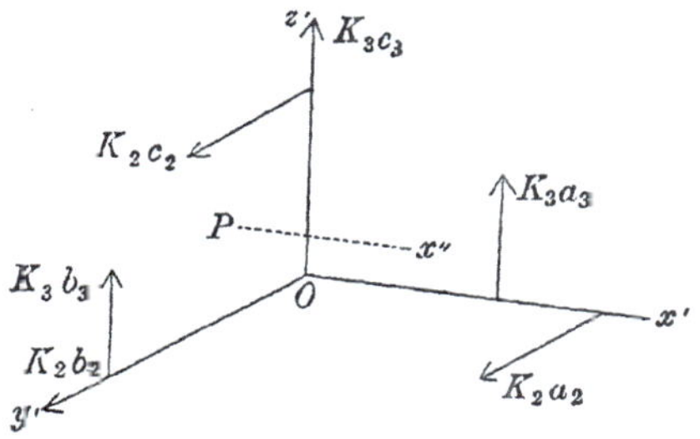

Another proof. We may also obtain these results very simply without resolving the couples. Let the arms OB, OC of the couples be taken as unity so that the forces G, H acting at B and C are measured by the astatic moments K_2, K_3, Art. 10. The axes Ox', Oy', Oz' being the axes of reference, the coordinates of B and C are respectively a_2, b_2, c_2; a_3, b_3, c_3. Since K_2 acts parallel to Oy', its moment about Oz' is K_2a_2, and since K_3 acts parallel to Oz' its moment about Oy' is $-K_3a_3$. In the same way their moments about Ox' are K_3b_3 and $-K_2c_2$. Equating these to the moment of R acting along Px'' and of Γ we have the same results as before.

40. When the body is rotated about Ox', the direction cosines a_2, a_3 are invariable. It follows that the straight line whose position is determined by the equations (2) and (3) is fixed relatively to the forces. Hence we infer, that, *when the body is rotated about an axis passing through the central point and parallel to the principal force, Poinsot's axis always coincides with a straight line fixed in space.*

This straight line traces out a right circular cylinder in the body whose radius ρ is given by the equation

$$R^2\rho^2 = K_2^2a_2^2 + K_3^2a_3^2 \qquad (4).$$

This cylinder is fixed in the body and moves with it. In one complete revolution of the body each generator in turn passes through the straight line fixed in space and becomes the Poinsot's axis for that position of the body.

Referring to the figure of Art. 38, the axis of this cylinder cuts the sphere of reference in R. We may also imagine the sphere of such size that the cylinder envelopes it along the circular boundary of the figure. In the figure the direction of the force R and the generators of the cylinder are supposed to be perpendicular to the plane of the paper.

As the body turns round OR as its axis, the dotted part of the figure remains fixed in space while the part indicated by the continuous lines moves round R.

Let a plane through the axis of the cylinder and the straight line fixed in space cut the sphere in the arc RP. Let RP produced backwards cut the circle GH in P'. Then the position of P or P' may be found from the equations

$$\tan GP = \tan GP' = \frac{\zeta'}{\eta'} = \frac{K_3a_3}{K_2a_2} \qquad (5).$$

In every position of the body Poinsot's central axis is a straight line drawn through P perpendicular to the plane of the circle GH. Here P is distinguished from P' by the sign of either η' or ζ' as given by the equations (2) and (3).

It follows from these results, that all the straight lines, each of which would be a Poinsot's axis if the body were properly placed, may be classified as the generators of a system of right circular cylinders. The axes of these cylinders pass through the central point and are always parallel to the direction of the principal force.

Conversely, *a straight line being given in the body, it may be required (when possible) to place the body in such a position that the straight line may be a Poinsot's axis.* To effect this, we turn the body about the central point until the given straight line is parallel to the principal force. If a_1, a_2, a_3 are the direction cosines of the given straight line referred to the principal axes of the body at the central point, then, in this position of the body, a_1, a_2, a_3 are also the direction cosines of the principal force. If the distance of the given straight line from the central point does not satisfy equation (4) the straight line cannot be a Poinsot's axis. If however the equation is satisfied, we turn the body round the principal force as an axis of rotation through the angle GP determined by equation (5), or, which is the same thing, we turn the body until the given straight line passes through the point η', ζ' in the plane $y'z'$ determined by the equations (2), (3). The body has then been placed in the required position. When the straight line fixed in the body has been made parallel to the principal force the body may be inverted, so that the given straight line is again parallel to the force but points in the opposite direction. If the condition (4) is satisfied in one case, it is satisfied in the other. Thus if the construction yield one position in which the given straight line is a Poinsot's axis, it will yield another.

41. In every position of the body the couple-moment of Poinsot's axis is given by

$$\Gamma = K_3 \cos Gz - K_2 \cos Hy$$
$$= K_3 (\cos\psi \sin\phi + \sin\psi \cos\phi \cos\theta)$$
$$+ K_2 (\sin\psi \cos\phi + \cos\psi \sin\phi \cos\theta),$$

by using the spherical formulæ for the triangles GIz and HIy. This may be written in the form

$$\Gamma = \Gamma_0 \sin(\phi - \phi_0) \text{.....................} (6),$$

where Γ_0 is the maximum value of Γ, and $\phi = \phi_0$ determines the position of the body when the couple-moment is zero. We easily find

$$\tan\phi_0 = -\frac{K_2 + K_3\cos\theta}{K_2\cos\theta + K_3}\tan\psi \quad \text{.........(7)},$$

$$\left.\begin{aligned}\Gamma_0^2 &= (K_2 + K_3\cos\theta)^2\sin^2\psi + (K_2\cos\theta + K_3)^2\cos^2\psi \\ &= \frac{(K_2 + K_3a_1)^2a_3^2 + (K_2a_1 + K_3)^2a_2^2}{a_3^2 + a_2^2}\end{aligned}\right\} \text{...(8)}.$$

Make the arc $MN_0 = \phi_0$, then the arc $N_0G = \phi - \phi_0$ and $\Gamma = \Gamma_0 \sin N_0G$. As the body rotates about the axis OR, both M and N_0 move with it. When $\phi - \phi_0 = 0$ or π, the point N_0 coincides with either P' or P; the couple-moment vanishes and the system is equivalent to a single resultant. As the body is turned from either of these opposite positions through any angle the couple Γ increases and its magnitude varies as the sine of the angle of rotation. The couple reaches a maximum in either of the positions given by $\phi - \phi_0 = \pm\frac{1}{2}\pi$ and then decreases again. Thus there are in general two positions of the body in which the couple-moment Γ has a given value, and two more in which it has the same value with an opposite sign.

42. We may interpret this result in a slightly different manner. We may ascribe to each generator a certain couple-moment Γ peculiar to itself, which becomes the couple-moment when the body is so placed that that generator is a Poinsot's axis. Make $MN_1 = MN_0 + GP$, then for any generator of the cylinder, say the one which passes through P, we have $\Gamma = \Gamma_0 \sin N_1P$.

It will be useful to state this result in words. *Through the line of action of R draw two planes, one passing through the two generators whose couple-moments are each zero, and the other arbitrary and cutting the cylinders in two other generators. If Γ be the couple-moment for these last two generators and χ the angle between the planes, then $\Gamma = \Gamma_0 \sin\chi$ where Γ_0 is given by either of the forms in equation* (8).

43. In what precedes it has been supposed that both the direction and the line of action of the principal force R are given in the body. In this case the body can only be rotated about Ox' as an axis. If the direction of R is not given, but only its line of action, the body can also be inverted by rotating it through two right angles about an axis perpendicular to Ox'. To avoid complicating the figure it will be more convenient to effect this last change by rotating the forces in the opposite direction, each about its point of application, so that the angles between their directions remain unaltered.

The effect of this inversion is easily seen to be, that the positive directions of x' and of one of the two y', z' are reversed. As it will be convenient that they should have the same positive directions in space as before, we shall represent the effect of the inversion by changing the signs of the force R and of that of one of the astatic moments K_2, K_3. The sign of the couple-moment Γ about Poinsot's axis also must be changed (even if its magnitude remains unaltered) when the positive direction of x in space is to be the same after inversion as before.

One result of these changes is that the arc $P'P$ (Art. 40) takes up another position (say $Q'Q$, not drawn in the figure of Art. 38) making the same angle with GR as before, but on the other side. The angle ϕ_0 and the couple Γ_0 are also changed. Thus the positions in which Poinsot's couple vanishes are changed by the inversion of the body.

44. *To find the equation of Poinsot's axis referred to the principal axes at the central point.*

Following the notation already described in Art. 39, the equations of Poinsot's axis referred to the axes of the forces are
$$R\eta' = -K_2a_2, \qquad R\zeta' = -K_3a_3 \text{..............(1)},$$
and the couple-moment Γ is given by $\Gamma = K_3b_3 - K_2c_2$........(2).
Transforming these to the axes fixed in the body we obviously have
$$R(b_1\xi + b_2\eta + b_3\zeta) = -K_2a_2,$$
$$R(c_1\xi + c_2\eta + c_3\zeta) = -K_3a_3.$$
Eliminating ξ, η, ζ in turn, and remembering that each constituent of the determinant of transformation in Art. 38 is equal to its minor, we have
$$\left.\begin{aligned} R(-\eta a_3 + \zeta a_2) &= -K_2a_2c_1 + K_3a_3b_1 \\ R(-\zeta a_1 + \xi a_3) &= -K_2a_2c_2 + K_3a_3b_2 \\ R(-\xi a_2 + \eta a_1) &= -K_2a_2c_3 + K_3a_3b_3 \end{aligned}\right\} \text{............(3)}.$$
These may also be written in the form
$$\left.\begin{aligned} R(-\eta a_3 + \zeta a_2) - \Gamma a_1 &= -K_2b_3 + K_3c_2 \\ R(-\zeta a_1 + \xi a_3) - \Gamma a_2 &= \qquad\quad -K_3c_1 \\ R(-\xi a_2 + \eta a_1) - \Gamma a_3 &= K_2b_1 \end{aligned}\right\} \text{.........(4)}.$$
Any two of these are the equations to Poinsot's axis when the relative positions of the body and the forces are given by the direction cosines a_1, &c. They are also the equations of the fixed generator of the circular cylinder, Art. 40.

Adding together the squares of the equations (3), we obtain the equation of the cylinder traced out by Poinsot's axis as the body is turned round Ox'. This cylinder is easily seen to be a right circular cylinder and its radius ρ is given by
$$R^2\rho^2 = K_2^2a_2^2 + K_3^2a_3^2 \text{.................(5)},$$
as already proved in Art. 40.

When the body is so placed that the forces reduce to a single resultant, the equations (4) may be put into a more convenient form. Since $\Gamma=0$, the first of those equations reduces to

$$\left.\begin{aligned} R(-\eta a_3+\zeta a_2) &= -K_2b_3+K_3c_2 \\ 0 &= -K_2c_2+K_3b_3 \end{aligned}\right\}.$$

also, by (2)

Subtracting the squares, we have

$$R^2(-\eta a_3+\zeta a_2)^2=(b_3^2-c_2^2)(K_2^2-K_3^2).$$

Let us seek the intersection of the single resultant with the plane of xy; putting therefore $\zeta=0$, the two first of equations (4) become

$$R^2a_3^2\frac{\eta^2}{K_3^2-K_2^2}=c_2^2-b_3^2,\qquad R^2a_3^2\frac{\xi^2}{K_3^2}=c_1^2 \text{ ......(6).}$$

A straight line drawn through the point thus determined parallel to the force R is the single resultant.

Adding these equations together and remembering that

$$b_3^2+a_3^2=1-c_3^2=c_1^2+c_2^2,$$

we have, after division by a_3^2,
$$\frac{\eta^2}{K_3^2-K_2^2}+\frac{\xi^2}{K_3^2}=\frac{1}{R^2} \text{ ....(7).}$$

This is the equation of a focal conic, Art. 36. The single resultant therefore intersects the focal conic in the plane of xy. In the same way, it intersects that in the plane of xz. We thus arrive at a *theorem due to Minding*, viz. that *when the body is so placed that the forces are equivalent to a single resultant, the line of action of that resultant is a focal line.* A fuller consideration of this mode of proof and of Minding's theorem will be found a little further on.

An apparent exception arises when either $a_3=0$ or $a_2=0$. Supposing that $a_3=0$ the equations (3) become $Ra_2\zeta=-K_2a_2c_1$, $Ra_1\zeta=K_2a_2c_2$.
Since $-c_2=a_1b_3-a_3b_1$, we have $\Gamma=K_3b_3-K_2c_2=(K_3+K_2a_1)b_3=0.$
Thus either $b_3=0$ or $K_3+K_2a_1=0$. Joining the former to $\Gamma=0$, we have $c_2=0$. The latter is impossible if K_3 is greater than K_2; if K_3 is less than K_2 the focal conic (7) is a hyperbola and the single resultant is parallel to an asymptote. Thus in both cases the single resultant intersects the focal conic.

Ex. 1. Show that the single resultant intersects the plane of the imaginary focal conic in the conic
$$\frac{\eta^2}{K_2^2}+\frac{\zeta^2}{K_3^2}=\frac{1}{R^2}\left(\frac{1}{a_1^2}-1\right).$$
This conic is fixed in the body when a_1 is given.

Ex. 2. Show that the circular cylinder (5) intersects the plane of xy in the conic whose equation is
$$R^2\{\xi^2+\eta^2-(\xi a_1+\eta a_2)^2\}=K_2^2a_2^2+K_3^2a_3^2.$$

46. The direction of the principal force R, and a point ξ, η, ζ on a generator of the circular cylinder being given referred to the principal axes of the body, it is required to find the couple-moment about that generator when the body is so placed that the generator is a Poinsot's axis.

For the sake of brevity let us write

$$-\eta a_3+\zeta a_2=p, \quad -\zeta a_1+\xi a_3=q, \quad -\xi a_2+\eta a_1=r.$$

Multiplying the second and third of equations (4) Art. 44 by $K_2^2 a_2$ and $K_3^2 a_3$ respectively we have

$$K_2^2 a_2(Rq-\Gamma a_2)+K_3^2 a_3(Rr-\Gamma a_3)=K_2K_3(-K_2a_2c_1+K_3a_3b_1)=K_2K_3Rp.$$

The couple-moment Γ is therefore given by

$$(K_2^2a_2^2+K_3^2a_3^2)\,\Gamma=R\,(K_2^2a_2q+K_3^2a_3r-K_2K_3p)\ldots\ldots\ldots\ldots\ldots(1).$$

If the line of action of R only is given and the force may act either way along it, we obtain another value of Γ by inverting either the body or the forces. If Γ' be the couple-moment after inversion we have by Art. 43

$$(K_2^2a_2^2+K_3^2a_3^2)\,\Gamma'=R\,(K_2^2a_2q+K_3^2a_3r-K_2K_3p)\ \ldots\ldots\ldots\ldots\ldots(2).$$

The force R then acts along the negative direction of its line of action.

We may write (1) in the form

$$(K_2^2a_2^2+K_3^2a_3^2)\,\Gamma=R\{-(K_3^2-K_2^2)\,a_2a_3\xi+K_3(K_3a_1+K_2)\,a_3\eta-K_2(K_2a_1+K_3)\,a_2\zeta\}\ldots(3).$$

We therefore see that the plane through the line of action of R and the two generators whose couple-moments are zero (Art. 41) is

$$-(K_3^2-K_2^2)\,a_2a_3\xi+K_3\,(K_3a_1+K_2)\,a_3\eta-K_2\,(K_2a_1+K_3)\,a_2\zeta=0\ldots\ldots\ldots(4).$$

Conversely, *when the magnitude of the couple Γ is given, either of the equations (1) or (3) enables us to find the generators which have the given moment Γ when the body is so placed that one of them is a Poinsot's axis.* When Γ is given, either of these equations represents a plane intersecting the circular cylinder (5) in two straight lines which are parallel to the principal force. These are the generators required; see also Art. 41. If we change the sign of Γ we obtain another plane, parallel to the former, giving two other generators, each of whose couple-moments has the given magnitude but an opposite sign. These four are obviously symmetrically arranged round the principal force.

Another construction for Poinsot's axis and moment is indicated in the following examples.

Ex. 1. A straight line OQ is drawn through the central point O perpendicular to the plane containing the force R and its corresponding fixed generator. Prove that p, q, r are the coordinates of the point Q in which this straight line cuts the circular cylinder. Prove also that Q is one of the poles of the great circle represented by PP' in the figure of Art. 38.

Ex. 2. Let OS be the straight line whose direction cosines are proportional to $-K_2K_3,\ K_2^2a_2,\ K_3^2a_3$, when referred to the principal axes of the body at the central point O; thus OS is fixed in the body when the position of OR is given. If ϕ be the angle contained by the lines OQ, OS, prove that

$$\frac{\Gamma}{\cos\phi}=\left\{\frac{K_2^2K_3^2+K_2^4a_2^2+K_3^4a_3^2}{K_2^2a_2^2+K_3^2a_3^2}\right\}^{\frac{1}{2}}$$

Show also that the straight line OS lies in the plane containing the force R and the two generators whose couple-moments are zero.

46. If the magnitude of the couple-moment Γ is given as well as the line of action of R, we may obtain other cylinders which will intersect the right cylinder already found in the corresponding Poinsot's axes.

The first of equations (4) Art. 44 is

$$R\,(-\eta a_3+\zeta a_2)-\Gamma a_1=-K_2b_3+K_3c_2,$$

and

$$\Gamma=-K_2c_2+K_3b_3.$$

Hence subtracting the squares, as in Art. 44,

$$\{R(-\eta a_3+\zeta a_2)-\Gamma a_1\}^2-\Gamma^2=(b_3^2-c_2^2)(K_2^2-K_3^2).$$

Now by Art. 38, $b_3^2-c_2^2=c_1^2-a_3^2$, hence, substituting for c_1^2 from the second of equations (4), we have

$$\frac{\{R(-\eta a_3+\zeta a_2)-\Gamma a_1\}^2-\Gamma^2}{K_3^2-K_2^2}+\frac{\{R(-\zeta a_1+\xi a_3)-\Gamma a_2\}^2}{K_3^2}=a_3^2 \quad\text{.........(1)}.$$

Again $b_3^2-c_2^2=a_2^2-b_1^2$, substituting for b_1^2 from the third of equations (4), we have

$$-\frac{\{R(-\eta a_3+\zeta a_1)^2-\Gamma a_1\}^2-\Gamma^2}{K_3^2-K_2^2}+\frac{\{R(-\xi a_2+\eta a_1)-\Gamma a_3\}^2}{K_2^2}=a_2^2 \quad\text{......(2)}.$$

Lastly, the last two of equations (4) give

$$\frac{\{R(-\zeta a_1+\xi a_3)-\Gamma a_2\}^2}{K_3^2}+\frac{\{R(-\xi a_2+\eta a_1)-\Gamma a_3\}^2}{K_2^2}=1-a_1^2 \quad\text{.........(3)}.$$

The three surfaces (1) (2) and (3) are cylinders, for the equation to any one of them shows that an expression of the first degree in ξ, η, ζ is some function of another expression of the first degree. Also the axis of each cylinder is parallel to the straight line $\xi/a_1=\eta/a_2=\zeta/a_3$, i.e. the axis of each is parallel to the line of action of the force R.

It may be noticed that the direction cosines b_1, b_2, b_3; c_1, c_2, c_3 have been eliminated so that the equations to these cylinders contain only the principal force R, the direction cosines of R and Poinsot's couple Γ.

47. Supposing that the coordinates (ξ, η, ζ) of some point on the cylindrical locus (5) are given, and that the line of action of the force R is also known, any one of the equations (1), (2), (3), of Art. 46 may be regarded as a quadratic to find the couple-moment when the body is so placed that the corresponding generator is a Poinsot's axis.

If we seek the corresponding equations when the forces are inverted we change the signs of R, Γ and one of the K's (Art. 43). But these changes leave the quadratics unaltered. Thus the two values of Γ given by any one of these quadratics correspond to the two directions in which R can act along the same given line of action.

Ex. The given point (ξ, η, ζ) being supposed to be on the circular cylinder, prove that the three quadratics (1) (2) (3) of Art. 46 reduce to the same, viz.

$$\Gamma^2(K_2^2a_2^2+K_3^2a_3^2)-2R\Gamma(K_2^2a_2q+K_3^2a_3r)+R^2(K_2^2q^2+K_3^2r^2)=K_2^2K_3^2(a_2^2+a_3^2).$$

Prove also that the roots of this quadratic are given by

$$\Gamma(K_2^2a_2^2+K_3^2a_3^2)=R(K_2^2a_2q+K_3^2a_3r\mp K_2K_3p)$$

where p, q, r have the meanings specified in Art. 45.

48. Minding's Theorem. By joining any one of the three cylinders (1), (2), (3) to the circular cylinder we have sufficient equations to find the generators which can have a given couple-moment and are also parallel to any given straight line. It will often be more convenient to use the intersections of the cylinders with one of the coordinate planes. Thus putting $\zeta=0$, the cylinder (1) cuts the plane of xy in the conic

$$\frac{(R\eta a_3+\Gamma a_1)^2-\Gamma^2}{K_3^2-K_2^2}+\frac{(R\xi a_3-\Gamma a_2)^2}{K_3^2}=a_3^2 \quad\text{.......................(1)}.$$

When the forces are equivalent to a single resultant we have $\Gamma=0$ and in that case equation (1) reduces to the focal conic

$$\frac{\eta^2}{K_3^2-K_2^2}+\frac{\xi^2}{K_3^2}=\frac{1}{R^2} \quad\text{..........(2)}.$$

The single resultant therefore intersects the focal conic in the plane of xy. Similarly it intersects that in the plane of xz. See Art. 44.

49. *Conversely*, let a straight line intersect both focal conics, then by Art. 37 it is a generator of the circular cylinder. If the direction cosines of this straight line are a_1, a_2, a_3, the corresponding couple-moment Γ is given by the quadratic (1) of Art. 48.

This quadratic gives two values of Γ. Multiplying (2) by $R^2a_3^2$ and subtracting the result from (1) we find that one root is $\Gamma = 0$ and that the other is given by

$$(K_2^2a_2^2 + K_3^2a_3^2)\,\Gamma = 2Ra_3\,\{K_2^2\xi a_2 + K_3^2(\eta a_1 - \xi a_2)\} \text{ .... } (3).$$

The result is that the couple-moment for the generator is zero for one of the two directions in which the force R can act along that generator.

These two values of Γ follow also from equations (1) and (2) Art. 45, for when the value of Γ given by (1) is zero, the value given by (2) agrees with that shown in equation (3) of this article.

Finally, we see that *if any straight line can be the line of action of a single resultant force that line must intersect both the focal conics, and if a straight line intersect both the focal conics it can be the line of action of a single resultant if the body be properly placed.*

50. Ex. 1. The direction of the principal force R being given by the direction cosines a_1, a_2, a_3 referred to the principal axes at the central point show that each of the planes

$$\left(\frac{\xi}{a_1} - \frac{\eta}{a_2}\right) K_3^2 \pm \left(\frac{\eta}{a_2} - \frac{\zeta}{a_3}\right) \frac{K_2K_3}{a_1} + \left(\frac{\zeta}{a_3} - \frac{\xi}{a_1}\right) K_2^2 = 0$$

passes through the line of action of R and intersects the focal conics in four points, which are the corners of a parallelogram formed by the focal lines, two of which are parallel to the direction of R. Prove also that the focal lines parallel to the given direction of R are the corresponding single resultants.

This follows easily from Art. 45.

Ex. 2. If the body is so placed that the force R acts along an asymptote of the hyperbolic focal conic, prove (1) that the circular cylinder contains the elliptic focal conic on its surface; (2) that as the body is turned round OR Poinsot's axis lies in the plane containing R and parallel to the force H which corresponds to the greater astatic moment K_3; (3) that Poinsot's couple Γ is always zero as the body is turned round OR provided the force R acts in the proper direction, but is zero only when the plane of the hyperbolic conic contains the force H if R act in the other direction.

51. Relations of Poinsot's axis to the confocals. The manner in which the single resultant is connected with the confocals is given by Minding's theorem. We may also find the relations of Poinsot's axis with the same confocals in the general case in which the couple is not zero. To effect this we require the following lemma in solid geometry.

52. *Lemma.* Let the squares of the semi-axes of two confocals be $\alpha^2 + \lambda$, $\beta^2 + \lambda$, $\gamma^2 + \lambda$ and $\alpha^2 + \lambda'$, $\beta^2 + \lambda'$, $\gamma^2 + \lambda'$. Let the direction cosines of any straight line be (l, m, n) and its distance from the origin be ρ. If two planes at right angles can be drawn through the straight line to touch the two confocals, then

$$\rho^2 + \alpha^2l^2 + \beta^2m^2 + \gamma^2n^2 = \alpha^2 + \beta^2 + \gamma^2 + \lambda + \lambda'.$$

It follows that when the confocals are given the left-hand side is constant for all straight lines.

Let (l', m', n'), (l'', m'', n'') be the direction cosines of the tangent planes, and p, p' the lengths of the perpendiculars on them. Then

$$p^2 = (\alpha^2 + \lambda)\,l'^2 + (\beta^2 + \lambda)\,m'^2 + (\gamma^2 + \lambda)\,n'^2,$$

$$p'^2 = (\alpha^2 + \lambda')\,l''^2 + (\beta^2 + \lambda')\,m''^2 + (\gamma^2 + \lambda')\,n''^2.$$

Noticing that $\rho^2 = p^2 + p'^2$ we find by addition

$$\rho^2 = \alpha^2 (l'^2 + l''^2) + \beta^2 (m'^2 + m''^2) + \gamma^2 (n'^2 + n''^2) + \lambda + \lambda'.$$

Hence since $l^2 + l'^2 + l''^2 = 1$ &c., we have

$$\rho^2 + \alpha^2 l^2 + \beta^2 m^2 + \gamma^2 n^2 = \alpha^2 + \beta^2 + \gamma^2 + \lambda + \lambda'.$$

53. Let us now apply this Lemma to any generator of the cylinder. Let α, β, γ be the semi-axes of the imaginary focal conic, then, by Art. 36,

$$\alpha^2 = 0, \quad \beta^2 = -K_2^2/R^2, \quad \gamma^2 = -K_3^2/R^2.$$

The values of λ, λ' are the squares of the semi-major axes of the two confocals; let these be represented by M_1/R^2 and M_1'/R^2 as in Art. 35. The direction cosines of any generator are (a_1, a_2, a_3) and its distance ρ from the central point is given by $R^2\rho^2 = K_2^2 a_2^2 + K_3^2 a_3^2$. Hence, substituting, the left-hand side of the equation in the Lemma reduces to zero. We therefore have $M_1^2 + M_1'^2 = K_2^2 + K_3^2$.

If therefore any two planes at right angles are drawn through a possible Poinsot's axis and two confocals are drawn to touch these planes, the sum of the squares of the semi-major axes of these confocals is constant. This constant when multiplied by R^2 is the sum of the squares of the astatic moments of the principal couples at the central point.

From this we may deduce as a corollary a theorem discovered by Darboux.

Let a plane be drawn through any possible Poinsot's axis to touch one of the focal conics, then a perpendicular plane through the same axis will touch another focal conic.

For in the limit these conics may be regarded as the bounding rims of two flat confocals whose semi-major axes are respectively K_2/R and K_3/R.

54. Ex. 1. If a possible Poinsot's axis touch two confocals prove that the sum of the squares of their semi-major axes is equal to $K_2^2 + K_3^2$ after division by R^2.

If a straight line touch two confocals, and tangent planes are drawn at the points of contact, these planes are known to be at right angles. If we apply the general theorem in Art. 53 to these two tangent planes, the result follows at once.

Ex. 2. If a possible Poinsot's axis intersect one of the focal conics prove that it must intersect the other also.

For suppose it intersects the plane of xy in the elliptic focal conic, it may be regarded as touching the confocal surface whose semi-major axis is K_3/R. Hence it also touches the confocal surface whose semi-major axis is K_2/R (by the last example), i.e. it intersects the plane of xz in the hyperbolic focal conic.

Reduction to Three and to Four Forces.

55. We have seen that the forces of any astatic system may be reduced to two couples and a single force. This representation of the forces, though very simple in its character, may not always be convenient. These couples and the force have an intimate relation to the central point and central plane, and the positions of this point and plane may not suit the circumstances of the problem we wish to consider.

We shall now examine some other representations of an astatic system. We shall show that the forces may be reduced to *three forces* which act at three arbitrary points in the central plane.

These points however must not in general lie in one straight line. We shall show that the forces of the system may also be reduced to *four forces* which act at any four points fixed in the body at which we may find it convenient to apply them. The four points must not in general lie in one plane.

We can see another advantage of these representations of the forces. For the points of application may be regarded as the corners of a triangle or tetrahedron of reference. We are thus enabled to use the systems of coordinates called trilinear and tetrahedral with considerable effect.

56. *To show that all the forces of any system may be reduced to three forces which act at three points lying in the central plane.*

Following the same notation as in Art. 9, let the forces of the system be P_1, P_2, &c. and let M_1, M_2, ... be their points of application. Let these be referred to any axes Ox, Oy, Oz, either rectangular or oblique, which are fixed relatively to the body. Let the coordinates of M_1, M_2, &c. be (x_1, y_1, z_1), (x_2, y_2, z_2), &c. Let Ox', Oy', Oz', be another system of axes, not necessarily rectangular, to which we may refer the forces. These are fixed relatively to the forces. Let the components of the forces along these be (X'_1, Y'_1, Z'_1), (X'_2, Y'_2, Z'_2), &c.

Consider the system of parallel forces X'_1, X'_2, &c. All these are astatically equivalent to a single force $\Sigma X'$ acting at their centre of parallel forces. In the same way the two other systems of parallel forces, viz. Y'_1, Y'_2 &c. and Z'_1, Z'_2 &c. are equivalent to $\Sigma Y'$ and $\Sigma Z'$ each acting at its own centre of parallel forces in directions parallel to y' and z' respectively. These forces we may represent by F, G, H, and their points of application by A, B, C. The centre of parallel forces is known to possess the astatic quality. If then we move the arbitrary axes Ox', Oy', Oz' in any manner about the origin, keeping their inclination to each other unaltered, *the system will yet be equivalent to the same three forces F, G, H acting at the same three points A, B, C in directions always parallel to the axes Ox', Oy', Oz'.*

To find the coordinates of these points we may therefore consider any one position of the forces and the body. In this position let X, Y, Z be the components of any force P resolved along the axes Ox, Oy, Oz. Then

$$X' = lX + l'Y + l''Z, \quad Y' = mX + m'Y + m''Z, \quad Z' = \&c.$$

where (l, m, n), (l', m', n'), (l'', m'', n''), are the direction ratios of the axes (x, y, z) referred to (x', y', z').

Let $(\bar{x}, \bar{y}, \bar{z})$ be the coordinates of A, then

$$\bar{x}_1 = \frac{\Sigma X'x}{\Sigma X'} = \frac{l\Sigma Xx + l'\Sigma Yx + l''\Sigma Zx}{l\Sigma X + l'\Sigma Y + l''\Sigma Z}$$

with similar values for $\bar{y}_1$ and $\bar{z}_1$. Taking the same notation as in Art. 10 we write $\Sigma Xx = X_x$ &c., $\Sigma X = X_0$ &c. We thus have

$$\left.\begin{aligned} F\bar{x}_1 &= lX_x + l'Y_x + l''Z_x \\ F\bar{y}_1 &= lX_y + l'Y_y + l''Z_y \\ F\bar{z}_1 &= lX_z + l'Y_z + l''Z_z \\ F &= lX_0 + l'Y_0 + l''Z_0 \end{aligned}\right\} \quad \text{...................(1).}$$

Hence it appears that the point A lies on the plane

$$\begin{vmatrix} \xi & X_x & Y_x & Z_x \\ \eta & X_y & Y_y & Z_y \\ \zeta & X_z & Y_z & Z_z \\ 1 & X_0 & Y_0 & Z_0 \end{vmatrix} = 0 \quad \text{.....................(2).}$$

In the same way the points B and C also lie on this plane.

57. We notice that the directions of the axes Ox', Oy', Oz', are perfectly arbitrary except that they cannot all lie in one plane. We may therefore obtain an infinite variety of triangles ABC with corresponding forces at the corners. Any one of these may be called an *astatic triangle*, and the points A, B, C, may be called *astatic points*.

We may obviously make the inclinations of the forces F, G, H to each other whatever we please, though of course the position of the triangle ABC is dependent on our choice of these inclinations. It is generally most convenient to make the forces F, G, H act in directions at right angles to each other.

We have seen that when we want to find the positions of A, B, C we may consider the body to have some fixed position relative to the forces. For this position X_x &c. are all constant whatever the positions of the axes x', y', z' may be. The equation (2) therefore gives, as the locus of the points A, B, C, a plane fixed in the body. We also see that the locus is a unique plane except when all the coefficients are zero. An independent and elementary proof that the plane ABC is unique has been given in Art. 13.

Comparing the equation (2) with that found in Art. 25 we notice that this plane is the same as that already called the *central plane.* It follows that all the astatic triangles lie on the central plane.

58. *To find the central plane and one astatic triangle with rectangular forces.*

The theorem proved in Art. 56 supplies us with a useful method of finding the position of the central plane. To effect this we resolve all the forces of the system into any three directions we may find convenient. Taking the forces in these three directions separately we have three sets of parallel forces. We then find the centre of parallel forces of each set by any method we may find convenient. We thus arrive at three points which we call A, B, C. The plane through A, B, C is the central plane. We have also found one astatic triangle.

Suppose the system referred to rectangular axes Ox, Oy, Oz and consider any position of the body relative to the forces. All the x-components form a system of parallel forces which may be collected into a single astatic force $\Sigma X = F$ acting at a point A whose coordinates are

$$\bar{x}_1 = \frac{\Sigma Xx}{\Sigma X} \qquad \bar{y}_1 = \frac{\Sigma Xy}{\Sigma X} \qquad \bar{z}_1 = \frac{\Sigma Xz}{\Sigma X}.$$

In the same way the y-components may be collected into a force $\Sigma Y = G$ acting at a point B whose coordinates are

$$\bar{x}_2 = \frac{\Sigma Yx}{\Sigma Y} \qquad \bar{y}_2 = \frac{\Sigma Yy}{\Sigma Y} \qquad \bar{z}_2 = \frac{\Sigma Yz}{\Sigma Y}.$$

The z-components may be similarly treated.

These three points lie on the central plane. The forces F, G, H act in directions at right angles to each other and their magnitudes have been found.

If the principal force is finite, the axes may always be so chosen that $\Sigma X, \Sigma Y, \Sigma Z$ are not zero. If the principal force is zero, the coordinates of the three points are either infinite or take an indeterminate form; and in this case the central plane is either at an infinite distance or is indeterminate in position. Thus whenever there is a central plane this construction may be used to find it.

59. Referring to the table of elementary couples given in Art. 10 these expressions for $(\bar{x}, \bar{y}, \bar{z})$ &c. give a new interpretation to those symbols. It has been shown in Art. 10 that the *constituents in any row* of that table are the components of the corresponding couples. It has now been proved that the *constituents in any column* are proportional to the coordinates of an astatic point with rectangular forces, Art. 57.

60. *To reduce all the forces of any system to four forces which act at four given points not all in one plane.*

Let A, B, C, D be any four points fixed in the body. These we shall regard as the corners of the tetrahedron of reference.

Let P_1, P_2, &c., be any forces acting on the body and let M_1, M_2, &c. be their points of application. We propose to replace each of these by four forces acting at the corners A, B, C, D parallel to the original direction of the force. Consider DA, DB, DC to be a system of oblique axes, let ξ, η, ζ, be the coordinates of any point M and let $DA = a$, $DB = b$, $DC = c$. Then by Art. 7 the forces acting at A, B, C, D are respectively

$$P\xi/a, \quad P\eta/b, \quad P\zeta/c, \quad P - P\xi/a - P\eta/b - P\zeta/c.$$

Now ζ/c is equal to the ratio of the perpendiculars drawn from M and C on the face ABC, and this ratio is the tetrahedral coordinate of M, Representing the four *tetrahedral* coordinates of M by $\alpha, \beta, \gamma, \delta$, and remembering that their sum is unity we see that the four forces at the corners A, B, C, D, are respectively $P\alpha$, $P\beta$, $P\gamma$, $P\delta$.

We therefore have the following working rule. *Any force P acting at the point whose tetrahedral coordinates are $\alpha, \beta, \gamma, \delta$ may be replaced by four parallel forces acting at the corners of the tetrahedron of reference whose magnitudes are respectively $P\alpha$, $P\beta$, $P\gamma$, $P\delta$.*

The several forces acting at each corner may now be compounded together. The result is that any system of forces can be replaced by four forces, one at each corner of the tetrahedron.

61. We may prove in the same way that a force P acting at any point M in the plane ABC may be replaced by three parallel forces respectively equal to $P\alpha$, $P\beta$, $P\gamma$, and acting at A, B, C, where α, β, γ, are the areal coordinates of M referred to the triangle ABC.

We may also deduce this result from the general theorem for a tetrahedron. We notice that tetrahedral coordinates become areal when the point considered lies in a coordinate plane. We may therefore disregard the coordinate δ and treat the tetrahedral coordinates α, β, γ, as if they were areal.

62. *To show that the system can be reduced to three forces acting at any three points in the central plane which form a triangle.*

Let the system be reduced to three forces acting at the corners A, B, C of some astatic triangle; then this triangle lies in the central plane. Let A', B', C', be any three points in the same plane, but not in a straight line, and let D' be a fourth point not in that plane. Regarding $A'B'C'D'$ as the tetrahedron of reference we shall transfer the forces from A, B, C to the corners of this tetrahedron.

To find the force at D', we multiply each force by its δ coordinate. Since this coordinate is zero for each of the points A, B, C, the resultant force at D' is zero.

63. **Transformation of Triangles.** *One astatic triangle ABC and the rectangular forces F, G, H at its corners being given, it is required to transfer this representation to any other triangle $A'B'C'$ and to find the rectangular forces F', G', H' at its corners.*

Let axes drawn through any point O parallel to either of these sets of forces be called the axes of those forces. We thus have two sets of rectangular axes. Let their mutual direction cosines be given in the usual way by the diagram.

	F'	G'	H'
F	l	m	n
G	l'	m'	n'
H	l''	m''	n''

Then any force F may be resolved into the components Fl, Fm, Fn, acting respectively parallel to the axes of F', G', H'. Treating the forces G, H in the same way we have $F' = Fl + Gl' + Hl''$,

$$G' = Fm + Gm' + Hm'', \qquad H' = \&c.$$

We also have

$$F = F'l + G'm + H'n, \quad G = F'l' + G'm' + H'n', \quad H = \&c.$$

The point of application of the force F' is the centre of the parallel forces Fl, Gl', Hl'' which act at A, B, C. Thus the point A' at which F' acts is the centre of gravity of three weights (positive or negative) proportional to Fl, Gl', Hl'' placed at the corners A, B, C of the given triangle. By properly choosing these ratios we can place the corner A' at any point we please.

The areal coordinates of the corners of either triangle referred to the other can also be found very simply by using the theorem of Art. 61. Let $(\alpha_1, \beta_1, \gamma_1)$, $(\alpha_2, \beta_2, \gamma_2)$, $(\alpha_3, \beta_3, \gamma_3)$ be the *areal* coordinates of the points A', B', C' referred to the given triangle ABC. If we transfer the forces F', G', H' back again to the triangle ABC, the three forces at A will be $F'\alpha_1$, $G'\alpha_2$, $H'\alpha_3$. But these are the components of F. The forces at B, C, may be similarly found.

Hence
$$F'\alpha_1 = Fl \qquad F'\beta_1 = Gl' \qquad F'\gamma_1 = Hl'',$$
$$G'\alpha_2 = Fm \qquad G'\beta_2 = Gm' \qquad G'\gamma_2 = Hm'',$$
$$H'\alpha_3 = Fn \qquad H'\beta_3 = Gn' \qquad H'\gamma_3 = Hn''.$$

By choosing the nine direction cosines in any way which their mutual relations permit we can use these formulæ to transform from one triangle to another.

If the forces of the two triangles are oblique we regard (l, m, n), (l', m', n'), (l'', m'', n''), as the direction ratios of F, G, H referred to the axes F', G', H'. The direction ratios of F', G', H' referred to the axes of F, G, H, are proportional to the minors of (l, l', l'') &c. If these direction ratios be $(\lambda, \lambda', \lambda'')$ (μ, μ', μ'') (ν, ν', ν'') we have
$$F = F'\lambda + G'\mu + H'\nu, \quad G = \&c., \quad H = \&c.,$$
instead of the expressions given above. With this exception all the other equations in this article apply to oblique forces.

64. The imaginary focal Conic. Let us suppose that the forces of the two triangles ABC, $A'B'C'$ are rectangular. The nine direction cosines are connected by relations such as $lm + l'm' + l''m'' = 0$ &c. Hence the coordinates of A', B', C' are connected by the three equations

$$\frac{\alpha_1\alpha_2}{F^2} + \frac{\beta_1\beta_2}{G^2} + \frac{\gamma_1\gamma_2}{H^2} = 0, \qquad \frac{\alpha_2\alpha_3}{F^2} + \frac{\beta_2\beta_3}{G^2} + \frac{\gamma_2\gamma_3}{H^2} = 0, \qquad \frac{\alpha_3\alpha_1}{F^2} + \frac{\beta_3\beta_1}{G^2} + \frac{\gamma_3\gamma_1}{H^2} = 0 \ldots(1).$$

If therefore A' be taken at any point (α, β, γ), both B' and C' must lie on the straight line
$$\frac{\alpha_1\alpha}{F^2} + \frac{\beta_1\beta}{G^2} + \frac{\gamma_1\gamma}{H^2} = 0 \ldots\ldots(2),$$
where α, β, γ are current coordinates. Taking B' anywhere on this line, then C' is found as the intersection of two straight lines.

This straight line (2) is evidently the polar line of $(\alpha_1, \beta_1, \gamma_1)$, with regard to the imaginary conic
$$\frac{\alpha^2}{F^2} + \frac{\beta^2}{G^2} + \frac{\gamma^2}{H^2} = 0 \ldots\ldots(3).$$

Thus the three astatic points are always at the corners of a self-conjugate triangle with regard to this conic.

The statical property of this conic is that each side of every astatic triangle with rectangular forces is the polar line of the opposite corner. But as two different conics cannot have the polar lines of every point the same in each conic, it follows that this conic is unique. Whatever astatic triangle ABC we take as the triangle of reference, the conic given by this equation is the same.

65. Ex. 1. Show that, whatever astatic triangle with rectangular forces is taken as the triangle of reference, the quantities

(1) $F^2 + G^2 + H^2$, (2) $FGH\Delta$, (3) $a^2G^2H^2 + b^2H^2F^2 + c^2F^2G^2$,

are invariable, where a, b, c are the sides, Δ the area of the triangle, and F, G, H the forces.

We have also the invariant property that the centre of gravity of three weights, proportional to F^2, G^2, H^2, placed at the corners is the same for all triangles.

Ex. 2. Show that, whatever astatic triangle with oblique forces is taken as the triangle of reference, the quantities

(1) $F^2+G^2+H^2+2FG\cos\gamma+2GH\cos\alpha+2HF\cos\beta$

(2) $FGH\Delta\mu$

(3) $$a'^2G'H'\{F'^2(\cos\alpha-\cos\beta\cos\gamma)-F'H'(\cos\beta-\cos\gamma\cos\alpha) \\ -F'G'(\cos\gamma-\cos\alpha\cos\beta)-G'H'\sin^2\alpha\}+\&c.+\&c.$$

are invariable, where α, β, γ are the mutual inclinations of the forces and

$$\mu=1-\cos^2\alpha-\cos^2\beta-\cos^2\gamma+2\cos\alpha\cos\beta\cos\gamma.$$

We notice that μ is six times the volume of the tetrahedron formed by unit lines drawn from any point parallel to the forces. It follows that μ cannot vanish unless the astatic forces are parallel to one plane.

Ex. 3. A system of forces is equivalent to a force R, acting at a point O, and two couples, whose astatic moments are K_2, K_3, and whose astatic arms are placed along the rectangular axes OY, OZ, the forces of the couples being perpendicular to each other and to the force R, see Art. 32. If these are transferred to an astatic triangle $A'B'C'$ situated in the plane yz, the coordinates of the corners being (η_1, ζ_1), (η_2, ζ_2), (η_3, ζ_3) and the rectangular forces F', G', H', prove that

$$\begin{array}{lll} F'=Rl & F'\eta_1=K_2l' & F'\zeta_1=K_3l'' \\ G'=Rm & G'\eta_2=K_2m' & G'\zeta_2=K_3m'' \\ H'=Rn & H'\eta_3=K_2n' & H'\zeta_3=K_3n'' \end{array}$$

where l, m, n &c. are the nine direction cosines of F', G', H', as in Art. 63.

If the forces F', G', H' are all equal, prove that the sum of the distances of the three corners from each of the axes of y and z is zero.

66. To find the Central Point. *The astatic triangle ABC with rectangular forces F, G, H being given, show that the central point is the centre of gravity of three weights proportional to F^2, G^2, H^2 placed at the corners.*

This follows easily from the theorem proved in Art. 30. We multiply each force, such as F, by the resolved part of all the forces along it, i.e. by F; the product is F^2. The rule asserts that the central point is the centre of gravity of the three products F^2, G^2, H^2, placed at the points of application of F, G, H.

Ex. If the forces F, G, H of an astatic triangle are not rectangular prove that the central point is the centre of gravity of three weights proportional to

$$F(F+G\cos\gamma+H\cos\beta),\quad G(F\cos\gamma+G+H\cos\alpha),\quad H(F\cos\beta+G\cos\alpha+H)$$

placed at the corners, where α, β, γ are the angles between the forces (G, H), (H, F), (F, G).

This result follows at once from the general theorem given in Art. 30.

67. *The central point coincides with the centre of the imaginary conic.* To find the centre of the conic we follow the rule given in treatises on Conics. Differentiating the equation of the conic (Art. 64) with regard to the areal coordinates α, β, γ separately, and equating the results, we find that α, β, γ are proportional to F^2, G^2, H^2. The result follows at once.

68. *The imaginary conic being given, it is required to find the central lines and the principal moments of the system.*

Let the system of forces be reduced to its simplest form (Art. 32), i.e. let the forces be represented by a force R acting at the central point O together with two astatic couples whose arms are placed along the central lines Oy, Oz. Let the astatic moments be K_2, K_3.

Consider the origin O as one corner of an astatic triangle and produce the arms of the couples to very distant points B and C, replacing the forces by two others, viz. G and H, both very small. Then OBC is an infinitely large astatic triangle with *rectangular forces*. Let $OB=b$, $OC=c$, then $bG=K_2$ and $cH=K_3$, also $F=R$. We shall now use this triangle to find the equation to the imaginary conic by the formula given in Art. 64.

Let η, ζ be the Cartesian coordinates referred to the rectangular axes Oy, Oz of any point. Let α, β, γ be the areal coordinates of the same point referred to the infinitely large triangle OBC. Then $\alpha=1$, $\beta=\eta/b$, $\gamma=\zeta/c$. The conic

$$\frac{\alpha^2}{F^2}+\frac{\beta^2}{G^2}+\frac{\gamma^2}{H^2}=0$$

therefore reduces to

$$\frac{\eta^2}{K_2{}^2}+\frac{\zeta^2}{K_3{}^2}+\frac{1}{R^2}=0.$$

We therefore infer (1) that the centre of the imaginary conic is the central point, (2) the principal diameters are the central lines of the system, (3) that the lengths of the principal semidiameters are $K_2\sqrt{-1}/R$ and $K_3\sqrt{-1}/R$.

Referring to Art. 36, we see that the imaginary conic is the same as the imaginary focal conic.

69. Ex. 1. If ABC be an astatic triangle with rectangular forces show that either central line makes an angle θ with the side BC where

$$a\tan 2\theta=\frac{4\Delta F^2(H^2b\cos C-G^2c\cos B)}{a^2G^2H^2+H^2F^2b^2\cos 2C+F^2G^2c^2\cos 2B},$$

and Δ is the area of the triangle.

Ex. 2. If a triangle having its orthocentre at the central point be projected orthogonally on the central plane, prove that the projection is a possible astatic triangle with rectangular forces, provided the self-conjugate circle projects into the real conic

$$\frac{\eta^2}{K_2{}^2}+\frac{\zeta^2}{K_3{}^2}=\frac{1}{R^2}.$$

70. Transformation of tetrahedra. The forces being referred to one tetrahedron as $ABCD$, it is required to refer them to any other tetrahedron as $A'B'C'D'$.

If the coordinates of the corners of the first tetrahedron with regard to the second are known, the transference may be effected at once by using the rule given in Art. 60. But if the coordinates of the second tetrahedron with regard to the first are given, we may proceed in the following manner.

Let the tetrahedral coordinates of $A'B'C'D'$ referred to the first tetrahedron be given by the diagram, and let the whole determinant be Δ. Then the coordinates of A referred to the second tetrahedron are the minors of the several terms in the row opposite A after division by Δ. The coordinates of B are the minors of the terms in the row opposite B after division by Δ, and so on.

	A'	B'	C'	D'
A	a_1	b_1	c_1	d_1
B	a_2	b_2	c_2	d_2
C	a_3	b_3	c_3	d_3
D	a_4	b_4	c_4	d_4

The coordinates of the corners of the first tetrahedron are now known and the transference may be effected as before.

71. Ex. 1. If one corner as D be changed to D' without altering the opposite face show that the direction of the force at D' is parallel to the force at D, and that their magnitudes are inversely proportional to the distances of D and D' from the unchanged face. See the rule in Art. 60.

If D' lie in the plane BDC show that the force at A is unaltered.

Ex. 2. The forces at the corners of a tetrahedron $ABCD$ are F, G, H, L respectively; it is required to find the central plane, the angles between the forces being given.

Let the cosine of the angle between two forces F, G be represented by $\cos FG$ and so on. Let f, g, h, l be the minors of the four constituents in the leading diagonal of the determinant.

$$\begin{vmatrix} 1, & \cos FG, & \cos FH, & \cos FL \\ \cos FG, & 1, & \cos GH, & \cos GL \\ \cos FH, & \cos GH, & 1, & \cos HL \\ \cos FL, & \cos GL, & \cos HL, & 1 \end{vmatrix}.$$

Then the central plane divides any side as AB in a point P such that

$$\frac{F \cdot AP}{G \cdot BP} = \pm \sqrt{\frac{f}{g}}.$$

Resolve the force F into three others F_1, F_2, F_3 acting parallel to G, H, L. Consider the three sets of parallel forces, viz. (G, F_1), (H, F_2), (L, F_3). We may collect each into its own centre of parallel forces and thus obtain three points on the central plane, Art. 58. The central plane therefore cuts AB in a point P where $F_1 \cdot AP = G \cdot BP$. But since F_1, F_2, F_3 are in equilibrium with $-F$, we have by Art. 48 of Vol. I., $F_1^2/F^2 = g/f$. The result follows at once.

72. If the forces F, G, H, of an astatic triangle ABC are rectangular and of finite magnitude, and if the area ABC is not zero, prove that the system cannot be reduced to fewer than three forces.

If possible let the forces be reduced to two, P and Q, and let these act at D and E in the plane of the triangle. Let p, q, r be the perpendiculars from A, B, C on DE. Turn the forces about their points of application until the force at A is perpendicular to the plane ABC, then the forces at B and C act in that plane. Taking moments about DE we have $Fp=0$. Similarly $Gq=0$, $Hr=0$. But this is impossible if the area of the triangle is not zero.

That the points of application D, E must lie in the plane ABC follows from Art. 57, for DE may be regarded as one side of an astatic triangle, the third force being zero. We may also prove this in an elementary manner. Place the body so that the direction of the force P is parallel to the plane of ABC, while the other Q is not parallel; this is possible provided P and Q are not parallel to each other. Then, as in Art. 13, taking the plane of ABC as that of xy we have Z_z the same for the three forces F, G, H and the two P, Q. The ordinate of E is therefore zero. In the same way the ordinate of D is zero.

If the forces P and Q are parallel to each other, they cannot form a couple because their components parallel to F, G and H are not zero. They can therefore be reduced to a single force. Proceeding as above we easily shew that its point of application lies in the triangle; thence we deduce as before that the area of ABC is zero.

That the three forces F, G, H cannot be reduced to two, P, Q, also follows from the invariants of an astatic triangle. Regarding DE as one side of the triangle, the third force being zero, we see that the second invariant of Art. 65 is zero. It follows that $FGH\Delta$ is also zero, which is impossible unless either the area Δ or one of the forces F, G, H is zero.

73. *To investigate the condition that the forces of an astatic system can be reduced to two forces.*

We have seen in Art. 57 that the forces of the system can be reduced to three forces, viz. X_0, Y_0, Z_0, acting at three points A, B, C whose coordinates (x_1, y_1, z_1) (x_2, y_2, z_2) (x_3, y_3, z_3) are given by

$$\begin{array}{lll} X_0x_1 = X_x & X_0y_1 = X_y & X_0z_1 = X_z, \\ Y_0x_2 = Y_x & Y_0y_2 = Y_y & Y_0z_2 = Y_z, \\ Z_0x_3 = Z_x & Z_0y_3 = Z_y & Z_0z_3 = Z_z. \end{array}$$

We shall suppose in the first instance that the principal force is not zero, and that the axes are so chosen that X_0, Y_0, Z_0 are all finite.

If the three points A, B, C lie in a straight line we may make a further reduction. We can replace each of these forces by two other forces parallel to it and of proper magnitude, acting at any two points M_1, M_2, which lie in the straight line. By compounding the three forces at M_1, and also those at M_2, the whole system can be reduced to two forces. In order therefore that the system of forces may be reducible to two forces it is *sufficient* that the three points A, B, C should lie in a straight line.

It is also *necessary*, for otherwise the system is equivalent to an astatic triangle with rectangular forces. Now by Art. 72 such a system cannot be reduced to two forces unless either the triangle is evanescent or one at least of the forces X_0, Y_0, Z_0, is zero.

If the three points A, B, C lie in a straight line a plane can be drawn through that straight line and the origin. Hence

$$\begin{vmatrix} X_x, & X_y, & X_z \\ Y_x, & Y_y, & Y_z \\ Z_x, & Z_y, & Z_z \end{vmatrix} = 0.$$

The projection of these points on any coordinate plane must also lie in a straight line. We therefore have

$$\begin{vmatrix} X_0, & X_y, & X_z \\ Y_0, & Y_y, & Y_z \\ Z_0, & Z_y, & Z_z \end{vmatrix} = 0, \qquad \begin{vmatrix} X_x, & X_0, & X_z \\ Y_x, & Y_0, & Y_z \\ Z_x, & Z_0, & Z_z \end{vmatrix} = 0, \qquad \begin{vmatrix} X_x, & X_y, & X_0 \\ Y_x, & Y_y, & Y_0 \\ Z_x, & Z_y, & Z_0 \end{vmatrix} = 0.$$

The second of these four equations expresses the fact that A, B, C lie in a plane perpendicular to that of yz, the third that they lie in a plane perpendicular to that of xz, and so on.

Since no two of these four planes coincide, except when the points A, B, C lie in a coordinate plane, any two of the last three equations are sufficient to express the fact that the three points A, B, C lie in a straight line except when the three force components are zero.

These determinants are the coefficients of the several terms in the equation to the central plane. That plane is therefore indeterminate.

Expressions for these determinants in terms of the forces, without the intervention of coordinate axes, have been given in Art. 31.

74. *To find the equivalent forces.* We have seen that they may be made to act at any two points M_1, M_2 which lie on the straight line ABC. The equation of this straight line is evidently $\frac{\xi - x_1}{x_2 - x_1} = \frac{\eta - y_1}{y_2 - y_1} = \frac{\zeta - z_1}{z_2 - z_1}$. This straight line is called the *central line of the two forces.*

If two forces, not parallel to each other, are together astatically equivalent to two other forces, we may prove in an elementary manner that the four points of application lie in one straight line.

Let P_1, P_2 acting at M_1, M_2 be equivalent to Q_1, Q_2 acting at N_1, N_2. Make P_1 act parallel to N_1N_2 and take moments about N_1N_2. It immediately follows that M_2 lies on N_1N_2. Similarly M_2 lies on N_1N_2. *Thus the central line is fixed in the body.*

Take any two distinct points M_1, M_2 on the central line. Let the coordinates of the points thus chosen be (f, g, h) and (f', g', h'). Let (F, G, H), (F', G', H') be the components of the forces at these two points. The forces will then be known when we have found (F, G, H) and (F', G', H').

Since this system of two forces is equivalent to the given system, the twelve elements must be the same for each system (Art. 12).

We therefore have

$$\begin{array}{llll} X_x = Ff + F'f', & X_y = Fg + F'g', & X_z = Fh + F'h', & X_0 = F + F' \\ Y_x = Gf + G'f', & Y_y = Gg + G'g', & Y_z = Gh + G'h', & Y_0 = G + G' \\ Z_x = Hf + H'f', & Z_y = Hg + H'g', & Z_z = Hh + H'h' & Z_0 = H + H'. \end{array}$$

Any six of these equations determine F, G, H; F', G', H' when f, g, h and f', g', h are given.

75. *To show that whatever points are chosen on the central line, the forces at those points are always parallel to the same plane.*

Supposing the system to be already reduced to two forces P_1, P_2 acting at some two points M_1, M_2, let us replace these by two other forces Q_1, Q_2 acting at any other points N_1, N_2 on the central line. The force Q_1 is the resultant of two forces which act parallel to P_1 and P_2; it is therefore parallel to any plane to which P_1 and P_2 are both parallel. In the same way the force Q_2 is parallel to the same plane.

It should also be noticed that the resultant of the two forces P_1, P_2, when transferred parallel to themselves to act at the same point, is a force fixed in direction and magnitude.

76. Referring to the determinantal conditions given in Art. 73, we see that *if we substitute ξ, η, ζ for the terms in any row* in the first determinant (repeated here in the margin) we have the equation of the plane containing the origin and the central line of the two resultant forces.

$$\begin{vmatrix} X_x, & X_y, & X_z \\ Y_x, & Y_y, & Y_z \\ Z_x, & Z_y, & Z_z \end{vmatrix} = 0$$

If however we substitute ξ, η, ζ for the terms in any column of the same determinantal equation, we have the equation of the plane to which the two resultant forces are parallel whatever be their points of application.

The first of these theorems follows at once from the values of x_1, &c. given in Art. 73. The second is easily proved by substituting in the terms of the first and second columns the values of X_x &c. given in Art. 74, and in the third column ξ, η, ζ. After an obvious reduction and division by $fg'-f'g$, the equation reduces to the form shown in the margin, which is the plane required. There is no exceptional case when the divisor vanishes, for the equation to the plane then takes the form $0=0$.

$$\begin{vmatrix} F, & F', & \xi \\ G, & G', & \eta \\ H, & H', & \zeta \end{vmatrix} = 0$$

77. We have hitherto assumed that X_0, Y_0, Z_0 are all finite. The case in which any one or any two are zero may be treated as a limiting case and the corresponding conditions may be derived from those obtained when X_0, Y_0, Z_0 have finite but *general* values. As long as the conditions thus obtained are not nugatory they will be the conditions required. If however the principal force R is zero, the three components X_0, Y_0, Z_0 vanish for all axes and the reasoning in Art. 73 fails from the beginning.

The equations of Art. 74 supply a method of arriving at the conditions that the given forces can be reduced to two forces without making any assumption about the principal force. The body being in any position, let the components of the two forces be, as before (F, G, H), (F', G', H'), and let their points of application be (f, g, h), (f', g', h'). The required conditions may then be deduced from the twelve equations given in Art. 74. It is evident by simple inspection that the four determinantal equations given in Art. 73 are satisfied.

If the principal force is zero and the system can be reduced to two forces, those two forces must be equal and opposite, i.e. they must form a couple. Let $\pm F$, $\pm G$, $\pm H$ be the resolved parts of the forces of this couple, (f, g, h) (f', g', h') the coordinates of the extremities of its astatic arm. Then equating the nine finite elements of the system to those of the couple we have

$$\begin{aligned} X_x&=F(f'-f), & X_y&=F(g'-g), & X_z&=F(h'-h) \\ Y_x&=G(f'-f), & Y_y&=G(g'-g), & Y_z&=G(h'-h) \\ Z_x&=H(f'-f), & Z_y&=H(g'-g), & Z_z&=H(h'-h). \end{aligned}$$

The necessary and sufficient conditions that the system should be equivalent to two forces are therefore that (X_x, Y_x, Z_x), (X_y, Y_y, Z_y), (X_z, Y_z, Z_z), should be each proportional to the direction cosines of one straight line. This straight line is parallel to the forces of the couple.

78. Ex. 1. Show that any force F acting at a point A may be replaced by forces P_1, P_2 acting parallel to F at any two points M_1, M_2 such that AM_1M_2 is a straight line. Show also that these forces are

$$P_1=F\frac{AM_2}{AM_2-AM_1} \text{ and } P_2=F\frac{AM_1}{AM_1-AM_2}.$$

Ex. 2. Two given forces P_1, P_2, acting at the points M_1, M_2, are changed into two forces Q_1, Q_2 which are at right angles to each other, and act at two other points N_1, N_2 in the straight line M_1M_2. If y_1, y_2 are the distances of N_1, N_2 from the central point of the forces P_1, P_2, prove that $R^4y_1y_2=-(P_1P_2D\sin\theta)^2$ where $R^2=P_1^2+P_2^2+2P_1P_2\cos\theta$, D is the distance M_1M_2 and θ is the inclination of the forces P_1, P_2 to each other. It follows that the product y_1y_2 is the same for all equivalent rectangular forces.

Ex. 3. In all transformations of two forces P_1, P_2 into two others in which the points of application remain on the same straight line, the quantities

(1) $P_1^2 + P_2^2 + 2P_1P_2 \cos\theta$,

(2) $P_1P_2 D \sin\theta$,

(3) $P_1(P_1 + P_2\cos\theta)\,x_1 + P_2(P_1\cos\theta + P_2)\,x_2$,

are invariable, where x_1, x_2 are the distances of the points of application M_1 M_2 from any fixed point on the central line, D is the distance M_1M_2 and θ is the angle made by the forces with each other.

Ex. 4. A system consists of two forces P_1, P_2 acting at M_1, M_2 and the inclination of the forces to each other is θ. Show that (1) the central point O is the centre of gravity of weights proportional to $P_1(P_1 + P_2\cos\theta)$ and $P_2(P_1\cos\theta + P_2)$ placed at M_1, M_2. (2) The central ellipsoid at O is two parallel planes perpendicular to M_1M_2. (3) The principal axes at O are M_1M_2 and any two perpendicular straight lines.

79. *To determine the conditions that the forces of an astatic system reduce to a single force.*

Let the single force be P_1, let it act at the point (x_1, y_1, z_1) and let its components be X_1, Y_1, Z_1. Comparing the elements at any base we have

$$X_x = X_1x_1, \qquad X_y = X_1y_1, \qquad X_z = X_1z_1, \text{ \&c.}$$

Hence we see that the constituents in any column of any of the four determinants of Art. 73 bear to each other the ratios (X_1, Y_1, Z_1) of the components of the single force and that these ratios must be the same for every column.

We also notice that the constituents in any row of any of the four determinants bear to each other the ratios (x_1, y_1, z_1) or $(1, y_1, z_1)$ &c. of the coordinates of the point of application.

We have twelve elementary equations and six arbitrary quantities (X_1, Y_1, Z_1), (x_1, y_1, z_1) leaving six conditions to be satisfied by the elements of the system.

Since $X_0 = X_1$, &c., it is clear that the single equivalent force is equal and parallel to the principal force, Art. 11. Also, since the coordinates of the central point depend on the twelve elements, it is evident that the central points of the two equivalent systems coincide, Art. 28. Thus it follows that the point of application of the equivalent single force is the central point of the system.

NOTES.

NOTE A, Art. 149. **Green's theorem.** We may deduce from equation (5) of Art. 149 *an extension of Gauss' theorem*, Art. 106. Let P, Q, R be the components of a vector I and let $V=1$, so that, by (1), $U=0$. We then have

$$\int I\cos i d\sigma = \int\left(\frac{dP}{dx}+\frac{dQ}{dy}+\frac{dR}{dz}\right)dv \quad\text{.........................(1).}$$

If therefore the components P, Q, R of a vector satisfy the condition

$$\frac{dP}{dx}+\frac{dQ}{dy}+\frac{dR}{dz}=0 \quad\text{..................................(2)}$$

the surface integral or flux of the vector taken through any closed surface is zero. It is of course obvious that when I as in Gauss' theorem represents the force due to an attracting body, $P=dV/dx$ &c., where V is now the potential of the body, and (2) becomes Laplace's equation.

I. Let two surfaces S, S' be bounded by the same rim. Let that side of either *be called the positive side* towards which the normals are drawn.

Since these surfaces enclose a space the surface integral of the vector taken over both surfaces is zero, provided the normals are drawn all outwards or all inwards, i.e. provided their positive sides are opposed to each other. Reversing the directions of the normals for one surface, it follows that *the surface integrals for two surfaces with the same rim or boundary are equal* provided their positive sides are the same.

II. Let a curve, such that the direction of the vector I at any point of the curve is a tangent, be called *a vectorial curve* (Art. 47). Let **a tube or filament** be formed by drawing vectorial curves through any small closed curve, as in Art. 126. Let σ, σ' be the areas of the normal sections at any two points P, P'.

By the extension of Gauss' theorem just proved, the surface integral of the vector over the boundary of the tube PP' is zero. The surface integral taken over the whole space PP', as in Art. 127, is $I'\sigma' - I\sigma$ where I, I' are the magnitudes of the vector at the bounding sections. Hence when the vector is such that *its components satisfy the equation* (2), *the flux across every section of a vectorial filament is the same.*

III. It is shown in Art. 149 that in some cases a volume integral can be replaced by a surface integral. We may also show that in some cases *a surface integral can be replaced by a line integral taken round the rim of the surface.*

Let X, Y, Z be the components of a vector whose line integral is to be taken round a closed curve. Let S be a continuous surface bounded by this curve as its rim. Let P, Q, R be the components of another vector related to X, Y, Z by the equations

$$P=\frac{dZ}{dy}-\frac{dY}{dz},\quad Q=\frac{dX}{dz}-\frac{dZ}{dx},\quad R=\frac{dY}{dx}-\frac{dX}{dy} \quad\text{..................(3).}$$

The theorem to be proved is that the surface integral of the vector (P, Q, R) taken over the surface S is equal to the line integral of the vector (X, Y, Z) taken round

the rim. Let (l, m, n) be the direction cosines of the normal to dS, the theorem then asserts that $\int(Pl+Qm+Rn)\,dS=\int(Xdx+Ydy+Zdz)$....................(4).
That side of S is called the positive side towards which the normals (l, m, n) are drawn. The line integral is to be taken clock-wise when viewed from the positive side.

If we construct an infinitely small sphere whose centre C is at (xyz), the components of the vector (X, Y, Z) at the point $x+\xi$, $y+\eta$, $z+\zeta$ are by Taylor's theorem $X'=X+\frac{dX}{dx}\xi+\frac{dX}{dy}\eta+\frac{dX}{dz}\zeta$, $Y'=Y+\frac{dY}{dx}\xi+$&c., $Z'=$&c.

The sum of the moments of the vector round a parallel to the axis of z drawn through C, taken for every element of volume dv of the sphere, is

$$\int(Y'\xi-X'\eta)\,dv=\tfrac{1}{2}vk^2\left(\frac{dY}{dx}-\frac{dX}{dy}\right),$$

where $\frac{1}{2}vk^2$ has been written for the equal integrals $\int\xi^2dv$, $\int\eta^2dv$. It is obvious that in a sphere $\int\xi\eta dv=0$, $\int\xi\zeta dv=0$, &c. $=0$.

It follows that if (X, Y, Z) are the components of one vector, (P, Q, R) are the components of another vector connected with the former at every point *by a geometrical relation which is independent of all coordinates.*

We shall now prove that the theorem (4) *is true for any area which is so small that it may be regarded as plane.* Taking the plane of xy to contain the area, we have $\int RdS=\iint\left(\frac{dY}{dx}-\frac{dX}{dy}\right)dx\,dy=\int(Ydy+Xdx)$,

where the third expression follows from the second by an integration between limits in the manner described in Art. 149. Thus, if AB, drawn parallel to x, cut the rim in A, B, $\iint\frac{dY}{dx}dx\,dy=\int(Y_B-Y_A)\,dy$. But at B, dy is positive and at A dy is negative, hence taking the integral round the rim and therefore giving dy its proper sign, this becomes $\int Ydy$. Since $l=0$, $m=0$, $n=1$ and $dz=0$, this equation asserts that the flux of the vector (P, Q, R) parallel to the positive direction of the axis of z is equal to the line integral round the rim taken clock-wise.

To prove the theorem for a surface of finite size we add the results obtained for each element of area. Let two adjacent elements meet along the arc AB. When integrating round each element we pass over AB in opposite directions so that the signs of dx, dy, dz in one integration are opposite to those in the other. The sum of the integrals may therefore be found by integrating round both elements as if they were one, omitting the arc AB. The same reasoning applies to all the elements and the sum of the line integrals may be found by integrating round the rim.

The surface integrals of the vector (P, Q, R) taken over two surfaces bounded by the same rim are each equal to the same line integral. Hence *the surface integral of the vector* (P, Q, R) *for any closed surface is zero.* This also follows at once from the extension of Gauss' theorem, for the vector (P, Q, R) as defined by (3) evidently satisfies the condition (2).

The following results show how some volume integrals can be replaced by surface integrals.

(1) The volume of a solid enclosed by a surface S is $\frac{1}{3}\int r\cos\phi d\sigma$ where $d\sigma$ is an element of the surface, and ϕ is the angle the outward normal at $d\sigma$ makes with the radius vector produced. [Gauss.]

(2) The potential at the origin, of the solid (if of unit density) is $\frac{1}{2}\int\cos\phi d\sigma$. [Smith's Prize, 1871.]

(3) The integral $\int \cos\phi d\sigma/r^2$ is 4π or 0 according as the origin is inside or outside S.

(4) The x component of attraction is $\int \cos\phi' d\sigma/r$ where ϕ' is the angle the normal at $d\sigma$ makes with x. [Gauss.]

In Arts 358, 360, 362 and Note M there are some examples of surface integrals replaced by line integrals.

NOTE B, Art. 190. **Potential of a thin circular ring.** When the law of force is the inverse square of the distance, Dickson puts the potential at any point R into the form

$$V_2 = \frac{M}{\frac{1}{2}(\rho+\rho')}\frac{2K}{\pi},$$

where $\frac{1}{2}(\rho+\rho')$ is the mean of the greatest and least distances of R from the ring, M is the mass, and K is the complete elliptic integral of the first kind to modulus OP/a. See the figure of Art. 185.

NOTE C, Art. 211. **Attraction of a solid ellipsoid.** In the text the potential at an internal point P is found first and the axial components of force are deduced by differentiation. The following method of finding the components of force is so simple as to deserve attention.

Through P we pass an ellipsoid concentric with and similar to the boundary of the solid. The attraction at P of the portion of the solid external to this ellipsoid has been proved to be zero in Art. 68. It is therefore necessary only to find the attraction at P of the portion of the solid bounded by this ellipsoid. The problem is thus reduced to that of finding the attraction of an ellipsoid at a point on its surface. Let the semi-axes of this ellipsoid be ma, mb, mc.

We now construct an elementary cone whose vertex is P and whose base is an element Q of the surface. If $d\omega$ be the solid angle of the cone, its attraction at P is $\int \rho r^2 d\omega dr/r^2$ taken between the limits $r=0$ and $r=r$. The attraction is therefore $\rho r d\omega$.

The axial components of the attraction of the whole ellipsoid at P are therefore

$$X = -\rho\int r\lambda d\omega, \quad Y = -\rho\int r\mu d\omega, \quad Z = -\rho\int r\nu d\omega \quad\quad (1),$$

where (λ, μ, ν) are the direction cosines of QP and the integrations are to be taken so as to include all the elementary cones which lie on one side of the tangent plane at P.

Let (ξ, η, ζ) be the coordinates of P when referred to the centre. Since Q lies on the ellipsoid we have

$$\frac{(\xi-\lambda r)^2}{m^2a^2} + \frac{(\eta-\mu r)^2}{m^2b^2} + \frac{(\zeta-\nu r)^2}{m^2c^2} = 1 \quad\quad (2).$$

Since the point (ξ, η, ζ) lies on the surface this gives

$$r = 2\left(\frac{\xi\lambda}{a^2} + \frac{\eta\mu}{b^2} + \frac{\zeta\nu}{c^2}\right)\Big/\left(\frac{\lambda^2}{a^2} + \frac{\mu^2}{b^2} + \frac{\nu^2}{c^2}\right) \quad\quad (3).$$

This value of r has to be substituted in the expressions (1) and the integrations effected. As the radius vector turns round P, it is evident by (3) that no values of λ, μ, ν make r imaginary. Since the value of r determined by λ, μ, ν differs only in sign from that determined by $-\lambda, -\mu, -\nu$, the equation (3) represents the surface twice over. Since the signs of X, Y, Z depend on the signs of the *products* $r\lambda$, $r\mu$, $r\nu$, it is clear that if we integrate the equations (1) taking *all* positions of the radius vector and not merely those on one side of the tangent plane, we shall obtain in each case twice the required attraction. We therefore have

$$X = -\rho\int\left(\frac{\xi\lambda^2}{a^2} + \frac{\eta\lambda\mu}{b^2} + \frac{\zeta\lambda\nu}{c^2}\right)d\omega\Big/\left(\frac{\lambda^2}{a^2} + \frac{\mu^2}{b^2} + \frac{\nu^2}{c^2}\right),$$

where (λ, μ, ν) have all possible values. It is obvious that the term containing the product $\lambda\mu$ disappears on integration, for the elements corresponding to (λ, μ) and $(\lambda, -\mu)$ destroy each other. In the same way the term containing the product $\lambda\nu$ disappears. We therefore have

$$X = -\rho\xi\int\frac{\frac{\lambda^2}{a^2}d\omega}{\frac{\lambda^2}{a^2}+\frac{\mu^2}{b^2}+\frac{\nu^2}{c^2}},\qquad Y=-\rho\eta\int\frac{\frac{\mu^2}{b^2}d\omega}{\frac{\lambda^2}{a^2}+\frac{\mu^2}{b^2}+\frac{\nu^2}{c^2}},\qquad Z=\&c.$$

These may be written in the form

$$X=-A\rho\xi,\qquad Y=-B\rho\eta,\qquad Z=-C\rho\zeta\ \ldots\ldots\ldots\ldots(4).$$

We notice that *the constants A, B, C are functions of the ratios of the axes* and are therefore the same for all similar ellipsoids.

The integrals given above for A, B, C may also be written in the form

$$A=\int\frac{x^2}{a^2}d\omega,\qquad B=\int\frac{y^2}{b^2}d\omega,\qquad C=\int\frac{z^2}{c^2}d\omega\ \ldots\ldots\ldots\ldots(5),$$

where the integration extends over the whole surface of the ellipsoid. It easily follows that $A+B+C=4\pi \qquad Aa^2+Bb^2+Cc^2=\int r^2 d\omega \ldots\ldots\ldots\ldots(6)$, where r is the radius vector of the bounding ellipsoid drawn from the centre as origin.

The potential is seen by an easy integration to be $V=\frac{1}{2}\rho\{D-A\xi^2-B\eta^2-C\zeta^2\}$, where $D=\int r^2 d\omega$, since $\frac{1}{2}\rho D$ must evidently be the potential at the centre.

NOTE D, Art. 218. **Other laws of force.** The potential of a thin homogeneous homoeoid at an internal point $(\xi\eta\zeta)$ when the force varies as the inverse κth power of the distance can be found, free from all signs of integration, when κ is an even integer >2. Let μp be the surface density at any point Q, where p is the perpendicular from the centre on the tangent plane at Q. The potential is

$$V=\frac{2\pi\mu}{(\kappa-1)(\kappa-3)}\left(\frac{2}{E}\right)^{\kappa-3}\left\{1+\frac{1}{2^2}\frac{E\nabla}{\kappa-4}+\frac{1}{2^4}\frac{E^2\nabla^2}{1\,.\,2\,.\,(\kappa-4)(\kappa-5)}+\&c.\right\}P^{\frac{1}{2}(\kappa-4)},$$

where $$\nabla=a^2\frac{d^2}{d\xi^2}+b^2\frac{d^2}{d\eta^2}+c^2\frac{d^2}{d\zeta^2},\qquad P=\frac{\xi^2}{a^4}+\frac{\eta^2}{b^4}+\frac{\zeta^2}{c^4},$$

and $$E=1-\frac{\xi^2}{a^2}-\frac{\eta^2}{b^2}-\frac{\zeta^2}{c^2}.$$

The general term is $$\frac{1}{2^{2f}}\frac{L(\kappa-4-f)}{L(f)\,L(\kappa-4)}E^f\nabla^f \text{ and } L(n)=1\,.\,2\,.\,3\ldots.n.$$

The series has $\frac{1}{2}(\kappa-2)$ terms. Thus for the law of the inverse fourth power it reduces to the first term; for the law of the inverse sixth power, there are two terms and so on.

At an external point P' whose coordinates are ξ', η', ζ', we have

$$V'=\frac{abc}{a'b'c'}\frac{2\pi\mu}{(\kappa-1)(\kappa-3)}\left(\frac{2}{E'}\right)^{\kappa-3}\left\{1+\frac{1}{2^2}\frac{E'\nabla'}{\kappa-4}+\&c.\right\}P'^{\frac{1}{2}(\kappa-4)},$$

where $$\nabla'=\frac{a'^4}{a^2}\frac{d^2}{d\xi'^2}+\frac{b'^4}{b^2}\frac{d^2}{d\eta'^2}+\&c.,\qquad P'=\frac{a^2\xi'^2}{a'^6}+\frac{b^2\eta'^2}{b'^6}\ \&c.$$

$$E'=1-\frac{a^2\xi'^2}{a'^4}-\&c.=\epsilon^2\left(\frac{\xi'^2}{a'^4}+\frac{\eta'^2}{b'^4}+\frac{\zeta'^2}{c'^4}\right).$$

Here a', b', c' are the semi-axes of the confocal drawn through P', and $\epsilon^2=a'^2-a^2=\&c.$ It should be noticed that the differentiations implied in the operator ∇' are to be performed on (ξ', η', ζ') on the supposition that a', b', c' are constant. The potential at an external point may be deduced from that at an internal point by a method which is practically one of inversion. See Art. 203. [*Phil. Trans.* 1895.]

NOTE E, Art. 250. **Heterogeneous ellipsoid.** When the attracted point P lies within the substance of the ellipsoid, a little more explanation may be added. Through P we describe an ellipsoid similar to the external surface of the given body. Let it be defined by $m=n$ as in Art. 241. The potential of the inner portion at P is

$$V_1=\Sigma\int_0^{n^2} dm^2 \int_{\lambda_1}^{\infty} du\,(1-m^2)^{\kappa-1}\,(m^2-1+R)^n F(u).$$

Now λ satisfies $\dfrac{\xi^2}{n^2a^2+\lambda}+\&c.=1$ (Art. 204) and since P lies on the ellipsoid (n), $\dfrac{\xi^2}{n^2a^2}+\&c.=1$. It follows that $\lambda=0$ and therefore, since $\lambda=\lambda_1 m^2$, $\lambda_1=0$ (Art. 249).

Next consider the shell outside the ellipsoid (n). As explained in Art. 240, we put $\lambda_1=0$ and integrate from $m^2=n^2$ to $m^2=1$. We have therefore

$$V_2=\Sigma\int_{n^2}^{1} dm^2 \int_0^{\infty} du\,(1-m^2)^{\kappa-1}\,(m^2-1+R)^n F(u).$$

Adding V_1 and V_2 we have $\quad V=\Sigma\displaystyle\int_0^1 dm^2\int_0^{\infty} du\,[\&c.].$

The order of the integrations may evidently be reversed, and the argument may be continued as in Art. 250, and in the result we have $\lambda=0$.

NOTE F, Art. 264. **Other laws of force.** When the law of force is the inverse κth power of the distance we require the expansion of $1/R^{\kappa-1}$. There are two ways of extending Legendre's series.

First we may continue to make the expansion in powers of h and put

$$(1-2ph+h^2)^{-\frac{1}{2}(\kappa-1)}=1+Q_1h+Q_2h^2+\dots$$

If $\kappa-1$ is an odd integer, say equal to $2m+1$, we have

$$Q_n=\frac{1}{1.3.5\dots(2m-1)}\frac{d^m}{dp^m}P_{m+n}.$$

If $\kappa-1$ is an even integer, say equal to $2m+2$, we have

$$Q_n=\frac{1}{2.4.6\dots2m}\frac{d^m}{dp^m}\frac{\sin(n+m+1)\theta}{\sin\theta},$$

where $p=\cos\theta$. The four most important theorems relating to the function Q_n are given in Art. 282, Ex. 3.

Secondly, we may retain Legendre's functions of p as the coefficients, but cease to expand in powers of h. We then have when κ is even and greater than 2

$$\frac{1}{(1-2ph+h^2)^{\frac{1}{2}(\kappa-1)}}=\Sigma P_nh^n\frac{\psi(h)}{(1-h^2)^{\kappa-3}}.$$

There is a similar expansion when κ is odd and >1, except that P_n is replaced by $\sin(n+1)\theta/\sin\theta$ and that the coefficients of the function $\psi(h)$ are different.

The function $\psi(h)$ is an integral rational function of h containing only even powers, the highest being $h^{\kappa-4}$. Thus the function does not increase in complexity as n increases, but has always the same number of terms.

When the body considered is a thin spherical surface or a circular ring, h is the ratio of the radius to the distance of the attracted particle. Thus $\psi(h)$ is constant for an integration over the surface of any portion of a sphere or along the circumference of the ring.

When the law of force is the inverse fourth power $\kappa=4$ and $\psi(h)=(2n+1)$; when the law is the inverse sixth, $\kappa=6$, and

$$1.3\psi(h)=(2n+1)\{-(2n-1)h^2+2n+3\}.$$

The general value of $\psi(h)$ is given in the *Proceedings of the Mathematical Society*, vol. XXVI., 1895, page 481.

NOTE G, Art. 281. **Legendre's theorem.** There is another proof of the theorem $\int P_n^2 dp = 2/(2n+1)$ which is in general use. We have

$$\frac{1}{1-2ph+h^2} = (P_0 + P_1 h + P_2 h^2 + \ldots)^2$$

We multiply both sides of this equation by dp and integrate between the limits -1 and $+1$. We then have, by Art. 278,

$$\int \frac{dp}{(1+h^2)-2hp} = \int (P_0^2 + P_1^2 h^2 + P_2^2 h^4 + \ldots) \, dp.$$

Integrating the left-hand side, we have

$$\{\log(1+h) - \log(1-h)\}/h = \Sigma \int P_n^2 h^{2n} dp.$$

Both series being convergent, we find the value of $\int P_n^2 dp$ by equating the coefficients of h^{2n} on each side.

We may deduce some other interesting results from the equation of differences

$$(n+1) P_{n+1} - (2n+1) p P_n + n P_{n-1} = 0.$$

Multiplying both sides by P_κ and integrating between the limits -1 and $+1$, we have
$$(2n+1) \int p P_n P_\kappa dp = (n+1) \int P_{n+1} P_\kappa dp + n \int P_{n-1} P_\kappa dp.$$

It follows from Art. 278 that $\int p P_n P_\kappa dp$ is zero except when κ and n differ by unity. In that case we have $\int p P_n P_{n+1} dp = \dfrac{2(n+1)}{(2n+1)(2n+3)}$ as in page 219.

In the same way we may show that $\int p^2 P_n P_\kappa dp$ is zero except when κ and n are equal or differ by 2. In these cases

$$\int p^2 P_n^2 dp = \frac{2(2n^2+2n-1)}{(2n-1)(2n+1)(2n+3)}, \qquad \int p^2 P_n P_{n+2} dp = \frac{2(n+1)(n+2)}{(2n+1)(2n+3)(2n+5)},$$

where the limits of the integrals are -1 to $+1$.

We may, by successive induction, deduce from the equation of differences, that

$$P_m P_n = \Sigma \frac{A(m-r) A(r) A(n-r)}{A(m+n-r)} \frac{2n+2m-4r+1}{2n+2m-2r+1} P_{m+n-2r},$$

where Σ expresses summation from $r=0$ to the lesser of the two quantities m, n.

Also
$$A(m) = \frac{1.3.5\ldots(2m-1)}{1.2.3\ldots m}, \qquad \therefore A(m) = \frac{m+1}{2m+1} A(m+1).$$

We may interpret $A(m)$, when m is zero or a negative integer, by supposing this relation to hold generally, so that putting $m=0$ we have $A(0)=1$. Similarly $A(-1)=0$, and hence, when m is any negative integer, $A(m)=0$.

In the series r is supposed to vary from $r=0$ to either m or n. If however r is taken beyond these limits, for instance if $r=-1$ or $m+1$, then (in consequence of the property of the function A just stated) the coefficient of the corresponding term is zero. Hence practically we may consider r to be unrestricted in value.

We notice that in this expansion the suffixes of P are all even or all odd according as $m+n$ is even or odd. If then we multiply by P_l and integrate the product between the limits -1 and $+1$, we have $\int P_l P_m P_n dp = 0$ if $l+m+n$ is odd (Art. 278).

Supposing $l+m+n$ to be even, it follows (by subtracting the even number $2l$) that $m+n-l$ is also even and that there may be a term on the right-hand side in which the suffix is given by $m+n-2r=l$. This term, after multiplication by P_l, supplies the integral $\int P_l^2 dp$ and is not zero. We then find

$$\int_{-1}^{+1} P_l P_m P_n dp = \frac{2}{2s+1} \frac{A(s-l) A(s-m) A(s-n)}{A(s)},$$

where $s=\frac{1}{2}(l+m+n)$. *In order that this integral may not be zero, no one of the quantities l, m, n must be greater than the sum of the other two, and $l+m+n$ must be an even integer.*

The reader may consult a paper by the late Prof. J. C. Adams in the *Proceedings of the Royal Society* 1878, No. 185. The value of the integral is also given by Ferrers as an example on page 156 of his *Treatise on Spherical Harmonics*, 1877.

By using the results referred to in Art. 292 we also find

$$\tfrac{1}{2}\int P_l \frac{d^m P_{m+n}}{dp^m}\,dp = \frac{m(m+1)\dots(m+s-1)}{1.2\dots s}\cdot\frac{1.3.5\dots(2m+2n-2s+1)}{1.3.5\dots(2n-2s+1)},$$

when $l=n-2s$. When $n-l$ is odd or $l>n$, the integral is zero.

$$\tfrac{1}{2}\int p^\kappa P_m\,dp = \frac{1.2.3\dots\kappa}{2.4\dots(\kappa-m)\,1.3.5\dots(\kappa+m+1)},$$

when $\kappa>m$. The integral is zero if $\kappa-m$ is odd, or if $\kappa<m$. In both integrals the limits are -1 to $+1$.

When the law of force is the inverse κth power of the distance, the equation of differences takes the form

$$(n+1)\,Q_{n+1} - p\,(2n+\kappa-1)\,Q_n + (n+\kappa-2)\,Q_{n-1} = 0,$$

as explained in Art. 282, Ex. 3. We may use this equation in a similar manner to find $\int\phi(p)\,Q_n{}^2dp$ and $\int\phi(p)\,p^2Q_n{}^2dp$ where $\phi(p)=(1-p^2)^{\frac{1}{2}(\kappa-2)}$.

NOTE H, Art. 288. **Laplace's theorem.** Laplace deduces the equation $\int Y_mY_n d\omega=0$ from the equation (7) of Art. 284. What follows is an extension of his method, *Mécanique Céleste*, livre troisième 12. Let us write (7) in the form

$$-\frac{d}{d\mu}\left(b\frac{dY_m}{d\mu}\right)-\frac{d}{d\phi}\left(c\frac{dY_m}{d\phi}\right)=pY_m \quad\dots\dots\dots(1),$$

$$\therefore -\frac{d}{d\mu}\left(b\frac{dY_n}{d\mu}\right)-\frac{d}{d\phi}\left(c\frac{dY_n}{d\phi}\right)=p'Y_n \quad\dots\dots\dots(2),$$

where $\qquad b=1-\mu^2,\quad c=1/(1-\mu^2),\quad p=m(m+1),\quad p'=n(n+1)$.

Multiplying these equations by Y_n, Y_m respectively and subtracting, we find

$$(p-p')\int Y_mY_n d\omega=\iint\left\{Y_m\frac{d}{d\mu}\left(b\frac{dY_n}{d\mu}\right)-Y_n\frac{d}{d\mu}\left(b\frac{dY_m}{d\mu}\right)\right\}d\mu\,d\phi+\&c.$$

Integrating by parts, we find that the unintegrated parts cancel, we therefore have

$$(p-p')\int Y_mY_n d\omega=\int\left[b\left(Y_m\frac{dY_n}{d\mu}-Y_n\frac{dY_m}{d\mu}\right)\right]d\phi$$
$$+\int\left[c\left(Y_m\frac{dY_n}{d\phi}-Y_n\frac{dY_m}{d\phi}\right)\right]d\mu \quad\dots\dots\dots(3),$$

where the quantities in square brackets are to be taken between limits, the first between $\mu=\pm1$, the second from $\phi=0$ to 2π.

Now $b=1-\mu^2$, if therefore Y_m, Y_n and their differential coefficients with regard to μ are finite all over the sphere, the first integral is zero.

The range of ϕ from 0 to 2π, carries a point P round the sphere on a small circle to the point from which P started. If then the quantities c, Y_m, Y_n, and their differential coefficients with regard to ϕ are "*one valued*" on the sphere, the second quantity in square brackets is the same at both limits, and the second integral is zero.

It follows that if p and p' are unequal (that is, if neither $m=n$ nor $m+n=-1$) the integral $\int Y_mY_n d\omega=0$.

If we generalise (7) and write it in the form

$$aY_m-\frac{d}{d\mu}\left(b\frac{dY_m}{d\mu}\right)-\frac{d}{d\phi}\left(c\frac{dY_m}{d\phi}\right)-\frac{d}{d\mu}\left(e\frac{dY_m}{d\phi}\right)-\frac{d}{d\phi}\left(e\frac{dY_m}{d\mu}\right)=pAY_m\dots(4),$$

where a, b, c, e and A are given finite functions of μ, ϕ, but not of p, while p is a given function of n, the function Y_m is not now a Laplace's function, but the equation

$$(p-p')\iint Y_m Y_n A d\mu d\phi = 0 \quad \text{.........................(5)}$$

will in certain cases be true. This may be proved by the same reasoning as before. The unintegrated parts cancel and the integrated parts vanish provided (1) b and e are zero when $\mu = \pm 1$, (2) Y_m, Y_n and their differential coefficients with regard to μ and ϕ are finite one-valued functions of μ, ϕ. Other cases in which the integrated parts are zero will suggest themselves to the reader and need not be particularised here.

We may also extend the theorem to the case in which *the integration is effected only over the area within some closed curve* drawn on the sphere, provided Y_m, Y_n are such that $\frac{1}{Y_m}\frac{dY_m}{d\mu} = \frac{1}{Y_n}\frac{dY_n}{d\mu}$, $\frac{1}{Y_m}\frac{dY_m}{d\phi} = \frac{1}{Y_n}\frac{dY_n}{d\phi}$ at all points of the boundary. For example, the equation (5) is true if Y_m and Y_n vanish at all points of the boundary.

The equation (5) is also true if both Y_m and Y_n satisfy the condition

$$-\left(\frac{b\sin\psi}{\sin^2\theta} + e\cos\psi\right)\frac{dY}{d\theta} + \left(c\sin\theta\cos\psi + \frac{e}{\sin\theta}\sin\psi\right)\frac{dY}{d\phi} = \lambda Y$$

at all points of the boundary, where ψ is the angle the arc θ makes with the elementary arc of the boundary and λ is an arbitrary function of θ, ϕ but not of m or n. When $b = 1/c = 1-\mu^2$ and $e=0$, $\lambda = 0$, this implies that the space variation of Y perpendicular to the boundary is zero.

NOTE I, Art. 329. **Magnetic sphere.** The expression in the text for the potential applies obviously to an external point. At an internal point, the potential, by the same rule, is equal to $\frac{4}{3}\pi Ir\cos\theta$. This also follows at once from the result given in the next article for an ellipsoid. The force due to a uniformly magnetised solid sphere at an internal point P is therefore $-\frac{4}{3}\pi I$. The direction, when taken positively, is opposite to the direction of magnetisation, and tends to demagnetise the body.

NOTE K, Art. 342. **Magnetic forces.** Kelvin, when speaking of the two definitions of resultant force in a crevasse (1) tangential and (2) perpendicular to the lines of magnetisation, sometimes calls the former "*the polar definition*" and the latter "*the electromagnetic definition*" (Reprint &c. Art. 517). This latter force is called "*the magnetic induction*" by Maxwell and this phrase has been generally adopted in the text. A slight modification has however been made in Art. 342 and on a few other occasions when the change seemed to make the meaning of the context clearer. Maxwell's phrase is not entirely unobjectionable and it is much to be desired that some short term could be generally agreed to.

NOTE L, Art. 345. **The magnetic induction.** At a point *outside a magnetic body* the magnetic force and the magnetic induction are the same. It follows that their components satisfy Laplace's equation, and we have

$$\frac{dX}{dx} + \frac{dY}{dy} + \frac{dZ}{dz} = 0, \qquad \frac{dX_1}{dx} + \frac{dY_1}{dy} + \frac{dZ_1}{dz} = 0 \quad \text{..................(1)}.$$

At a point *inside a magnetic body*, we have by substitution (Art. 345)

$$\frac{dX_1}{dx} + \frac{dY_1}{dy} + \frac{dZ_1}{dz} = \left(\frac{dX}{dx} + \&c.\right) + 4\pi\left(\frac{dA}{dx} + \&c.\right).$$

Since the magnetic force is by definition that due to Poisson's two distributions, the sum of the terms in the first bracket on the right-hand side is equal to $4\pi\rho$ (Arts. 105, 41). The sum of the terms in the second bracket is $-4\pi\rho$ (Art. 339). We therefore have

$$\frac{dX}{dx}+\frac{dY}{dy}+\frac{dZ}{dz}=4\pi\rho, \quad \frac{dX_1}{dx}+\frac{dY_1}{dy}+\frac{dZ_1}{dz}=0 \quad \text{..................(2)}.$$

It follows that the components (X, Y, Z) of the magnetic force satisfy different differential equations according as the point under consideration is external or internal. The *components of the magnetic induction satisfy the same equation* (viz. Laplace's equation) *whether the point is inside or outside.*

Since the equation satisfied by the components of the magnetic induction is the same as the condition (2) given in Note A, page 356, it follows immediately that *the surface integral of the magnetic induction taken through any closed surface is zero.* This surface may be wholly within or wholly without or partly within and partly without the magnetic body. See also Art. 488.

It also follows that the surface integrals of the magnetic induction taken through any two surfaces having the same rim are equal. See Note A.

NOTE M, Art. 358. **Vector potential.** Since the surface integral of the magnetic induction depends on the closed rim and not on the form of the surface (Note L, page 363), it should be possible to find the induction through a closed curve, without constructing a surface to act as a diaphragm.

This is effected by finding a vector A whose components F, G, H satisfy the equations

$$X_1=\frac{dH}{d\eta}-\frac{dG}{d\zeta}, \qquad Y_1=\frac{dF}{d\zeta}-\frac{dH}{d\xi}, \qquad Z_1=\frac{dG}{d\xi}-\frac{dF}{d\eta} \quad \text{......(3)},$$

where (X_1, Y_1, Z_1) are the components of the induction at a point P whose coordinates are (ξ, η, ζ). Then, as proved in Note A, page 356, the induction through any closed surface is equal to the line integral of the vector (FGH) round the rim. *This new vector is called* by Maxwell *the vector potential of magnetic induction.* [See his *Electricity*, Art. 405.]

The relations (3) are satisfied at an external point for *a simple lamellar shell of unit strength* by taking

$$F=\int\frac{dx}{R}, \qquad G=\int\frac{dy}{R}, \qquad H=\int\frac{dz}{R} \quad \text{..................(4)},$$

where the integration extends round the rim of the shell, and R is the distance of an element of the rim (xyz) from a point $(\xi\eta\zeta)$ in space. This follows at once from the values of X, Y, Z given in Art. 358.

Example. Prove that for a simple magnetic shell of strength m, in the form of a small circle of radius a and centre O, the vector potential at a point P is approximately

$$A=\frac{m\pi a^2 p}{r^3}\left\{1-\frac{3}{2}\frac{a^2}{r^2}+\frac{15}{8}\frac{a^2p^2}{r^4}\right\},$$

where $r=OP$ and p is the distance of P from the axis of the shell. [Coll. Ex. 1896.]

To prove this we take the plane of the circle as the plane of xy, the centre as origin and the plane of xz to contain P. We then have $x=a\cos\phi$, $y=a\sin\phi$, $z=0$ and $R^2=r^2-2ap\cos\phi+a^2$. Substituting in (4) and expanding the denominator in powers of a/r, we see that $F=0$. Rejecting all odd powers of $\cos\phi$ in the expansion for G we find at once that G has the value given in the enunciation.

We must refer this to axes of x, y which are independent of the position of P if we wish to use equations (3). We then have

$$F=-A\eta/p, \qquad G=A\xi/p, \qquad H=0,$$

where $(\xi, \eta, 0)$ are the coordinates of P and $p^2=\xi^2+\eta^2$.

For an elementary lamellar shell, the vector potential is $A=M\sin\theta/r^2$, *where* $r=OP$, θ *is the angle r makes with the axis Oz and* $M=\pi a^2 m$. *The direction of the vector is perpendicular to the plane POz and its positive direction is clockwise round Oz.*

For an elementary magnet whose moment is M, centre O, and axis the axis of z, we assume the magnitude of the vector to be $M\sin\theta/r^2$ and its direction to be as just described. The components are then evidently $F=-\dfrac{M\eta}{r^3}$, $G=\dfrac{M\xi}{r^3}$, $H=0$. Since the potential of an elementary magnet is $M\cos\theta/r^2$, it is not difficult to see that the equations (3) are satisfied.

To find the components of the vector potential of a small magnet when the direction cosines of the axis are λ, μ, ν, we resolve the magnet into $M\lambda$, $M\mu$, $M\nu$. The F component of $M\lambda$ is zero, those of $M\mu$, $M\nu$ are $M\mu\zeta/R^3$ and $-M\nu\eta/R^3$ respectively. The F component for a magnetic body at P is therefore

$$F'=\iiint I\,\frac{\mu\zeta-\nu\eta}{R^3}\,dx\,dy\,dz,$$

where R is the distance of any point (xyz) of the body from the point P in space whose coordinates are (ξ, η, ζ) and $M=Idv$, Art. 326.

NOTE N, Art. 397. **Electrified sphere.** The figure has been drawn by Dickson to show the distribution of electrical density on the surface of a sphere under the influence of a point-charge at S (where $OS=10$, $OA=6$). Let a radius vector from the centre O cut the curve drawn inside the circle in P, the circle itself in Q, and

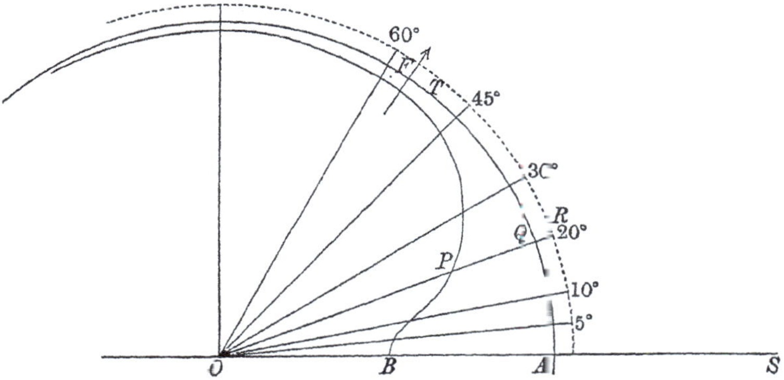

the dotted circle outside in R. The length PQ then represents the density of the (negative) charge at any point Q of the sphere, when uninsulated; while the length QR would represent the uniform density of an equal (positive) charge freely distributed on the sphere, when the point-charge at S is absent and the sphere insulated. Consequently, if the sphere be initially uncharged and at zero potential, and if the point-charge be then brought to S, $QR - PQ$ will represent the (positive) density at the point Q. This density will be negative from A to F, at which latter point the total density is zero. If the whole figure be rotated about OS, F will trace out the line of no force. For the data given, the angle FOS is about $56\frac{1}{2}°$, and if the tangent from S touch the circle at T, the angle SOT will be about $53\frac{1}{8}°$.

NOTE P, Art. 486. **Discontinuity.** The result in Ex. 8 is interesting as it exhibits a discontinuity. The difficulty thus introduced would disappear if we supposed the value of K to be continuous but to change rapidly from K to K'. See some brief remarks on this subject in chap. XIII. of the second volume of the Author's treatise on *Rigid Dynamics* (Art. 620 of the fifth edition).

INDEX TO ATTRACTIONS.

The numbers refer to the articles.

The numbers refer to the articles.

The numbers refer to the articles.

The numbers refer to the articles.

The numbers refer to the articles.

The numbers refer to the articles.

The numbers refer to the articles.

INDEX TO THE BENDING OF RODS.

The numbers refer to the articles.

INDEX TO ASTATICS.

The numbers refer to the articles.

The numbers refer to the articles.

CAMBRIDGE: PRINTED BY JOHN CLAY, M.A. AT THE UNIVERSITY PRESS.